Advanced Production Decline Analysis and Application

Advanced Production Decline Analysis and Application

Edited by

Hedong Sun

石油工业出版社
PETROLEUM INDUSTRY PRESS

ELSEVIER

AMSTERDAM • BOSTON • HEIDELBERG • LONDON
NEW YORK • OXFORD • PARIS • SAN DIEGO
SAN FRANCISCO • SINGAPORE • SYDNEY • TOKYO
Gulf Professional Publishing is an Imprint of Elsevier

G P
P

Gulf Professional Publishing is an imprint of Elsevier
225 Wyman Street, Waltham, MA 02451, USA
The Boulevard, Langford Lane, Kidlington, Oxford, OX5 1GB, UK

British Library Cataloguing-in-Publication Data
A catalogue record for this book is available from the British Library

Library of Congress Cataloging-in-Publication Data
A catalog record for this book is available from the Library of Congress

ISBN: 978-0-12-802411-9

For information on all Gulf Professional Publishing publications
visit our website at http://store.elsevier.com/

Typeset by Thomson Digital

Contents

Companion Website for this Book:
http://booksite.elsevier.com/9780128024119/index.php

Preface

Over the past 20 years, advanced production decline analysis (APDA) evolved fast with the improvement of wellhead pressure (WHP) measurement and flow metering techniques. Today, it has become a well-established technique. In field applications, APDA is essential and beneficial for performance monitoring, and plays a major role in the fine description of reservoirs and the analysis of field development. In the past decade, the APDA-based production analysis software has been widely applied in various oil and gas fields. Unfortunately, this technique and its theory have never been systematically and thoroughly formulated in any existing work. The publication of the book entitled "Advanced Production Decline Analysis and Application" will break the situation.

This book considers the advanced production decline analysis by way of manual matching for a vertical well centered in a closed circular reservoir. Based on the APDA concepts, several production decline methods, such as Arps, Fetkovich, Blasingame, Agarwal-Gardner, Normalized Pressure Integral (NPI), Transient, Long-term Linear Flow and Dynamic Material Balance, are discussed thoroughly, including their principles, processes, cases, and application. The plotting and analysis of APDA curves are introduced in detail for complex reservoir, such as closed circular composite reservoir, two-layered reservoir, and dual-porosity reservoir. In the appendix, the theoretical curves of Blasingame method and the calculation procedure for normalized pseudo-pressure of gas well are presented, in order that the readers can understand how the type curves are plotted and what the "black box" of analysis software contains. Accordingly, this book is a relatively complete and systematic work concerning advanced production decline analysis for oil and gas wells.

This book integrates the author's achievements and experience in his long-term research, so it is meaningful both theoretically and practically. Its publication will be helpful to the promotion and application of APDA, thereby further identifying the performance variation of oil and gas fields and enhancing the field development. Besides, it will have positive and significant effects on the development of reservoir engineering personnel.

Academicians of Chinese Academy of Engineering
Dakuang Han

中国工程院院士

2013 年 7 月 31 日

About the author

Hedong Sun, PhD, SPE member, born in 1973, is a senior engineer, and earned his PhD degree from Xi'an Jiaotong University in 2004. Since 2004, he has been a Research Engineer in Research Institute of Petroleum Exploration and Development (RIPED)-Langfang Branch, which is the R&D center of China National Petroleum Corporation (CNPC). Hedong has over 18 years of reservoir engineering experience in well test analysis and production analysis. He has been one of the academic leaders of reservoir engineering of RIPED-Langfang Branch since 2008. In 2013, he was appointed as a technical expert of Well Testing Analysis and Productivity Evaluation in RIPED-Langfang Branch. He is the reviewer for five journals, including ACTA Petrolei Sinica and Well testing etc. He has published over 40 papers in peer-reviewed journals and SPE conferences. He has published two books – entitled *Modern Well Test Analysis and Deliverability Analysis of Complex Gas Reservoir* (2012) and *Advanced Production Decline Analysis and Application* (2013).

Introduction

Advanced production decline analysis (APDA), or production analysis or rate transient analysis, has become a hotspot in reservoir engineering in recent years. Based on transient filtration theory, the technique can provide novel typical curves by way of reservoir engineering and modern well test analysis. It is also used to analyze the daily production data and quantify the reservoir parameters, percolation characteristics and OOIP (OGIP) with type curve matching method.

APDA involves four kinds of methods, including (1) empirical method, e.g., Arps; (2) classical analysis method, e.g., Fetkovich; (3) log-log type curve matching analysis, e.g., Blasingame; and (4) reservoir engineering method, e.g., FMB. This technique, together with Lifecycle Modern Well Test Analysis, has become one of the main methods for dynamic reservoir description, and APDA-based analysis software has been widely used in oil and gas fields. However, to the best of our knowledge, there is no book that provides a systematic introduction on this technique.

Under this background, this book entitled *Advanced Production Decline Analysis and Application* is launched. It is compiled with reference to the previous research results, and in combination with the author's experience in dynamic reservoir description. This book keeps a foothold by carrying out advanced production decline analysis manually. As Professor Nengqiang Liu noted, a famous well test expert of CNPC logging, the modern well test interpretation process is basically the recapitulation of manual operation on the computer, so the operator must learn manual operation first, which will provide great help for thoroughly understanding the programs, instructions, and procedures and providing the best interpretation using computer software freely. The same is true for the APDA technique.

This book presents the APDA technique to the reservoir engineering professionals, serving as a modest spur to induce them to effectively learn the overseas advanced technologies and improve the reservoir production analysis. In view of the basic concept of APDA, this book thoroughly and systematically elaborates the basic principles, analysis process and cases of APDA methods, including Arps, Fetkovich, Blasingame, Agarwal–Gardner, NPI, Transient, Long-term Linear Flow and FMB methods, when they are used in a vertical well in closed circular reservoir. Combining with the field cases, this book also explains the integrated application of APDA process in the practical production. Besides, the plotting and analysis of APDA curves with Blasingame method are briefly introduced for complex reservoir, such as radial composite reservoir and two-layered reservoir. Online downloadable computer programs such as Arps, Fetkovich, Blasingame method and the calculation programs for normalized pseudo-pressure of gas well are available, so that the readers can understand how the type curves are plotted or the pertinent APDA models and theoretical curves are derived and plotted based on this book. (Details provided in http://booksite.elsevier.com/9780128024119/ or http://team.agoil.cn/sunhedong/.)

The completion of the manuscript requires the help of a large number of people. I hereby express my heart-felt gratitude to my postgraduate supervisor, Professor Chengtai Gao of Xi'an Shiyou University, doctoral supervisor, Professor Fangde Zhou of Xi'an Jiaotong University, as well as postdoctoral supervisor, Yuewu Liu, research fellow of Institute of Mechanics, CAS. It is their guidance that attracted me to the road of scientific research and encouraged me to make achievements in reservoir engineering. I am also grateful to Professor Xiaodong Wang of China University of Geosciences, as well as Professor Junbin Chen and Jia'en Lin of Xi'an Shiyou University, for their assistance during the compilation of this book. I also owe my sincere thanks to Professor Tongwen Jiang, Zhongqian Zhu, and Wenqing Pan of PetroChina Tarim Oilfield Company; project directors as Chunshu Luo, Ying Shi, Xiangjiao Xiao, Xingliang Deng, and Jianping Yang; as well as Jiwu Fan, director of Sulige Research Center, PCOC; Bin Wang, director of Research Institute of Exploration and Development, PetroChina Xinjiang Oilfield Company; Hu Sun and Jianting Duan of CCDC Changqing Downhole Technology Company; Lianchao Jia of PCOC No. 2 Gas Recovery Plant, etc., for their assistance and help in the project. My gratitude then goes to Doctor Weiyang Wang, Mingliang Luo, Huijuan Chen of China University of Petroleum (Huadong), who provided substantial help for literature delivery. I'm also indebted to Professor Xizhe Li and Jianjun Chen of Institute of Petroleum Exploration and Development (RIPED)-Langfang Branch; Jialiang Lu, Daojiang Long, Yujin Wan, and Yongxin Han of Gas Development Institute. Their guidance and encouragement propelled me to successfully complete the task. Thanks also go to Wen Cao and Xifei Yang, etc. for their assistance and help during the editing, proofreading, and publishing of this book.

The work of this book was supported in part by China Postdoctoral Science Foundation (No. 2011M500403), China Postdoctoral Science Foundation – the fifth special (No. 2012T50140), and Youth Innovation Fund of Research Institute of Petroleum Exploration and Development (RIPED) (No. 2009A1715).

Due to the limited level of knowledge and experience, the author could not avoid inappropriate statements in this book. Your comments and criticism are thereby warmly welcomed.

Thank you to the Elsevier staff who worked on the book, most notably Nicky Carter and Simon Tian.

December 2, 2014
Hedong Sun

Fundamentals of Advanced Production Decline Analysis

1

Advanced production decline analysis (APDA), or rate transient analysis or production analysis, is a procedure to process and interpret the daily production data of wells for obtaining parameters of such wells or reservoirs. This chapter introduces the history of APDA based on filtration theory, its similarities to and differences with well test analysis. In addition, this chapter also introduces several concepts related to the APDA.

1.1 History of Advanced Production Decline Analysis (APDA)

At the middle and later stages of reservoir development, daily production data of a well becomes the focus for reservoir analysis. They can be used to forecast the most probable well life, evaluate well production in the future, and determine the interwell communication relation and infill potential. Currently, the production decline analysis technique consists of the conventional Arps (1945) method, classical Fetkovich (1980) type curve matching method, modern Palacio and Blasingame (1993) and Agarwal et al. (1998) type curve matching methods and FMB (1998) reservoir engineering method.

Extrapolating the characteristic trend of some variables of a well can be helpful for our jobs. As to a well, the simplest and the most easily available variable is its production. If the flow rate versus time or cumulative production curve is plotted and extrapolated, the ultimate cumulative production can be obtained. The trend or mathematical relations indicated by the entire rate history of a well can be used to forecast the production performance in the future, which is referred to as the conventional Arps (1945) decline curve analysis method. This method magnificently describes the production decline laws of well at a constant bottom hole flowing pressure (BHFP) and in the completely boundary-dominated flow period. The greatest advantage of this method is that formation parameters are not necessarily obtained. On the other hand, it is not suitable for data analysis from the transient flow stage.

A variety of interpretations may occur for the data of one well or one reservoir, mostly resulting from the experiences of appraisers or the difference of appraisal targets. Just as pointed out by Ramsay (1968), "Some new papers contributing to the decline curve analysis were published in 1964-1968, but there was hardly any new technique." Slider (1968) developed a matching method applicable to the production-time data, which is similar to the log–log type curve matching method in well test analysis and uncovers a new direction for decline curve analysis. Because this method was quick and easy, Ramsay extensively used it to determine the distribution of decline exponent b in the appraisal of more than 200 wells. Gentry (1972) plotted three Arps decline curves on one chart to match the decline data of wells, where the dimensionless time was defined the same as with the Fetkovich (1980) method, and the dimensionless production was the reciprocal of relevant variables in Fetkovich method.

Advanced Production Decline Analysis and Application

Arps type curve can only be used to analyze the data of a boundary-dominated flow period. Fetkovich (1980), on the basis of homogeneous bounded formation transient filtration theory, introduced the transient flow formula in well test analysis to the decline analysis, so that the Arps type curve is extended to the transient flow period prior to boundary-dominated flow, and the transient rate decline curve and the Arps rate decline equation are organically combined. In this way, the production decline laws and the effect of boundary are intuitively shown, and a set of relatively complete log–log production decline curve matching analysis method similar to well test analysis is developed. The greatest advantage of the method is its ability to reliably determine whether the production is in a transient flow period or in a boundary-dominated flow period.

Both Arps and Fetkovich methods assume that the BHFP is constant to analyze the production data without considering the change of gas pressure–volume–temperature (PVT) charateristics with pressure. Palacio and Blasingame (1993) introduced the pseudo-pressure normalized production ($q/\Delta p_p$) and the material balance pseudo-time t_{ca} to develop the type curve, which considered the production at variable BHFP and the gas PVT changing with formation pressure.

Agarwal et al. (1998) used the relations of pseudo-pressure normalized production ($q/\Delta p_p$), material balance pseudo-time t_{ca}, and dimensionless parameters in well test analysis to develop the Agarwal-Gardner production decline analysis. Owing to the different definitions of dimensionless quantity, the early part of the curve is more discrete than the Blasingame chart and thus is in favor of reducing the ambiguity of matching analysis.

Both Blasingame and Agarwal-Gardner methods used the pseudo-pressure normalized production ($q/\Delta p_p$) and the material balance pseudo-time t_{ca} to create type curve, while the NPI (normalized pressure integral) method (Blasingame et al., 1989) used the production normalized pressure integral to analyze the data available, which was not affected by the scatter of data.

Palacio and Blasingame (1993) and Agarwal et al. (1998) type curve matching analysis methods introduced pseudo-time (or material balance pseudo-time) and production normalized pseudo-pressure (pseudo-pressure normalized production) to deal with variable BHFP, variable rate, and change of gas PVT with pressure. They used the flow rate integral, flow rate integral derivative, cumulative production–time, and flow rate–cumulative production type curves as the auxiliary matching analysis curves to reduce the ambiguity of interpretation results.

Her-Yuan and Teufel (2000) developed the method on the basis of Fetkovich's findings, and presented the linear flow characteristic curve usually occurring in low-permeability tight gas reservoir. Wattenbarger and El-Banbi (1998) and his students combined the linear flow model and the curve matching analysis method in well test analysis to present the analysis method for long-term linear flow production data of gas well in low-permeability tight gas reservoirs. Pratikno et al. (2003) developed the type curve and analysis method of a vertical fracture well. Yong-Xin Han (2006) also made helpful research on the long-term linear flow of low-permeability fracture wells.

Mattar et al. (1998, 2006) and Agarwal et al. (1998) suggested using the "flow (dynamic) material balance" method to analyze the production data, and conducted detailed discussion on the calculation of material balance time. This method is simple and easy. Mattar and Anderson (2003) believes that there is no one universal production data analysis method that can meet all types of reservoirs, and the best way to eliminate analysis errors is to synthetically use all analysis methods and consider flowing pressure data.

Over nearly a century, the APDA technique has evolved with several advances, including target to be analyzed, that is, from purely production data to both flow rate and pressure data; analytic model, that is, from no model to both analytical model and numerical model; analytic method, that is, from the empirical Arps method to the log-log method represented by Blasingame; applicable conditions, that is, from simple constant pressure production data to variable pressure and variable rate data; and the estimation parameters, that is, from only cumulative production to many parameters such as formation permeability, skin factor, dynamic reserves and drainage area, as well as interwell communication and infill potential.

1.2 Similarities and Differences between Production Decline Analysis and Well Test Analysis

As to dynamic reservoir description, APDA and well test analysis are combined to appraise the reservoir where the well is located, with the high precision pressure data acquired from transient well test and the dynamic data like pressure and flow rate obtained in production test and actual production, and based on understandings obtained from static geologic data. The parameters to be appraised include reservoir permeability, skin factor, dynamic reserves, drainage radius, fault sealing, and advancing range of edge water. As two major techniques for dynamic reservoir description, the APDA and the well test analysis have both specific and common features. They should be well combined and constrained with each other, so as to minimize the uncertainties of parameters interpretation. The similarities and differences between them are shown in Table 1.1.

Both of them are based on the classical filtration theory, by using the curve-matching method to get parameters and applying numerical solution method by means of building models for cases like complicated boundary, multiphase flow, and multiwell interference. However, they are different with respect to the precision of appraisal data adopted in the workflow; that is, production decline analysis can be conducted using only the flow rate and pressure data calculated for each day, while well test analysis is conducted using the high precision transient pressure test data. The data source of different quality decides the reliability of appraisal results. That is, the production decline analysis technique adopts daily test flow rate and pressure data, which are of a great quantity but low precision, especially the BHFP data that are mostly converted from the WHP, thereby leading to some errors. In contrast, the well test analysis technique adopts transient pressure test data, which are available in great quantities and have high precision.

1.3 Basic Concepts

1.3.1 Wellbore Storage Effect

As the wellbore has the ability to store fluid and flow or shut in operation is carried out at wellhead, the changes in wellhead flow rate and bottom hole sandface rate are different,

Table 1.1 Comparison of the PA and WTA (after Oliver et al., 2013)

Item	Well test analysis (WTA)		Production analysis (PA)
Theoretical basis	Filtration theory		Filtration theory
Time range	Traditional pressure buildup analysis	Hours, days, weeks	Days, weeks, months, years
	Full life cycle analysis	Hours, days, weeks, months, years	
Periods of interest	Traditional pressure buildup analysis	Mostly shut-ins Clean productions possible	Producing phases but buildups may be included
	Full life cycle analysis	Full life cycle	
Data sources	Pressure buildup well test, formation test, permanent pressure gauge		Measured/allocated production Surface pressure readings Permanent gauges (PDG)
Reservoir areas of interest	Whatever volume of investigation during the test and/or the shut-in		Drainage area of well or well group
Generation	Manual matching	MDH, Horner	Arps
	Classical curve matching	Ramey	Fetkovich
	Modern curve matching	Bourdet	Blasingame
Main flow regime of interest	Requirement	Radial flow	Pseudo-steady state, boundary-dominated flow
	Capability	High	General
Results	Traditional pressure buildup	Formation flow capacity, skin, boundary, etc.	Drainage area and reserves, formation flow capacity, skin factor, etc.
	Full life cycle	Also drainage area and reserves	
Validity	Traditional pressure buildup	Higher	Higher
	Full life cycle	Very high	

as shown in Figure 1.1. At the initial stage of open flow, the wellhead flow rate mainly results from the pressure relief and expansion of fluid in the wellbore, whereas the formation does not supply any fluid to the wellbore. As time goes on, the sandface flow rate increases gradually. After a period of time, the surface flow rate equals the bottom hole flow rate. At the initial stage of shut in, the surface flow rate is zero, while the reservoir goes on supplying fluid to the wellbore. As time goes on, the sandface flow rate reduces, and after a period of time, the sandface flow rate becomes zero.

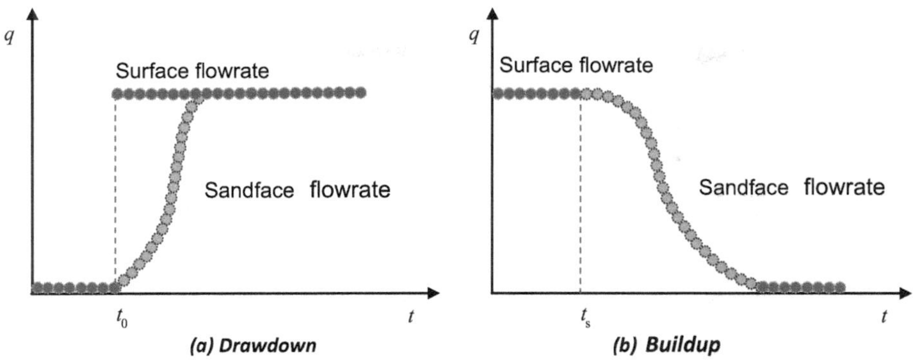

Figure 1.1 Wellbore storage

The wellbore storage effect is expressed by the wellbore storage coefficient, C, and is defined as the increase of fluid volume in the wellbore corresponded by the increase of unit BHFP. That is, we have

$$C = \frac{\Delta V_{\mathrm{w}}}{\Delta p} \tag{1.1}$$

1.3.2 Skin Effect

Due to contamination or improvement in drilling, as well as completion and stimulation on the reservoir, the permeability of the formation near the wellbore can change and lead to additional resistance. Let us assume that a layer of skin is stuck on the borehole wall, and the additional resistance generated by the fluid flowing through it just equals the additional resistance resulting from the permeability variation of formation near the wellbore. Based on the concept of skin, it may be believed that the permeability of formation near the wellbore does not change. The mathematical manipulation difficulty, which results from the change in permeability of formation near the wellbore, is avoided. The resistance resulting from the skin is expressed by the skin factor S, and under basic SI unit,

$$\Delta p_{\mathrm{skin}} = \frac{q\mu}{2\pi Kh} S \tag{1.2}$$

When $S > 0$, the additional resistance pressure change is positive. When $S < 0$, the additional resistance pressure change is negative. As to steady flow of circular reservoir, we have

$$p_{\mathrm{e}} - p_{\mathrm{wf}} = \Delta p_{\mathrm{ideal}} + \Delta p_{\mathrm{skin}} = \frac{q\mu}{2\pi Kh}\left[\ln\left(\frac{r_{\mathrm{e}}}{r_{\mathrm{w}}}\right) + S\right] \tag{1.3}$$

From Eq. (1.3), we have

$$q = \frac{2\pi Kh(p_e - p_{wf})}{\mu\left[\ln\left(\frac{r_e}{r_w}\right) + S\right]} \tag{1.4}$$

As to unsteady flow, we have

$$\begin{aligned} p_i - p_{wf} &= \Delta p_{ideal} + \Delta p_{skin} = \frac{q\mu}{4\pi Kh}\left[-Ei\left(-\frac{r_w^2}{4\eta t}\right) + 2S\right] \\ &= \frac{q\mu}{4\pi Kh}\left[\ln\left(\frac{4\eta t}{e^{\gamma} r_w^2}\right) + 2S\right] \end{aligned} \tag{1.5}$$

1.3.3 Apparent Wellbore Radius

Apparent wellbore radius can be introduced along with skin factor. As to steady flow of circular reservoir, assuming that the permeability of formation near the wellbore does not change, the well still produces fluid at flow rate q, and the wellbore radius is the apparent wellbore radius r_{wa} instead of the real wellbore radius r_w. According to Eq. (1.3), we have

$$q = \frac{2\pi Kh(p_e - p_{wf})}{\mu\left[\ln\left(\frac{r_e}{r_w}\right) + S\right]} = \frac{2\pi Kh(p_e - p_{wf})}{\mu \ln\left(\frac{r_e}{r_{wa}}\right)} \tag{1.6}$$

where

$$r_{wa} = r_w e^{-S} \tag{1.7}$$

When $S < 0$, $r_{wa} > r_w$, it corresponds to the increase of wellbore radius; when $S = 0$, $r_{wa} = r_w$, it means no change in wellbore radius; when $S > 0$, $r_{wa} < r_w$, it corresponds to the decrease of wellbore radius. The skin factor is related to the range and permeability of altered zone near the wellbore. Let the radius of altered zone be r_s, the pressure be p_s, and the permeability be K_s, and we have

$$\begin{aligned} q &= \frac{2\pi Kh(p_e - p_s)}{\mu \ln\left(\frac{r_e}{r_s}\right)} = \frac{2\pi K_s h(p_s - p_{wf})}{\mu \ln\left(\frac{r_s}{r_w}\right)} = \frac{2\pi Kh(p_e - p_{wf})}{\mu\left[\ln\left(\frac{r_e}{r_s}\right) + \frac{K}{K_s}\ln\left(\frac{r_s}{r_w}\right)\right]} \\ &= \frac{2\pi Kh(p_e - p_{wf})}{\mu\left[\ln\left(\frac{r_e}{r_w}\right) + \left(\frac{K}{K_s} - 1\right)\ln\left(\frac{r_s}{r_w}\right)\right]} \end{aligned} \tag{1.8}$$

we have

$$S=\left(\frac{K}{K_s}-1\right)\ln\left(\frac{r_s}{r_w}\right) \tag{1.9}$$

If the wellbore has been damaged, $K_s < K$, S is positive, and the larger the difference between K_s and K, the larger the damage zone is and the larger the skin factor S. If stimulation has been implemented in the wellbore, $K_s > K$, the skin factor S is negative, and the larger the stimulation radius r_s, the smaller the skin factor S. If the wellbore is neither damaged nor stimulated, $K_s = K$, and the skin factor, $S = 0$.

1.3.4 Infinite Formation Line-Source Solution

As there is an infinite acting radial flow period, at the initial stage of well test, the formation can be regarded as infinite. After the wellbore diameter has been neglected, the well can approximately be regarded as line source. Given that the well starts to produce fluid at $t = 0$ at a constant rate of q, the inner boundary condition approximates as

$$\left(r\frac{\partial p}{\partial r}\right)\bigg|_{r\to 0}=\frac{\mu q}{2\pi Kh} \tag{1.10}$$

It is known from the fluid flow in porous medium that the solution of definite solution problem Eq. (1.14) to Eq. (1.16) and Eq. (1.10) is

$$p_i - p(r,t) = -\frac{q\mu}{4\pi Kh}Ei\left(-\frac{r^2}{4\eta t}\right) \tag{1.11}$$

Ei is an exponential integration function, where

$$Ei(-x) = -\int_x^\infty \frac{e^{-u}}{u}du \approx \ln\left(e^\gamma x\right) = \ln x + 0.5772 \quad when\, x < 0.01 \tag{1.12}$$

where γ is the Euler constant, being 0.5772. If $\frac{r^2}{4\eta t} < 0.01$, we have

$$\begin{aligned} p_i - p &= \frac{q\mu}{4\pi Kh}\left[\ln\left(\frac{4\eta t}{e^\gamma r^2}\right)\right] = \frac{q\mu}{4\pi Kh}\ln\left(\frac{2.25\eta t}{r^2}\right) \\ &= \frac{q\mu}{4\pi Kh}\left[\ln\left(\frac{4\eta t}{r^2}\right) - 0.5772\right] \end{aligned} \tag{1.13}$$

As to well test procedure, the approximate formula of Eq. (1.13) can be used several seconds after the commencement of well test, which can be known from the following calculations.

Let $\eta = 1000\ \mathrm{cm}^2/\mathrm{s}$, $r_\mathrm{w} = 10\ \mathrm{cm}$, according to

$$x = \frac{r_\mathrm{w}^2}{4\eta t} < 0.01$$

We have

$$t > \frac{r_\mathrm{w}^2}{4\eta \times 0.01} = \frac{10 \times 10}{4 \times 0.01 \times 1000} = 2.5\mathrm{s}$$

Therefore, as to $r = r_\mathrm{w}$, the approximate formula of Eq. (1.13) can be used three seconds after the commencement of well test. However, as to the point far away from the wellbore, for example, another offset well, because r is very large, only when t reaches a certain level, the approximate formula of Eq. (1.13) can be used, which accounts for the inability of conducting semi-log analysis using the data obtained at the initial stage of an interference test. In an actual well test analysis, semilog analysis generally cannot be carried out using the data obtained at the earlier time, which have not resulted from the commission of the approximate formula of Eq. (1.13) but is due to the case that the wellbore storage effect has no means of keeping the sandface rate unchanged.

1.3.5 Steady State, Pseudo-Steady State, and Boundary-Dominated Flows

1. **Diffusivity equation**

   Assume that the formation is homogeneous, isopachous, and isotropic, the reservoir is filled with slightly compressible fluid with a constant compressibility, the flow behavior is isothermal and follows Darcy's law, and that the gravity and capillary force are negligible. Under these fundamental assumptions, the definite solution problem of a single well producing at a constant rate can be described as follows under basic SI.

   That is, we have the flow equation,

$$\eta\left[\frac{1}{r}\frac{\partial}{\partial r}\left(r\frac{\partial p}{\partial r}\right)\right] = \frac{\partial p}{\partial t} \tag{1.14}$$

   where $\eta = \dfrac{K}{\varphi\mu C_\mathrm{t}}$.

   To solve the above partial differential equation, it is required to give initial and boundary conditions, namely,

   Initial condition is

$$p\,|_{t=0} = p_\mathrm{i} \tag{1.15}$$

   Outer boundary condition is

infinite reservoir is

$$p|_{r\to\infty} = p_{\text{i}} \tag{1.16}$$

constant pressure is

$$p|_{r=r_{\text{e}}} = p_{\text{i}} \tag{1.17}$$

and closed boundary is

$$\left.\frac{\partial p}{\partial r}\right|_{r=r_{\text{e}}} = 0 \tag{1.18}$$

Inner boundary condition, where
constant rate at bottom hole

$$\left.\frac{\partial p}{\partial r}\right|_{r=r_{\text{w}}} = \frac{\mu q}{2\pi K h r_{\text{w}}} \tag{1.19}$$

constant pressure at bottom hole

$$p|_{r=r_{\text{w}}} = p_{\text{wf}} \tag{1.20}$$

and constant rate at wellhead

$$\left.\frac{\partial p}{\partial r}\right|_{r=r_{\text{w}}} = \frac{\mu}{2\pi K h r_{\text{w}}}\left(qB + C\frac{dp_{\text{wf}}}{dt}\right) \tag{1.21}$$

The above problem during the flow period can be divided into three stages:
Infinite acting radial flow, which refers to the period of time after the flow period when the pressure wave spreads outwards but does not reach the boundary. At this time, the boundary has not yet been touched, and the formation can be regarded as an infinite formation. As to a large reservoir, the infinite acting radial flow period is long, while as to a micro-reservoir, the infinite acting radial flow period is short.
Transition stage, which refers to the period of time after the flow period when the pressure wave spreads outwards and has reached the boundary, while the change in pressure in the vicinity of the outer boundary has not yet been stabilized.
Steady state or pseudo-steady state flow, which refers to the period of time after the pressure variation all over the formation has been stabilized. As to the state of constant pressure outer boundary, the pressure all over the reservoir does not change, while as to the state of closed outer boundary, the pressure all over the reservoir uniformly drops with time.

2. **Steady state flow**
If a well has been producing at a constant rate for a long time, and if the pressure distribution around the whole wellbore remains constant, the flow state is termed as steady state flow. The time of commencement of steady state flow is denoted by t_{ss}. When $t > t_{\text{ss}}$, the pressure at any point of the formation does not change with time, that is, $\frac{\partial p}{\partial t} = 0$.

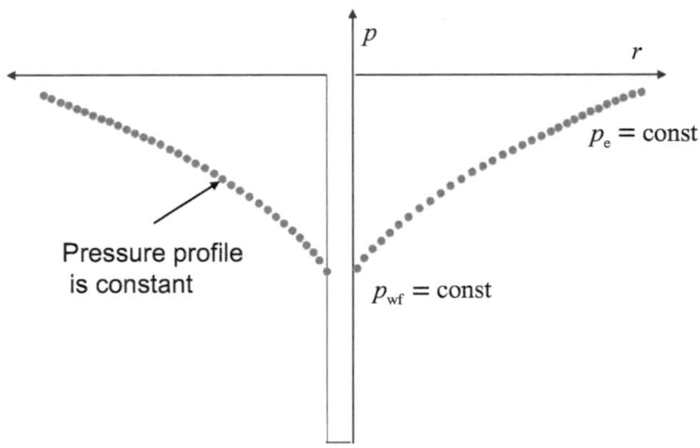

Figure 1.2 Pressure distribution during steady state flow

As to the natural water drive reservoir with significant surrounding water or the water flooding reservoir, the oil–water boundary may possibly form "constant pressure boundary," making the well production stay at a steadyflow state or approximately reach a steadyflow state. The pressure distribution in the vicinity of such a well is shown in Figure 1.2. It must be pointed out that as to gas wells, such a steadyflow state generally would not occur, and the so-called "constant pressure boundary" would not exist either, because of a large difference of gas and water viscosities, and it is usually characterized by a composite model in which the outer region becomes poor.

3. **Pseudo-steady state flow**
The so-called pseudo-steady state flow virtually is a kind of transient flow. After the wells in a closed reservoir have been producing at a stable flow rate for a certain time, the pressure spreads to all the boundaries around. Thereafter, the pressure at all points of the closed reservoir will drop at the same speed and then a pseudo-steady state flow occurs, as shown in Figure 1.3.

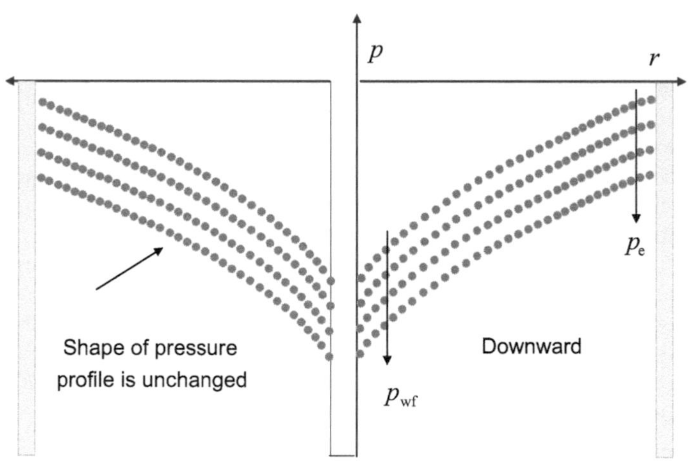

Figure 1.3 Pressure distribution during pseudo-steady state flow

If the time of commencement of pseudo-steady state flow is t_{ss}, when $t > t_{pss}$, at any point in the formation, $\frac{\partial p}{\partial t} = Const$. Observed from Figure 1.3, the shape of the pressure distribution curve remains unchanged at this time, where the pressure distribution curves of different times parallel to each other and only their heights are different.

After having entered the pseudo-steady state, the pressure all over the reservoir uniformly drops with time, therefore, $\frac{\partial p}{\partial t} = Const$. Based on the material balance principle, we have

$$V_p C_t \frac{dp}{dt} = -q \tag{1.22}$$

where

$$V_p = \pi r_e^2 \varphi h, \eta = \frac{K}{\varphi \mu C_t}$$

We have

$$\frac{1}{r}\frac{d}{dr}\left(r\frac{dp}{dr}\right) = \frac{1}{\eta}\frac{\partial p}{\partial t} = -\frac{\mu q}{\pi K h r_e^2} \tag{1.23}$$

Integrating Eq. (1.23), we have

$$p_e(t) - p_{wf}(t) = \frac{q\mu}{2\pi Kh}\left[\ln\left(\frac{r_e}{r_w}\right) - \frac{1}{2}\right] \tag{1.24}$$

$$\bar{p}(t) - p_{wf}(t) = \frac{q\mu}{2\pi Kh}\left[\ln\left(\frac{r_e}{r_w}\right) - \frac{3}{4}\right] \tag{1.25}$$

The $\bar{p}$ in Eq. (1.25) is the average formation pressure. The expression for the transition stage solution of closed circular reservoir is extremely complicated, which is rarely used and thus omitted in the book. For more information, please refer to the book *Reservoir Pressure Buildup and Flow Test in Wells* (Matthews and Russell, 1967).

4. Boundary-Dominated Flow

As to wells producing in a closed reservoir, when the pressure spreads to all boundaries around, if they still produce fluid at a constant BHFP, with the depletion of producing energy, the flow rate continuously drops, and the pressure inside the closed reservoir will also continuously drop. However, what is different from the pseudo-steady state is that the pressure drop speed at each point inside the reservoir is changing, as shown in Figure 1.4. In this way, the pseudo-steady state flow is a particular case for the boundary-dominated flow; that is, it can occur only at constant rate.

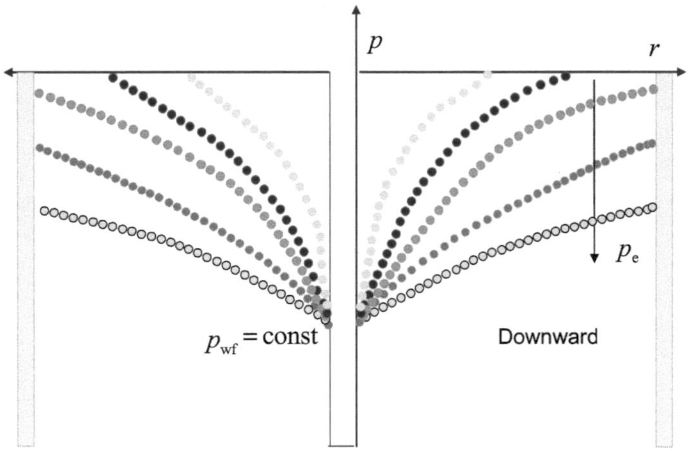

Figure 1.4 Pressure distribution during boundary-dominated flow

1.3.6 Investigation Radius

The investigation radius of a well refers to the range of pressure variation introduced to the reservoir after the change in well yield. The propagation time and propagation length meet the following equation.

$$r_i = 2\sqrt{\eta t_m}, r_i = 2\sqrt{\frac{Kt_m}{\varphi\mu C_t}} \tag{1.26}$$

For example, let $\eta = 1000\ \text{cm}^2/\text{s}$, $r_e = 500$ m, then the time propagated to the boundary is

$$t = \frac{r_i^2}{4\eta} = \frac{(50000)^2}{4\times 1000} = 6.25\times 10^5\,\text{s} = 7.23\,\text{d}$$

1.3.7 Dimensionless Variables

The dimensions of physical quantity in reservoir engineering are listed in Table 1.2.

The advantages of using a dimensionless physical quantity are as follows. The results are suitable for different unit systems, the number of parameters and variables are reduced, the problem is simplified, and the nature of physical problem is better shown. The dimensionless physical quantities used in the well test analysis and production decline analysis mainly are as follows

$$r_D = \frac{r}{r_w} \quad r_{eD} = \frac{r_e}{r_w} \quad t_D = \frac{\alpha_t Kt}{\varphi\mu C_t r_w^2} \quad t_{DA} = \frac{\alpha_t Kt}{\varphi\mu C_t A}$$

$$p_D = \frac{Kh}{\alpha_p q\mu B}(p_i - p) \quad p_{wD} = \frac{Kh}{\alpha_p q\mu B}(p_i - p_{wf}) \quad C_D = \frac{\alpha_c C}{\varphi C_t h r_w^2}$$

Table 1.2 Physical quantity dimension in reservoir engineering

Physical quantity	Symbol	Dimension
Formation permeability	K	$[L^2]$
Flow rate	q	$[L^3T^{-1}]$
Pressure	p	$[M/T^2L]$
Wellbore storage coefficient	C	$[T^2L/M]$
Viscosity	μ	$[M/TL]$
Time	t	$[T]$
Radial distance from a well, formation thickness	r, h	$[L]$

where, α_p, α_t, and α_c represent conversion factors of unit systems. The units and conversion factors of different unit systems are shown as follows.

As the dimensions of physical quantity are independent of the system of units, Table 1.3 can be used conveniently for unit conversion. With the above definitional equation of dimensionless quantity, the basic flow equation Eq. (1.14) is converted into

$$\frac{1}{r_D}\frac{\partial}{\partial r_D}\left(r_D\frac{\partial p_D}{\partial r_D}\right)=\frac{\partial p_D}{\partial t_D} \tag{1.27}$$

The Dupuit flow rate Eq. (1.6) is converted into

$$p_{eD}=\ln r_{eD} \tag{1.28}$$

The line-source solution of infinite formation Eq. (1.11) is converted into

$$p_D\left(r_D,t_D\right)=-\frac{1}{2}Ei\left(-\frac{r_D^2}{4t_D}\right) \tag{1.29}$$

Table 1.3 Physical quantity unit of reservoir engineering

Unit system		Basic SI	Legal SI	Imperial field unit
Length	***r, h, L***	**m**	**m**	**ft**
Time	t	s	h	h
Pressure	p	Pa	MPa	psi
Permeability	K	m^2	mD	md
Oil Rate	q_o	m^3/s	m^3/d	STB/D
Gas Rate	q_g	m^3/s	$10^4 m^3/d$	Mscf/D
Viscosity	μ	Pa·s	mPa·s(= cP)	cP
Conversion factor	α_p	$1/2\pi$	1.842	141.2
	α_t	1	0.0036	2.637×10^{-4}
	α_c	$1/2\pi$	$1/2\pi$	0.8936
	α_m	$1/\pi$	3.683×10^4	50312

$$p_{\mathrm{wD}} = p_{\mathrm{D}}(1, t_{\mathrm{D}}) + S = -\frac{1}{2} Ei\left(-\frac{1}{4t_{\mathrm{D}}}\right) + S$$
$$= \frac{1}{2}\ln\left(\frac{4t_{\mathrm{D}}}{e^{\gamma}}\right) + S = \frac{1}{2}(\ln t_{\mathrm{D}} + 0.80907) + S \tag{1.30}$$

The solution of closed circular reservoir at the pseudo-steady state stage is converted into

$$p_{\mathrm{D}} = \frac{2t_{\mathrm{D}}}{r_{\mathrm{eD}}^2} + \ln r_{\mathrm{eD}} - \frac{3}{4} \tag{1.31}$$

It is often required to carry out unit conversion in reservoir engineering calculation and well test analysis, which is difficult for many operators. Therefore, cases are given in the book to show how to correctly conduct unit conversion.

In case 1, the Dupuit equation is converted from basic SI into legal SI.

Under basic SI, the Dupuit equation is

$$q = \frac{2\pi K h \Delta p}{\mu \ln\left(\frac{r_{\mathrm{e}}}{r_{\mathrm{w}}}\right)} \tag{1.32}$$

For the sake of unit conversion, the original unit and the new unit are written out firstly at both sides of the equation as follows

$$q\frac{[\mathrm{m}^3/\mathrm{s}]}{[\mathrm{m}^3/\mathrm{d}]}[\mathrm{m}^3/\mathrm{d}] = \frac{2\pi K\frac{[\mathrm{m}^2]}{[\mathrm{m\,D}]}[\mathrm{m\,D}]h[\mathrm{m}]\Delta p\frac{[\mathrm{Pa}]}{[\mathrm{MPa}]}[\mathrm{MPa}]}{\mu\frac{[\mathrm{Pa.s}]}{[\mathrm{mPa.s}]}[\mathrm{mPa.s}]\ln\left(\frac{r_{\mathrm{e}}}{r_{\mathrm{w}}}\right)}$$

Then, substituting the real data for the new/old unit ratio, we have

$$q\frac{1}{86400}[\mathrm{m}^3/\mathrm{d}] = \frac{2\pi K\frac{1}{10^{15}}[\mathrm{m\,D}]h[\mathrm{m}]\Delta p\frac{10^6}{1}[\mathrm{MPa}]}{\mu\frac{1}{1000}[\mathrm{mPa.s}]\ln\left(\frac{r_{\mathrm{e}}}{r_{\mathrm{w}}}\right)}$$

At last, rearranging to get the equation under the new system of units, we have

$$q = \frac{1}{1.842}\frac{Kh\Delta p}{\mu \ln\left(\frac{r_e}{r_w}\right)}$$

In case 2, t_D is converted from basic SI into legal SI.

Under basic SI, the dimensionless time is defined as $t_D = \frac{Kt}{\varphi\mu C_t r_w^2}$, and then

$$t_D = \frac{K\frac{[\text{m}^2]}{[\text{mD}]}[\text{mD}]t\frac{[\text{s}]}{[\text{h}]}[\text{h}]}{\varphi\mu\frac{[\text{Pa.s}]}{[\text{mPa.s}]}[\text{mPa.s}]C_t\frac{[\text{MPa}]}{[\text{Pa}]}[\text{MPa}^{-1}]r_w^2[\text{m}^2]}$$

$$= \frac{K\frac{1}{10^{15}}[\text{mD}]t\frac{3600}{1}[\text{h}]}{\varphi\mu\frac{1}{1000}[\text{mPa.s}]C_t\frac{1}{10^6}[\text{MPa}^{-1}]r_w^2[\text{m}^2]} = \frac{0.0036Kt}{\varphi\mu C_t r_w^2}$$

1.3.8 Shape Factor and Pseudo-Steady State Arrival Time

It can be inferred from the above results that the solution of closed circular reservoir at pseudo-steady state is

$$p_D = \frac{2t_D}{r_{eD}^2} + \ln r_{eD} - \frac{3}{4} \tag{1.33}$$

Let $A = \pi r_e^2$, $t_{DA} = t_D r_w^2 / A$, and $C_A = 4\pi e^{\frac{3}{2}} / e^{\gamma}$. Eq. (1.33) can be written as

$$p_D = \frac{2t_D}{r_{eD}^2} + \ln r_{eD} - \frac{3}{4} = 2\pi t_{DA} + \frac{1}{2}\ln\left(\frac{4A}{e^{\gamma}C_A r_w^2}\right) \tag{1.34}$$

where A is the reservoir area and C_A is shape factor.

This is the solution for ideal state without skin. If with skin, Eq. (1.34) is converted into

$$p_{wD} = p_D + S = 2\pi t_{DA} + \frac{1}{2}\ln\left(\frac{4A}{e^{\gamma}C_A r_w^2}\right) + S \tag{1.35}$$

Although Eq. (1.35) is derived from a closed circular reservoir, it is applicable to ordinary bounded reservoir as long as A and C_A are considered as the area and shape factor of the bounded reservoir respectively. The shape factor C_A is decided by the shape of bounded reservoir and the location of the well, and it varies depending on reservoirs. As to bounded reservoirs with different shapes and sizes, the pseudo-steady state arrival time is different. See Table 1.4 for shape factor and pseudo-steady state arrival time of different shapes of reservoirs. Prior to this, we need to know the shape of reservoir and the location of well, so as to select the approximate state from the table to read its C_A value and $(t_{DA})_{pss}$ value for analysis.

Table 1.4 Shape factor C_A with different shape and size (after Dietz, 1965)

No.	Reservoir shape	Shape factor C_A		
		ln C_A	C_A	Start time of pseudo-steady state t_{DA}>
1	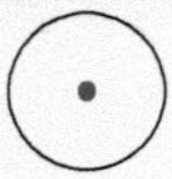	3.45	31.62	0.1
2	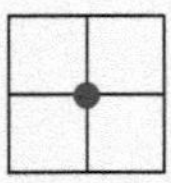	3.43	30.90	0.1
3	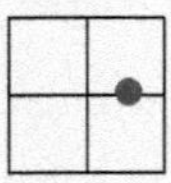	2.56	12.99	0.7
4	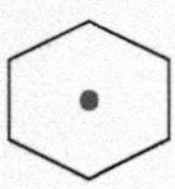	3.45	31.60	0.1
5	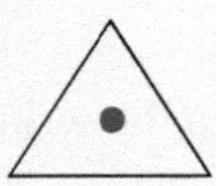	3.32	27.60	0.2
6	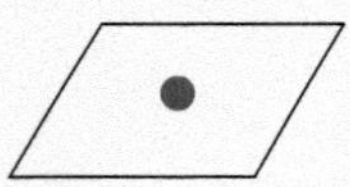	3.3	27.10	0.2
7	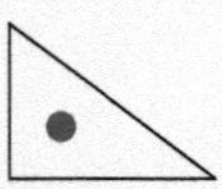	3.09	21.90	0.4
8	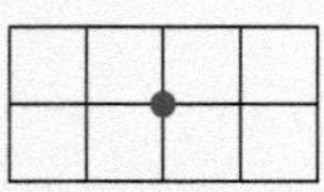	3.08	21.84	0.3
9	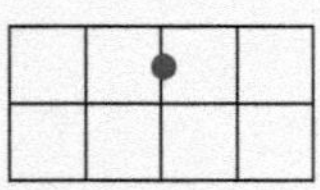	2.38	10.84	0.4
10	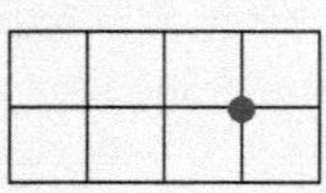	1.51	4.51	1.5
11	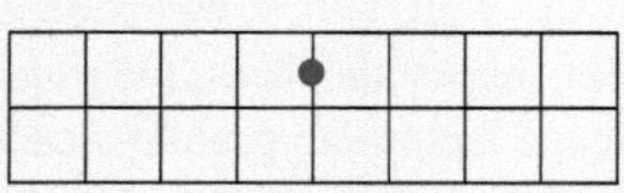	0.99	2.69	0.8
12	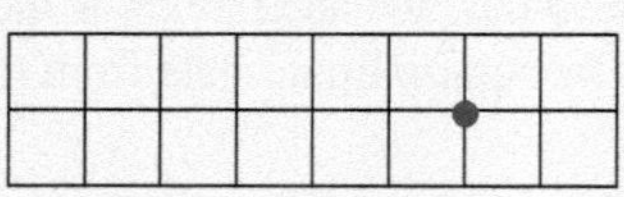	-1.46	0.232	4.0

1.3.9 Superposition Principle

1. **Multiwell superposition**

   Assuming that there are N wells in the infinite formation and each well is produced at a constant rate, the pressure change at any point of the formation is the sum of pressure change at this point resulting from the independent production of each well. Then,

$$\Delta p = p_i - p(r,t) = -\frac{\mu}{4\pi Kh}\sum_{j=1}^{N} q_j Ei\left[-\frac{r_j^2}{4\eta(t-t_j)}\right] = \frac{\mu}{4\pi Kh}\sum_{j=1}^{N} q_j \ln\left[\frac{4\eta(t-t_j)}{e^{\gamma} r_j^2}\right] \tag{1.36}$$

   where t_j is the starting time of production of well j and r_j is the distance from this point to well j.

2. **Variable-rate superposition of single well**

   If the flow rate has been changing slightly, the single well rate change can be approximately expressed by segment. In each interval of time, the flow rate changes little, which can be approximately regarded as a constant. As shown in Figure 1.5, divide the production into n intervals, and let $q_0 = 0$, $t_0 = 0$. The flow rate is q_i $(i = 1,2,3,\ldots,n)$ in the interval i, that is,

$$q = q_1,\quad 0 \le t \le t_1$$
$$q = q_2,\quad t_1 \le t \le t_2$$
$$q = q_3,\quad t_2 \le t \le t_3$$
$$\ldots$$
$$q = q_{n-1},\quad t_{n-2} \le t \le t_{n-1}$$
$$q = q_n,\quad t_{n-1} \le t \le t_n$$

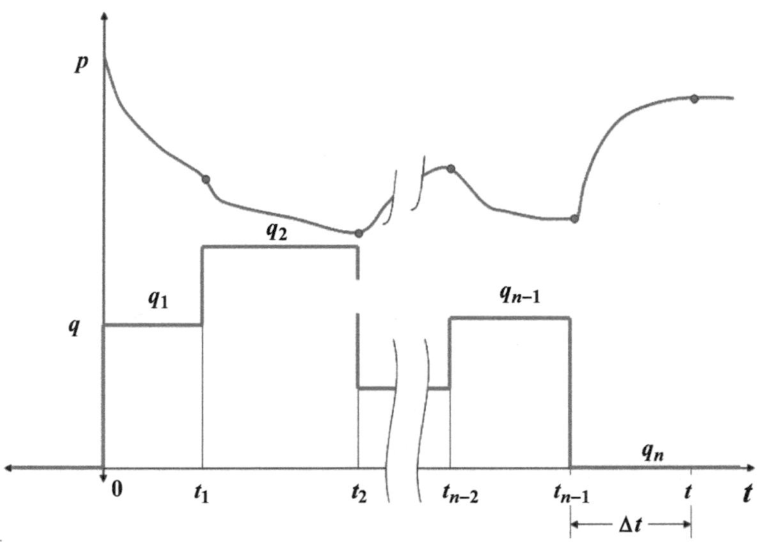

Figure 1.5 Variable rate and pressure history

At the interval n, $t \geq t_{n-1}$, then

$$p_i - p = \sum_{j=1}^{n} \frac{\mu\left(q_j - q_{j-1}\right)}{2\pi Kh}\left\{-\frac{1}{2}Ei\left[-\frac{r^2}{4\eta\left(t - t_{j-1}\right)}\right]\right\}$$
$$\approx \sum_{j=1}^{n} \frac{\mu\left(q_j - q_{j-1}\right)}{4\pi Kh}\left[\ln\frac{4\eta\left(t - t_{j-1}\right)}{e^{\gamma} r^2}\right] \tag{1.37}$$

$$\frac{2\pi Kh}{\mu}\left(p_i - p\right) = \sum_{j=1}^{n}\left(q_j - q_{j-1}\right)p_D\left(r_D, t_D - t_{Dj-1}\right) \tag{1.38}$$

Let r = r_w, that is, $r_D = 1$. The BHFP expression can be derived from the above equation as follows

$$\frac{2\pi Kh}{\mu}\left(p_i - p_{wf}\right) = \sum_{j=1}^{n}\left(q_j - q_{j-1}\right)p_D\left(1, t_D - t_{Dj-1}\right) + q_n S \tag{1.39}$$

3. **Image method**

If there is a well in the vicinity of no-flow boundary or fault, the fault can be taken as a mirror for the production to symmetrically allocate an image well. Because of symmetry, the fault looks like a streamline, and the flow is turned to a state in which two wells are produced in an infinite formation, as shown in Figure 1.6. Therefore, we have

$$p_i - p_{wf} = \frac{\mu q}{2\pi Kh}\left[p_D\left(1, t_D\right) + p_D\left(2L_D, t_D\right) + S\right] \tag{1.40}$$

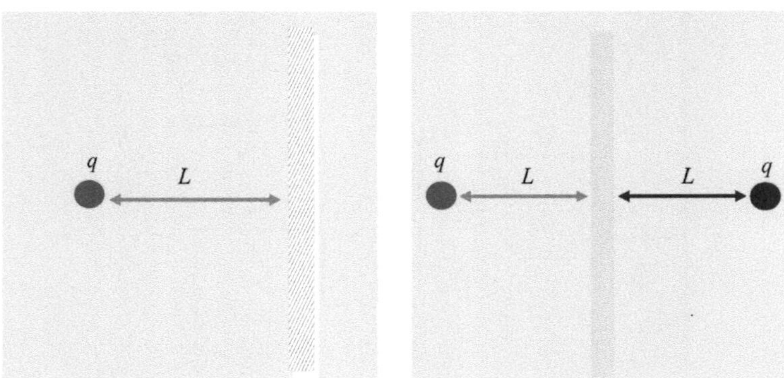

Figure 1.6 Single-fault geologic and mathematic model (after Ehlig-Economides, 2000)

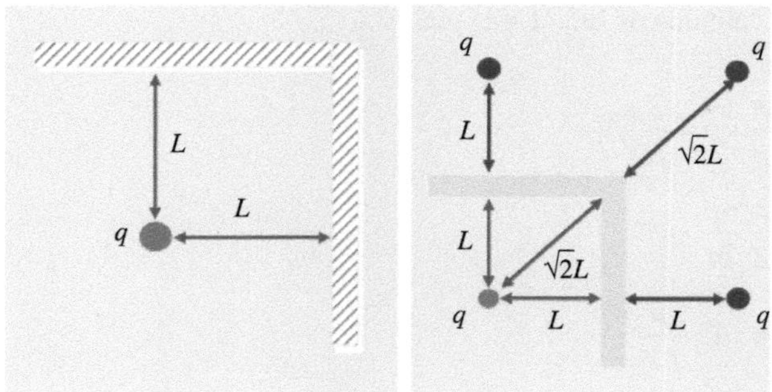

Figure 1.7 Ninety-degree intersecting fault geologic and mathematic model (after Ehlig-Economides, 2000)

where $L_D = L/r_w$. L is the distance from well to fault.

If a well is located in the central of included angle of two orthogonal faults, mapping is done by taking the two orthogonal faults as mirrors, and four wells are formed. In this way, the issue is turned into a state in which four wells are produced in an infinite formation, and the result is their sum, as shown in Figure 1.7.

If a well is intercalated between two parallel faults, mapping is performed repeatedly by taking the two parallel faults as mirrors, and a linear well array is formed. In this way, the issue is turned into a state in which infinite wells are produced in an infinite formation, and the result is an infinite series.

If a well is located at the center of a rectangular reservoir, mapping repeatedly by taking the four sides as mirrors, infinite identical rectangles overspread in the infinite formation are formed, and there is a well in each rectangle. In this way, the issue is turned into a state in which infinite wells are produced in an infinite formation, and the result is a dual infinite series.

1.3.10 Pseudo-Pressure, Pseudo-Time

Assuming that the gas reservoir is homogeneous and the flow follows Darcy's law, the flow equation is

$$\frac{\partial}{\partial t}\left(\frac{p}{Z}\right) = \frac{K}{\varphi}\nabla \cdot \left(\frac{p}{\mu Z}\nabla p\right) \tag{1.41}$$

Because both viscosity μ and real gas deviation factor Z are functions of pressure, the equation is nonlinear. In order to linearize it, the pseudo-pressure (Al-Hussainy and Ramey, 1966) is introduced as

$$m(p) = \int_{p_0}^{p} \frac{2p}{\mu Z} dp \tag{1.42}$$

where p_0 is a reference pressure taken randomly.

Differential calculus of Eq. (1.42) yields

$$\nabla m = \frac{\partial m}{\partial p}\nabla p = \frac{2p}{\mu Z}\nabla p$$

$$\frac{\partial m}{\partial t} = \frac{2p}{\mu Z}\frac{\partial p}{\partial t}$$

$$\frac{\partial}{\partial t}\left(\frac{p}{Z}\right) = \mu C_{\mathrm{t}}\frac{p}{\mu Z}\frac{\partial p}{\partial t}$$

Substituting it into Eq. (1.41) yields

$$\nabla^2 m = \frac{\varphi \mu C_{\mathrm{t}}}{K}\frac{\partial m}{\partial t} \tag{1.43}$$

Pseudo-time (Agarwal, 1979) is

$$t_{\mathrm{a}} = \int_0^t \frac{\mu_{\mathrm{i}} C_{\mathrm{ti}}}{\mu C_{\mathrm{t}}} dt \tag{1.44}$$

Diffusivity is

$$\eta = \frac{K}{\varphi \mu_{\mathrm{i}} C_{\mathrm{ti}}}$$

Eq. (1.43) yields

$$\nabla^2 m = \frac{1}{\eta}\frac{\partial m}{\partial t_{\mathrm{a}}} \tag{1.45}$$

For radial flow, Eq. (1.45) yields

$$\frac{1}{r}\frac{\partial}{\partial r}\left(r\frac{\partial m}{\partial r}\right) = \frac{1}{\eta}\frac{\partial m}{\partial t_{\mathrm{a}}} \tag{1.46}$$

Eq. (1.46) has the same form as flow equation of oil well. Therefore, after the pseudo-pressure and pseudo-time are introduced, the equation is linearized, but also the flow problem of gas reservoir can be transformed into that of oil reservoir. For the sake of utilizing the known solution related to the flow problem of oil reservoir, the dimensionless pseudo-pressure is defined.

$$p_{\mathrm{D}} = \frac{KhT_{\mathrm{sc}}\Delta m}{\alpha_m p_{\mathrm{sc}} q_{\mathrm{sc}} T} \tag{1.47}$$

The pseudo-pressure defined by Eq. (1.47) is different from pressure in respect of dimension, resulting in much inconvenience. Therefore, the normalized pseudo-pressure (Meunier et al., 1987) is introduced as follows

$$p_\mathrm{p} = p_\mathrm{i} + \frac{\mu_\mathrm{i} Z_\mathrm{i}}{p_\mathrm{i}} \int_{p_\mathrm{i}}^{p} \frac{p}{\mu Z} dp \tag{1.48}$$

Obviously, p_p possesses the dimension of pressure, and if $p = p_\mathrm{i}$, we have $p_\mathrm{p} = p_\mathrm{i}$. The dependence of pseudo-pressure and pressure can be figured out using the numerical integration method and the relation of μ, Z, and pressure. In this case, if pseudo-pressure is given, the pressure can be derived, and vice versa.

When the pressure is higher than 14.0 MPa, $\frac{p}{\mu Z} \approx Const.$ From Eq. (1.42), we have

$$m = \frac{2p_\mathrm{i}}{\mu_\mathrm{i} Z_\mathrm{i}} p \tag{1.49}$$

Here, the difference between pseudo-pressure and pressure is a constant coefficient, and the pseudo-pressure analysis can be transformed into pressure analysis.

When the pressure is lower than 14.0 MPa, $\mu Z \approx$ const. From Eq. (1.42), we have

$$m = \frac{1}{\mu_\mathrm{i} Z_\mathrm{i}} p^2 \tag{1.50}$$

Here, the difference between pseudo-pressure and pressure square is a constant coefficient, and the pseudo-pressure analysis is transformed into pressure square analysis.

1.3.11 Material Balance Time

The material balance time is $t_\mathrm{c} = Q/q$. Its geometric meaning is shown in Figure 1.8; that is, to build the equivalent relation between variable rate and constant rate production.

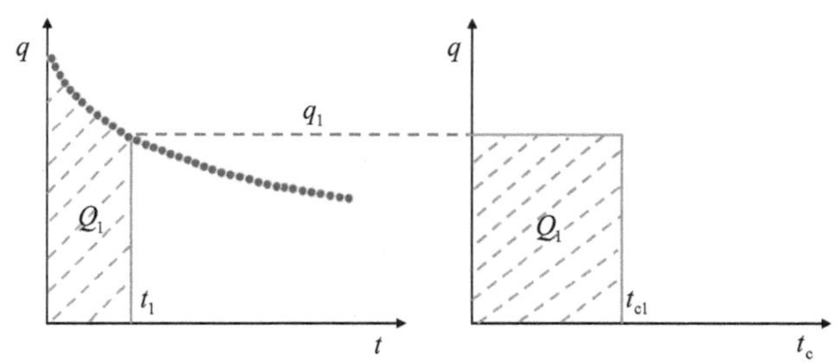

Figure 1.8 Material balance time

1.3.12 Flow Regime Identification

As to different types of reservoirs, the flow has diverse shapes on the log–log plot at different flow regimes, which mainly includes radial flow, linear flow, bilinear flow, spherical or hemispherical flow, and pseudo-radial flow.

The above types of flow regimes may result from a variety of causes. Radial flow is the most common transient flow state, which occurs in the homogeneous and isopachous reservoir after a well has been perforated wholly and put into production. As to vertically fractured well and horizontal well, observed from a position far away from them, they can be regarded as a vertical well with an enlarged influence range. After a long-term production, pseudo-radial flow would occur. Linear flow is also a transientflow state, which often occurs in the following states, namely, linear flow formed by parallel impermeable boundary in the reservoir, linear flow or bilinear flow formed by hydraulically fractured vertical fractures, linear flow formed by horizontal wells, and vertical linear flow formed by high-permeability streaks in low-permeability tight reservoirs. As to imperfect perforation, incomplete perforation, or RFT formation test, etc., spherical and hemispherical flows can be formed as shown in Table 1.5.

1.3.13 Type Curve Matching Principle

The fundamental principle of type curve matching and the workflow are illustrated below by taking log–log analysis of well test as an example. In respect to dimensionless physical quantity, the dimensionless pressure and dimensionless time are defined, respectively. If logarithm is taken for these two parameters, we have

$$\lg t_{\mathrm{D}} = \lg\Delta t + \lg\left(\frac{\alpha_{\mathrm{t}} K}{\varphi\mu C_{\mathrm{t}} r_{\mathrm{w}}^2}\right) = \lg\Delta t + \lg C_1 \tag{1.51}$$

$$\lg p_{\mathrm{D}} = \lg\Delta p + \lg\left(\frac{Kh}{\alpha_{\mathrm{p}} q\mu B}\right) = \lg\Delta p + \lg C_2 \tag{1.52}$$

where

$$C_1 = \frac{\alpha_{\mathrm{t}} K}{\varphi\mu C_{\mathrm{t}} r_{\mathrm{w}}^2} \qquad C_2 = \frac{Kh}{\alpha_{\mathrm{p}} q\mu B}$$

In other words, a dimensionless curve is different from a dimensional curve only by a constant on the log–log paper. If p_{D}~$t_{\mathrm{D}}/C_{\mathrm{D}}$ and Δp~Δt curves are plotted, respectively, on the log–log paper, the formation parameters can be determined based on match points. This is the fundamental principle of log–log matching.

Table 1.5 Different types of flow regimes

SN	Flow regime				
	Name	Slope	Region	p–t	Types
1	Wellbore	1.0	Early time	Δp–Δt	
2	Radial flow	0.0	1 Homogeneous reservoir		
			2 Vertically fractured well Late pseudo-radial flow		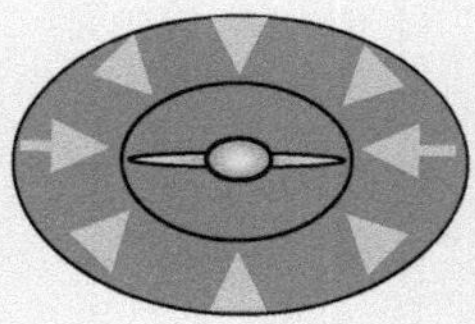
			3 Horizontal well Late pseudo-radial flow		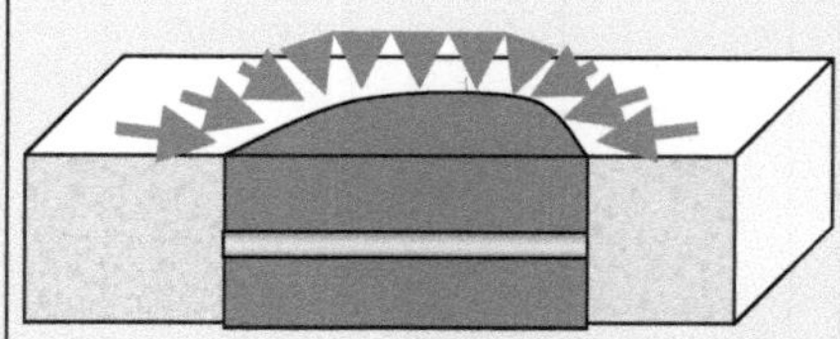
3	Hemi-spherical flow	-0.50	1 Few perforations open	Δp–$(\Delta t)^{-0.5}$	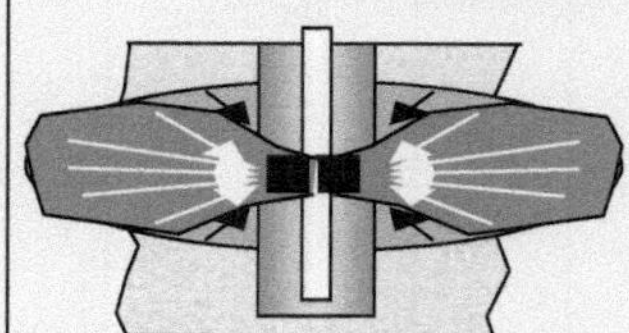
			2 Small part of zone perforated		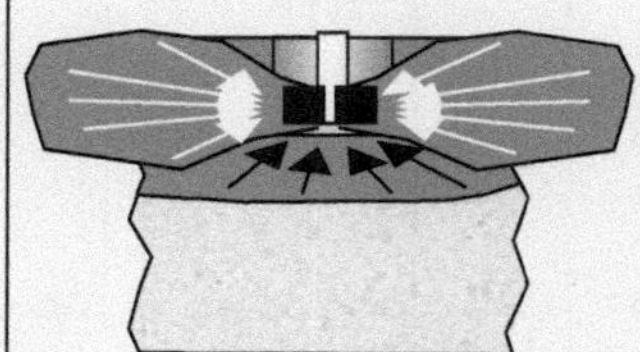
			3 Certain wireline testing tools		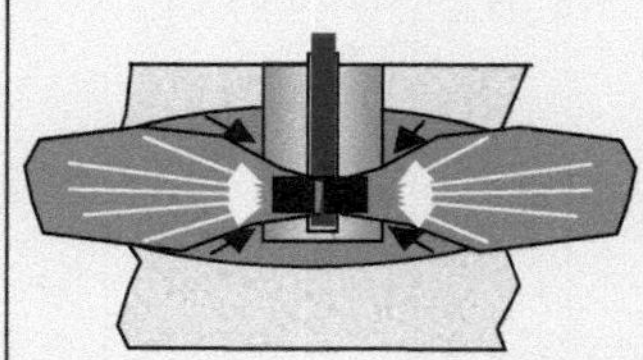

(Continued)

Table 1.5 **Different types of flow regimes** ***(cont.)***

SN	Flow regime				
	Name	**Slope**	**Region**	***p–t***	**Types**
4	Linear flow	0.50	1 Vertically fractured well	$\Delta p – (\Delta t)^{0.5}$	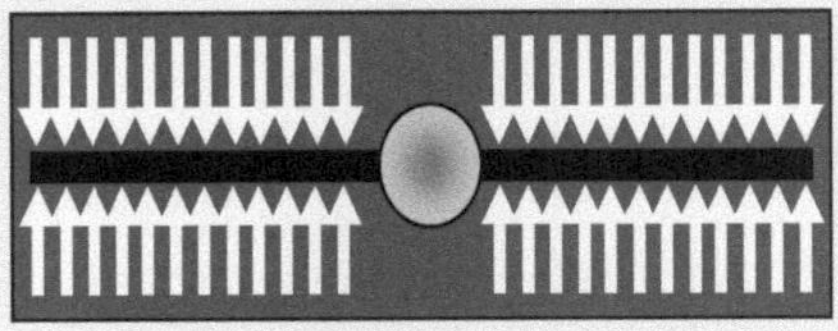
			2 Channel reservoir		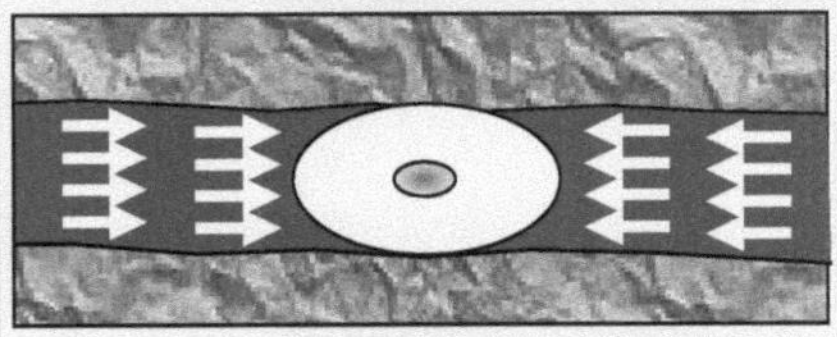
			3 Horizontal well Early linear flow		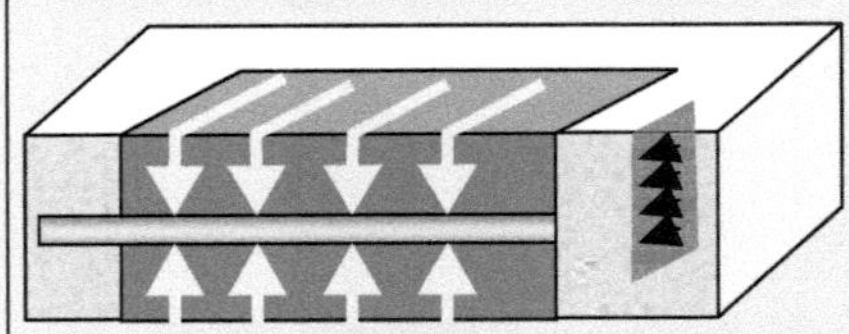
			4 Horizontal well Late linear flow		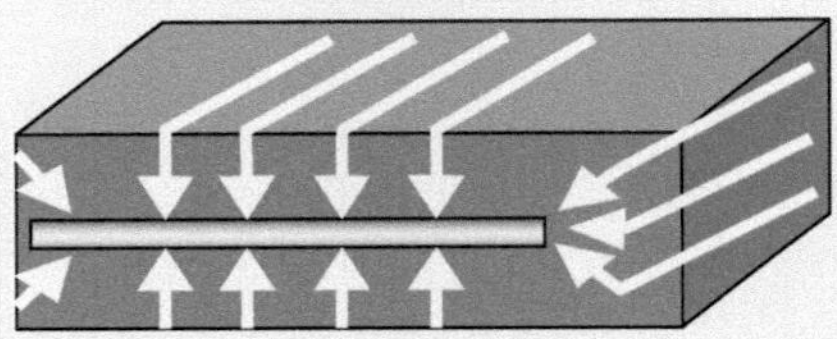
			5 High-permeability zone		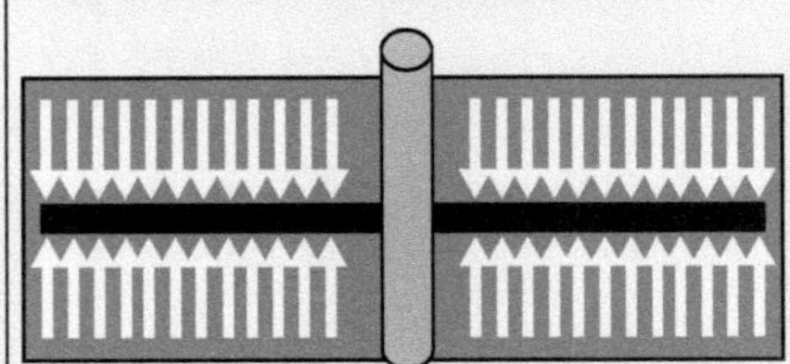
5	Bilinear flow	0.25	Vertically fractured well	$\Delta p – (\Delta t)^{0.25}$	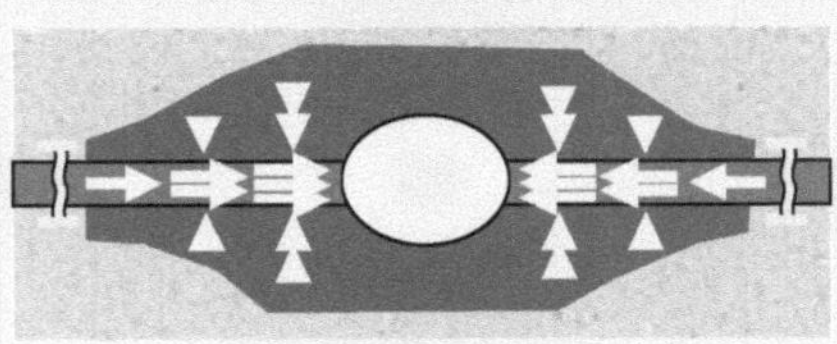
6	Pseudo-steady state (draw-down)	1.00	Late time	$\Delta p – \Delta t$	

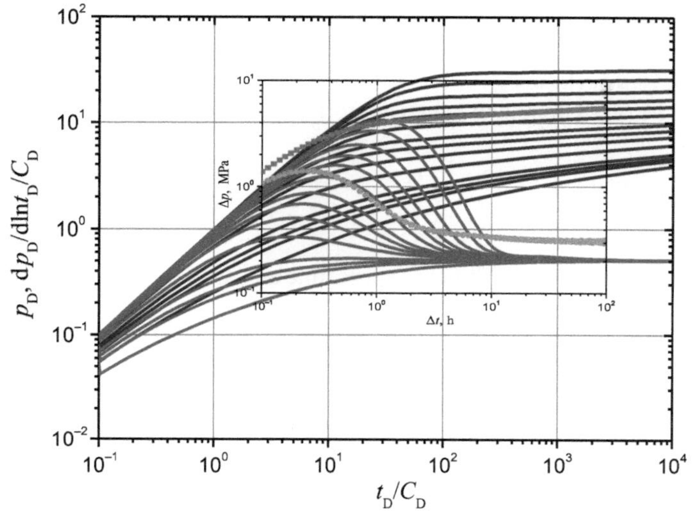

Figure 1.9 Log–log type curve matching—1

Firstly, plot p_D~t_D/C_D and Δp~Δt curves respectively on the log–log paper, as shown in Figure 1.9.

Shift the transparent log–log paper vertically and horizontally, allowing the observed pressure and derivative data to best match the pressure and derivative curves corresponding to a C_De^{2S} value, and record the match value of the curve as $(C_De^{2S})_M$, as shown in Figure 1.10.

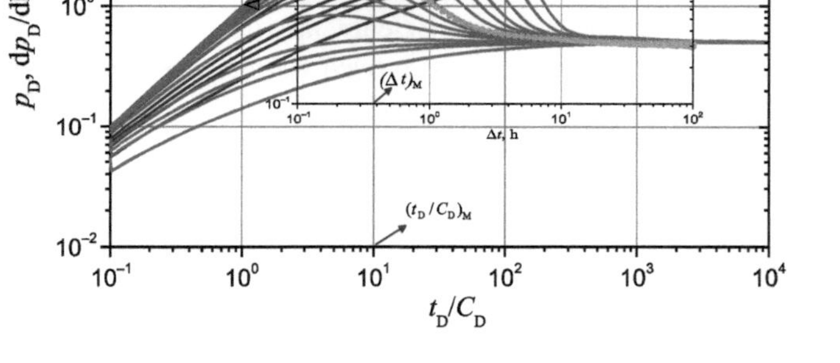

Figure 1.10 Log–log type curve matching—2

Randomly select a point on the transparent graph paper as the match point, and read the coordinate value of the point on the log–log paper and interpretation chart paper, $(\Delta p)_M$, $(\Delta t)_M$, $(p_D)_M$, $(t_D/C_D)_M$.

Based on the pressure match value, the formation flow capacity can be obtained as

$$Kh = \alpha_p q \mu B \left(\frac{p_D}{\Delta p} \right)_M$$

Based on the definitions of dimensionless time and dimensionless wellbore storage, we have

$$\frac{t_D}{C_D} = \frac{\alpha_t K t}{\varphi \mu C_t r_w^2} \times \frac{\varphi C_t h r_w^2}{\alpha_c C} = \frac{\alpha_t K h t}{\alpha_c C \mu}$$

Based on the time match value, the wellbore storage coefficient can be obtained as

$$C = \frac{\alpha_t K h}{\alpha_c \mu \left(\frac{t_D/C_D}{\Delta t} \right)_M}$$

Based on the read $(C_D e^{2S})_M$, the skin factor can be obtained as

$$S = 0.5 \ln \left[\frac{\left(C_D e^{2S} \right)_M}{C_D} \right]$$

This is the matching procedure for log–log analysis of homogeneous infinite formation, that is, to conduct pressure match and time match firstly, record the corresponding match points, and figure out the formation parameters. The log–log matching principle for the APDA is the same; apart from that, the definitions of dimensionless variables are different.

1.3.14 Recoverable Reserves

Based on the CNS "Oil and Gas Resource Extent Classification" (GBT 19492-2004), the recoverable reserves are classified into proved estimated ultimate recovery and proved initial reserves, as shown in Figure 1.11.

The possible reserves are the total volume of natural gas expected to withdraw ultimately from the discovered oil and gas reservoirs under the current technical conditions and government regulations, and it is also called ultimate recoverable reserves. The probable reserves refer to the total volume of natural gas estimated and capable of being economically exploited as per the current economic conditions (price, cost, additional capital, etc.) under the conditions of applied or affirmatively to be applied technologies.

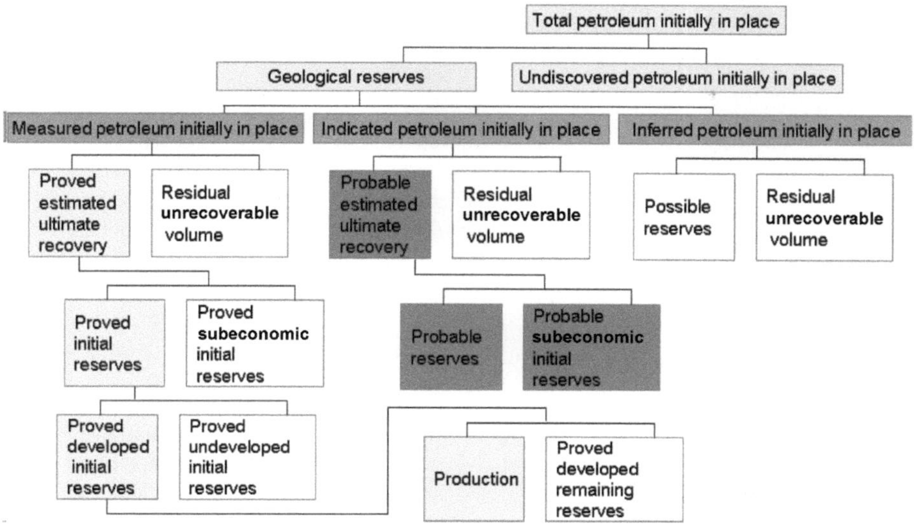

Figure 1.11 Reserves classification criteria of China (after GBT 19492, 2004)

1.3.15 Dynamic Reserve

Dynamic reserve is a generalized concept, identical with "movable reserves" and "geological reserves estimated by a dynamic method." Various formulations are listed in Table 1.6.

In summary, the dynamic reserves possess the following characteristics, including (1) targeting gas reservoir or single well; (2) movable theoretically, usually lower than reserves estimated by volumetric method; (3) obtained from production data; (4) containing both recoverable and unrecoverable reserves, a value in the middle of measured petroleum initially in place and technically recoverable reserves, and related to the current technical levels and well pattern; and (5) timeliness.

Table 1.6 **Definition of dynamic reserves (after Sun, 2012)**

Year	Author	Source		Formulation
1994	Ya-he Yang	Natural Gas Industry	Dynamic reserves estimation for low-permeability gas reservoir by pressure buildup data	Put forward the phrase of "dynamic reserves" for the first time but did not present its definite definition

(Continued)

Table 1.6 Definition of dynamic reserves (after Sun, 2012) *(cont.)*

Year	Author	Source		Formulation
1996	Lun-you Zhang	Natural Gas Exploration and Development	Concept of producible reserves and approach for determining economically recoverable reserves	In the developed geological reserves, under the conditions of existing technologies and stable exploitation mode of existing well pattern, when all the wells have been brought into production and the gas output and the formation pore pressure within the swept range have dropped to zero, the total volume of natural gas that can be flown out of the gas reservoir
1999	Yu-hong Hao	Well Testing and Production Technology	On the dynamic reserves of gas field	The gas in place capable of seepage or flow in the course of development
2002	Xi Feng	Natural Gas Industry	Analysis of dynamic reserves computational problems at early stages of heterogeneous gas reservoir development	The sum of volume converted to the standard conditions of gas ultimately capable of effectively flowing in the interconnected porosity volume of gas reservoir under the existing exploitation technical levels
2005	Ministry of Land and Resources of PRC	Geological and mineral industry standard of the People's Republic of China DZ/T 0217-2005 "Regulation of Petroleum Reserves Estimation"	Regulation of petroleum reserves estimation	When the geological reserves is difficult to be estimated by volumetric method, it should be estimated using dynamic method (material balance method, elastic two-phase process, etc.); the measured petroleum initially in place and the indicated

Table 1.6 Definition of dynamic reserves (after Sun, 2012) *(cont.)*

Year	Author	Source		Formulation
				petroleum initially in place should be classified based on the confidence of production and pressure data
2008	Ping Guo	Journal of Chongqing University of Science and Technology	Method study on dynamic reserves of gas well	Reserves figured out using dynamic approaches like material balance, transient well test, production decline, and production accumulation
2009	Xiao-gang Chen	Special Oil & Gas Reservoirs	Overview of prediction methods for dynamic reserves of gas reservoir	
2010	Shun-li He	Science Technology and Engineering	New method for estimating dynamic reserves of low-permeability gas reservoir	
2012	He-dong Sun	Modern Well testing Analysis and Productivity Evaluation of Complex Gas Reservoir	Chapter 5–Dynamic reserves assessment techniques	Under the existing technology and stable exploitation mode of existing well pattern, on the basis of production performance data such as flow rate and pressure of single well or gas reservoir, estimated using the gas reservoir engineering method—the cumulative gas production when the gas well flow rate drops to zero and the formation pore pressure within the swept range drops to 1 normal atmosphere.

Arps Decline Curves Analysis

2

The chapter presents the types, law, and theory of the Arps method, the methodology and cases of curve matching to estimate the recoverable reserves. Finally, the power function analysis method is introduced.

2.1 Arps Equations

As to wells with a long production history and producing at constant bottom hole flowing pressure (BHFP), Arps (1945) derived three types of production declines by the rate (cumulative production)-time relationship, including exponential decline, hyperbolic decline, and harmonic decline. This method is simple, without consideration of reservoir or well parameters, and can be applied to different types of reservoirs. However, this method is limited in two aspects. First, the ultimate recoverable reserves must be estimated under the assumption that the production conditions will keep unchanged in the future. Second, what is represented by the decline curve is the decline law at boundary dominated flow state; therefore, it cannot be used to analyze the data under a transient flow state.

2.1.1 Exponential Decline

The concept of decline rate will be first introduced. It refers to the flow rate changed in unit time, often denoted by D, as shown in Eq. (2.1).

$$D = -\frac{\Delta q/q}{\Delta t} \tag{2.1}$$

As to exponential decline, the rate–time relationship is

$$q = q_i e^{-Dt} \tag{2.2}$$

The rate–cumulative production relationship is

$$Q = \int_0^t q\,dt = \int_0^t q_i e^{-Dt}\,dt = \frac{q_i - q}{D} \tag{2.3}$$

On the semi-log plot, lg $q \sim t$ is a straight line, that is, D is a constant. On the Cartesian plot, the rate and cumulative production also represents as a straight line, as shown in Figure 2.1. Because of the simple linear relation, it is easy to identify and use the exponential decline law.

Advanced Production Decline Analysis and Application

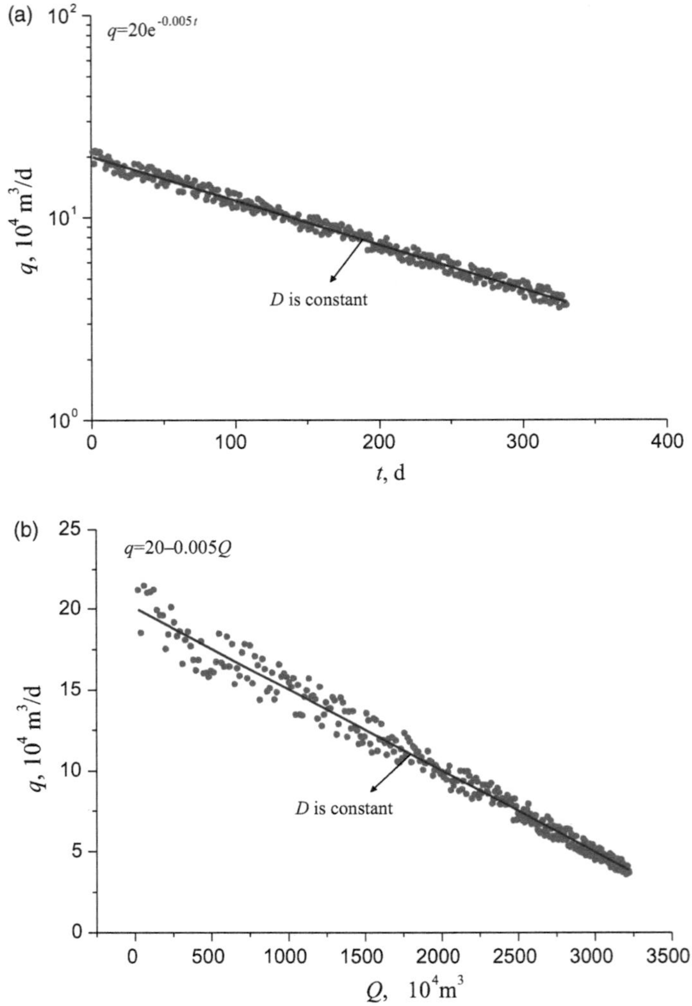

Figure 2.1 Exponential decline curve. (a) Rate–time curve; (b) Rate–cumulative production curve

The decline rate is used to figure out the ultimate produced fluid under exponential decline case. Initial flow rate q_f at the time of t_f is determined by

$$q_f = q_i e^{-Dt_f} \tag{2.4}$$

Let the flow rate at the time of abandonment t_{ab} be q_{ab}. Based on Eq. (2.3), the cumulative production between time t_f and t_{ab} can be obtained from

$$Q_f = \frac{q_f - q_{ab}}{D} \tag{2.5}$$

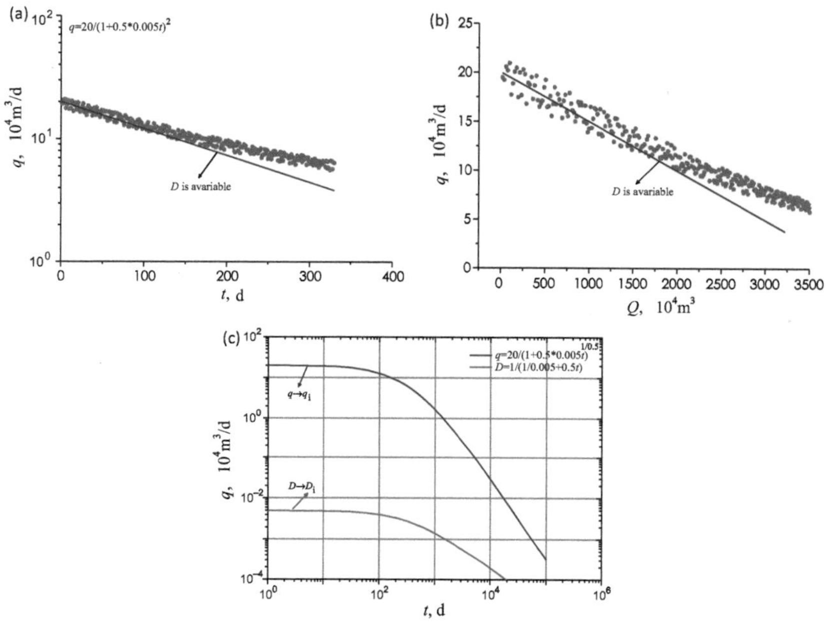

Figure 2.2 Hyperbolic decline curves. (a) Rate–time curve; (b) Rate–cumulative production curve; (c) D–time curve

Therefore, the ultimate recoverable volume is

$$EUR = N_p + Q_f = N_p + \frac{q_f - q_{ab}}{D} \tag{2.6}$$

Based on Eq. (2.4), the production time reaching abandonment can also be estimated as

$$\Delta t = t_{ab} - t_f = \frac{1}{D}\ln\left(\frac{q_f}{q_{ab}}\right) \tag{2.7}$$

2.1.2 *Hyperbolic Decline*

As to hyperbolic decline, lg $q \sim t$ is no longer a straight line on the semilog plot; that is, D is a variable, as shown in Figure 2.2. The decline exponent b is introduced to describe such variation as in Eq. (2.8): $b = 0$ corresponds to exponential decline, and $b = 1$ corresponds to harmonic decline; the smaller the b, the faster the decline is.

$$q = \frac{q_i}{(1+bD_i t)^{1/b}} (b \text{ is constant}, 0 < b < 1) \tag{2.8}$$

If Eq. (2.8) is integrated, the cumulative production–time relationship within a period of time can be derived by

$$Q=\int_{t_0}^{t} q dt=\int_{t_0}^{t} \frac{q_i}{\left(1+bD_i t\right)^{1/b}} dt \tag{2.9}$$

Based on the mathematical manual, the indefinite integral of the above form is

$$\int (ax+b)^{\mu} dx=\frac{1}{a(\mu+1)}(ax+b)^{\mu+1}+C_i \qquad (\mu \neq -1) \tag{2.10}$$

Therefore,

$$\int \frac{q_i}{\left(1+bD_i t\right)^{1/b}} dt=\frac{q_i}{bD_i\,(1-1/b)}\left(1+bD_i t\right)^{1-1/b}=\frac{q_i}{D_i\,(b-1)}\left(1+bD_i t\right)^{1-1/b} \tag{2.11}$$

The limits of integration t_0 and t are substituted into Eq. (2.11), we have

$$\int_{t_0}^{t} \frac{q_i}{\left(1+bD_i t\right)^{1/b}} dt=\frac{q_i}{(b-1)\,D_i}\left[\frac{1}{\left(1+bD_i t\right)^{1/b-1}}-\frac{1}{\left(1+bD_i t_0\right)^{1/b-1}}\right] \tag{2.12}$$

When $t_0 = 0$,

$$\int_{0}^{t} \frac{q_i}{\left(1+bD_i t\right)^{1/b}} dt=\frac{q_i}{(b-1)\,D_i}\left[\frac{1}{\left(1+bD_i t\right)^{1/b-1}}-1\right]=\frac{q_i}{(1-b)\,D_i}\left[1-\frac{1}{\left(1+bD_i t\right)^{1/b-1}}\right] \tag{2.13}$$

Eq. (2.8) is rearranged and substituted into Eq. (2.13); then, the cumulative production–time relationship can be derived as

$$\int_{0}^{t} \frac{q_i}{\left(1+bD_i t\right)^{1/b}} dt=\frac{q_i}{(1-b)\,D_i}\left[1-\frac{1}{\left(q_i/q\right)^{(1/b-1)b}}\right]=\frac{q_i}{(1-b)\,D_i}\left[1-\left(\frac{q}{q_i}\right)^{1-b}\right] \tag{2.14}$$

The decline rate is used to work out the ultimate produced fluid under hyperbolic decline case. Initial flow rate and the flow rate at the time of abandonment are predicted as

$$q_f=\frac{q_i}{\left(1+bD_i t_f\right)^{1/b}} \tag{2.15}$$

$$q_{ab}=\frac{q_i}{\left(1+bD_i t_{ab}\right)^{1/b}} \tag{2.16}$$

When $t_0 = t_f$ and $t = t_{ab}$, the cumulative production Q_f of the forecast period can be derived from Eq. (2.12)

$$Q_f = \int_{t_f}^{t_{ab}} \frac{q_i}{(1+bD_i t)^{1/b}} dt = \frac{q_i}{(b-1)D_i}\left[\frac{1}{(1+bD_i t_{ab})^{1/b-1}} - \frac{1}{(1+bD_i t_f)^{1/b-1}}\right] \tag{2.17}$$

Eq. (2.15) and Eq. (2.16) are rearranged and substituted into Eq. (2.17). Then,

$$\int_{t_f}^{t_{ab}} \frac{q_i}{(1+bD_i t)^{1/b}} dt = \frac{q_i^b}{(1-b)D_i}\left(q_f^{1-b} - q_{ab}^{1-b}\right) \tag{2.18}$$

The ultimate recoverable volume is

$$EUR = N_p + Q_f = N_p + \frac{q_i^b}{(1-b)D_i}\left(q_f^{1-b} - q_{ab}^{1-b}\right) \tag{2.19}$$

Based on Eq. (2.15) and Eq. (2.16), the production time reaching abandonment can also be estimated

$$\Delta t = t_{ab} - t_f = \frac{\left(\frac{q_i}{q_{ab}}\right)^b - \left(\frac{q_i}{q_f}\right)^b}{bD_i} \tag{2.20}$$

2.1.3 Harmonic Decline

Harmonic decline is the special form of hyperbolic decline. In Eq. (2.8), b = 1 corresponds to harmonic decline

$$q = \frac{q_i}{(1+D_i t)} \tag{2.21}$$

If Eq. (2.21) is integrated, the cumulative production-time relationship within a period of time can be derived

$$Q = \int_{t_0}^{t} q dt = \int_{t_0}^{t} \frac{q_i}{(1+D_i t)} dt \tag{2.22}$$

Based on the mathematical manual, the indefinite integral of the above form is

$$\int (ax+1)^{-1} dx = \frac{1}{a}\ln(ax+1) + C_i \tag{2.23}$$

Therefore,

$$\int \frac{q_i}{1+D_i t}\,dt = \frac{q_i}{D_i}\ln(1+D_i t)+C_i \tag{2.24}$$

The limits of integration t_0 and t are substituted into Eq. (2.24). Then,

$$\int_{t_0}^{t} \frac{q_i}{1+D_i t}\,dt = \frac{q_i}{D_i}\ln\left(\frac{1+D_i t}{1+D_i t_0}\right) \tag{2.25}$$

When $t_0 = 0$, the cumulative production–time relationship can be derived as

$$Q = \int_0^t \frac{q_i}{1+D_i t}dt = \frac{q_i}{D_i}\ln(1+D_i t) = \frac{q_i}{D_i}\ln\frac{q_i}{q} \tag{2.26}$$

It is known from Eq. (2.21) and Eq. (2.26) that the reciprocal of rate of harmonic decline $1/q{\sim}t$ is linear, and the cumulative production $Q{\sim}\ln q$ is also linear, as shown in Figure 2.3.

The decline rate is used to figure out the ultimate produced fluid under the harmonic decline case. Initial flow rate and the flow rate at the time of abandonment are predicted as

$$q_f = \frac{q_i}{1+D_i t_f} \tag{2.27}$$

$$q_{ab} = \frac{q_i}{1+D_i t_{ab}} \tag{2.28}$$

When $t_0 = t_f$ and $t = t_{ab}$, the cumulative production Q_f of the forecast period can be derived from Eq. (2.25)

$$Q_f = \int_{t_f}^{t_{ab}} \frac{q_i}{1+D_i t}\,dt = \frac{q_i}{D_i}\ln\left(\frac{1+D_i t_{ab}}{1+D_i t_f}\right) = \frac{q_i}{D_i}\ln\left(\frac{q_f}{q_{ab}}\right) \tag{2.29}$$

The ultimate recoverable volume is

$$EUR = N_p + Q_f = N_p + \frac{q_i}{D_i}\ln\left(\frac{q_f}{q_{ab}}\right) \tag{2.30}$$

Based on Eq. (2.27) and Eq. (2.28), the production time reaching abandonment can be estimated by

$$\Delta t = t_{ab} - t_f = \frac{q_i}{D_i}\left(\frac{1}{q_{ab}} - \frac{1}{q_f}\right) \tag{2.31}$$

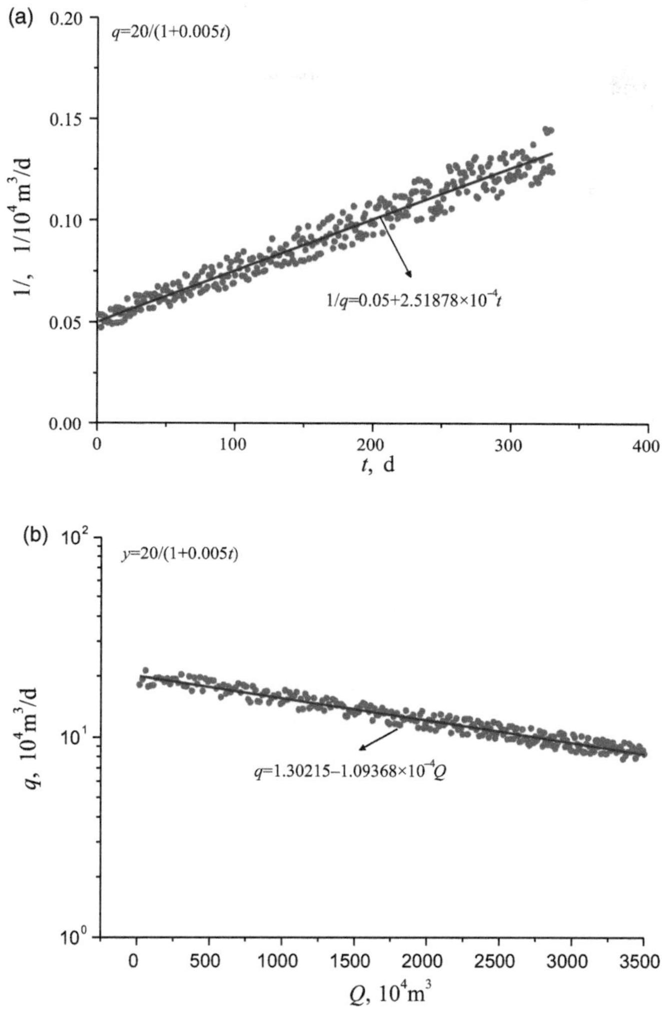

Figure 2.3 Harmonic decline curves. (a) Rate–time curve; (b) Rate–cumulative production curve

2.1.4 Comparison of Three Decline Types

The rate, cumulative production, and time relationships of three decline curves are listed in Table 2.1. Given the same parameters, trends of three decline curves are shown in Figure 2.4. It is observed from Figure 2.4 that the exponential decline is the fastest, but the forecast cumulative production is the lowest; the harmonic decline is the slowest, but the forecast cumulative production is the highest; whereas the hyperbolic decline lies in between them.

As shown in Figure 2.5, if the data within 150 days are used for matching, excellent fitting curves can be obtained respectively from the three methods, but prodigious

Table 2.1 **Arps equations**

Basic characteristics	Decline type	Rate–time	Rate–cumulative production	$D_i t$
$b=0$	Exponential	$q=q_i e^{-D_i t}$	$Q=(q_i-q)/D_i$	$D_i t=\ln(q_i/q)$
$0<b<1$	Hyperbolic	$q=q_i(1+bD_i t)^{-1/b}$	$Q=\frac{q_i}{(1-b)D_i}\left[1-\left(\frac{q}{q_i}\right)^{1-b}\right]$	$D_i t=\frac{(q_i/q)^b-1}{b}$
$b=1$	Harmonic	$q=q_i(1+D_i t)^{-1}$	$Q=q_i\ln(q_i/q)/D_i$	$D_i t=(q_i/q)-1$

deviation will occur in the cumulative production trend prediction. Therefore, when the Arps method is used to conduct the analysis, the production time should be long enough.

2.1.5 Modified Hyperbolic Decline

To avoid the issue that the results of hyperbolic decline might be too positive, Robertson (1988) introduced the modified hyperbolic decline formula, as shown in Eq. (2.32). This method divides the conventional hyperbolic decline into two stages at a given time, namely, the first stage conforming to the hyperbolic decline regularity and the second stage conforming to the exponential decline regularity, as shown in Figure 2.6.

$$q=q_i\frac{(1-\beta)^b e^{-Dt}}{\left(1-\beta e^{-Dt}\right)^b} \qquad (\beta \text{ is constant}, 0<\beta<1) \tag{2.32}$$

where β is a transition coefficient (which is dimensionless).

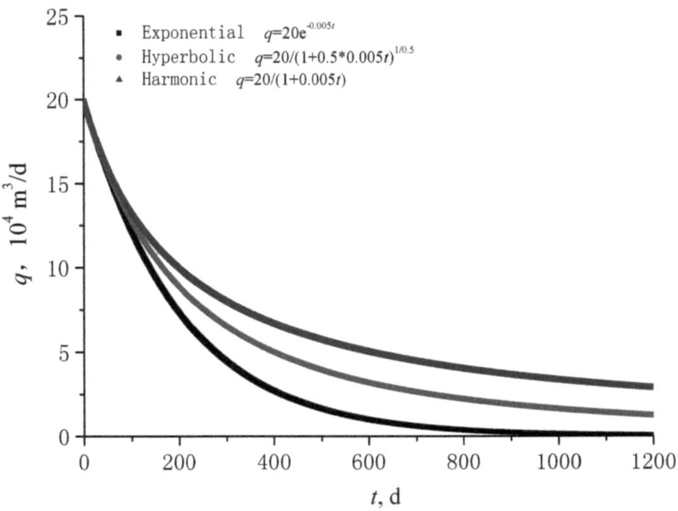

Figure 2.4 Arps decline curve shapes for a Cartesian rate–time plot

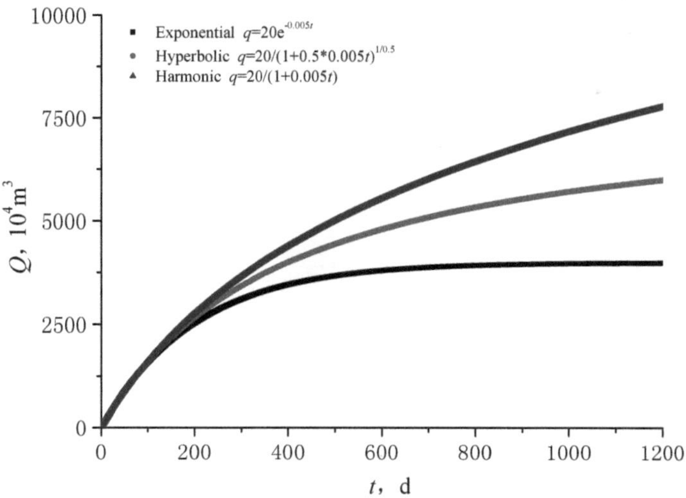

Figure 2.5 Arps decline curve shapes for a Cartesian cumulative production–time plot

The first stage of modified hyperbolic decline conforms to the hyperbolic decline and changes into exponential decline at a given time. Here, the decline rate is denoted by D_{lim}, and the flow rate change at different stages is expressed by the following equation

$$D = \frac{D_i}{1 + bD_i t} \qquad D > D_{\text{lim}} \tag{2.33}$$

$$q = \frac{q_i}{\left(1 + bD_i t\right)^{1/b}} \qquad D > D_{\text{lim}} \tag{2.34}$$

$$q = q_{\text{lim}} e^{-D_{\text{lim}}(t - t_{\text{lim}})} \qquad D \le D_{\text{lim}} \tag{2.35}$$

Based on Eq. (2.33) and Eq. (2.34), the flow rate at the transition point q_{lim} can be derived by

$$q_{\text{lim}} = q_i \left(\frac{D_{\text{lim}}}{D}\right)^{1/b} \tag{2.36}$$

Based on Eq. (2.34), the time at the transition point t_{lim} can be derived by

$$t_{\text{lim}} = \frac{\left(q_i / q_{\text{lim}}\right)^b - 1}{bD_i} \tag{2.37}$$

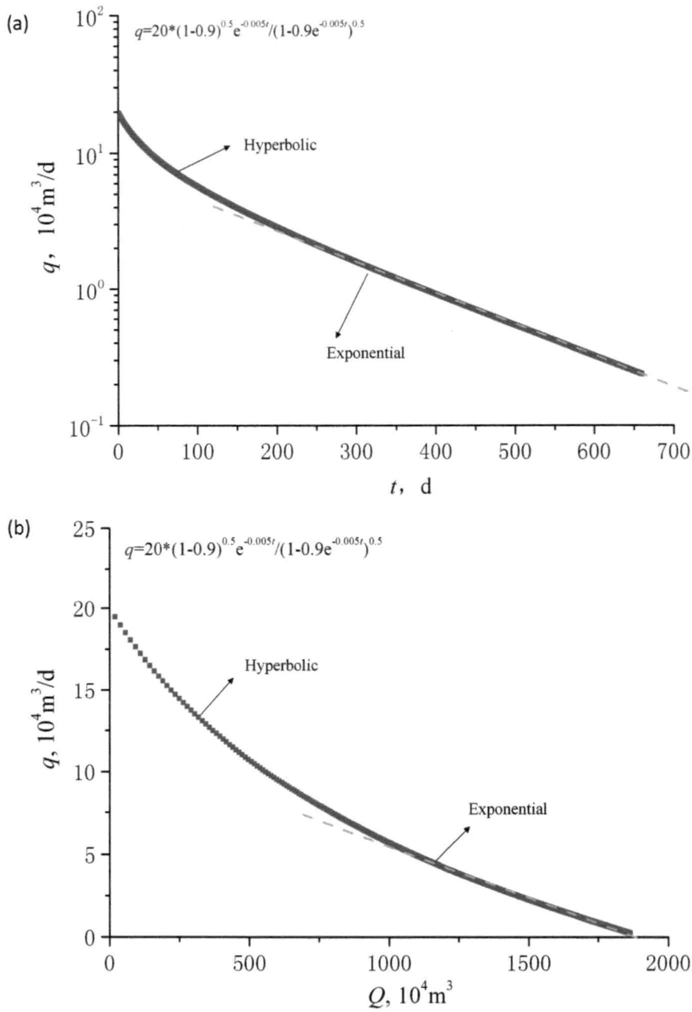

Figure 2.6 Modified hyperbolic decline curves. (a) Rate–time curve; (b) Rate–cumulative production curve

The cumulative production is

$$Q = \frac{q_i^b}{(1-b)D_i}\left(q_i^{1-b} - q^{1-b}\right) \quad \text{when } D > D_{\text{lim}} \tag{2.38}$$

$$Q = \frac{q_i^b}{(1-b)D_i}\left(q_i^{1-b} - q_{\text{lim}}^{1-b}\right) + \frac{q_{\text{lim}} - q}{D_{\text{lim}}} \quad \text{when } D \le D_{\text{lim}} \tag{2.39}$$

When $b = 1$, the hyperbolic decline becomes harmonic decline, and the cumulative production equation is

$$Q = \frac{q_i}{D_i} \ln\left(q_i / q\right) \quad \text{when } D > D_{lim} \tag{2.40}$$

$$Q = \frac{q_i}{D_i} \ln\left(q_i / q_{lim}\right) + \frac{q_{lim} - q}{D_{lim}} \quad \text{when } D \le D_{lim} \tag{2.41}$$

If Eq. (2.32) is directly integrated, the cumulative production equation can also be derived. Based on Eq. (2.10), we have

$$\int q_i \frac{(1-\beta)^b e^{-Dt}}{\left(1-\beta e^{-Dt}\right)^b} dt = \frac{q_i (1-\beta)^b \left(1-\beta e^{-Dt}\right)^{1-b}}{\beta D(1-b)} + C_i \tag{2.42}$$

Therefore,

$$Q = \int_0^t q dt = \frac{q_i (1-\beta)}{\beta D(1-b)} \left[\left(\frac{1-\beta}{1-\beta e^{-Dt}} \right)^{b-1} - 1 \right] \tag{2.43}$$

When $b = 1$, based on Eq. (2.23), Eq. (2.32) is directly integrated. Then,

$$\int q_i \frac{(1-\beta) e^{-Dt}}{\left(1-\beta e^{-Dt}\right)} dt = \frac{q_i (1-\beta) \ln\left(1-\beta e^{-Dt}\right)}{\beta D} + C_i \tag{2.44}$$

Therefore,

$$Q = \int_0^t q dt = \frac{q_i (1-\beta)}{\beta D} \ln\left(\frac{1-\beta e^{-Dt}}{1-\beta} \right) \tag{2.45}$$

Like the hyperbolic decline analysis, when $b \neq 1$, the ultimate recoverable volume is

$$EUR = N_p + \frac{q_i^b}{(1-b) D_i} \left(q_f^{1-b} - q_{lim}^{1-b} \right) + \frac{q_{lim} - q_{ab}}{D_{lim}} \tag{2.46}$$

When $b = 1$, the ultimate recoverable volume is

$$EUR = N_p + \frac{q_i}{D_i} \ln\left(\frac{q_f}{q_{lim}} \right) + \frac{q_{lim} - q_{ab}}{D_{lim}} \tag{2.47}$$

2.1.6 *Fluid in Place and Drainage Area*

If the ultimate recoverable volume of a single well is given, the reserves and area can be estimated based on the material balance principle. As to oil well, there is

$$N = \frac{\alpha_1 EUR}{C_t \Delta p} \tag{2.48}$$

$$A = \frac{\alpha_2 N B_o}{\varphi h S_o} \tag{2.49}$$

As to gas well, based on the material balance of volumetric gas reservoir, we have

$$\frac{p}{Z} = \left(\frac{p}{Z}\right)_i \left(1 - \frac{EUR}{G}\right) \tag{2.50}$$

there is

$$G = \frac{EUR}{\left(1 - \frac{p}{Z} \bigg/ \frac{p_i}{Z_i}\right)} \tag{2.51}$$

$$A = \frac{\alpha_3 Z_i T p_{sc} G}{\varphi h S_g p_i T_{sc}} \tag{2.52}$$

The α_1, α_2, and α_3 in Eq. (2.48), Eq. (2.49), and Eq. (2.52) are the conversion factors under different unit systems, respectively.

2.2 Theoretical Fundamentals of Arps Production Decline

Arps developed the three production decline equations based on his experience. However, he as well as most of later researchers did not conduct any research related to filtration theory. Bingyu Ji (1995), starting from the relative permeability curve and material balance principle, derived three Arps decline equations and demonstrated that the rate decline equation of waterdrive reservoir depends mainly on the characteristics of relative permeability to oil curves. As to the gas well producing at constant BHFP, the flow rate follows the exponential decline rules.

2.2.1 *Flow Rate Equation of Waterdrive Reservoir*

Based on the filtration mechanics principle, in the case of maintaining the formation pressure by water injection, the flow rate of an oil well under basic SI is expressed by

$$q = \frac{2\pi K K_{ro}(S_w) h \Delta p}{\mu_o B\left[\ln\left(\frac{r_e}{r_w}\right) - \frac{3}{4} + S\right]} \tag{2.53}$$

Let

$$\alpha = \frac{2\pi K h \Delta p}{\mu_o B\left[\ln\left(\frac{r_e}{r_w}\right) - \frac{3}{4} + S\right]}$$

We have

$$q = \alpha K_{ro}(S_w) \tag{2.54}$$

Under the condition of voidage balance,

$$\frac{dS_w}{dt} = \frac{qB}{V\varphi} \tag{2.55}$$

Substitute Eq. (2.54) into Eq. (2.55), separating and integrating gives

$$\int_{S_{wc}}^{S_w} \frac{dS_w}{K_{ro}(S_w)} = \int_0^t \frac{\alpha B}{V\varphi} dt \tag{2.56}$$

In the case of keeping the current constant pressure production mode unchanged, as long as the relative permeability curve $K_{ro}(S_w) \sim S_w$ of the well is known, the different forms of rate decline equations can be developed based on Eq. (2.54) and Eq. (2.56).

2.2.2 *Exponential Decline Equation of Waterdrive Reservoir*

Assume that the relative permeability to oil is a linear form, that is,

$$K_{ro}(S_w) = a_1 - b_1 S_w \tag{2.57}$$

Substituting it into Eq. (2.56), we have

$$\int_{S_{wc}}^{S_w} \frac{dS_w}{a_1 - b_1 S_w} = \int_0^t \frac{\alpha B}{V\varphi} dt \tag{2.58}$$

Based on Eq. (2.23), integrating Eq. (2.58), we have

$$\ln\left(a_1 - b_1 S_w\right) = -\frac{\alpha B b_1 t}{V\varphi} + \ln\left(a_1 - b_1 S_{wc}\right) \tag{2.59}$$

Substituting Eq. (2.57) and Eq. (2.59) into Eq. (2.54), we have

$$q = \alpha e^{\ln\left(a_1 - b_1 S_{wc}\right)} e^{-\frac{\alpha B b_1 t}{V\varphi}} \tag{2.60}$$

Eq. (2.60) is the exponential decline equation. Respectively, the initial flow q_i and decline rate D_i are

$$q_i = \alpha e^{\ln\left(a_1 - b_1 S_{wc}\right)} \tag{2.61}$$

$$D_i = \frac{\alpha B b_1}{V\varphi} \tag{2.62}$$

2.2.3 Hyperbolic Decline Equation of Waterdrive Reservoir

Assume that the relative permeability to oil is in a form of power function, that is,

$$K_{ro}\left(S_w\right) = a_1\left(1 - S_w\right)^{b_1} \qquad b_1 \neq 1 \tag{2.63}$$

Substituting Eq. (2.63) into Eq. (2.56), we have

$$\int_{S_{wc}}^{S_w} \frac{dS_w}{a_1\left(1 - S_w\right)^{b_1}} = \int_0^t \frac{\alpha B}{V\varphi} dt \tag{2.64}$$

Based on Eq. (2.10), integrating Eq. (2.64), we have

$$\left(1 - S_w\right)^{1-b_1} = a_1\left(b_1 - 1\right)\frac{\alpha B}{V\varphi} t + \left(1 - S_{wc}\right)^{1-b_1} \tag{2.65}$$

Substituting Eq. (2.65) into Eq. (2.54), we have

$$\begin{aligned} q &= \alpha a_1\left(1 - S_w\right)^{b_1} = \alpha a_1\left[a_1\left(b_1 - 1\right)\frac{\alpha B}{V\varphi} t + \left(1 - S_{wc}\right)^{1-b_1}\right]^{\frac{b_1}{1-b_1}} \\ &= \alpha a_1\left(1 - S_{wc}\right)^{b_1}\left[1 + \frac{a_1\left(b_1 - 1\right)\frac{\alpha B}{V\varphi} t}{\left(1 - S_{wc}\right)^{1-b_1}}\right]^{\frac{b_1}{1-b_1}} \end{aligned} \tag{2.66}$$

Obviously,

$$\alpha a_1 \left(1 - S_{\mathrm{wc}}\right)^{b_1} = q_{\mathrm{i}} \tag{2.67}$$

Eq. (2.66) is converted to

$$q = q_{\mathrm{i}} \left[1 + \frac{a_1 \left(b_1 - 1\right) \dfrac{\alpha B}{V\varphi} t}{\left(1 - S_{\mathrm{wc}}\right)^{1-b_1}} \right]^{\frac{b_1}{1-b_1}} = q_{\mathrm{i}} \left(1 + Et\right)^N \tag{2.68}$$

where

$$E = \frac{a_1 \left(b_1 - 1\right) \dfrac{\alpha B}{V\varphi}}{\left(1 - S_{\mathrm{wc}}\right)^{1-b_1}} \qquad N = \frac{b_1}{1 - b_1}$$

Based on the definition of decline rate, we have

$$D = -\frac{1}{q}\frac{dq}{dt} = -\frac{1}{q_{\mathrm{i}} \left(1 + Et\right)^N} \times q_{\mathrm{i}} NE \left(1 + Et\right)^{N-1} = -\frac{NE}{\left(1 + Et\right)} \tag{2.69}$$

So the initial decline rate is

$$D_{\mathrm{i}} = -NE \tag{2.70}$$

Substituting Eq. (2.70) into Eq. (2.68), we have

$$q = q_{\mathrm{i}} \left(1 - \frac{D_{\mathrm{i}}}{N} t\right)^N \tag{2.71}$$

Let

$$b = -\frac{1}{N}$$

Eq. (2.71) is converted to

$$q = \frac{q_{\mathrm{i}}}{\left(1 + bD_{\mathrm{i}} t\right)^{1/b}} \tag{2.72}$$

Eq. (2.72) is the hyperbolic decline equation.

2.2.4 Harmonic Decline Equation of Water Drive Reservoir

Assume that the relative permeability to oil is an exponential function form, that is

$$K_{ro}\left(S_w\right)=a_1 e^{-b_1 S_w} \tag{2.73}$$

Substituting Eq. (2.73) into Eq. (2.56), we have

$$\int_{S_{wc}}^{S_w}\frac{dS_w}{a_1 e^{-b_1 S_w}}=\int_0^t \frac{\alpha B}{V\varphi}dt \tag{2.74}$$

Integrating Eq. (2.74), we have

$$e^{b_1 S_w}=a_1 b_1 \frac{\alpha B}{V\varphi}t+e^{b_1 S_{wc}} \tag{2.75}$$

Substituting Eq. (2.75) into Eq. (2.54), we have

$$q=\alpha a_1 e^{-b_1 S_w}=\frac{\alpha a_1}{a_1 b_1 \frac{\alpha B}{V\varphi}t+e^{b_1 S_{wc}}}=\frac{\alpha a_1 e^{-b_1 S_{wc}}}{1+a_1 b_1 \frac{\alpha B}{V\varphi}e^{-b_1 S_{wc}}t} \tag{2.76}$$

Obviously,

$$\alpha a_1 e^{-b_1 S_w}=q_i$$

Let

$$D_i=a_1 b_1 \frac{\alpha B}{V\varphi}e^{-b_1 S_{wc}}$$

Eq. (2.76) is converted to

$$q=\frac{q_i}{1+D_i t} \tag{2.77}$$

Eq. (2.72) is the harmonic decline equation.

2.2.5 Decline Equation of Bounded Elastic Drive–Constant Pressure Production

The material balance equation of bounded elastic drive oil reservoir is expressed as

$$N_p B=C_t N B_i\left(p_i-p\right) \tag{2.78}$$

Differentiating Eq. (2.78), we have

$$\frac{dN_p}{dt}=\frac{C_t N B_i}{B}\frac{d(p_i-p)}{dt} \tag{2.79}$$

The left of Eq. (2.79) is the instantaneous producing rate of the oil well; then Eq. (2.79) is expressed as

$$q=-\frac{C_t N B_i}{B}\frac{dp}{dt} \tag{2.80}$$

Based on the rate equation of steady flow Eq. (2.53), we have

$$q=J(p_i-p_{wf}) \tag{2.81}$$

where J is termed as production index, and it can be expressed as

$$J=\frac{2\pi K K_{ro}(S_w)h}{\mu_o B\left[\ln\left(\frac{r_e}{r_w}\right)-\frac{3}{4}+S\right]} \tag{2.82}$$

As to constant BHFP production state, the differential of Eq. (2.81) is

$$\frac{dq}{dt}=J\frac{dp}{dt} \tag{2.83}$$

Substituting Eq. (2.83) into Eq. (2.80), we have

$$q=-\frac{C_t N B_i}{JB}\frac{dq}{dt} \tag{2.84}$$

Separating and integrating Eq. (2.84), we have

$$\int_{q_i}^{q}\frac{dq}{q}=-\int_{0}^{t}\frac{JB}{C_t N B_i}dt \tag{2.85}$$

$$q=q_i e^{-\frac{JB}{C_t N B_i}t}=q_i e^{-D_i t} \tag{2.86}$$

Therefore, when the bounded elastic drive oil well is produced at a constant pressure, the decline law conforms to the exponential form; the decline rate is adversely proportional to the producing geological reserves and proportional to the productivity index.

2.2.6 Decline Equation of Multilayer Oil Well—Constant Pressure Production

Assume that N pay zones are opened in the oil well, there is no exchange of fluid between layers, the oil well is producing at constant BHFP, the difference of initial formation pressure can be ignored, and all layers meet the material balance condition of bounded elastic drive reservoir. Then, the exponential decline equation of layer j can be derived as per the method similar to single layer.

$$q_j = q_{ji} e^{-\frac{J_j B}{C_t N_j B_i} t} = q_{ji} e^{-D_{ji} t} \quad (2.87)$$

As to the N-layer oil well producing at constant BHFP and free of interlayer exchange, the single well cumulative production decline equations is

$$q = \sum_{j=1}^{N} q_j = \sum_{j=1}^{N} q_{ji} e^{-D_{ji} t} \quad (2.88)$$

Therefore, under constant BHFP, at the time of development by elastic energy, the flow rate of multilayer well shows multiexponential decline.

As to a carbonate oil well communicating multiple fractures and caves, the decline law is similar to that of a multilayer state without cross-flow between layers, and it changes as per the multi-exponential superposition.

2.2.7 Decline Equation of Gas Well in Volumetric Gas Reservoir—Constant Pressure Production

The flow equation of real gas is linearized by introducing pseudo-pressure , as given below.

$$m(p) = 2\int_0^p \frac{p}{\mu Z} dp = 2\int_0^p \frac{1}{\mu C_g} d\left(\frac{p}{Z}\right) \quad (2.89)$$

After production having entered into boundary-dominated flow stage, the LIT deliverability equation expressed in pseudo-pressure form is

$$\Delta m(p) = Aq + Bq^2 \quad (2.90)$$

The LIT deliverability equation coefficients A and B are

$$A = \frac{12.69T}{Kh} \ln\left(\frac{2.2458 A_g}{C_A r_w^2}\right) \quad (2.91)$$

$$B = \frac{12.69T}{Kh} D \quad (2.92)$$

As to volumetric gas reservoir, the material balance equation is

$$\frac{p}{Z}=\left(\frac{p}{Z}\right)_{\mathrm{i}}\left(1-\frac{G_{\mathrm{p}}}{G}\right) \tag{2.93}$$

Differentiating Eq. (2.89), Eq. (2.90), and Eq. (2.93), respectively, we have

$$dm=\frac{2}{\mu C_{\mathrm{g}}}d\left(\frac{p}{Z}\right) \tag{2.94}$$

$$dm=(A+2Bq)dq \tag{2.95}$$

$$d\left(\frac{p}{Z}\right)=-\left(\frac{p}{Z}\right)_{\mathrm{i}}\frac{q}{G}dt \tag{2.96}$$

Combining Eq. (2.94), Eq. (2.95), and Eq. (2.96), we have

$$\left(\frac{A}{q}+2B\right)dq=-2\left(\frac{p}{Z}\right)_{\mathrm{i}}\frac{1}{G\mu C_{\mathrm{g}}}dt \tag{2.97}$$

Integrating Eq. (2.97), we have

$$\ln\left(\frac{q}{q_{\mathrm{i}}}\right)+\frac{2Bq_{\mathrm{i}}}{A}\left(\frac{q}{q_{\mathrm{i}}}-1\right)=-2\left(\frac{p}{\mu C_{\mathrm{g}}Z}\right)_{\mathrm{i}}\frac{1}{AG}t_a \tag{2.98}$$

where

$$t_{\mathrm{a}}=\int_0^{\mathrm{t}}\frac{(\mu C_{\mathrm{g}})_{\mathrm{i}}}{\mu C_{\mathrm{g}}}dt \tag{2.99}$$

If $B = 0$, the decline equation is reduced to the Fraim and Wattenbarger (1987) model; we have

$$q=q_{\mathrm{i}}e^{-2\left(\frac{p}{\mu C_{\mathrm{g}}Z}\right)_{\mathrm{i}}\frac{1}{AG}t_{\mathrm{a}}} \tag{2.100}$$

That is, the decline law of gas well in a volumetric gas reservoir producing at a constant pressure conforms to exponential decline, and the decline rate is influenced by factors like OGIP, geometry, and formation pressure.

2.3 Arps Type Curves

After the production of a well starts to decline, it is required to, based on the obtained production data, use different methods to judge the decline type, determine the decline parameters, and build its relevant empirical formula. Currently, such methods usually include graphology, trial and error, and curve matching, etc. The type curve matching analysis method is mainly introduced in the section.

2.3.1 Gentry–Arps Type Curve

Gentry (1972) introduced the concepts of dimensionless rate q_{Dd}, dimensionless time t_{Dd} and dimensionless cumulative production N_{pDd}, plotted three types of Arps decline curves on the q_{Dd}~t_{Dd} and q_{Dd}~N_{pDd} semilog coordinates, and thus unified the curve form and built the Arps-type curves.

On the Gentry chart, the dimensionless times of three Arps decline types are defined as follows

$$t_{Dd} = D_i t \tag{2.101}$$

As to the exponential decline, the dimensionless rate is defined as

$$q_{Dd} = \frac{q_i}{q} = e^{t_{Dd}} \tag{2.102}$$

The dimensionless cumulative production is defined as

$$N_{pDd} = \frac{Q}{q_i t} = \frac{q_i - q}{D_i q_i t} = \frac{q_{Dd} - 1}{q_{Dd} t_{Dd}} \tag{2.103}$$

As to the hyperbolic decline, the dimensionless rate is defined as

$$q_{Dd} = \left(1 + b t_{Dd}\right)^{1/b} \; (b \text{ is constant}, 0<b<1) \tag{2.104}$$

The dimensionless cumulative production is defined as

$$N_{pDd} = \frac{Q}{q_i t} = \frac{1}{(1-b) t_{Dd}} \left[1 - \left(\frac{1}{q_{Dd}}\right)^{1-b}\right] \tag{2.105}$$

As to the harmonic decline, the dimensionless rate is defined as

$$q_{Dd} = 1 + t_{Dd} \tag{2.106}$$

The dimensionless cumulative production is defined as

$$N_{pDd} = \frac{Q}{q_i t} = \frac{q_i}{D_i q_i t} \ln \frac{q_i}{q} = \frac{1}{t_{Dd}} \ln q_{Dd} \tag{2.107}$$

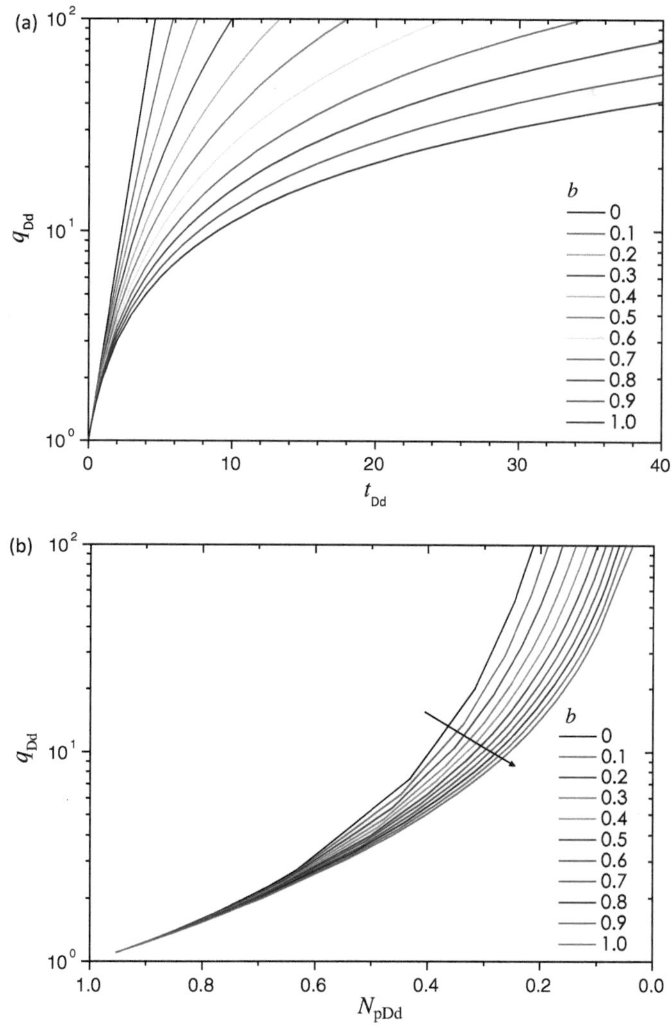

Figure 2.7 Gentry–Arps decline curves (after Gentry, 1972). (a) Dimensionless rate and time; (b) Dimensionless rate and dimensionless cumulative production

With the above seven sets of dimensionless variables, the type curve is plotted as Figure 2.7. It is observed that the curve form is independent of the initial flow q_i and the initial decline rate D_i, and it can be applied to the decline analysis of all wells.

The matching procedure of the method is similar to the well test type curve matching procedure, that is, plotting the $q_i/q \sim t$ dependence of real data, matching with the chart (a) in Figure 2.7, until the real data can best fit a certain type curve; at this time, it can read the b value on the type curve, and judge the decline type; by means of reading the (t, q_{Dd}) values of the match points $(t_{Dd}, q_{Dd})_M$ and the corresponding measured curves, the decline rate $D_i = \frac{(t_{Dd})_M}{t}$ can be derived. Both dimensionless rate and dimensionless cumulative production curves can be used as the verification curve of the matched results.

In the production data analysis, after b, q_i, and D_i are determined using the type curve matching, the corresponding equations in Section 2.1 can be used to estimate the single well recoverable reserves and production. In the actual analysis, the value of b has a great effect on estimating the recoverable reserves. The b value of a typical gas well varies from 0.4 to 0.5. In the course of actual matching, the case when $b > 1$ may also occur as a result of the following: 1) wrong interpretation, and $b < 1$ may also have a better fitting; 2) the data are still at transient stage, and the boundary-dominated flow stage has not yet been reached; and 3) such a case may occur in fractured gas well.

2.3.2 Arps Dimensionless Decline Flow Rate Type Curve

The definition of dimensionless time is

$$t_{Dd} = D_i t \tag{2.108}$$

As to the exponential decline, the dimensionless rate is defined as

$$q_{Dd} = \frac{q}{q_i} = e^{-t_{Dd}} \tag{2.109}$$

As to the hyperbolic decline, the dimensionless rate is defined as

$$q_{Dd} = \left(1 + bt_{Dd}\right)^{-1/b} \text{ (b is constant, 0<b<1)} \tag{2.110}$$

As to the harmonic decline, the dimensionless rate is defined as

$$q_{Dd} = \frac{1}{\left(1 + t_{Dd}\right)} \tag{2.111}$$

With the above four sets of dimensionless variables, the type curve is plotted as Figure 2.8.

It can be observed from Figure 2.8 that when $t_{Dd} < 0.3$, all curves are superposed and cannot be distinguished. Therefore, all data when $t_{Dd} < 0.3$ can come down to an exponential decline curve, which is independent of b value.

2.3.3 Arps Dimensionless Decline Cumulative Production Curve

The dimensionless cumulative production N_{pDd} is defined as

$$N_{pDd} = \int_0^{t_{Dd}} q_{Dd}(\tau) d\tau \tag{2.112}$$

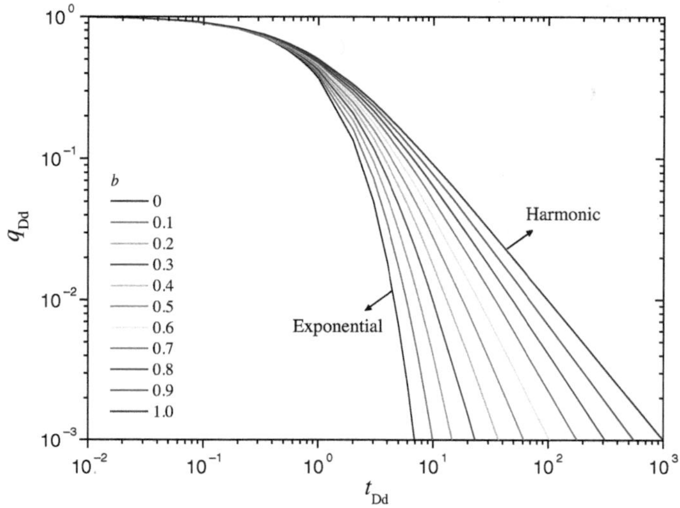

Figure 2.8 Arps dimensionless rate vs. dimensionless time curves

The dimensionless cumulative production N_{pDd} for different types of decline is defined as

$$N_{pDd} = \int_0^{t_{Dd}} q_{Dd}(\tau)d\tau = \int_0^{t_{Dd}} e^{-\tau}d\tau = 1 - e^{-t_{Dd}} \tag{2.113}$$

or

$$N_{pDd} = 1 - q_{Dd} \tag{2.114}$$

As to the exponential decline, substituting Eq. (2.109) into Eq. (2.112), and we have

$$N_{pDd} = \int_0^{t_{Dd}} (1 + b\tau)^{-1/b} d\tau = \frac{1}{b-1}\left[\left(1 + bt_{Dd}\right)^{(1-1/b)} - 1\right] \tag{2.115}$$

Based on Eq. (2.10), we have

$$N_{pDd} = \frac{1}{b-1}\left[q_{Dd}^{(1-b)} - 1\right] \tag{2.116}$$

As to the harmonic decline, substitute Eq. (2.111) into Eq. (2.112), based on Eq. (2.23), and we have

$$N_{pDd} = \int_0^{t_{Dd}} (1 + \tau)^{-1} d\tau = \ln\left(1 + t_{Dd}\right) \tag{2.117}$$

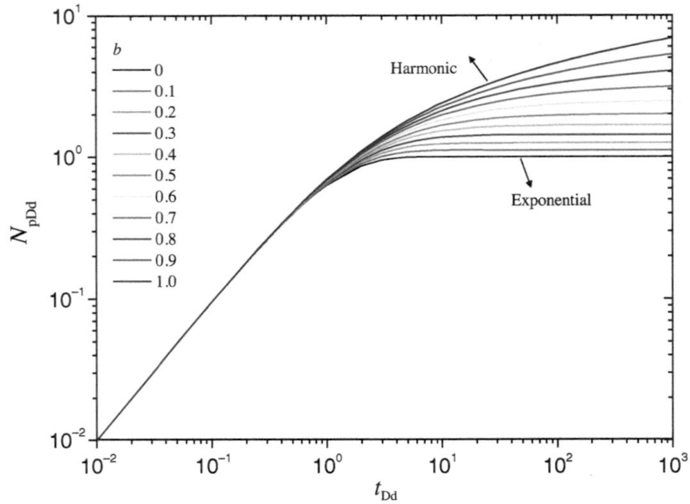

Figure 2.9 Arps dimensionless cumulative production vs. dimensionless time curves

Substituting Eq. (2.111) into Eq. (2.117), and we have

$$N_{pDd} = \ln(1 + t_{Dd}) = -\ln q_{Dd} \tag{2.118}$$

The dimensionless cumulative production type curves are shown in Figure 2.9.

2.3.4 Arps Dimensionless Decline Flow Rate Integral Type Curve

Dimensionless decline rate integral function q_{Ddi} is defined as

$$q_{Dd\,i} = \frac{N_{pDd}}{t_{Dd}} = \frac{1}{t_{Dd}} \int_0^{t_{Dd}} q_{Dd}(\tau)d\tau \tag{2.119}$$

As to exponential decline, based on Eq. (2.113), we have

$$q_{Dd\,i} = \frac{1 - e^{-t_{Dd}}}{t_{Dd}} \tag{2.120}$$

Based on Eq. (2.114), we have

$$q_{Dd\,i} = \frac{1 - q_{Dd}}{t_{Dd}} \tag{2.121}$$

As to hyperbolic decline, based on Eq. (2.115), we have

$$q_{\mathrm{Ddi}} = \frac{1}{(b-1)\,t_{\mathrm{Dd}}}\left[\left(1+bt_{\mathrm{Dd}}\right)^{(1-1/b)}-1\right] \tag{2.122}$$

Based on Eq. (2.116), we have

$$q_{\mathrm{Ddi}} = \frac{1}{(b-1)\,t_{\mathrm{Dd}}}\left[q_{\mathrm{Dd}}^{(1-b)}-1\right] \tag{2.123}$$

As to harmonic decline, based on Eq. (2.117), we have

$$q_{\mathrm{Ddi}} = \frac{\ln\left(1+t_{\mathrm{Dd}}\right)}{t_{\mathrm{Dd}}} \tag{2.124}$$

Based on Eq. (2.118), we have

$$q_{\mathrm{Ddi}} = -\frac{\ln q_{\mathrm{Dd}}}{t_{\mathrm{Dd}}} \tag{2.125}$$

The dimensionless decline flow rate integral type curves are shown in Figure 2.10.

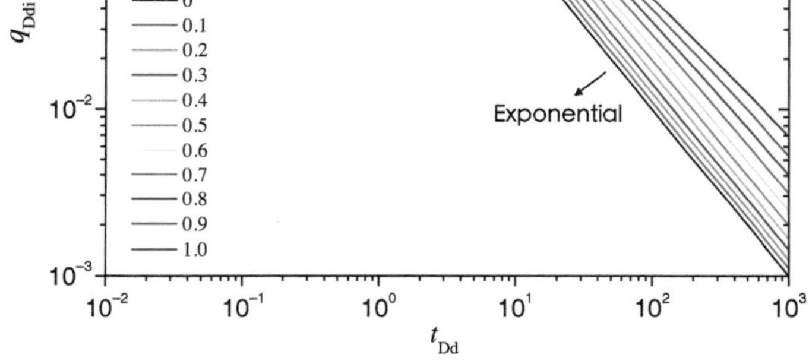

Figure 2.10 Arps dimensionless decline flow rate integral vs. dimensionless time curves

2.3.5 Arps Dimensionless Decline Flow Rate Integral Derivative Type Curve

Dimensionless decline flow rate integral derivative q_{Ddid} is defined as

$$q_{\mathrm{Dd\,id}} = -\frac{dq_{\mathrm{Dd\,i}}}{d\ln t_{\mathrm{Dd}}} = -t_{\mathrm{Dd}}\frac{dq_{\mathrm{Dd\,i}}}{dt_{\mathrm{Dd}}} = -t_{\mathrm{Dd}}\frac{d\left(N_{\mathrm{pDd}}/t_{\mathrm{Dd}}\right)}{dt_{\mathrm{Dd}}} \tag{2.126}$$

Substituting Eq. (2.112) into Eq. (2.126), we have

$$q_{\mathrm{Dd\,id}} = -t_{\mathrm{Dd}}\frac{d}{dt_{\mathrm{Dd}}}\left[\frac{1}{t_{\mathrm{Dd}}}\int_0^{t_{\mathrm{Dd}}} q_{\mathrm{Dd}}(\tau)d\tau\right] \tag{2.127}$$

Based on the mathematical manual, we have

$$d\left(\frac{u}{v}\right) = \frac{vdu - udv}{v^2} \tag{2.128}$$

Rearranging Eq. (2.127) yields

$$\begin{aligned} q_{\mathrm{Ddid}} &= -t_{\mathrm{Dd}}\frac{d}{dt_{\mathrm{Dd}}}\left[\frac{1}{t_{\mathrm{Dd}}}\int_0^{t_{\mathrm{Dd}}} q_{\mathrm{Dd}}(\tau)d\tau\right] \\ &= -t_{\mathrm{Dd}}\times\left[\frac{t_{\mathrm{Dd}}d\int_0^{t_{\mathrm{Dd}}} q_{\mathrm{Dd}}(\tau)d\tau - \int_0^{t_{\mathrm{Dd}}} q_{\mathrm{Dd}}(\tau)d\tau}{t_{\mathrm{Dd}}^2}\right] = \frac{1}{t_{\mathrm{Dd}}}\int_0^{t_{\mathrm{Dd}}} q_{\mathrm{Dd}}(\tau)d\tau - q_{\mathrm{Dd}}(t_{\mathrm{Dd}}) \end{aligned} \tag{2.129}$$

Substituting Eq. (2.119) into Eq. (2.129), and we have

$$q_{\mathrm{Ddid}} = q_{\mathrm{Ddi}} - q_{\mathrm{Dd}} \tag{2.130}$$

As to the exponential decline, based on Eq. (2.120), we have

$$q_{\mathrm{Dd\,id}} = \frac{1 - e^{-t_{\mathrm{Dd}}}}{t_{\mathrm{Dd}}} - q_{\mathrm{Dd}} \tag{2.131}$$

Based on Eq. (2.121), we have

$$q_{\mathrm{Dd\,id}} = \frac{1 - q_{\mathrm{Dd}}}{t_{\mathrm{Dd}}} - q_{\mathrm{Dd}} \tag{2.132}$$

As to the hyperbolic decline, based on Eq. (2.122), we have

$$q_{\mathrm{Dd\,id}} = \frac{1}{(b-1)\,t_{\mathrm{Dd}}}\left[\left(1+bt_{\mathrm{Dd}}\right)^{(1-1/b)}-1\right]-q_{\mathrm{Dd}} \tag{2.133}$$

Based on Eq. (2.123), we have

$$q_{\mathrm{Dd\,id}} = \frac{1}{(b-1)\,t_{\mathrm{Dd}}}\left[q_{\mathrm{Dd}}^{(1-b)}-1\right]-q_{\mathrm{Dd}} \tag{2.134}$$

As to the harmonic decline, based on Eq. (2.124), we have

$$q_{\mathrm{Dd\,id}} = \frac{1}{t_{\mathrm{Dd}}}\ln\left(1+t_{\mathrm{Dd}}\right)-q_{\mathrm{Dd}} \tag{2.135}$$

Based on Eq. (2.125), we have

$$q_{\mathrm{Dd\,id}} = -\frac{1}{t_{\mathrm{Dd}}}\ln q_{\mathrm{Dd}}-q_{\mathrm{Dd}} \tag{2.136}$$

The dimensionless decline flow rate integral derivative type curves are shown in Figure 2.11.

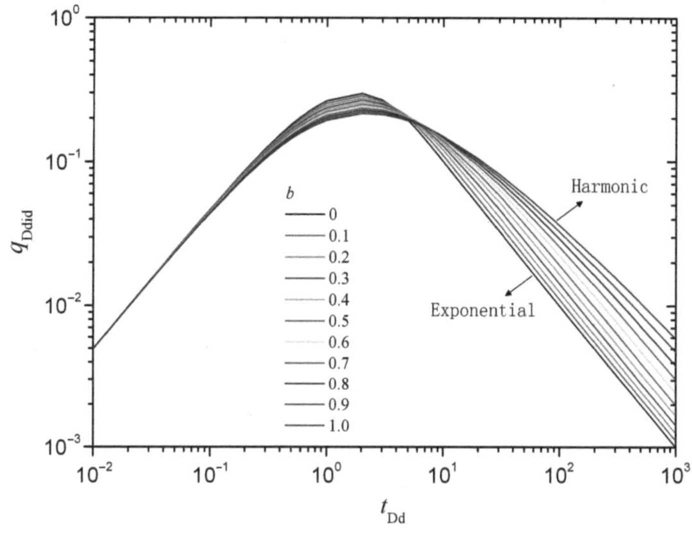

Figure 2.11 Arps dimensionless decline flow rate integral derivative vs. dimensionless time curves

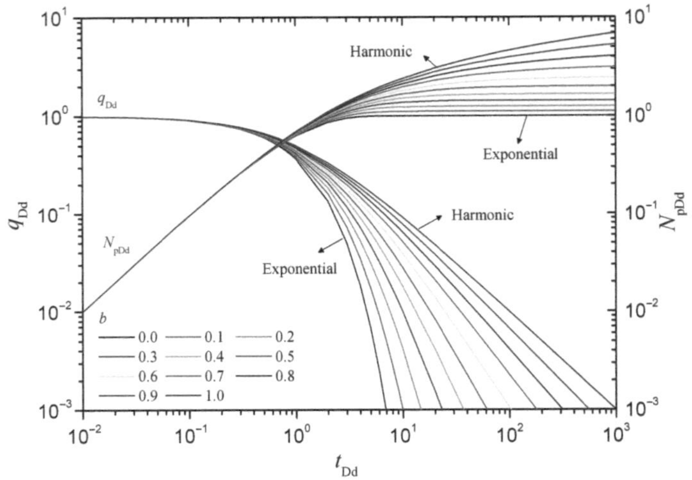

Figure 2.12 Arps decline curve 1

2.3.6 Arps Type Curve

With the concept of building the pressure and pressure derivative type curve in well test analysis, if both the dimensionless rate curve and the dimensionless cumulative production curve are superimposed, the Arps type curve can be obtained, as shown in Figure 2.12. The type curves, together with the transient part of the Fetkovich curve, constitute the Fetkovich type curve. The matching procedure and analytic cases will be introduced in detail in Chapter 3.

If the dimensionless rate integral curve and the dimensionless rate integral derivative curve are superimposed, as shown in Figure 2.13, the curve is the right part of the Palacio–Blasingame chart (1993).

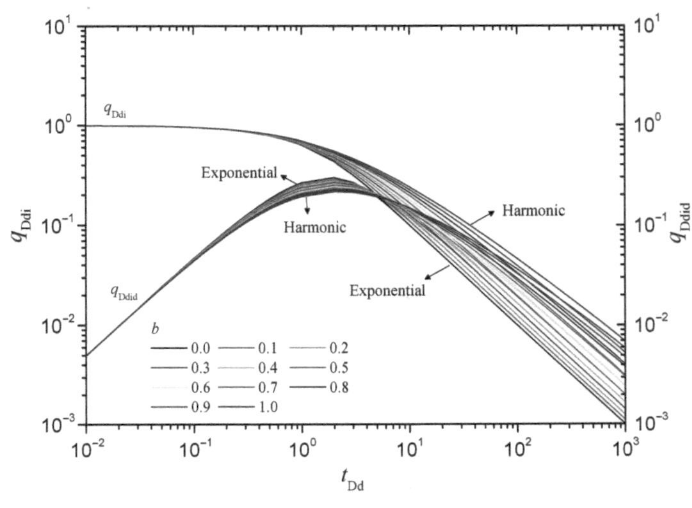

Figure 2.13 Arps decline curve 2

2.4 Power Function Analysis

As mentioned above, when the Arps method is used for analysis, $b > 1$ could occur and result in larger deviation of reserves forecast, partly because the flow is still at transient flow stage and has not formed boundary-dominated flow. IlK et al. (2008) and Mattar and Moghadam (2009) proposed a new strategy to use the power function method to predict the reserves of a low-permeability gas reservoir.

2.4.1 IlK Method

Spivey et al. (1992) presented the general expression of Arps decline rate D and hyperbolic decline exponent b, as shown below

$$D = -\frac{dq}{qdt} = -\frac{dq}{dN_{\mathrm{p}}} \tag{2.137}$$

$$b = \frac{d}{dt}\left(\frac{1}{D}\right) = q\frac{d}{dN_{\mathrm{p}}}\left(\frac{1}{D}\right) \tag{2.138}$$

IlK et al. (2008) suggested using the attenuation power function method to characterize the parameter D, which is defined as

$$D = D_{\infty} + D_1 t^{-(1-n)} \tag{2.139}$$

where
D_{∞} is the decline rate as t approaches infinity,
D_1 is the decline rate intercept at $t = 1$ day, and
n is the time exponent.

Eq. (2.139) indicates that the decline rate D at transient flow stage, transition stage, and boundary-dominated flow stage can be expressed by an attenuation power function. On the log-log paper, the D function is a straight line with slope $n-1$ at the early stage, and tend to a constant at late stage, as shown in Figure 2.14. This results indicate that the impact of a small period of time D_{∞} at the early stage on D is negligible, whereas in the large period of time at a late stage, the time-dependent items gradually attenuate, and D_{∞} becomes the main control factor.

If Eq. (2.139) is substituted into Eq. (2.55) to separate variables and integrate, we have

$$q = \hat{q}_{\mathrm{i}} e^{-D_{\infty}t - \frac{D_1}{n}t^n} = \hat{q}_{\mathrm{i}} e^{-D_{\infty}t - \hat{D}_{\mathrm{i}}t^n} \tag{2.140}$$

where
$\hat{q}_{\mathrm{i}}$ is the rate intercept that is q at $t = 0$
$\hat{D}_{\mathrm{i}}$ is $\frac{D_1}{n}$

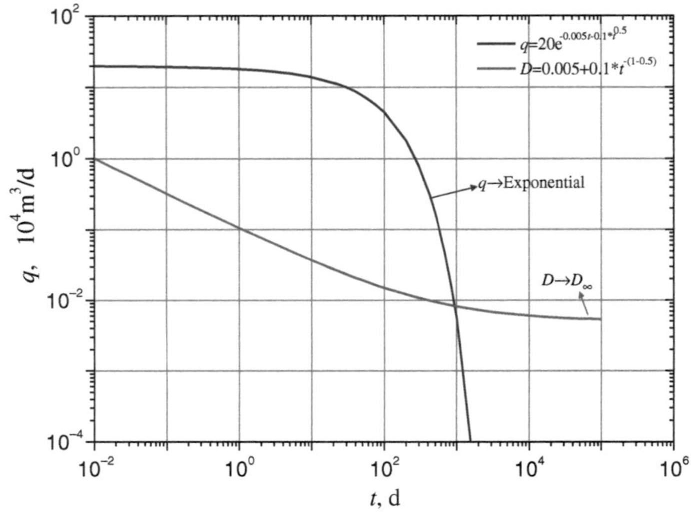

Figure 2.14 IlK power function method (after IlK et al., 2008)

The power function method formally is similar to an exponential decline, which solves the problem of b value attenuating with time in the Arps method, and can satisfactorily match the production data prior to boundary-dominated flow.

2.4.2 Mattar Method

For a tight gas well, it is difficult to reach a pseudo-steady state or boundary-dominated flow stage, and cannot meet the applied conditions of an Arps decline method; therefore, the case when $b > 1$ may occur frequently, and the OGIP could be overestimated while carrying out reserves evaluation. There are four unknown parameters in Eq. (2.140), and significant uncertainty exists at the time of carrying out matching analysis. On the basis of IlK et al. (2008) research, Mattar and Moghadam (2009) reduced the IlK equation aiming at the flow stages, such as linear flow, radial flow, and boundary-dominated flow.

As to the case with only linear flow–boundary-dominated flow stage, Wattenbarger and El-Banbi (1998) derived the solution for the transient flow stage of long linear flow, as given below

$$\frac{1}{q_{\mathrm{D}}} = \frac{\pi}{2}\sqrt{\pi t_{\mathrm{Dx_f}}} \tag{2.141}$$

Substitute Eq. (2.141) into Eq. (2.137), and we have

$$D = -\frac{dq_{\mathrm{D}}}{q_{\mathrm{D}}dt_{\mathrm{Dx_f}}} = 0.5t_{\mathrm{Dx_f}}^{-1} \tag{2.142}$$

After having entered into the boundary-dominated flow stage, the flow rate follows the exponential decline regularity, and D is a constant.

$$q = q_i e^{-D_i t} \tag{2.143}$$

As to the linear flow–radial flow–boundary-dominated flow cases, Mattar and Moghadam (2009) pointed out that at radial flow stage, Eq. (2.140) can be reduced to

$$q = \hat{q}_i e^{-D_\infty t - \frac{D_1}{n} t^n} = \hat{q}_i e^{-\hat{D}_i t^{0.13}} \tag{2.144}$$

or the more general form

$$q = \hat{q}_i e^{-\hat{D}_i t^n}, (0 < n < 0.13) \tag{2.145}$$

The analytic process is as follows:

1. Based on the measured data, use Eq. (2.137) to calculate the instantaneous decline rate D, and plot the lg D~lg t curve on the log–log paper.
2. Conduct linear regression on the datum points, allow the slope to be (–1.13 to –1), and pay attention to increasing the late time data weight at the time of regression. The time index is $n = 1 + slope$.
3. The intercept at $t = 1$ d equals to $n\hat{D}_i$ and can be used to calculate $\hat{D}_i$.
4. On the lg q~t semilog plot, adjust $\hat{q}_i$ to match the measured data with the $q = \hat{q}_i e^{-\hat{D}_i t^n}$ datum points.
5. Based on the equation of boundary-dominated flow starting time $t_{BDF} = \frac{\alpha \varphi \mu A}{\pi p K}$, use $q = \frac{q_i}{(1 + 0.5 D_i t)^2}$ to forecast for the later stage.

2.5 Case Analysis of Arps Method

A fractured gas well in a closed circular gas reservoir is producing at a constant BHFP. The initial formation pressure is 30 MPa, the formation temperature is 80°C, the formation net thickness is 10 m, the porosity is 0.1, the closed boundary is 500 m, the OGIP is 2.0×10^8 m^3, the half-length of fracture is 60 m, the natural gas relative density is 0.6, the permeability estimation from well test is 1.0 mD, the skin factor is 0, the BHFP is 10 MPa, the production period is 800 days, and the cumulative gas production is $10{,}954 \times 10^4$ m^3, as shown in Table 2.2.

Then, the Arps decline analysis method is used to estimate the ultimate recovery percentage and remaining production period of the well. Assume that the flow rate at the time of abandonment is 1.0×10^4 m^3.

2.5.1 Identification of Decline Law

Firstly plot the rate-time and rate-cumulative production curves, as shown in Figure 2.15. Based on the curve characteristics, it is believed that the curve of the well can be matched either by exponential decline or by harmonic decline.

Table 2.2 Historical production data of well A

Time, d	Rate, $10^4\ m^3/d$	Cumulative production, $10^4\ m^3$	Time, d	Rate, $10^4\ m^3/d$	Cumulative production, $10^4\ m^3$	Time, d	Rate, $10^4\ m^3/d$	Cumulative production, $10^4\ m^3$
1	88.9	89	220	17.8	6239	520	6.6	9561
2	65.7	155	230	16.1	6405	530	6.8	9629
3	64.1	219	240	15.9	6567	540	6.7	9697
4	60.1	279	250	15.9	6724	550	6.7	9762
5	55.9	335	260	14.9	6876	560	6.2	9825
6	52.6	387	270	14.2	7023	570	6.3	9887
7	51.0	438	280	14.3	7166	580	6.1	9947
8	51.0	489	290	13.3	7303	590	5.7	10005
9	49.9	539	300	13.6	7438	600	5.9	10063
10	49.9	589	310	12.7	7567	610	5.5	10119
20	42.4	1019	320	12.6	7692	620	5.3	10173
30	38.9	1412	330	11.6	7814	630	5.5	10226
40	35.1	1779	340	11.4	7930	640	5.3	10278
50	32.6	2126	350	11.0	8042	650	4.8	10328
60	31.7	2457	360	11.0	8153	660	4.7	10376
70	32.2	2780	370	10.2	8261	670	4.9	10424
80	29.6	3084	380	9.8	8366	680	4.7	10471
90	29.3	3373	390	10.4	8468	690	4.4	10516
100	27.7	3654	400	9.3	8567	700	4.3	10560
110	26.7	3919	410	9.1	8663	710	4.3	10604
120	25.8	4173	420	9.6	8758	720	4.0	10646
130	24.0	4416	430	9.3	8849	730	3.9	10687
140	22.2	4650	440	8.3	8938	740	4.1	10728
150	21.8	4879	450	8.7	9024	750	3.7	10767
160	20.5	5092	460	8.5	9106	760	3.9	10806
170	19.8	5302	470	7.8	9187	770	3.9	10844
180	19.0	5504	480	8.1	9266	780	3.6	10882
190	19.9	5699	490	7.8	9344	790	3.7	10918
200	19.0	5885	500	7.7	9418	800	3.6	10954
210	17.6	6065	510	7.1	9490			

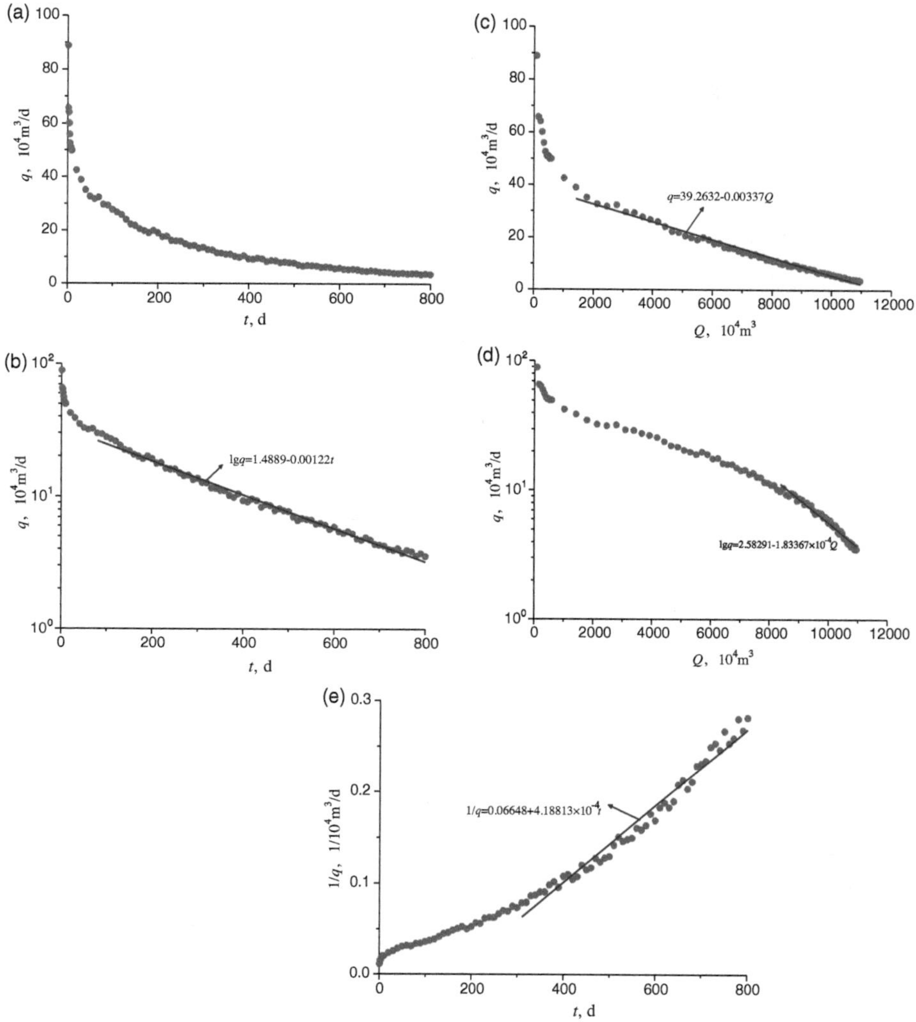

Figure 2.15 Production data curve of well A. (a) Cartesian rate-time plot; (b) Semilog rate–time plot; (c) Cartesian rate–cumulative production plot; (d) Semilog rate–cumulative production plot; (e) Cartesian rate countdown–time plot

2.5.2 Arps Exponential Decline Analysis

Based on the linear regression results of Figure 2.15 (b), we have

$$\lg q = 1.4889 - 0.00122t \quad (2.146)$$

Based on the change of base formula, Eq. (2.146) can be expressed as

$$q = 30.82e^{-0.0028t} \quad (2.147)$$

Based on Table 2.2, the current production rate is 3.6×10^4 m^3/d, and the flow rate at the time of abandonment is $q_{ab} = 1.0 \times 10^4$ m^3/d. Based on Eq. (2.5), the cumulative production Q_f is estimated as follows

$$Q_f = \frac{q_f - q_{ab}}{D} = \frac{3.6 - 1.0}{0.0028} = 928.6 \times 10^4 \text{ m}^3 \tag{2.148}$$

Based on Eq. (2.6), the ultimate recoverable volume is

$$EUR = N_p + Q_f = 10954 + 928.6 = 11882.6 \times 10^4 \text{ m}^3 \tag{2.149}$$

Based on Eq. (2.4), the production time reaching the abandonment is

$$\Delta t = \frac{1}{D} \ln\left(\frac{q_f}{q_{ab}}\right) = \frac{1}{0.0028} \ln\left(\frac{3.6}{1.0}\right) = 457.5\text{d} \tag{2.150}$$

2.5.3 Arps Harmonic Decline Analysis

Based on the linear regression results of Figure 2.15 (e), we have

$$\frac{1}{q} = 0.06648 + 4.18813 \times 10^{-4} t \tag{2.151}$$

Based on Eq. (2.21), we have

$$q_i = \frac{1}{0.06648} = 15.0 \times 10^4 \text{ m}^3/\text{d} \tag{2.152}$$

$$D_i = \frac{4.18813 \times 10^{-4}}{0.06648} = 0.0063 \text{ d}^{-1} \tag{2.153}$$

$$q = \frac{q_i}{1 + D_i t} = \frac{15.0}{1 + 0.0063t} \tag{2.154}$$

Based on Table 2.2, the current production rate is 3.6×10^4 m^3/d, and the flow rate at the time of abandonment is $q_{ab} = 1.0 \times 10^4$ m^3/d. Based on Eq. (2.29), the cumulative production Q_f is estimated as follows

$$Q_f = \frac{q_i}{D_i} \ln \frac{q_f}{q_{ab}} = \frac{15.0}{0.0063} \ln \frac{3.6}{1.0} = 3049.8 \times 10^4 \text{ m}^3 \tag{2.155}$$

The ultimate recoverable volume is

$$EUR = N_p + Q_f = 10954 + 3049.8 = 14003.8 \times 10^4 \text{ m}^3 \tag{2.156}$$

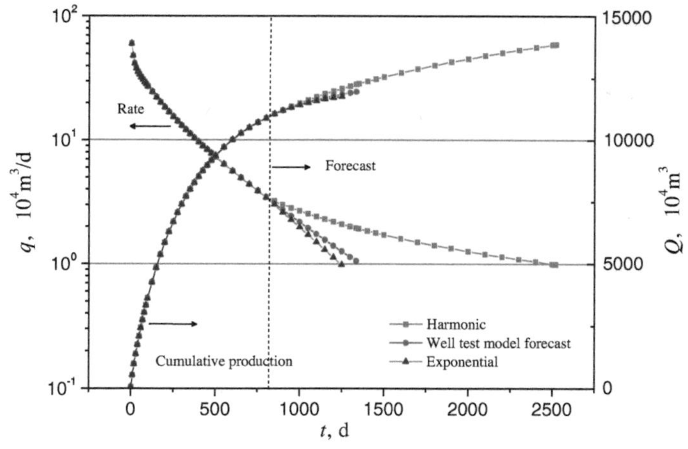

Figure 2.16 Production forecast of well A

Based on Eq. (2.31), the production time reaching the abandonment is

$$\Delta t = \frac{q_\mathrm{i}}{D_\mathrm{i}}\left(\frac{1}{q_\mathrm{ab}} - \frac{1}{q_\mathrm{f}}\right) = \frac{15.0}{0.0063}\left(\frac{1}{1.0} - \frac{1}{3.6}\right) = 1719.6\,\mathrm{d} \tag{2.157}$$

2.5.4 Result Correlation and Analysis

The prediction model is built with the known gas reservoir parameters. When the flow rate at the time of abandonment is 1.0×10^4 m^3/d, the theoretically predicted results and the exponential decline and harmonic decline predicted results are shown in Figure 2.16. It is seen that the prediction results of exponential decline and model method are basically the same, while the harmonic decline predicted results are obviously higher.

Fetkovich Decline Curves Analysis

3

Arps type curves could only be used to analyze the data in the period of boundary-dominated flow. Fetkovich introduced the idea of log-log type curve analysis to production analysis for both transient flow period and boundary-dominated flow period. Boundary-dominated flow for constant pressure production is similar to pseudo-steady state flow for constant rate production. Fetkovich type curve analysis was intended to be a rapid way to estimate performance when a well is producing at a constant bottom hole flowing pressure (BHFP), which is the same as what people usually do in well test analysis. From Fetkovich type curves, future performance can be forecast along with estimates of OOIP (OGIP) and ultimate recovery.

In this chapter, we mainly present the theoretical basis of Fetkovich production decline analysis method which is fit for the vertical well with closed circular boundary, as well as the type curve plotting and the curve matching method. Furthermore, we also introduce several cases with applications using the above method.

3.1 Solution for a Well in a Closed Circular Reservoir: Constant Pressure Production

We consider the flow of a slightly compressible fluid of constant viscosity in a cylindrical reservoir in which the outer boundary is closed. The well is located at the center of the cylinder and the fluid is produced at a constant BHFP. Initially, the pressure is uniform throughout the reservoir. The skin and wellbore storage effect is not considered. The reservoir radius, reservoir thickness, initial reservoir pressure, wellbore radius, porosity, total compressibility, permeability, fluid viscosity, volume factor, and flow rate are denoted by r_e, h, p_i, r_w, φ, C_t, K, μ, B, and q, respectively.

The basic partial differential equations for the system under investigation are given by

$$\frac{1}{r}\frac{\partial}{\partial r}\left(r\frac{\partial p}{\partial r}\right)=\frac{\varphi\mu C_t}{K}\frac{\partial p}{\partial t} \tag{3.1}$$

$$p(r,0)=p_\mathrm{i} \tag{3.2}$$

$$p(r_\mathrm{w},t)=p_\mathrm{wf} \tag{3.3}$$

$$\left.\frac{\partial p}{\partial r}\right|_{r=r_\mathrm{e}}=0 \tag{3.4}$$

Advanced Production Decline Analysis and Application

$$\left(r\frac{\partial p}{\partial r}\right)_{r=r_{\mathrm{w}}}=\frac{q\mu B}{2\pi Kh} \tag{3.5}$$

In dimensionless variables

$$\frac{1}{r_{\mathrm{D}}}\frac{\partial}{\partial r_{\mathrm{D}}}\left(r_{\mathrm{D}}\frac{\partial p_{\mathrm{D}}}{\partial r_{\mathrm{D}}}\right)=\frac{\partial p_{\mathrm{D}}}{\partial t_{\mathrm{D}}} \tag{3.6}$$

$$p_{\mathrm{D}}\left(r_{\mathrm{D}},0\right)=0 \tag{3.7}$$

$$p_{\mathrm{D}}\left(1,t_{\mathrm{D}}\right)=1 \tag{3.8}$$

$$\left.\frac{\partial p_{\mathrm{D}}}{\partial r_{\mathrm{D}}}\right|_{r_{\mathrm{D}}=r_{\mathrm{eD}}}=0 \tag{3.9}$$

$$q_{\mathrm{D}}=-\left(r_{\mathrm{D}}\frac{\partial p_{\mathrm{D}}}{\partial r_{\mathrm{D}}}\right)_{r_{\mathrm{D}}=1} \tag{3.10}$$

The dimensionless variables are defined as follows

$$p_{\mathrm{D}}=\frac{p_{\mathrm{i}}-p}{p_{\mathrm{i}}-p_{\mathrm{wf}}}\quad t_{\mathrm{D}}=\frac{Kt}{\varphi\mu C_{\mathrm{t}}r_{\mathrm{w}}^{2}}\quad r_{\mathrm{D}}=\frac{r}{r_{\mathrm{w}}}\quad r_{\mathrm{eD}}=\frac{r_{\mathrm{e}}}{r_{\mathrm{w}}}\quad q_{\mathrm{D}}=\frac{qB\mu}{2\pi Kh\left(p_{\mathrm{i}}-p_{\mathrm{wf}}\right)}$$

Taking the Laplace transform of Eq. (3.6) through Eq. (3.10), we have

$$\frac{d^{2}\bar{p}_{\mathrm{D}}}{dr_{\mathrm{D}}^{2}}+\frac{1}{r_{\mathrm{D}}}\frac{d\bar{p}_{\mathrm{D}}}{dr_{\mathrm{D}}}=s\,\bar{p}_{\mathrm{D}} \tag{3.11}$$

$$\bar{p}_{\mathrm{D}}\left(r_{\mathrm{D}},0\right)=0 \tag{3.12}$$

$$\bar{p}_{\mathrm{D}}\left(1,s\right)=\frac{1}{s} \tag{3.13}$$

$$\left.\frac{d\bar{p}_{\mathrm{D}}}{dr_{\mathrm{D}}}\right|_{r_{\mathrm{D}}=r_{\mathrm{eD}}}=0 \tag{3.14}$$

$$\bar{q}_{\mathrm{D}} = -\left(\frac{\partial \bar{p}_{\mathrm{D}}}{\partial r_{\mathrm{D}}}\right)_{r_{\mathrm{D}}=1} \tag{3.15}$$

The solution of Eq. (3.11), in the Laplace space, is given as follows

$$\bar{p}_{\mathrm{D}} = A_0 I_0\left(r_{\mathrm{D}}\sqrt{s}\right) + B_0 K_0\left(r_{\mathrm{D}}\sqrt{s}\right) \tag{3.16}$$

According to the inner boundary condition Eq. (3.13), we have

$$A_0 I_0\left(\sqrt{s}\right) + B_0 K_0\left(\sqrt{s}\right) = \frac{1}{s} \tag{3.17}$$

According to the outer boundary condition Eq. (3.14), we have

$$A_0 I_1\left(r_{\mathrm{eD}}\sqrt{s}\right) - B_0 K_1\left(r_{\mathrm{eD}}\sqrt{s}\right) = 0 \tag{3.18}$$

According to Eq. (3.17) and Eq. (3.18), we have

$$A_0 = \frac{K_1\left(r_{\mathrm{eD}}\sqrt{s}\right)}{s\left[I_0\left(\sqrt{s}\right)K_1\left(r_{\mathrm{eD}}\sqrt{s}\right) + K_0\left(\sqrt{s}\right)I_1\left(\mathrm{r}_{\mathrm{eD}}\sqrt{s}\right)\right]} \tag{3.19}$$

$$B_0 = \frac{I_1\left(r_{\mathrm{eD}}\sqrt{s}\right)}{s\left[I_0\left(\sqrt{s}\right)K_1\left(r_{\mathrm{eD}}\sqrt{s}\right) + K_0\left(\sqrt{s}\right)I_1\left(r_{\mathrm{eD}}\sqrt{s}\right)\right]} \tag{3.20}$$

Based on Eq. (3.15), we have

$$\bar{q}_{\mathrm{D}} = \sqrt{s}\left[-A_0 I_1\left(\sqrt{s}\right) + B_0 K_1\left(\sqrt{s}\right)\right] \tag{3.21}$$

Substituting of Eq. (3.19) and Eq. (3.20) into Eq. (3.21) yields

$$\bar{q}_{\mathrm{D}} = \frac{1}{\sqrt{s}} \frac{\dfrac{K_1(\sqrt{s})}{I_0(\sqrt{s})} - \dfrac{K_1(r_{\mathrm{eD}}\sqrt{s})}{I_1(r_{\mathrm{eD}}\sqrt{s})}\dfrac{I_1(\sqrt{s})}{I_0(\sqrt{s})}}{\dfrac{K_0(\sqrt{s})}{I_0(\sqrt{s})} + \dfrac{K_1(r_{\mathrm{eD}}\sqrt{s})}{I_1(r_{\mathrm{eD}}\sqrt{s})}} \tag{3.22}$$

Table 3.1 $V(i)$ **with Stehfest parameter** $N = 8$

$V(1)$	$V(2)$	$V(3)$	$V(4)$	$V(5)$	$V(6)$	$V(7)$	$V(8)$
–0.3333	48.3333	–906	5464.6667	–14376.6667	18730	–11946.6667	2986.6667

Through the Stehfest numerical inverse transformation method (1976), we can calculate dimensionless flow rate in real space. The Laplace inverse transform method can be given by

$$V(i) = (-1)^{\frac{N}{2}+i} \sum_{k=\left[\frac{i+1}{2}\right]}^{\min\left(i,\frac{N}{2}\right)} \frac{k^{\frac{N}{2}}(2k)!}{\left(\frac{N}{2}-k\right)!(k)!(k-1)!(i-k)!(2k-i)!} \tag{3.23}$$

where

$$f(t) = \frac{\ln(2)}{t}\sum_{i=1}^{N} V(i)\bar{f}(s), \qquad s = i\frac{\ln(2)}{t}$$

The variable N could be one of the even numbers from 8 to 12, where 8 is frequently used. Theoretically, it is believed that the result will be better if N is relatively bigger; however, the actual result is worse because of the effect of round-off error (Table 3.1).

For a vertical well under production with constant BHFP centered in a closed circular reservoir, the relationship between dimensionless flow rate q_{D} and dimensionless time t_{D} is shown in Figure 3.1. In different reservoir sizes (i.e., different $r_{\mathrm{e}} / r_{\mathrm{w}}$), the curves coincide with each other during the early transient flow period. When it turns into the boundary-dominated flow region, the curves gradually disperse. As the radius of reservoir increases, the state of transient flow lasts longer and boundary-dominated flow occurs later.

3.2 Fetkovich Decline Curve Plotting

3.2.1 Fetkovich Dimensionless Variables

In the period of boundary-dominated flow, the initial flow rate q_{i} could be calculated by Dupuit formula, which can be given by

$$q_{\mathrm{i}} = \frac{2\pi Kh(p_{\mathrm{i}} - p_{\mathrm{wf}})}{\mu B\left[\ln\left(\frac{r_{\mathrm{e}}}{r_{\mathrm{wa}}}\right) - \frac{1}{2}\right]} \tag{3.24}$$

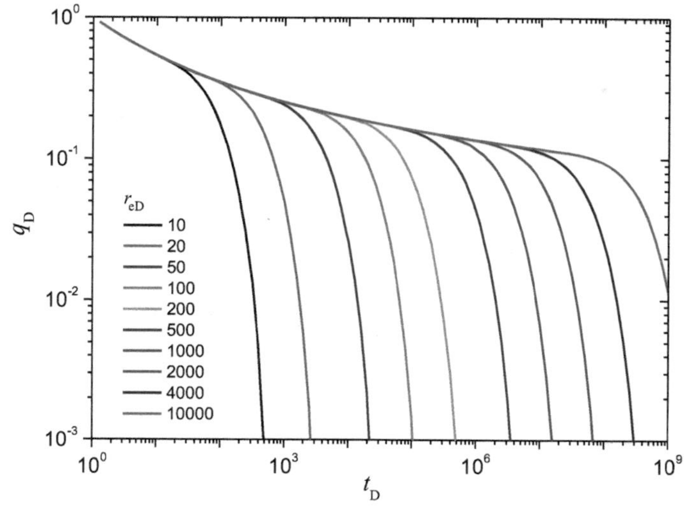

Figure 3.1 Dimensionless flow rate functions for finite outer boundary and constant pressure at inner boundary (after Hurst, 1934)

or

$$q_i = \frac{2\pi Kh(p_i - p_{wf})}{\mu B\left(\ln r_{eD} - \frac{1}{2}\right)} \tag{3.25}$$

$$q_{iD} = \frac{1}{\ln r_{eD} - \frac{1}{2}} \tag{3.26}$$

The dimensionless initial flow rate can be expressed as

$$q_{iD} = \frac{1}{\ln r_{eD} - \frac{1}{2}} \tag{3.26}$$

where r_{wa} is the apparent wellbore radius, which is given by

$$r_{wa} = r_w e^{-S} \tag{3.27}$$

Substituting Eq. (3.26) into Eq. (2.109) yields

$$q_{Dd} = \frac{q}{q_i} = q_D\left(\ln r_{eD} - \frac{1}{2}\right) = \alpha q_D \tag{3.28}$$

Fetkovich (1980) suggested that for a well centered in a closed circular reservoir with constant BHFP, the flow rate is conformed to the exponential decline law. When $r_{\mathrm{eD}}^2 > 10^3$ and $t_{\mathrm{D}} > 0.1\pi r_{\mathrm{eD}}^2$, the pseudo-steady state solution could be expressed by the formula below approximately.

$$p_{\mathrm{wD}}(t_{\mathrm{D}}) = \frac{2t_{\mathrm{D}}}{r_{\mathrm{eD}}^2} + \ln r_{\mathrm{eD}} - \frac{3}{4} \tag{3.29}$$

Van Everdingen and Hurst (1949) indicated that according to the superposition principle, the relationship between the pressure solution with constant flow rate and the flow rate solution with constant BHFP could be determined as follows

$$\bar{p}_{\mathrm{wD}}(s)\bar{q}_{\mathrm{D}}(s) = \frac{1}{s^2} \tag{3.30}$$

Substituting Eq. (3.30) into Eq. (3.29) yields

$$\begin{aligned} \bar{q}_{\mathrm{D}}(s) &= \frac{1}{s^2\left[\dfrac{2}{s^2 r_{\mathrm{eD}}^2} + \dfrac{1}{s}\left(\ln r_{\mathrm{eD}} - \dfrac{3}{4}\right)\right]} = \frac{1}{\dfrac{2}{r_{\mathrm{eD}}^2} + s\left(\ln r_{\mathrm{eD}} - \dfrac{3}{4}\right)} \\ &= \frac{1}{\left(\ln r_{\mathrm{eD}} - \dfrac{3}{4}\right)\left[\dfrac{2}{r_{\mathrm{eD}}^2\left(\ln r_{\mathrm{eD}} - \dfrac{3}{4}\right)} + s\right]} \end{aligned} \tag{3.31}$$

Taking the Laplace inverse transformation of Eq. (3.31), and according to the *Mathematics Handbook*, we find that the inverse transformation of the $\dfrac{a}{s+b}$ shape formula is ae^{-bt}, and then we could have

$$q_{\mathrm{D}} = \frac{1}{\left(\ln r_{\mathrm{eD}} - \dfrac{3}{4}\right)} e^{-\frac{2t_{\mathrm{D}}}{r_{\mathrm{eD}}^2\left(\ln r_{\mathrm{eD}} - \frac{3}{4}\right)}} \tag{3.32}$$

This result shows that the dimensionless flow rate solution with constant BHFP conforms to the exponential decline law. Based on Eq. (3.22), Fetkovich (1980) replaced 3/4 by 1/2 (tried 1/2, 5/8 and 3/4, and finally found that 1/2 had the most significant correlation) and defined the dimensionless decline time.

$$t_{\mathrm{Dd}} = t_{\mathrm{D}} \cdot \frac{1}{\dfrac{1}{2}\left(r_{\mathrm{eD}}^2 - 1\right)\left(\ln r_{\mathrm{eD}} - \dfrac{1}{2}\right)} = \beta t_{\mathrm{D}} \tag{3.33}$$

3.2.2 Fetkovich Dimensionless Decline Rate and Decline Cumulative Production Type Curves

Substituting dimensionless decline rate function q_{Dd} and dimensionless decline time function t_{Dd} as defined by Fetkovich into Eq. (3.6) through Eq. (3.10), we have

$$\frac{1}{r_{\mathrm{D}}}\frac{\partial}{\partial r_{\mathrm{D}}}\left(r_{\mathrm{D}}\frac{\partial p_{\mathrm{D}}}{\partial r_{\mathrm{D}}}\right)=\beta\frac{\partial p_{\mathrm{D}}}{\partial t_{\mathrm{Dd}}} \tag{3.34}$$

$$p_{\mathrm{D}}\left(r_{\mathrm{D}},0\right)=0 \tag{3.35}$$

$$p_{\mathrm{D}}\left(1,t_{\mathrm{Dd}}\right)=1 \tag{3.36}$$

$$\left.\frac{\partial p_{\mathrm{D}}}{\partial r_{\mathrm{D}}}\right|_{r_{\mathrm{D}}=r_{\mathrm{eD}}}=0 \tag{3.37}$$

$$q_{\mathrm{Dd}}=-\alpha\left(r_{\mathrm{D}}\frac{\partial p_{\mathrm{D}}}{\partial r_{\mathrm{D}}}\right)_{r_{\mathrm{D}}=1} \tag{3.38}$$

Then we transform the q_{D} and t_{D} into dimensionless variables q_{Dd} and t_{Dd} in Fetkovich decline curve. Taking the Laplace transformation of Eq. (3.34) through Eq. (3.38) yields

$$\frac{d^2\bar{p}_{\mathrm{D}}}{dr_{\mathrm{D}}^2}+\frac{1}{r_{\mathrm{D}}}\frac{d\bar{p}_{\mathrm{D}}}{dr_{\mathrm{D}}}=s\beta\,\bar{p}_{\mathrm{D}} \tag{3.39}$$

$$\bar{p}_{\mathrm{D}}\left(r_{\mathrm{D}},0\right)=0 \tag{3.40}$$

$$\bar{p}_{\mathrm{D}}(1,s)=\frac{1}{s} \tag{3.41}$$

$$\left.\frac{d\bar{p}_{\mathrm{D}}}{dr_{\mathrm{D}}}\right|_{r_{\mathrm{D}}=r_{\mathrm{eD}}}=0 \tag{3.42}$$

$$\bar{q}_{\mathrm{Dd}}=-\alpha\left(\frac{\partial\bar{p}_{\mathrm{D}}}{\partial r_{\mathrm{D}}}\right)_{r_{\mathrm{D}}=1} \tag{3.43}$$

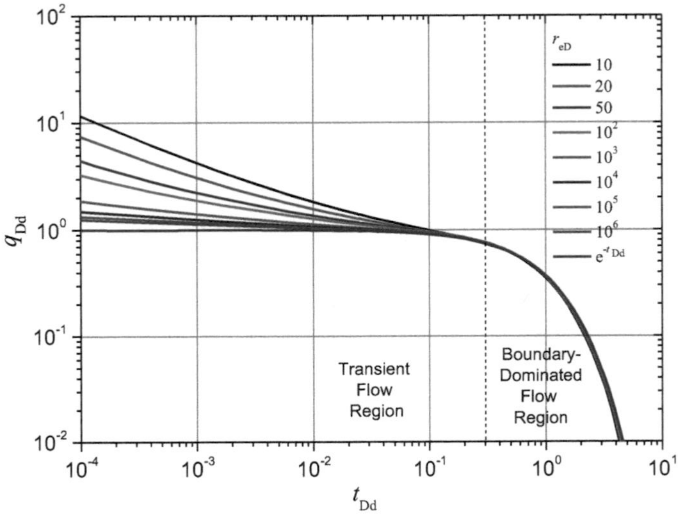

Figure 3.2 Fetkovich dimensionless flow rate q_{Dd} type curves (after Fetkovich, 1980)

The process for solving Eq. (3.39) is similar to the former. The Laplace space solution for constant BHFP for a well entered in a closed circular reservoir is

$$\bar{q}_{Dd} = \frac{\alpha\sqrt{\beta}}{\sqrt{s}} \frac{\dfrac{K_1(\sqrt{\beta s})}{I_0(\sqrt{\beta s})} - \dfrac{K_1(r_{eD}\sqrt{\beta s})}{I_1(r_{eD}\sqrt{\beta s})}\dfrac{I_1(\sqrt{\beta s})}{I_0(\sqrt{\beta s})}}{\dfrac{K_0(\sqrt{\beta s})}{I_0(\sqrt{\beta s})} + \dfrac{K_1(r_{eD}\sqrt{\beta s})}{I_1(r_{eD}\sqrt{\beta s})}} \tag{3.44}$$

For constant BHFP for a well entered in a closed circular reservoir, the relationship between Fetkovich dimensionless decline rate function q_{Dd} and dimensionless decline time t_{Dd} is shown in Figure 3.2. It can be seen that it is just contrary to what is shown in Figure 3.1. Before $t_{Dd} = 0.3$, it is the early transient flow period, which is controlled by r_{eD}. When $t_{Dd} > 0.3$, it turns into boundary-dominated flow period, and the curves coincide with each other and present an exponential decrease trend.

Replacing 3/4 by 1/2 in Eq. (3.32), we have

$$q_D\left(\ln r_{eD} - \frac{1}{2}\right) = e^{-\frac{2t_D}{r_{eD}^2\left(\ln r_{eD} - \frac{1}{2}\right)}} = q_{Dd} = e^{-t_{Dd}} \tag{3.45}$$

where t_{Dd} could be described as

$$t_{\mathrm{Dd}} = \frac{2t_{\mathrm{D}}}{\left(r_{\mathrm{eD}}^2 - 1\right)\left(\ln r_{\mathrm{eD}} - \frac{1}{2}\right)} \tag{3.46}$$

Substituting $t_{\mathrm{D}} = \frac{Kt}{\varphi\mu C_t r_{\mathrm{w}}^2}$ into Eq. (3.46), we have

$$t_{\mathrm{Dd}} = \frac{2K/\varphi\mu C_{\mathrm{t}} r_{\mathrm{w}}^2}{\left(r_{\mathrm{eD}}^2 - 1\right)\left(\ln r_{\mathrm{eD}} - \frac{1}{2}\right)} t \tag{3.47}$$

According to Eq. (2.108), we have

$$D_{\mathrm{i}} = \frac{2K/\varphi\mu C_{\mathrm{t}} r_{\mathrm{w}}^2}{\left(r_{\mathrm{eD}}^2 - 1\right)\left(\ln r_{\mathrm{eD}} - \frac{1}{2}\right)} \tag{3.48}$$

It could be found that D_{i} is the function of rock properties, fluid properties, and reservoir shape, and it has nothing to do with flow pressure.

Dimensionless decline cumulative production function, N_{pDd}, is described as

$$N_{\mathrm{pDd}} = \int_0^{t_{\mathrm{Dd}}} q_{\mathrm{Dd}}(\tau)d\tau \tag{3.49}$$

Laplace transformation of Eq. (3.49) yields

$$\bar{N}_{\mathrm{pDd}} = \frac{\bar{q}_{\mathrm{Dd}}(s)}{s} \tag{3.50}$$

Then substituting Eq. (3.45) into Eq. (3.50), we have

$$\bar{N}_{\mathrm{pDd}} = \frac{\alpha\sqrt{\beta}}{s\sqrt{s}} \cdot \frac{\frac{K_1(\sqrt{\beta s})}{I_0(\sqrt{\beta s})} - \frac{K_1(r_{\mathrm{eD}}\sqrt{\beta s})}{I_1(r_{\mathrm{eD}}\sqrt{\beta s})}\frac{I_1(\sqrt{\beta s})}{I_0(\sqrt{\beta s})}}{\frac{K_0(\sqrt{\beta s})}{I_0(\sqrt{\beta s})} + \frac{K_1(r_{\mathrm{eD}}\sqrt{\beta s})}{I_1(r_{\mathrm{eD}}\sqrt{\beta s})}} \tag{3.51}$$

The relationship between the dimensionless decline cumulative production and dimensionless decline time is shown in Figure 3.3, where the difference from a dimensionless decline rate curve is that the transient flow period ends at around $t_{\mathrm{Dd}} = 0.6$ and then boundary-dominated flow follows. Figure 3.4 shows dimensionless decline rate and dimensionless decline cumulative production type curves.

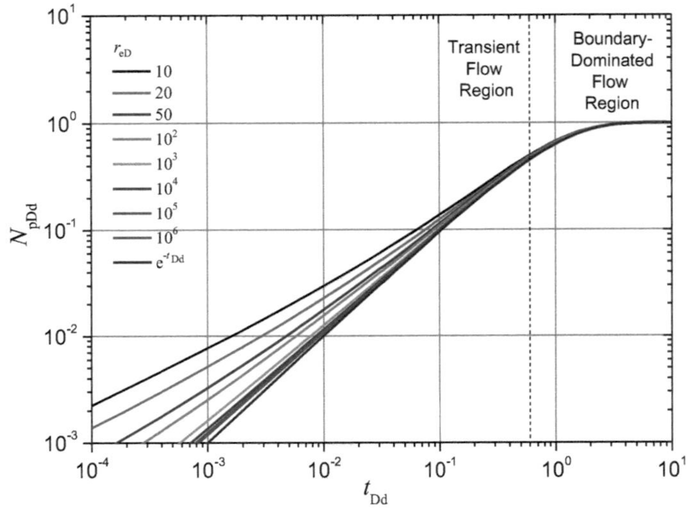

Figure 3.3 Fetkovich dimensionless decline cumulative production N_{pDd} type curves (after Fetkovich, 1980)

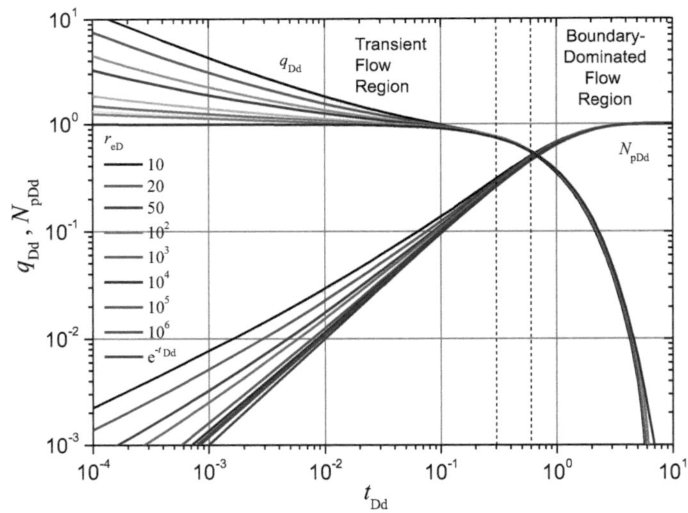

Figure 3.4 Fetkovich dimensionless flow rate q_{Dd} and dimensionless decline cumulative production N_{pDd} type curves (after Fetkovich, 1980)

3.2.3 Fetkovich–Arps Dimensionless Decline Rate and Decline Cumulative Production Type Curves

As discussed previously, the solution for constant BHFP for a well entered in a closed circular reservoir could be regarded as an Arps exponential decline

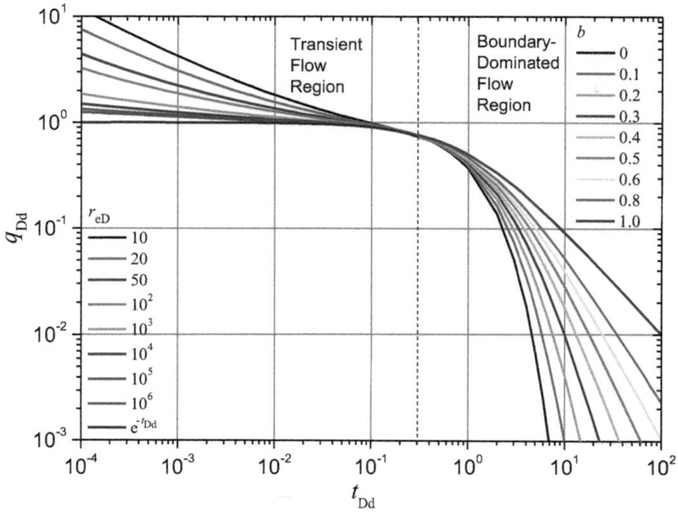

Figure 3.5 Fetkovich dimensionless flow rate q_{Dd} type curves (after Fetkovich, 1980)

curve in the later flow period, $q_{Dd} = e^{-t_{Dd}}$; it could be described by a general formula:

$$q_{Dd}(t_{Dd}) = \frac{1}{\left(1 + bt_{Dd}\right)^{1/b}} \tag{3.52}$$

Readers can refer to detailed information about the Arps decline curves as shown in Section 2.3. If following the definition by Fetkovich (1980), we could combine the dimensionless decline rate curves and Arps decline equations, and then we could get a new dimensionless decline rate curve, which is the so-called Fetkovich–Arps production curve.

As shown in Figure 3.5, Fetkovich–Arps decline rate curve could be divided into two parts. The left part when $t_{Dd} < 0.3$ represents the early and middle transient flow period, which is mainly affected by dimensionless reservoir drainage radius r_{eD}, and the decline rate curve shifts downwards with the increased r_{eD}, and the right part when $t_{Dd} > 0.3$ represents the boundary-dominated flow, which is the curve cluster controlled by Arps decline exponent b. The decline rate curve shifts to the right when b generally increases.

Figure 3.6 shows the dimensionless cumulative production curve, which is the result of the integration of the dimensionless decline rate curve. Figure 3.7 shows the result of dimensionless decline rate and dimensionless decline cumulative production. Given the production data, that is, the relationship between flow rate $q(t)$ or cumulative production N_p and producing time, the reservoir characteristics and the well performance could be evaluated through curve matching.

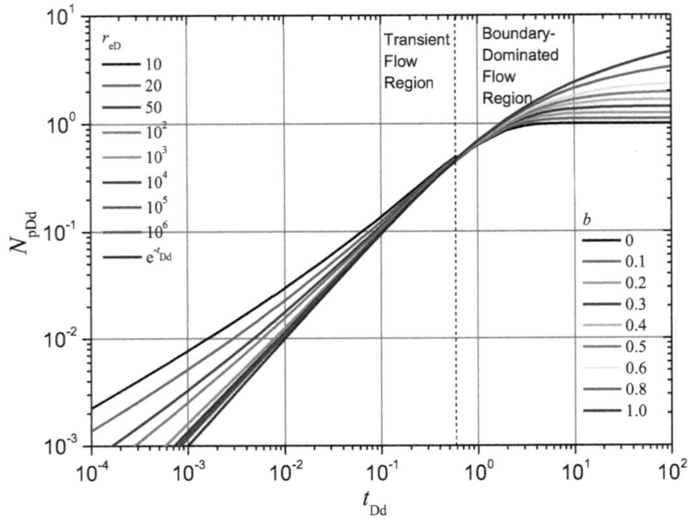

Figure 3.6 Fetkovich–Arps dimensionless decline cumulative production N_{pDd} type curves (after Fetkovich, 1980)

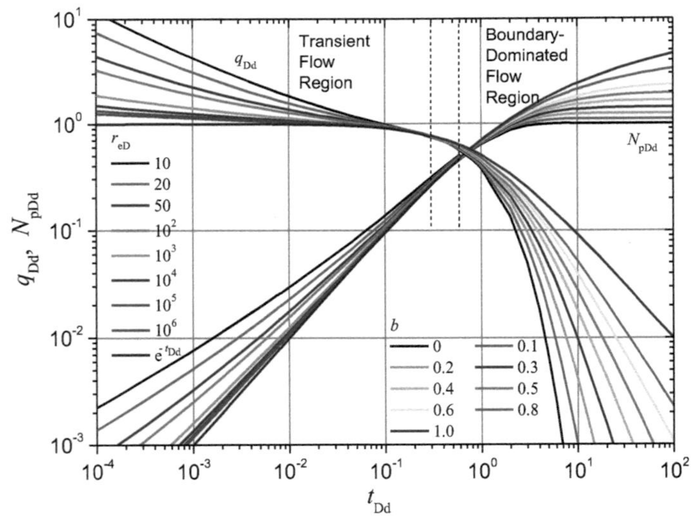

Figure 3.7 Fetkovich–Arps dimensionless flow rate q_{Dd} and dimensionless decline cumulative production N_{pDd} type curves (after Fetkovich, 1980)

3.2.4 Fetkovich–Arps Dimensionless Decline Rate Integral Type Curves

Blasingame et al. (1989) firstly introduced the concept of rate integral in the field of well test analysis, and Spivey et al. (1992) extended it into the production decline analysis. This method could reduce the effect of the data noise and allow the data to be more easily analyzed to some degree. Eq. (2.199) gives the dimensionless rate integral q_{Ddi} as defined by McCray (1990),

$$q_{\mathrm{Ddi}} = \frac{N_{\mathrm{pDd}}}{t_{\mathrm{Dd}}} = \frac{1}{t_{\mathrm{Dd}}} \int_0^{t_{\mathrm{Dd}}} q_{\mathrm{Dd}}(\tau) d\tau \tag{3.53}$$

It is the average flow rate since the well has come into producing. After acquiring the dimensionless cumulative production through Eq. (3.51), the dimensionless rate integral could be calculated on the basis of Eq. (3.53). In boundary-dominated flow period, the Arps exponential decline equations could be applied in the calculation of the dimensionless rate integral, that is, Eq. (2.120) through Eq. (2.125).

For exponential decline, we have

$$q_{\mathrm{Ddi}} = \frac{1 - e^{-t_{\mathrm{Dd}}}}{t_{\mathrm{Dd}}} \tag{3.54}$$

Or in terms of q_{Dd}, we have

$$q_{\mathrm{Ddi}} = \frac{1 - q_{\mathrm{Dd}}}{t_{\mathrm{Dd}}} \tag{3.55}$$

For hyperbolic decline, we have

$$q_{\mathrm{Ddi}} = \frac{1}{(b-1)t_{\mathrm{Dd}}} \left[\left(1 + bt_{\mathrm{Dd}}\right)^{(1-1/b)} - 1 \right] \tag{3.56}$$

Or in terms of q_{Dd}, we have

$$q_{\mathrm{Ddi}} = \frac{1}{(b-1)t_{\mathrm{Dd}}} \left[q_{\mathrm{Dd}}^{(1-b)} - 1 \right] \tag{3.57}$$

For harmonic decline, we have

$$q_{\mathrm{Ddi}} = \frac{\ln\left(1 + t_{\mathrm{Dd}}\right)}{t_{\mathrm{Dd}}} \tag{3.58}$$

Or in terms of q_{Dd}, we have

$$q_{\mathrm{Ddi}} = -\frac{\ln q_{\mathrm{Dd}}}{t_{\mathrm{Dd}}} \tag{3.59}$$

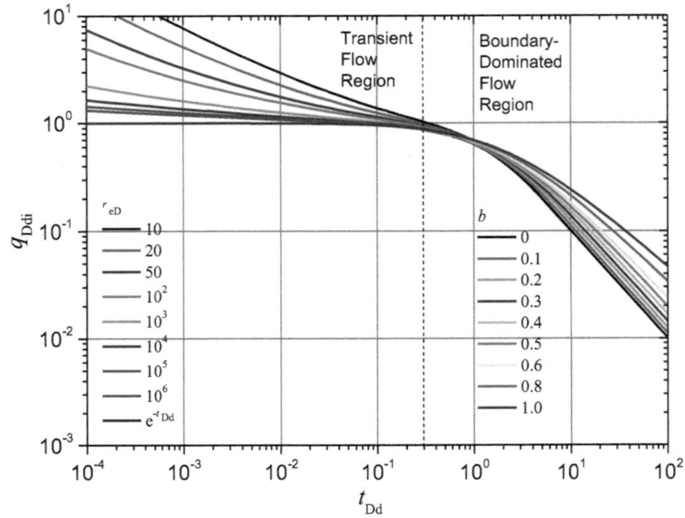

Figure 3.8 Fetkovich–Arps dimensionless flow rate integral type curves (after McCray, 1990)

The dimensionless decline rate integral curves are shown in Figure 3.8. The dimensionless decline rate and dimensionless decline rate integral curves are shown in Figure 3.9. It could be seen that the two kinds of curves have similar characteristics, and the latter kind has relatively a little upward bias.

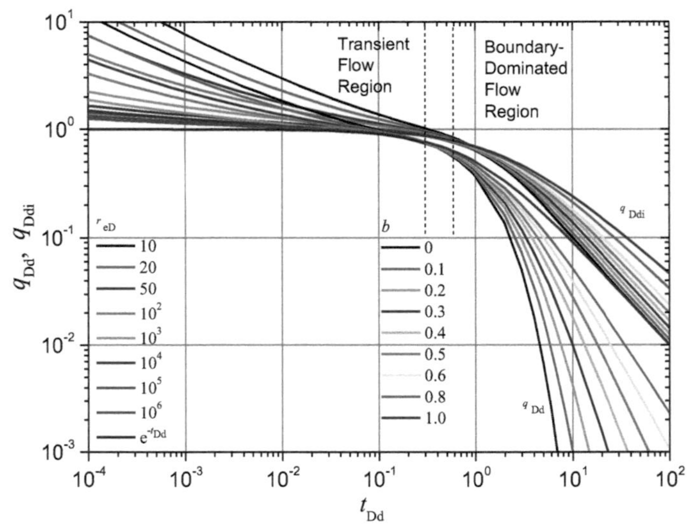

Figure 3.9 Fetkovich–Arps dimensionless flow rate q_{Dd} and dimensionless flow rate integral q_{Ddi} type curves (after McCray, 1990)

3.2.5 Fetkovich–Arps Dimensionless Decline Rate Integral Derivative Type Curves

The conception of decline rate integral derivative q_{Ddid} is just the same as the Arps method, which has been introduced in Chapter 2, and readers could refer to Section 2.3 for more information.

$$q_{\mathrm{Ddid}} = -\frac{dq_{\mathrm{Ddi}}}{d\ln t_{\mathrm{Dd}}} = -t_{\mathrm{Dd}}\frac{dq_{\mathrm{Ddi}}}{dt_{\mathrm{Dd}}} = -t_{\mathrm{Dd}}\frac{d\left(N_{\mathrm{pDd}}/t_{\mathrm{Dd}}\right)}{dt_{\mathrm{Dd}}} \tag{3.60}$$

$$q_{\mathrm{Ddid}} = q_{\mathrm{Ddi}} - q_{\mathrm{Dd}} \tag{3.61}$$

For exponential decline, we have

$$q_{\mathrm{Ddid}} = \frac{1-e^{-t_{\mathrm{Dd}}}}{t_{\mathrm{Dd}}} - q_{\mathrm{Dd}} \tag{3.62}$$

Or in terms of q_{Dd}

$$q_{\mathrm{Ddid}} = \frac{1-q_{\mathrm{Dd}}}{t_{\mathrm{Dd}}} - q_{\mathrm{Dd}} \tag{3.63}$$

For hyperbolic decline, we have

$$q_{\mathrm{Ddid}} = \frac{1}{(b-1)t_{\mathrm{Dd}}}\left[\left(1+bt_{\mathrm{Dd}}\right)^{(1-1/b)} - 1\right] - q_{\mathrm{Dd}} \tag{3.64}$$

Or in terms of q_{Dd}

$$q_{\mathrm{Ddid}} = \frac{1}{(b-1)t_{\mathrm{Dd}}}\left[q_{\mathrm{Dd}}^{(1-b)} - 1\right] - q_{\mathrm{Dd}} \tag{3.65}$$

For harmonic decline, we have

$$q_{\mathrm{Ddid}} = \frac{1}{t_{\mathrm{Dd}}}\ln\left(1+t_{\mathrm{Dd}}\right) - q_{\mathrm{Dd}} \tag{3.66}$$

Or in terms of q_{Dd}

$$q_{\mathrm{Ddid}} = -\frac{1}{t_{\mathrm{Dd}}}\ln q_{\mathrm{Dd}} - q_{\mathrm{Dd}} \tag{3.67}$$

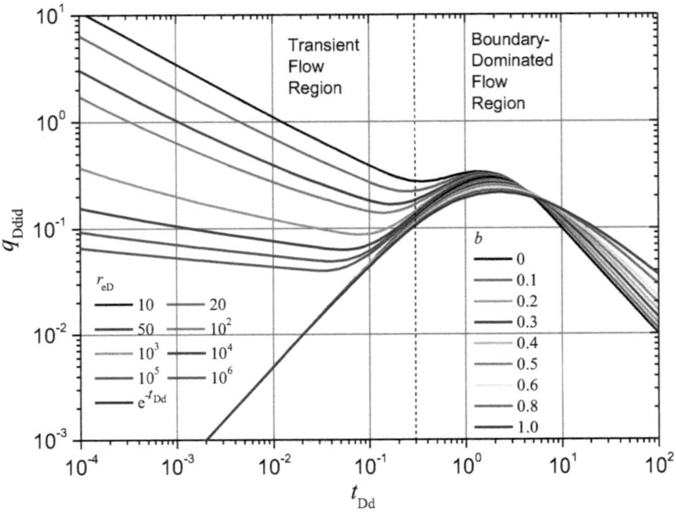

Figure 3.10 Fetkovich–Arps dimensionless decline rate integral derivative type curves (after McCray, 1990)

The decline rate integral derivative curves are shown in Figure 3.10. The decline rate and decline rate integral derivative curves are shown in Figure 3.11. The decline rate integral and decline rate integral derivative curves are shown in Figure 3.12.

From those figures, it can be seen that both the decline rate integral curve and the decline rate integral derivative curve is smooth and easy to analyze. However,

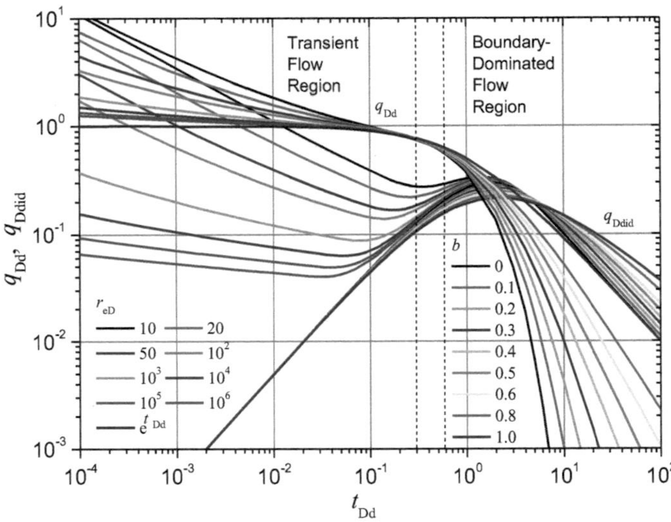

Figure 3.11 Fetkovich–Arps dimensionless decline rate and decline rate integral derivative type curves (after McCray, 1990)

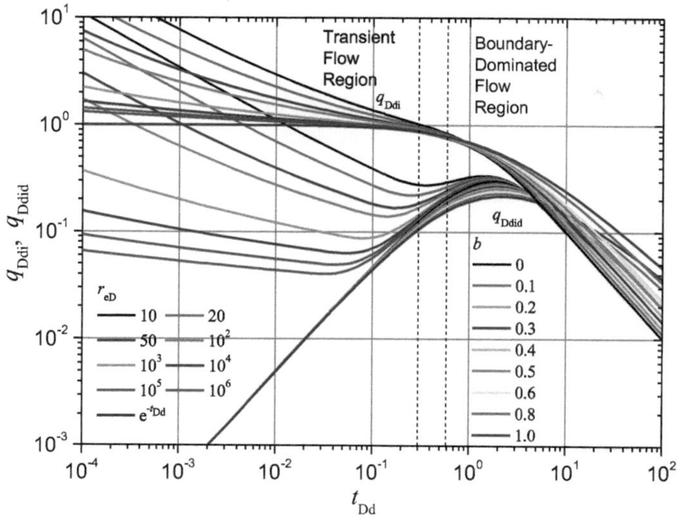

Figure 3.12 Fetkovich–Arps dimensionless decline rate integral and decline rate integral derivative type curves (after McCray, 1990)

the derivative curve is very sensitive to the error in the early data. A tiny error tends to result in a large accumulative deviation.

3.3 Decline Curve Analysis Using Fetkovich-Style Type Curves

3.3.1 Conventional Decline Curve Analysis

As discussed previously, the Fetkovich–Arps type curve consists of two parts, where the first half is with respect to the transient flow period, which depends on the value of r_{eD}, while the second half represents the Arps decline equation, which depends on the b value of the decline exponent.

Based on the curve matching analysis between field data and Fetkovich–Arps type curves, the Arps decline parameters q_i, D_i, and b could be determined by the first half of the type curve. And the value of r_{eD} could be determined by the second half of the curves. Then the permeability K, skin factor S, reservoir drainage radius r_e, single-well original gas in place, and the cumulative production before reaching the abandoned condition could all be determined.

The detailed process of the conventional Fetkovich–Arps curve matching is given below.

1. Draw the q~t curve on the log–log plot.
2. Similar to the well test analysis, matching the q~t plot with the theoretical type curves (Figure 3.7). Please refer to Section 1.3 for the matching principles.

3. Based on the matching results, record the value of dimensionless reservoir drainage radius r_{eD} and decline exponent b.
4. Choose any one of the matching points and record the actual matching point $(t,q)_M$ and the corresponding theoretical matching point $(t_{Dd},q_{Dd})_M$.
5. On the basis of the flow rate matching point, determine the initial flow rate $q_i = \left(\frac{q}{q_{Dd}}\right)_M$.
6. On the basis of the time matching point, determine the initial decline rate D_i.

$$D_i = \left(\frac{t_{Dd}}{t}\right)_M \tag{3.68}$$

7. On the basis of the value of r_{eD} and q_i, which were calculated in Step (3), determine the value of permeability according to Eq. (3.24).

$$K = \frac{\mu B\left(\ln r_{eD} - \frac{1}{2}\right)}{2\pi h(p_i - p_{wf})} q_i = \frac{\mu B\left(\ln r_{eD} - \frac{1}{2}\right)}{2\pi h(p_i - p_{wf})} \left(\frac{q}{q_{Dd}}\right)_M \tag{3.69}$$

8. Calculate the apparent wellbore radius according to Eq. (3.47) based on the time matching point

$$r_{wa} = \sqrt{\frac{2K/\varphi\mu C_t}{(r_{eD}^2 - 1)\left(\ln r_{eD} - \frac{1}{2}\right)} \left(\frac{t}{t_{Dd}}\right)_M} \tag{3.70}$$

And then determine the skin factor S

$$S = \ln \frac{r_w}{r_{wa}} \tag{3.71}$$

9. Determine the ultimate recoverable reserves.
According to related equations in Section 2.1, we can get the recoverable reserves before the abandoned condition and the corresponding production time. For the Arps exponential decline, the ultimate recoverable reserves could be estimated by Eq. (3.72)

$$EUR = N_p + Q_f = N_p + \frac{q_f - q_{ab}}{D_i} \tag{3.72}$$

And the time when the well must be abandoned could be calculated by Eq. (3.73)

$$\Delta t = t_{ab} - t_f = \frac{1}{D_i} \ln\left(\frac{q_f}{q_{ab}}\right) \tag{3.73}$$

For the Arps hyperbolic decline, the ultimate recoverable reserves could be estimated by Eq. (3.74)

$$EUR = N_p + Q_f = N_p + \frac{q_i^b}{(1-b)D_i}\left(q_f^{1-b} - q_{ab}^{1-b}\right) \tag{3.74}$$

The time when the well must be abandoned could be calculated by Eq. (3.75)

$$\Delta t = t_{ab} - t_f = \frac{\left(\frac{q_i}{q_{ab}}\right)^b - \left(\frac{q_i}{q_f}\right)^b}{bD_i} \tag{3.75}$$

For the Arps harmonic decline, the ultimate recoverable reserves could be estimated by Eq. (3.76)

$$EUR = N_p + Q_f = N_p + \frac{q_i}{D_i}\ln\left(\frac{q_f}{q_{ab}}\right) \tag{3.76}$$

The time when the well must be abandoned could be calculated by Eq. (3.77)

$$\Delta t = t_{ab} - t_f = \frac{q_i}{D_i}\left(\frac{1}{q_{ab}} - \frac{1}{q_f}\right) \tag{3.77}$$

10. Determine the reservoir pore volume.
According to the time matching point and flow rate matching point, we could determine the reservoir pore volume

$$V_p = \pi r_e^2 h\varphi \tag{3.78}$$

According to Eq. (3.47), we have

$$\left(\frac{t}{t_{Dd}}\right)_M = \frac{\left(r_{eD}^2 - 1\right)\left(\ln r_{eD} - \frac{1}{2}\right)\varphi\mu C_t r_w^2}{2K} \tag{3.79}$$

According to Eq. (3.25), we have

$$\left(\frac{q}{q_{Dd}}\right)_M = \frac{2\pi Kh(p_i - p_{wf})}{\mu B\left(\ln r_{eD} - \frac{1}{2}\right)} \tag{3.80}$$

Multiply Eq. (3.79) and Eq. (3.80), we have

$$\left(\frac{t}{t_{\mathrm{Dd}}}\right)_{\mathrm{M}}\left(\frac{q}{q_{\mathrm{Dd}}}\right)_{\mathrm{M}}=\frac{\pi\left(r_{\mathrm{e}}^{2}-r_{\mathrm{w}}^{2}\right)h\varphi C_{\mathrm{t}}}{B}(p_{\mathrm{i}}-p_{\mathrm{wf}})=\frac{V_{\mathrm{p}}C_{\mathrm{t}}}{B}(p_{\mathrm{i}}-p_{\mathrm{wf}}) \tag{3.81}$$

It could also be represented as

$$V_{\mathrm{p}}=\frac{B}{C_{\mathrm{t}}(p_{\mathrm{i}}-p_{\mathrm{wf}})}\left(\frac{t}{t_{\mathrm{Dd}}}\right)_{\mathrm{M}}\left(\frac{q}{q_{\mathrm{Dd}}}\right)_{\mathrm{M}} \tag{3.82}$$

11. Determine the drainage area
Determine the drainage area according to Eq. (3.78) and Eq. (3.83)

$$r_{\mathrm{e}}=\sqrt{\frac{V_{\mathrm{p}}}{\pi h\varphi}} \tag{3.83}$$

$$A=\frac{V_{\mathrm{p}}}{h\varphi} \tag{3.84}$$

12. Determine the reservoir reserves
After transforming the volume underground into the volume at surface, we can determine the original gas (oil) in place according to Eq. (3.82).

$$N=\frac{V_{\mathrm{p}}\left(1-S_{\mathrm{w}}\right)}{B_{\mathrm{i}}} \tag{3.85}$$

We can also calculate through Arps matching results. According to Eq. (2.14), we have

$$N=\int_{0}^{t_{\infty}}\frac{q_{\mathrm{i}}}{\left(1+bD_{\mathrm{i}}t\right)^{1/b}}dt=\frac{q_{\mathrm{i}}}{(1-b)D_{\mathrm{i}}}\left[1-\frac{1}{\left(1+bD_{\mathrm{i}}t_{\infty}\right)^{1/b-1}}\right]\approx\frac{q_{\mathrm{i}}}{(1-b)D_{\mathrm{i}}} \tag{3.86}$$

3.3.2 Normalized Decline Curve Analysis

We can replace the flow rate and cumulative production by normalized rate and normalized cumulative production, and repeat the same steps given above.

The normalized flow rate is given by

$$\frac{q}{\Delta p}=\frac{q}{p_{\mathrm{i}}-p_{\mathrm{wf}}} \tag{3.87}$$

The normalized cumulative production is given by

$$N_p = \int_0^t \frac{q}{p_i - p_{wf}} d\tau \tag{3.88}$$

For a gas well, we only need to replace the pressure by a normalized pseudo-pressure.

The process of normalized Fetkovich–Arps curves matching analysis is similar to conventional analysis, and the detailed process is given below.

1. Draw the $q/\Delta p \sim t$ curve on the log-log plot.
2. Matching the $q/\Delta p \sim t$ plot with the theoretical type curves (Figure 3.7).
3. Based on the matching results, record the value of dimensionless drainage reservoir radius r_{eD} and decline exponent b.
4. Choose any one of the matching points and record the actual matching point $(t, q/\Delta p)_M$ and the corresponding theoretical matching point $(t_{Dd}, q_{Dd})_M$.
5. On the basis of the flow rate matching point, determine the initial flow rate $q_i = \left(\frac{q/\Delta p}{q_{Dd}}\right)_M (p_i - p_{wf})$.
6. On the basis of the time matching point, determine the initial decline rate D_i.

$$D_i = \left(\frac{t_{Dd}}{t}\right)_M \tag{3.89}$$

7. On the basis of the values of r_{eD} and q_i, which were calculated in Step (3), determine the value of permeability according to Eq. (3.24).

$$K = \frac{\mu B\left(\ln r_{eD} - \frac{1}{2}\right)}{2\pi h(p_i - p_{wf})} q_i = \frac{\mu B\left(\ln r_{eD} - \frac{1}{2}\right)}{2\pi h}\left(\frac{q/\Delta p}{q_{Dd}}\right)_M \tag{3.90}$$

8. Determine r_{eD} based on the time matching point and Step (3), calculate the apparent wellbore radius according to Eq. (3.47), and then determine the skin factor S.

$$r_{wa} = \sqrt{\frac{2K/\varphi\mu C_t}{\left(r_{eD}^2 - 1\right)\left(\ln r_{eD} - \frac{1}{2}\right)}\left(\frac{t}{t_{Dd}}\right)_M} \tag{3.91}$$

$$S = \ln\frac{r_w}{r_{wa}} \tag{3.92}$$

9. Determine the ultimate recoverable reserves
   According to related equations in Section 2.1, we could get the recoverable reserves before the abandoned condition and the corresponding production time. For the

Arps exponential decline, the ultimate recoverable reserves could be predicted by Eq. (3.93)

$$EUR = N_{\mathrm{p}} + Q_{\mathrm{f}} = N_{\mathrm{p}} + \frac{\left(\frac{q}{\Delta p}\right)_{\mathrm{f}} (\Delta p)_{\mathrm{f}} - q_{\mathrm{ab}}}{D_{\mathrm{i}}} \tag{3.93}$$

And the time when the well must be abandoned could be calculated by

$$\Delta t = t_{\mathrm{ab}} - t_{\mathrm{f}} = \frac{1}{D_{\mathrm{i}}} \ln \left[\frac{\left(\frac{q}{\Delta p}\right)_{\mathrm{f}} (\Delta p)_{\mathrm{f}}}{q_{\mathrm{ab}}} \right] \tag{3.94}$$

For the Arps hyperbolic decline, the ultimate recoverable reserves could be estimated by Eq. (3.95)

$$EUR = N_{\mathrm{p}} + Q_{\mathrm{f}} = N_{\mathrm{p}} + \frac{q_{\mathrm{i}}^{b}}{(1-b)D_{\mathrm{i}}} \left[\left(\frac{q}{\Delta p}\right)_{\mathrm{f}}^{1-b} (\Delta p)_{\mathrm{f}}^{1-b} - q_{\mathrm{ab}}^{1-b} \right] \tag{3.95}$$

The time at which the well must be abandoned could be calculated by

$$\Delta t = t_{\mathrm{ab}} - t_{\mathrm{f}} = \frac{\left(\frac{q_{\mathrm{i}}}{q_{\mathrm{ab}}}\right)^{b} - \left[\frac{q_{\mathrm{i}}}{\left(\frac{q}{\Delta p}\right)_{\mathrm{f}} (\Delta p)_{\mathrm{f}}} \right]^{b}}{bD_{\mathrm{i}}} \tag{3.96}$$

For the Arps harmonic decline, the ultimate recoverable reserves could be estimated by

$$EUR = N_{\mathrm{p}} + Q_{\mathrm{f}} = N_{\mathrm{p}} + \frac{q_{\mathrm{i}}}{D_{\mathrm{i}}} \ln \left[\frac{\left(\frac{q}{\Delta p}\right)_{\mathrm{f}} (\Delta p)_{\mathrm{f}}}{q_{\mathrm{ab}}} \right] \tag{3.97}$$

The time at which the well must be abandoned could be calculated by

$$\Delta t = t_{\mathrm{ab}} - t_{\mathrm{f}} = \frac{q_{\mathrm{i}}}{D_{\mathrm{i}}} \left[\frac{1}{q_{\mathrm{ab}}} - \frac{1}{\left(\frac{q}{\Delta p}\right)_{\mathrm{f}} (\Delta p)_{\mathrm{f}}} \right] \tag{3.98}$$

10. Determine the reservoir pore volume.
According to the time matching point and flow rate matching point, we could determine the reservoir pore volume.

$$V_p = \pi r_e^2 h\varphi \tag{3.99}$$

According to Eq. (3.47), we have

$$\left(\frac{t}{t_{Dd}}\right)_M = \frac{\left(r_{eD}^2 - 1\right)\left(\ln r_{eD} - \frac{1}{2}\right)\varphi\mu C_t r_w^2}{2K} \tag{3.100}$$

According to Eq. (3.25), we have

$$\left(\frac{q/\Delta p}{q_{Dd}}\right)_M = \frac{2\pi Kh}{\mu B\left(\ln r_{eD} - \frac{1}{2}\right)} \tag{3.101}$$

Multiplying Eq. (3.79) and Eq. (3.80), we have

$$\left(\frac{t}{t_{Dd}}\right)_M \left(\frac{q/\Delta p}{q_{Dd}}\right)_M = \frac{\pi\left(r_e^2 - r_w^2\right)h\varphi C_t}{B} = \frac{V_p C_t}{B} \tag{3.102}$$

It could also be represented as

$$V_p = \frac{B}{C_t}\left(\frac{t}{t_{Dd}}\right)_M \left(\frac{q/\Delta p}{q_{Dd}}\right)_M \tag{3.103}$$

11. Determine the drainage area.
Determine the drainage area according to Eq. (3.99) and Eq. (3.103):

$$r_e = \sqrt{\frac{V_p}{\pi h\varphi}} \tag{3.104}$$

$$A = \frac{V_p}{h\varphi} \tag{3.105}$$

12. Determine the reservoir reserves.
Determine the reservoir reserves according to Eq. (3.103).

$$N = \frac{V_p\left(1 - S_w\right)}{B_i} \tag{3.106}$$

We can also calculate it through Arps matching results. According to Eq. (2.14), we have

$$N=\int_0^{t_\infty}\frac{q_i}{(1+bD_it)^{1/b}}dt=\frac{q_i}{(1-b)D_i}\left[1-\frac{1}{(1+bD_it_\infty)^{1/b-1}}\right]\approx\frac{q_i}{(1-b)D_i} \tag{3.107}$$

3.3.3 The Multiplicity of Fetkovich–Arps Decline Curve Analysis

In the process of curve matching by Fetkovich–Arps type curves, as shown in Figure 3.7, you may consider that the reservoir drainage area could be determined as long as the value r_{eD} has been got and the data matching in transient flow period has been achieved. Unfortunately, Mattar et al. (2012) indicated that it is incorrect. The reason is that in the early transient flow period, the pressure has not spread to the boundary of the reservoir. As a result, we could not match the points and determine the value of r_{eD} only based on the data from this flow period. We cannot calculate the value of r_{eD} until it turns into the boundary-dominated flow period. Pseudo-steady state flow is a kind of special flow shape of this period. If not, the matching result has a high uncertainty. The reason can be seen from Figure 3.13, that is, the curves in transient flow period that depend on r_{eD} are similar to each other, and one point could be fitted with several curves.

As a result, the Fetkovich–Arps decline analysis method is fit for the condition when the well is producing with constant BHFP, and the type curve could not be used until it reaches the period of boundary-dominated flow. The well drainage area could be determined by Eq. (3.83). If not, the matching point of r_{eD} will be multiple.

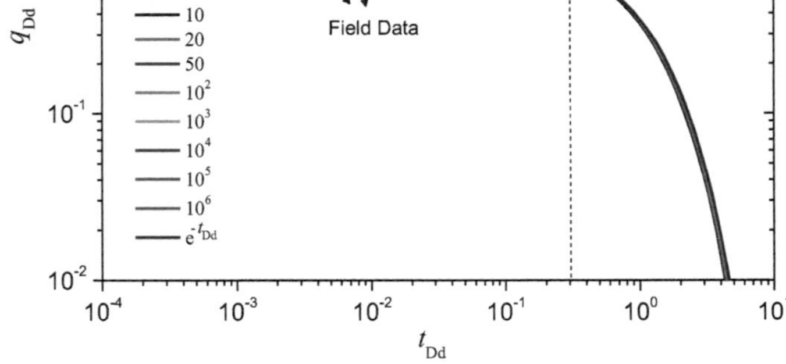

Figure 3.13 The multiplicity of Fetkovich–Arps decline curve analysis

Fetkovich–Arps method does not need flow pressure data. Essentially, it is kind of empirical and thus could be widely used. But the limitation is that the depletion analysis is usually multiple because of the similarity of the hyperbolic decline curve.

3.4 Example of Fetkovich–Arps Method

We still take the case in Section 2.3 as an example. The main parameters are as follows. There is a fractured gas well centered in a closed circular reservoir, producing with constant BHFP. The initial reservoir pressure is 30 MPa. The reservoir temperature is 80°C. The net pay thickness is 10 m, porosity is 0.1, reservoir radius is 500 m, original gas in place is 2.0×10^8 m^3, fracture half-length is 60 m, gas density ratio is 0.6, permeability estimation from well test analysis is 1.0 mD, skin factor is 0.0, BHFP is 10 MPa, production time is 800 days, cumulative gas production is 10954×10^4 m^3, fluid viscosity is 0.0228 mPa.s, rock compressibility is 4.35113×10^{-4} MPa^{-1}, gas compressibility is 0.0246 MPa^{-1} under original condition, abandoned flow rate is 1.0×10^4 m^3/d, and the current flow rate is 3.6×10^4 m^3/d. Detailed production data are shown in Table 2.2.

The solving process is as below.

1. Draw the $q \sim t$ curve and the $N_p \sim t$ curve on the log–log plot, as shown in Figure 3.14.
2. Match the production history of well A (Figure 3.14) with the theoretical type curves (Figure 3.7), and the results are shown in Figure 3.15.
3. Based on the matching results, record the value of dimensionless drainage reservoir radius r_{eD} and the decline exponent b, where $r_{eD} = 20$ and $b = 0.5$.
4. Choose any one of the matching points and record the actual matching point $(10,100)_M$ and the corresponding theoretical matching point $(0.078, 0.19)_M$.

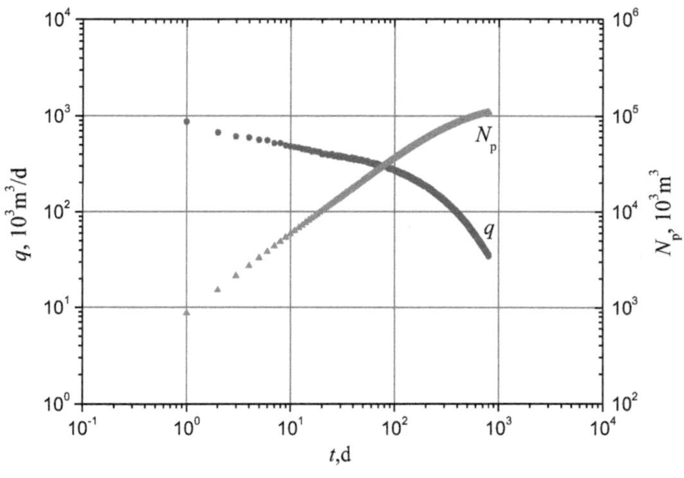

Figure 3.14 Production history of well A

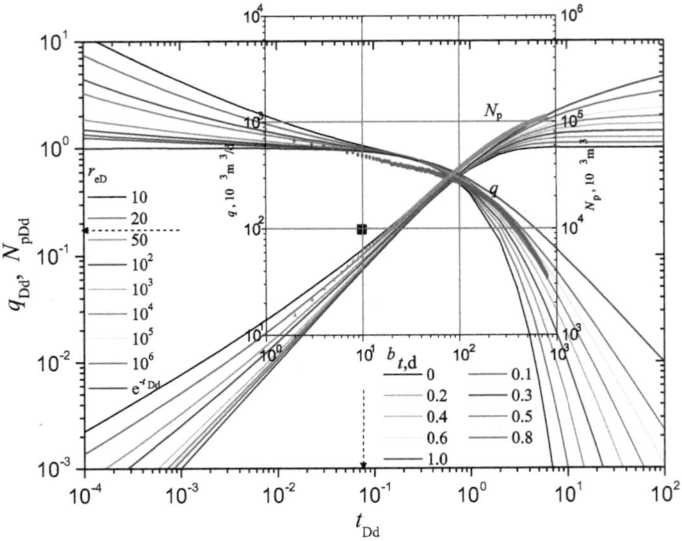

Figure 3.15 Log–log matching curves of Fetkovich–Arps method

5. On the basis of the production matching point, determine the initial flow rate

$$q_i = \left(\frac{q}{q_{Dd}}\right)_M = \left(\frac{100}{0.19}\right)_M = 526 \quad 10^3\,\text{m}^3/\text{d}$$

6. On the basis of the time matching point, determine the initial decline rate D_i

$$D_i = \left(\frac{t_{Dd}}{t}\right)_M = 0.078/10 = 0.0078\ \text{d}^{-1} \tag{3.108}$$

7. On the basis of the value of r_{eD} and q_i, determine the value of permeability according to Eq. (3.24). Note that Eq. (3.24) should be transformed as the pressure-square form under the SI unit system before use.

$$\begin{aligned} K &= \frac{p_{sc}(\mu Z)_{avg} T(\ln r_{eD} - 0.5)}{2.714\times10^{-5} h T_{sc}(p_i^2 - p_{wf}^2)}\left(\frac{q/10}{q_{Dd}}\right)_M \\ &= \frac{12.7355\times(\mu Z)_{avg} T(\ln r_{eD} - 0.5)}{h(p_i^2 - p_{wf}^2)}\left(\frac{q/10}{q_{Dd}}\right)_M \\ &= \frac{12.7355\times0.0187\times0.90\ \times353.15\times(\ln 20 - 0.5)}{10\times(30^2 - 10^2)}\times52.6 = 1.2\ \text{mD} \end{aligned} \tag{3.109}$$

8. Calculate the apparent wellbore radius and the skin factor S.

$$r_{\mathrm{wa}} = \sqrt{\frac{2K/\varphi\mu C_{\mathrm{t}}}{\left(r_{\mathrm{eD}}^2-1\right)\left(\ln r_{\mathrm{eD}}-\frac{1}{2}\right)}\left(\frac{t}{t_{\mathrm{Dd}}}\right)_{\mathrm{M}}}$$
$$= \sqrt{\frac{3.6\times10^{-3}\times2\times1.2/(0.1\times0.02\times0.0246)}{\left(20^2-1\right)(\ln 20-0.5)}\left(\frac{10\times24}{0.078}\right)} = 23.3\ \mathrm{m} \tag{3.110}$$

$$S = \ln\frac{r_{\mathrm{w}}}{r_{\mathrm{wa}}} = \ln\frac{0.1}{23.3} = -5.45 \tag{3.111}$$

As well A is a fractured well with vertical fractures, when converting skin factor into fracture half-length, we have

$$x_{\mathrm{f}} = 2r_{\mathrm{w}}e^{-S} = 2\times0.1\times e^{-(-5.45)} = 46.6\ \mathrm{m} \tag{3.112}$$

9. Determine the ultimate recoverable reserves.
The matching results show that the well is conformed to Arps hyperbolic decline. So we should substitute related parameters into Eq. (3.74), and then the ultimate recoverable reserves could be determined. The abandoned time could be calculated through Eq. (3.75)

$$EUR = N_{\mathrm{p}} + Q_{\mathrm{f}} = N_{\mathrm{p}} + \frac{q_{\mathrm{i}}^b}{(1-b)D_{\mathrm{i}}}\left(q_{\mathrm{f}}^{1-b} - q_{\mathrm{ab}}^{1-b}\right)$$
$$= 10954 + \frac{(52.6\times10)^{0.5}\left[(3.6\times10)^{0.5}-(1.0\times10)^{0.5}\right]}{(1-0.5)\times0.0078\times10} = 12623\times10^4\ \mathrm{m}^3 \tag{3.113}$$

$$\Delta t = t_{\mathrm{ab}} - t_{\mathrm{f}} = \frac{\left(\frac{q_{\mathrm{i}}}{q_{\mathrm{ab}}}\right)^b - \left(\frac{q_{\mathrm{i}}}{q_{\mathrm{f}}}\right)^b}{bD_{\mathrm{i}}} = \frac{\left(\frac{52.6}{1.0}\right)^{0.5} - \left(\frac{52.6}{3.6}\right)^{0.5}}{0.5\times0.0078} = 880\ \text{days} \tag{3.114}$$

10. Determine the reservoir pore volume.
In the SI unit system, Eq. (3.79) could be

$$\left(\frac{24t}{t_{\mathrm{Dd}}}\right)_{\mathrm{M}} = \frac{\left(r_{\mathrm{eD}}^2-1\right)\left(\ln r_{\mathrm{eD}}-\frac{1}{2}\right)\varphi\mu C_{\mathrm{t}}r_{\mathrm{w}}^2}{2\times3.6\times10^{-3}K} \tag{3.115}$$

According to Eq. (3.109), we have

$$K = \frac{12.7355 \times \mu ZT\left(\ln r_{\mathrm{eD}} - \frac{1}{2}\right)}{h(p_{\mathrm{i}}^2 - p_{\mathrm{wf}}^2)}\left(\frac{q/10}{q_{\mathrm{Dd}}}\right)_{\mathrm{M}} \tag{3.116}$$

Substituting Eq. (3.116) into Eq. (3.115), we have

$$V_{\mathrm{p}} = \frac{0.6912 \times ZT}{C_{\mathrm{t}}(p_{\mathrm{i}}^2 - p_{\mathrm{wf}}^2)} \times \left(\frac{t}{t_{\mathrm{Dd}}}\right)_{\mathrm{M}}\left(\frac{q}{q_{\mathrm{Dd}}}\right)_{\mathrm{M}} \tag{3.117}$$

$$V_{\mathrm{p}} = \frac{0.6912 \times 0.96 \times 353.15}{0.0246 \times (30^2 - 10^2)} \times \left(\frac{1}{0.0078}\right) \times (526.0) = 0.80 \times 10^6\,\mathrm{m}^3 \tag{3.118}$$

Note that the compressibility of the gas is so large that the system compressibility C_{t} is roughly equal to the gas compressibility C_{g}.

11. Determine the drainage area

$$r_{\mathrm{e}} = \sqrt{\frac{V_{\mathrm{p}}}{\pi h \varphi}} = \sqrt{\frac{0.8 \times 10^6}{\pi \times 10 \times 0.1}} = 506\ \mathrm{m} \tag{3.119}$$

12. Determine the reservoir reserves.

According to Eq. (3.85), we have

$$N = \frac{0.80 \times 10^6 \times (1 - 0.0)}{3.447 \times 10^{-4} \times 0.96 \times 353.15 / 30} = 2.05 \times 10^8\,\mathrm{m}^3 \tag{3.120}$$

According to Eq. (3.86), we have

$$N \approx \frac{q_{\mathrm{i}}}{(1-b)D_{\mathrm{i}}} = \frac{52.6}{(1-0.5) \times 0.0078} = 1.35 \times 10^8\,\mathrm{m}^3 \tag{3.121}$$

Blasingame Decline Curves Analysis 4

Both Arps and Fetkovich methods are based on constant bottom hole flowing pressure (BHFP) assumption and mainly consider the flow rate data, while neglecting the change of the gas pressure–volume–temperature (PVT) properties with reservoir pressure. However, the Blasingame method adopts the pseudo-pressure normalized rate ($q/\Delta p_{\mathrm{p}}$) and the material balance pseudo-time t_{ca} to build the corresponding type curve. This method takes into account the variable BHFP condition and the gas PVT properties, which are changing with reservoir pressure. In this chapter, we mainly introduce the theoretical basis of Blasingame decline curves analysis method, which is fit for the vertical well centered in a closed circular homogeneous reservoir as well as the type curve plotting. In addition, we also illustrate some applications using this method.

4.1 Pressure Distribution in a Closed Circular Reservoir: Constant Rate Production

Assume that there is a vertical well centered in a closed circular reservoir producing at a constant rate q. The reservoir radius is r_{e}. The BHFP, reservoir thickness, initial reservoir pressure, wellbore radius, reservoir porosity, total compressibility, permeability, fluid viscosity, and formation volume factor are denoted by p_{wf}, h, p_{i}, r_{w}, φ, C_{t}, K, μ, B, respectively. Besides, the skin effect is not considered. Under the basic SI unit system, the expressions Eq. (4.1) to Eq. (4.4) can be obtained.

$$\frac{1}{r}\frac{\partial}{\partial r}\left(r\frac{\partial p}{\partial r}\right)=\frac{\varphi\mu C_{\mathrm{t}}}{K}\frac{\partial p}{\partial t} \tag{4.1}$$

$$p(r,0)=p_{\mathrm{i}} \tag{4.2}$$

$$\left(r\frac{\partial p}{\partial r}\right)_{r=r_{\mathrm{w}}}=\frac{q\mu B}{2\pi Kh} \tag{4.3}$$

$$\left.\frac{\partial p}{\partial r}\right|_{r=r_{\mathrm{e}}}=0 \tag{4.4}$$

Advanced Production Decline Analysis and Application

In dimensionless variables

$$\frac{1}{r_D}\frac{\partial}{\partial r_D}\left(r_D\frac{\partial p_D}{\partial r_D}\right)=\frac{\partial p_D}{\partial t_D} \tag{4.5}$$

$$p_D(r_D,0)=0 \tag{4.6}$$

$$\left(r_D\frac{\partial p_D}{\partial r_D}\right)_{r_D=1}=-1 \tag{4.7}$$

$$\left.\frac{\partial p_D}{\partial r_D}\right|_{r_D=r_{eD}}=0 \tag{4.8}$$

The dimensionless variables are defined as follows

$$p_D=\frac{2\pi Kh(p_i-p_{wf})}{q\mu B}\quad t_D=\frac{Kt}{\varphi\mu C_t r_w^2}\quad r_D=\frac{r}{r_w}\quad r_{eD}=\frac{r_e}{r_w}$$

Laplace transformation of Eq. (4.5) through Eq. (4.8) yields

$$\frac{d^2\bar{p}_D}{dr_D^2}+\frac{1}{r_D}\frac{d\bar{p}_D}{dr_D}=s\,\bar{p}_D \tag{4.9}$$

$$\left.\frac{d\bar{p}_D}{dr_D}\right|_{r_D=1}=-\frac{1}{s} \tag{4.10}$$

$$\left.\frac{d\bar{p}_D}{dr_D}\right|_{r_D=r_{eD}}=0 \tag{4.11}$$

In the Laplace space, the solution of Eq. (4.9) is given as follows

$$\bar{p}_D=A_0I_0\left(r_D\sqrt{s}\right)+B_0K_0\left(r_D\sqrt{s}\right) \tag{4.12}$$

According to the inner boundary condition Eq. (4.10), we have

$$A_0I_1\left(\sqrt{s}\right)-B_0K_1\left(\sqrt{s}\right)=-\frac{1}{s\sqrt{s}} \tag{4.13}$$

According to the outer boundary condition Eq. (4.11), we have

$$A_0 I_1\left(r_{eD}\sqrt{s}\right) - B_0 K_1\left(r_{eD}\sqrt{s}\right) = 0 \tag{4.14}$$

According to Eq. (4.13) and Eq. (4.14), we have

$$A_0 = \frac{\dfrac{1}{s\sqrt{s}} \dfrac{K_1\left(r_{eD}\sqrt{s}\right)}{I_1\left(r_{eD}\sqrt{s}\right)}}{K_1\left(\sqrt{s}\right) - I_1\left(\sqrt{s}\right)\dfrac{K_1\left(r_{eD}\sqrt{s}\right)}{I_1\left(r_{eD}\sqrt{s}\right)}} \tag{4.15}$$

$$B_0 = \frac{\dfrac{1}{s\sqrt{s}}}{K_1\left(\sqrt{s}\right) - I_1\left(\sqrt{s}\right)\dfrac{K_1\left(r_{eD}\sqrt{s}\right)}{I_1\left(r_{eD}\sqrt{s}\right)}} \tag{4.16}$$

Substituting Eq. (4.15) and Eq. (4.16) into Eq. (4.12), we have

$$\bar{p}_D = \frac{1}{s\sqrt{s}}\left[\frac{\dfrac{K_1\left(r_{eD}\sqrt{s}\right)}{I_1\left(r_{eD}\sqrt{s}\right)} + \dfrac{K_0\left(\sqrt{s}\right)}{I_0\left(\sqrt{s}\right)}}{\dfrac{K_1\left(\sqrt{s}\right)}{I_0\left(\sqrt{s}\right)} - \dfrac{I_1\left(\sqrt{s}\right)}{I_0\left(\sqrt{s}\right)}\dfrac{K_1\left(r_{eD}\sqrt{s}\right)}{I_1\left(r_{eD}\sqrt{s}\right)}}\right] \tag{4.17}$$

The definition of pressure derivative is as follows

$$p_D' = \frac{dp_D}{dt_D} \tag{4.18}$$

The following expression is obtained through the Laplace transformation of Eq. (4.18)

$$\bar{p}_D' = s\,\bar{p}_D = \frac{1}{\sqrt{s}}\left[\frac{\dfrac{K_1\left(r_{eD}\sqrt{s}\right)}{I_1\left(r_{eD}\sqrt{s}\right)} + \dfrac{K_0\left(\sqrt{s}\right)}{I_0\left(\sqrt{s}\right)}}{\dfrac{K_1\left(\sqrt{s}\right)}{I_0\left(\sqrt{s}\right)} - \dfrac{I_1\left(\sqrt{s}\right)}{I_0\left(\sqrt{s}\right)}\dfrac{K_1\left(r_{eD}\sqrt{s}\right)}{I_1\left(r_{eD}\sqrt{s}\right)}}\right] \tag{4.19}$$

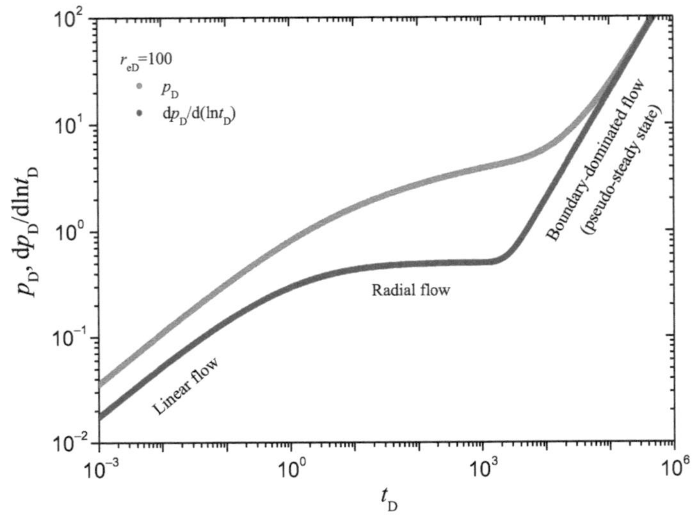

Figure 4.1 Pressure distribution in a closed circular reservoir: constant rate production

For a vertical well centered in a closed circular reservoir producing at a constant rate, the relationship between dimensionless pressure p_D and dimensionless time t_D is shown in Figure 4.1. The pressure curve presents several flow characteristics, including linear flow in the early period, radial flow in the middle period and boundary-dominated flow during the late period (pseudo-steady state).

Before the late period, Eq. (4.17) can be simplified as

$$\bar{p}_D = \frac{1}{s\sqrt{s}} \frac{K_0\left(\sqrt{s}\right)}{K_1\left(\sqrt{s}\right)} \tag{4.20}$$

In the early period, when $t_D < 0.01$, that is, the big s period, we have

$$K_v(s) = \sqrt{\frac{\pi}{2s}}\, e^{-s} \tag{4.21}$$

Substituting the asymptotic formula Eq. (4.21) into Eq. (4.20), we have

$$\bar{p}_D = \frac{1}{s\sqrt{s}} \tag{4.22}$$

According to the mathematics handbook, there is

$$L^{-1}\left[\frac{1}{s\sqrt{s}}\right] = 2\sqrt{\frac{t_D}{\pi}} \tag{4.23}$$

Performing the Laplace inversion transformation of Eq. (4.22), we have

$$p_D = 2\sqrt{t_D/\pi} \tag{4.24}$$

Eq. (4.24) reflects the linear flow characteristic in the early period.

In the big time period, $t_D/r_D^2 > 100$, we have

$$K_0(s) \approx -\left(\ln\frac{s}{2} + \gamma\right) \tag{4.25}$$

$$K_1(s) \approx \frac{1}{s} \tag{4.26}$$

Substituting Eq. (4.25) and Eq. (4.26) into Eq. (4.20), we have

$$\bar{p}_D = \frac{-\left(\ln\frac{\sqrt{s}}{2} + \gamma\right)}{s} = -\frac{\ln s}{2s} - \frac{\gamma - \ln 2}{s} \tag{4.27}$$

According to the mathematics handbook, there is

$$L^{-1}\left[\frac{\ln s}{s}\right] = -\ln t_D - \gamma \tag{4.28}$$

Performing the Laplace inversion transformation of Eq. (4.27), we have

$$p_D = \frac{1}{2}(\ln t_D + \gamma) - (\gamma - \ln 2) = \frac{1}{2}\ln\frac{4t_D}{e^{\gamma}} \tag{4.29}$$

Eq. (4.29) reflects the radial flow characteristic in the big time period.

Van Everdingen and Hurst (1949) presented the analytical solution of Eq. (4.17) in Euclidean space through Laplace transformation

$$\begin{aligned} p_D(r_D, t_D) = {} & \frac{2}{(r_{eD}^2 - 1)}\left(t_D + \frac{r_D^2}{4}\right) - \frac{r_{eD}^2 \ln r_D}{(r_{eD}^2 - 1)} - \frac{3r_{eD}^4 - 4r_{eD}^4 \ln r_{eD} - 2r_{eD}^2 - 1}{4(r_{eD}^2 - 1)^2} \\ & + \pi\sum_{n=1}^{\infty} e^{-z_n^2 t_D} \frac{J_1^2(z_n r_{eD})\left[J_1(z_n)Y_0(z_n r_D) - J_0(z_n r_D)Y_1(z_n)\right]}{z_n\left[J_1^2(z_n r_{eD}) - J_1^2(z_n)\right]} \end{aligned} \tag{4.30}$$

where Z_n is the root of the following characteristic equation

$$Y_1(z_n)J_1(z_n r_{eD}) - J_1(z_n)Y_1(z_n r_{eD}) = 0 \tag{4.31}$$

Under the pseudo-steady state, Eq. (4.30) can be simplified as

$$p_D \approx \frac{2t_D}{r_{eD}^2 - 1} + \ln r_{eD} - \frac{3}{4} \tag{4.32}$$

Van Everdingen and Hurst (1949) pointed out that according to the superposition principle, the relationship between the pressure solution at a constant production rate and the production rate solution under constant pressure turned out to be

$$\bar{p}_D(s)\bar{q}_D(s) = \frac{1}{s^2} \tag{4.33}$$

Substituting Eq. (4.17) into Eq. (4.33), we have

$$\bar{q}_D = \frac{\dfrac{K_1(\sqrt{s})}{I_0(\sqrt{s})} - \dfrac{I_1(\sqrt{s})}{I_0(\sqrt{s})}\dfrac{K_1(r_{eD}\sqrt{s})}{I_1(r_{eD}\sqrt{s})}}{\sqrt{s}\left[\dfrac{K_1(r_{eD}\sqrt{s})}{I_1(r_{eD}\sqrt{s})} + \dfrac{K_0(\sqrt{s})}{I_0(\sqrt{s})}\right]} \tag{4.34}$$

It is the production rate equation Eq. (3.22) under constant BHFP.

Plotting q_D (the production solution under constant BHFP) and $1/p_D$ (the reciprocal of p_D, which is the pressure solution at a constant rate) in the same coordinate system, as shown in Figure 4.2. The figure shows that in the transient flow period, the two

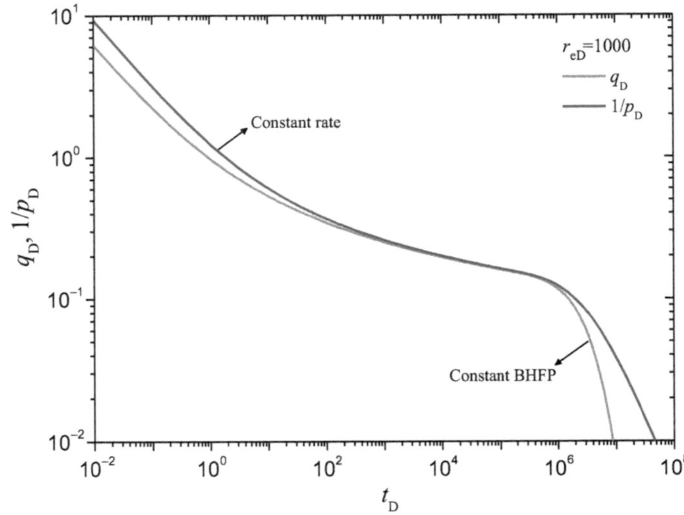

Figure 4.2 q_D, $1/p_D$ vs. t_D curve

curves are almost overlapped and dispersed in the boundary-dominated flow period. In this period, the q_D~t_D curve follows a exponential decline trend while the $1/(p_D-t_D)$ curve exhibits a harmonic decline trend.

4.2 Blasingame Type Curve Plotting

4.2.1 *Material Balance Time of Oil Well*

The material balance time is the ratio of current cumulative production and daily flow rate, which could be described as $t_c = N_p/q$. Its geometric meaning has been shown in Figure 1.8. Based on this equation, the equivalent relationship between producing with a constant and variable rate could be built. Once the pressure wave spreads to the boundary, the fluid flow will enter into the boundary-dominated flow period, and the flow rate is completely dictated by the elastic expansion of fluid and rock due to the reservoir pressure depletion.

Under the SI unit, according to the definition of compressibility factor, we have

$$q = -\frac{Ah\varphi C_t}{B}\frac{d\bar{p}}{dt} \tag{4.35}$$

Assuming that C_t is constant, the integration of the above equation is

$$\int_0^t q\,dt = -\frac{Ah\varphi C_t}{B}\int_{p_i}^{\bar{p}} d\bar{p} \tag{4.36}$$

$$N_p = \frac{Ah\varphi C_t}{B}\left(p_i - \bar{p}\right) \tag{4.37}$$

Eq. (4.37) could be given by

$$\frac{N_p}{A\varphi C_t} = \frac{h}{B}\left(p_i - \bar{p}\right) \tag{4.38}$$

And it could also be given by

$$\frac{2\pi Kh}{q\mu B}\left(p_i - \bar{p}\right) = \frac{2\pi K}{\mu A\varphi C_t}\frac{N_p}{q} \tag{4.39}$$

Defining $t_c = \frac{N_p}{q}$ as the material balance time function, we have

$$\frac{2\pi Kh}{q\mu B}\left(p_i - \bar{p}\right) = \frac{2\pi K}{\mu A\varphi C_t}t_c \tag{4.40}$$

If the dimensionless time is given by

$$t_{\mathrm{cD}} = \frac{K}{\varphi \mu C_{\mathrm{t}} A} t_{\mathrm{c}} \tag{4.41}$$

Substituting Eq. (4.41) into Eq. (4.40), we have

$$\frac{2\pi Kh}{q\mu B}\left(p_{\mathrm{i}} - \bar{p}\right) = 2\pi t_{\mathrm{cD}} \tag{4.42}$$

As Eq. (4.40) is based on the material balance equation, it is correct no matter what the flow state is or when it is, and no matter if it is under variable BHFP, constant flow rate, or variable flow rate condition.

For a closed circular homogenous reservoir, the Dupuit flow rate formula could be given by

$$q = \frac{2\pi Kh\left(\bar{p} - p_{\mathrm{wf}}\right)}{\mu B\left(\ln r_{\mathrm{eD}} - \frac{3}{4}\right)} \tag{4.43}$$

Similarly, for any bounded reservoir, the BHFP under pseudo-steady state could be

$$\bar{p} - p_{\mathrm{wf}} = \frac{q\mu B}{2\pi Kh}\left(\ln r_{\mathrm{eD}} - \frac{3}{4}\right) = \frac{q\mu B}{4\pi Kh}\ln\left(\frac{4A}{C_{\mathrm{A}} e^{\gamma} r_{\mathrm{w}}^2}\right) \tag{4.44}$$

Where A is the drainage area, C_{A} is the Dietz (1965) shape factor, and γ is the Euler's constant. For a closed circular reservoir, $C_A = 4\pi e^{\frac{3}{2}-\gamma} \approx 31.62$

Assuming that the transient flow enters into a pseudo-steady state immediately, we have

$$\bar{p} - p_{\mathrm{wf}} = \left(p_{\mathrm{i}} - p_{\mathrm{wf}}\right) - \left(p_{\mathrm{i}} - \bar{p}\right) \tag{4.45}$$

Substituting Eq. (4.44) and Eq. (4.42) into Eq. (4.45), we have

$$\frac{2\pi Kh\left(p_{\mathrm{i}} - p_{\mathrm{wf}}\right)}{q\mu B} = 2\pi t_{\mathrm{cD}} + \frac{1}{2}\ln\left(\frac{4A}{C_{\mathrm{A}} e^{\gamma} r_{\mathrm{w}}^2}\right) \tag{4.46}$$

Rearranging Eq. (4.46)

$$\frac{\left(p_{\mathrm{i}} - p_{\mathrm{wf}}\right)}{q} = \frac{\mu B}{Kh} t_{\mathrm{cD}} + \frac{\mu B}{4\pi Kh}\ln\left(\frac{4A}{C_{\mathrm{A}} e^{\gamma} r_{\mathrm{w}}^2}\right) \tag{4.47}$$

$$\frac{(p_{\mathrm{i}} - p_{\mathrm{wf}})}{q} = mt_{\mathrm{c}} + b_{\mathrm{pss}} \tag{4.48}$$

where,

$$m = \frac{\mu B}{Kh} \times \frac{K}{\varphi \mu C_{\mathrm{t}} A} = \frac{B}{\varphi A h C_{\mathrm{t}}} = \frac{1}{NC_{\mathrm{t}}} \tag{4.49}$$

$$b_{\mathrm{pss}} = \frac{\mu B}{4\pi Kh} \ln\left(\frac{4A}{C_{\mathrm{A}} e^{\gamma} r_{\mathrm{w}}^2}\right) \tag{4.50}$$

Inversing and rearranging Eq. (4.48), we have

$$\frac{q}{p_{\mathrm{i}} - p_{\mathrm{wf}}} b_{\mathrm{pss}} = \frac{1}{1 + \dfrac{m}{b_{\mathrm{pss}}} t_{\mathrm{c}}} \tag{4.51}$$

For a closed circular reservoir, we have

$$C_{\mathrm{A}} = 4\pi e^{\frac{3}{2} - \gamma} \tag{4.52}$$

Substituting Eq. (4.52) into Eq. (4.50), we have

$$b_{\mathrm{pss}} = \frac{\mu B}{2\pi Kh}\left(\ln r_{\mathrm{eD}} - \frac{3}{4}\right) \tag{4.53}$$

Rearranging the left of Eq. (4.51), we have

$$\frac{q}{p_{\mathrm{i}} - p_{\mathrm{wf}}} b_{\mathrm{pss}} = \frac{q}{p_{\mathrm{i}} - p_{\mathrm{wf}}} \frac{\mu B}{2\pi Kh}\left(\ln r_{\mathrm{eD}} - \frac{3}{4}\right) = \frac{q}{q_{\mathrm{i}}} = q_{\mathrm{Dd}} \tag{4.54}$$

Rearranging the right of Eq. (4.51), we have

$$\begin{aligned} t_{\mathrm{cDd}} &= \frac{m}{b_{\mathrm{pss}}} t_{\mathrm{c}} = \frac{B}{\varphi A h C_{\mathrm{t}}} \times \frac{1}{\dfrac{\mu B}{2\pi Kh}\left(\ln r_{\mathrm{eD}} - \dfrac{3}{4}\right)} t_{\mathrm{c}} = \frac{2K}{\varphi r_{\mathrm{e}}^2 C_{\mathrm{t}} \mu \left(\ln r_{\mathrm{eD}} - \dfrac{3}{4}\right)} t_{\mathrm{c}} \\ &= t_{\mathrm{cD}} \frac{1}{\dfrac{r_{\mathrm{eD}}^2}{2}\left(\ln r_{\mathrm{eD}} - \dfrac{3}{4}\right)} \approx t_{\mathrm{cD}} \frac{1}{\dfrac{1}{2}\left(r_{\mathrm{eD}}^2 - 1\right)\left(\ln r_{\mathrm{eD}} - \dfrac{1}{2}\right)} \end{aligned} \tag{4.55}$$

Apparently, Eq. (4.51) is just the same as Fetkovich harmonic decline equation

$$q_{\mathrm{Dd}} = \frac{1}{1 + t_{\mathrm{cDd}}} \tag{4.56}$$

The only difference between t_{cDd} and Fetkovich dimensionless decline time t_{Dd} is that the production time t is replaced by the material balance time t_{c}. Since the Fetkovich method is valid under the assumption that the reservoir shape is circular, the coefficient of Eq. (4.51) is $\ln\left(\frac{r_{\mathrm{e}}}{r_{\mathrm{w}}}\right) - \frac{3}{4}$. However, in the process of calculating, Fetkovich chose an offset of ½ instead of ¾ because he proved that the former could lead to better correlation (Fetkovich, 1980).

4.2.2 Material Balance Time of Gas Well

Meunier et al. (1987) developed the well analysis method under a normalized pseudo-pressure and pseudo-time. The advantage of this method is that the pseudo-pressure has the same dimension with real pressure; thus, all equations in oil well analysis could be introduced into gas well analysis.

Define the normalized pseudo-pressure as

$$p_{\mathrm{p}} = \left(\frac{\mu Z}{p}\right)_{\mathrm{i}} \int_0^p \frac{p}{\mu Z} dp \tag{4.57}$$

Define material balance pseudo-time as

$$t_{\mathrm{ca}} = \frac{(\mu C_{\mathrm{t}})_{\mathrm{i}}}{q} \int_0^t \frac{q}{\mu C_{\mathrm{t}}} dt \tag{4.58}$$

Define gas compressibility as

$$C_{\mathrm{g}} = \frac{1}{\rho}\frac{d\rho}{dp} = \frac{ZRT}{pM_{\mathrm{g}}}\frac{d}{dp}\left(\frac{pM_{\mathrm{g}}}{ZRT}\right) = \frac{Z}{p}\frac{d}{dp}\left(\frac{p}{Z}\right) \tag{4.59}$$

As to gas well, based on the material balance of the volumetric gas reservoir

$$\frac{p}{Z} = \left(\frac{p}{Z}\right)_{\mathrm{i}}\left(1 - \frac{G_{\mathrm{p}}}{G}\right) \tag{4.60}$$

Differentiating Eq. (4.60), we have

$$q = -\frac{Z_i G}{p_i}\frac{d}{dt}\left(\frac{p}{Z}\right) = -\frac{Z_i G}{p_i}\frac{d}{dp}\left(\frac{p}{Z}\right)\frac{dp}{dt} \tag{4.61}$$

Substituting Eq. (4.59) into Eq. (4.61), we have

$$q = -\frac{Z_i G C_g p}{Z p_i}\frac{dp}{dt} \tag{4.62}$$

Substituting Eq. (4.62) into Eq. (4.58), we have

$$t_{ca} = -\frac{G}{q}\left(\frac{\mu Z C_t}{p}\right)_i \int_{p_i}^{p} \frac{p}{Z\mu} dp \tag{4.63}$$

Rearranging Eq. (4.63), we have

$$t_{ca} = \frac{G C_{ti}}{q}\left(p_{p_i} - p_p\right) \tag{4.64}$$

Rearranging Eq. (4.64), we have

$$\frac{p_{p_i} - p_p}{q} = \frac{t_{ca}}{G C_{ti}} \tag{4.65}$$

Al-Hussainy and Ramey (1966) pointed out that once the single-phase gas flow enters into a pseudo-steady state, under the SI unit system, we have

$$\frac{p_p - p_{p_{wf}}}{q} = \frac{(\mu B)_i}{2\pi K h}\left[\frac{1}{2}\ln\left(\frac{4A}{C_A e^{\gamma} r_w^2}\right)\right] \tag{4.66}$$

Substituting Eq. (4.65) into Eq. (4.66), we have

$$\frac{p_{p_i} - p_{p_{wf}}}{q} = \frac{t_{ca}}{G C_{ti}} + \frac{(\mu B)_i}{2\pi K h}\left[\frac{1}{2}\ln\left(\frac{4A}{C_A e^{\gamma} r_w^2}\right)\right] \tag{4.67}$$

Rearranging Eq. (4.67), we have

$$\frac{\Delta p_p}{q} = m_a t_{ca} + b_{a,pss} \tag{4.68}$$

where

$$\Delta p_{\mathrm{p}} = p_{p_{\mathrm{i}}} - p_{p_{\mathrm{wf}}} \tag{4.69}$$

$$m_{\mathrm{a}} = \frac{1}{GC_{\mathrm{ti}}} \tag{4.70}$$

$$b_{\mathrm{a,pss}} = \frac{(\mu B)_{\mathrm{i}}}{2\pi Kh}\left[\frac{1}{2}\ln\left(\frac{4A}{C_{\mathrm{A}} e^{\gamma} r_{\mathrm{w}}^{2}}\right)\right] \tag{4.71}$$

Eq. (4.68) and Eq. (4.48) are identical in form. Inversing and rearranging Eq. (4.68), we have

$$\frac{q}{\Delta p_{\mathrm{p}}} b_{\mathrm{a,pss}} = \frac{1}{1+\dfrac{m_{\mathrm{a}}}{b_{\mathrm{a,pss}}} t_{\mathrm{ca}}} \tag{4.72}$$

The forms of Eq. (4.72) and Eq. (4.51) are identical. Rearranging Eq. (4.72), we have

$$q_{\mathrm{Dd}} = \frac{1}{1+t_{\mathrm{caDd}}} \tag{4.73}$$

where

$$q_{\mathrm{Dd}} = \frac{b_{\mathrm{a,pss}}}{\Delta p_{p}} \tag{4.74}$$

$$t_{\mathrm{caDd}} = \frac{m_{\mathrm{a}}}{b_{\mathrm{a,pss}}} t_{\mathrm{ca}} \tag{4.75}$$

For a closed circular reservoir,

$$\begin{aligned} t_{\mathrm{caDd}} &= \frac{m_{\mathrm{a}}}{b_{\mathrm{a,pss}}} t_{\mathrm{ca}} = \frac{K}{r_{\mathrm{eD}}^{2} r_{\mathrm{w}}^{2} \varphi C_{\mathrm{ti}} (\mu B)_{\mathrm{i}} \dfrac{1}{2}\left(\ln r_{\mathrm{eD}} - \dfrac{3}{4}\right)} t_{\mathrm{ca}} \\ &\approx \frac{t_{\mathrm{caD}}}{\dfrac{1}{2}\left(r_{\mathrm{eD}}^{2} - 1\right)\left(\ln r_{\mathrm{eD}} - \dfrac{1}{2}\right)} \end{aligned} \tag{4.76}$$

$$q_{\mathrm{Dd}} = \frac{b_{\mathrm{a,pss}}}{\Delta p_{\mathrm{p}}} = \frac{(\mu B)_{\mathrm{i}}}{2\pi Kh\Delta p_{\mathrm{p}}}\left(\ln r_{\mathrm{eD}} - \frac{3}{4}\right) \approx q_{\mathrm{D}}\left(\ln r_{\mathrm{eD}} - \frac{1}{2}\right) \tag{4.77}$$

The dimensionless flow rate decline functions of Eq. (4.73) for gas well and Eq. (4.56) for oil well are equivalent. The dimensionless flow rate decline function is a harmonic decline curve in the boundary-dominated flow period.

4.2.3 Material Balance Time Curve

Plotting the rate solution q_D under a constant BHFP and reciprocal of pressure solution at a constant rate $(1/p_D)$ in the coordinate system is shown in Figure 4.2. The figure indicates that the two curves almost overlap in the transient flow period and dispersed in the boundary-dominated flow period. Palacio and Blasingame (1993) replaced dimensionless time t_D with dimensionless material balance time t_{cD} and plotted the above curves again. The result shows that the two curves overlap, as shown in Figure 4.3. That is to say, the solutions of variable flow rate and constant flow rate are equivalent due to the introduction of material balance time t_{cD}. Therefore, the Blasingame method is applicable to the condition of a variable BHFP, but also the condition of a variable flow rate.

Define dimensionless cumulative production as

$$N_{pD} = \int_0^{t_D} q_D(\tau)d\tau \tag{4.78}$$

Performing the Laplace transformation of Eq. (4.78), we have

$$\bar{N}_{pD} = \frac{\bar{q}_D(s)}{s} \tag{4.79}$$

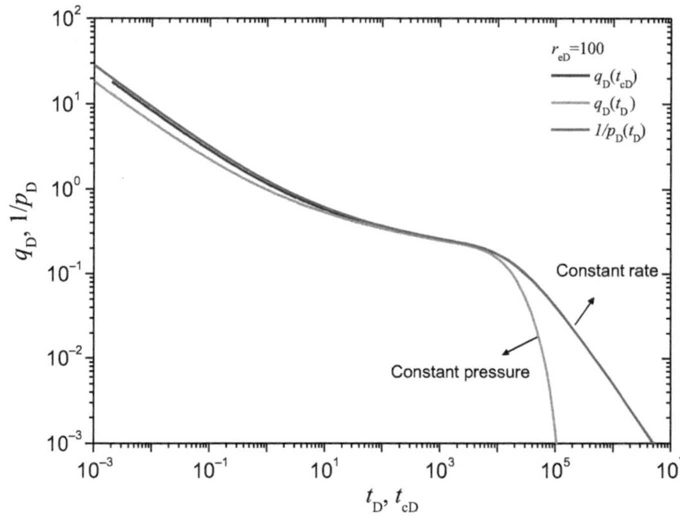

Figure 4.3 q_D and $1/p_D$ vs. t_D, q_D vs. t_{cD} curve (after Palacio and Blasingame, 1993)

Substituting Eq. (3.22) into Eq. (4.79), we have

$$\bar{N}_{\mathrm{pD}}=\frac{1}{s\sqrt{s}}\frac{\dfrac{K_1(\sqrt{s})}{I_0(\sqrt{s})}-\dfrac{K_1(r_{\mathrm{eD}}\sqrt{s})}{I_1(r_{\mathrm{eD}}\sqrt{s})}\dfrac{I_1(\sqrt{s})}{I_0(\sqrt{s})}}{\dfrac{K_0(\sqrt{s})}{I_0(\sqrt{s})}+\dfrac{K_1(r_{\mathrm{eD}}\sqrt{s})}{I_1(r_{\mathrm{eD}}\sqrt{s})}} \tag{4.80}$$

Define dimensionless material balance time as

$$t_{\mathrm{cD}}=\frac{N_{\mathrm{pD}}}{q_{\mathrm{D}}}=\frac{1}{q_{\mathrm{D}}}\int_0^{t_{\mathrm{D}}}q_{\mathrm{D}}(\tau)d\tau \tag{4.81}$$

4.2.4 Blasingame Dimensionless Decline Flow Rate Curve

Define dimensionless decline flow rate as

$$q_{\mathrm{Dd}}=\frac{b_{\mathrm{a,pss}}}{\Delta p_{\mathrm{p}}}=\frac{(\mu B)_{\mathrm{i}}}{2\pi Kh\Delta p_{\mathrm{p}}}\left(\ln r_{\mathrm{eD}}-\frac{3}{4}\right)\approx q_D\left(\ln r_{\mathrm{eD}}-\frac{1}{2}\right) \tag{4.82}$$

Define dimensionless material balance time as

$$t_{\mathrm{cDd}}=\frac{m}{b_{\mathrm{pss}}}t_{\mathrm{c}}=\frac{t_{\mathrm{cD}}}{\frac{1}{2}\left(r_{\mathrm{eD}}^2-1\right)\left(\ln r_{\mathrm{eD}}-\frac{1}{2}\right)} \tag{4.83}$$

$$t_{\mathrm{caDd}}=\frac{m_{\mathrm{a}}}{b_{\mathrm{a,pss}}}t_{\mathrm{ca}}=\frac{t_{\mathrm{caD}}}{\frac{1}{2}\left(r_{\mathrm{eD}}^2-1\right)\left(\ln r_{\mathrm{eD}}-\frac{1}{2}\right)} \tag{4.84}$$

Define dimensionless cumulative production as

$$N_{\mathrm{pDd}}=\int_0^{t_{\mathrm{Dd}}}q_{\mathrm{Dd}}(\tau)d\tau \tag{4.85}$$

Performing the Laplace transformation of Eq. (4.85), we have

$$\bar{N}_{\mathrm{pDd}}=\frac{\bar{q}_{\mathrm{Dd}}(s)}{s} \tag{4.86}$$

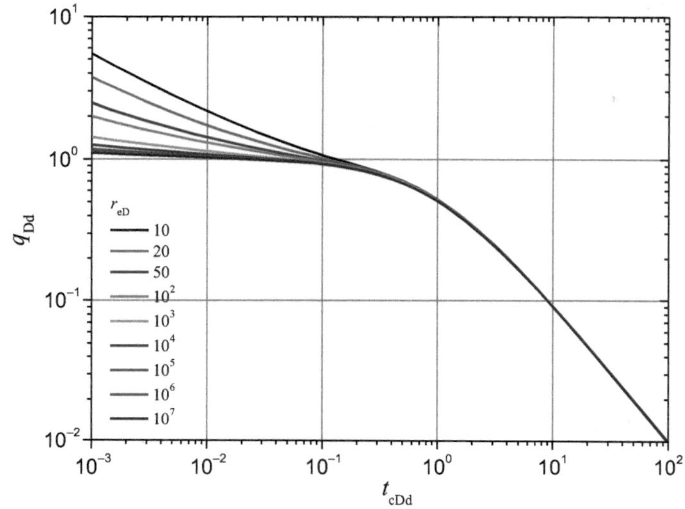

Figure 4.4 Blasingame normalized dimensionless decline flow rate curve: according to the solution of constant pressure

Substituting Eq. (3.44) into Eq. (4.86), we have

$$\bar{N}_{pDd} = \frac{\alpha\sqrt{\beta}}{s\sqrt{s}} \frac{\dfrac{K_1(\sqrt{\beta s})}{I_0(\sqrt{\beta s})} - \dfrac{K_1(r_{eD}\sqrt{\beta s})}{I_1(r_{eD}\sqrt{\beta s})}\dfrac{I_1(\sqrt{\beta s})}{I_0(\sqrt{\beta s})}}{\dfrac{K_0(\sqrt{\beta s})}{I_0(\sqrt{\beta s})} + \dfrac{K_1(r_{eD}\sqrt{\beta s})}{I_1(r_{eD}\sqrt{\beta s})}} \tag{4.87}$$

Define dimensionless material balance time as

$$t_{cDd} = \frac{N_{pDd}}{q_{Dd}} = \frac{1}{q_{Dd}} \int_0^{t_{Dd}} q_{Dd}(\tau) d\tau \tag{4.88}$$

The normalized dimensionless decline flow rate is shown in Figure 4.4. Because of the application of material balance time t_{cDd}, the latter part of Fetkovich-Arps type curves is converged to a harmonic decline curve.

Substituting t_{Dd} into Eq. (4.5) through Eq. (4.8), we have

$$\frac{1}{r_D}\frac{\partial}{\partial r_D}\left(r_D \frac{\partial p_D}{\partial r_D} \right) = \beta \frac{\partial p_D}{\partial t_{Dd}} \tag{4.89}$$

$$p_D(r_D, 0) = 0 \tag{4.90}$$

$$\left(r_{\mathrm{D}}\frac{\partial p_{\mathrm{D}}}{\partial r_{\mathrm{D}}}\right)_{r_{\mathrm{D}}=1}=-1 \tag{4.91}$$

$$\left.\frac{\partial p_{\mathrm{D}}}{\partial r_{\mathrm{D}}}\right|_{r_{\mathrm{D}}=r_{\mathrm{eD}}}=0 \tag{4.92}$$

The solution is

$$\bar{p}_{\mathrm{D}}=\frac{1}{s\sqrt{\beta s}}\left[\frac{\dfrac{K_1\left(r_{\mathrm{eD}}\sqrt{\beta s}\right)}{I_1\left(r_{\mathrm{eD}}\sqrt{\beta s}\right)}+\dfrac{K_0\left(\sqrt{\beta s}\right)}{I_0\left(\sqrt{\beta s}\right)}}{\dfrac{K_1\left(\sqrt{\beta s}\right)}{I_0\left(\sqrt{\beta s}\right)}-\dfrac{I_1\left(\sqrt{\beta s}\right)}{I_0\left(\sqrt{\beta s}\right)}\dfrac{K_1\left(r_{\mathrm{eD}}\sqrt{\beta s}\right)}{I_1\left(r_{\mathrm{eD}}\sqrt{\beta s}\right)}}\right] \tag{4.93}$$

According to Eq. (4.77), dimensionless decline flow rate can be given by

$$q_{\mathrm{Dd}}=\frac{\ln r_{\mathrm{eD}}-\dfrac{1}{2}}{L^{-1}\left[\bar{p}_{\mathrm{D}}\right]} \tag{4.94}$$

The normalized dimensionless decline flow rate type curves based on Eq. (4.94) are shown in Figure 4.5. The material balance time under a constant BHFP in Figure 4.4

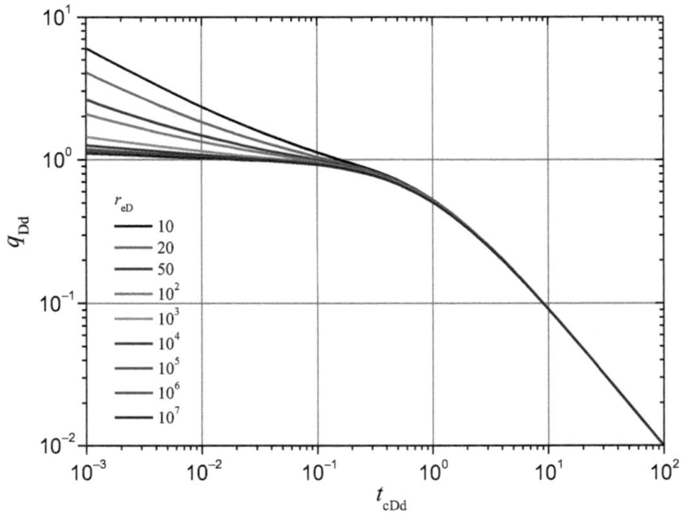

Figure 4.5 Blasingame normalized dimensionless decline flow rate curve: according to the solution of constant rate

is almost identical as the material balance time under a constant production rate in Figure 4.5.

4.2.5 Blasingame Normalized Dimensionless Decline Flow Rate Integral Curves

Blasingame et al. (1989) first introduced the concept of rate integration in the field of well test. This method can effectively reduce the noise of production data, thus facilitating the analysis. The dimensionless decline flow rate integral function q_{Ddi} could be given by

$$q_{\mathrm{Ddi}} = \frac{N_{\mathrm{pDd}}}{t_{\mathrm{Dd}}} = \frac{1}{t_{\mathrm{Dd}}} \int_0^{t_{\mathrm{Dd}}} q_{\mathrm{Dd}}(\tau)d\tau \tag{4.95}$$

The above equation represents the average value of the production data up to now. After acquiring the dimensionless cumulative production according to Eq. (4.87), the latter part of the production solution at a constant BHFP could be summarized as an exponential decline function and the dimensionless decline flow rate integral function tends to be 1.0. Meanwhile, the corresponding dimensionless time $t_{\mathrm{Dd}} \approx 6.5$. If converted to the material balance time, the curve could no longer represent the well production characteristic as shown in Figure 4.6.

By substituting Eq. (4.94) into Eq. (4.95), the dimensionless decline flow rate integral curve is acquired, as shown in Figure 4.7. It is worth keeping in mind that the

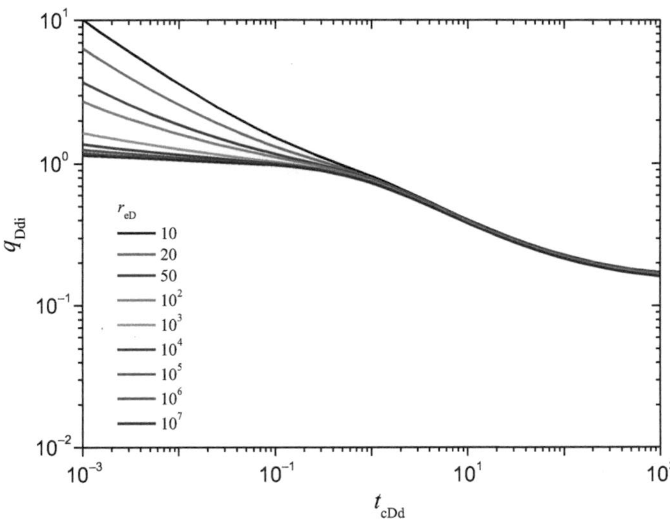

Figure 4.6 Blasingame normalized dimensionless decline flow rate integral curve: according to the solution of constant pressure

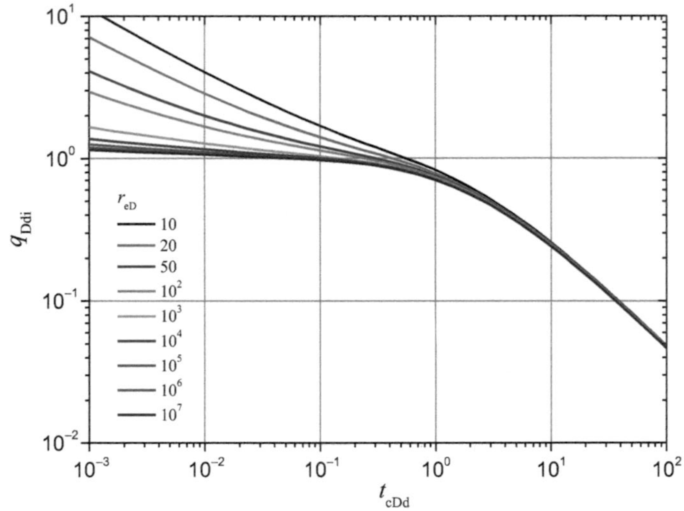

Figure 4.7 Blasingame normalized dimensionless decline flow rate integral curve: according to the solution of constant rate

value of N_{pDd} at the initial time should be calculated according to Eq. (4.87), and then the trapezoidal integration method can be used to calculate the integration.

4.2.6 Blasingame Normalized Dimensionless Decline Flow Rate Integral Derivative Curve

The definition of dimensionless decline flow rate integral derivative q_{Ddid} is just the same as the Arps method, introduced in Chapter 2, as given below

$$q_{Dd\,id} = -\frac{dq_{Dd\,i}}{d\ln t_{Dd}} = -t_{Dd}\frac{dq_{Dd\,i}}{dt_{Dd}} = -t_{Dd}\frac{d\left(N_{pDd}/t_{Dd}\right)}{dt_{Dd}} \tag{4.96}$$

In the boundary-dominated flow period

$$q_{Ddid} = q_{Ddi} - q_{Dd} \tag{4.97}$$

Please refer to the detailed deriving process in Section 2.3. The normalized dimensionless decline flow rate integral derivative curves are shown in Figure 4.8.

4.2.7 Blasingame Type Curves

In the Blasingame type curves, three kinds of rate–material balance time curves can be plotted, including the normalized dimensionless decline flow rate curve, the

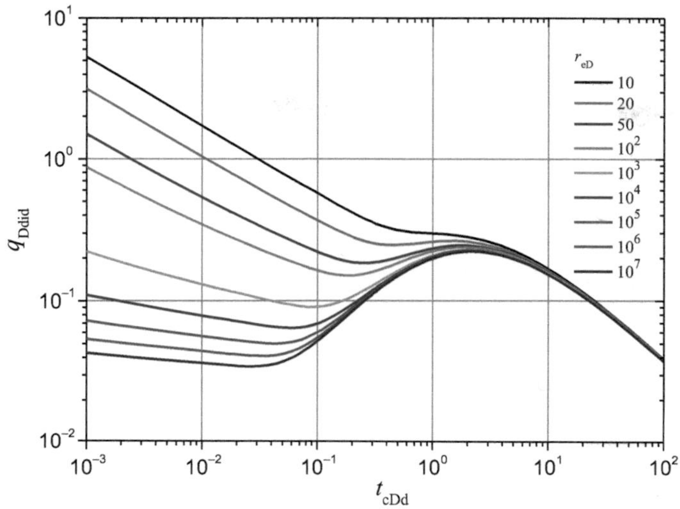

Figure 4.8 Blasingame normalized dimensionless decline flow rate integral derivative curve: according to the solution of constant rate

normalized dimensionless decline flow rate integral curve, and the dimensionless decline flow rate integral derivative curve, as shown in Figure 4.9.

Any of the three curves or any combination thereof can be used in the process of matching the field data with type curves. Through type curve matching, we can calculate the permeability K, skin factor S, drainage radius r_e, and original oil in place

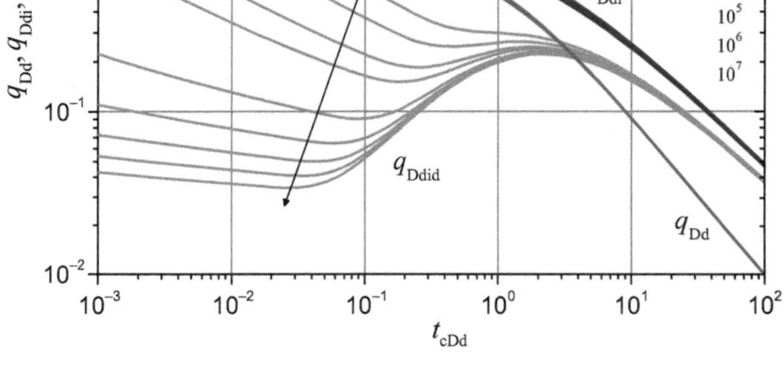

Figure 4.9 Blasingame type curves: according to the solution of constant rate (after Palacio and Blasingame, 1993)

(OOIP), etc. The advantage of the Blasingame method is that it follows the principle of "integration first, then derivation," which can reduce the noise of field data and facilitate the identification of the production characteristics. However, as the flow rate integration of the early data points is very sensitive to errors, a tiny error in the early data points may lead to a large deviation of the flow rate integral and the flow rate integral derivative.

4.3 Decline Curves Analysis Using Blasingame-Style Type Curves

4.3.1 Type Curves Matching Procedure—Oil Well

Similar to the Fetkovich-Arps normalized matching analysis procedure, the Blasingame curve matching analysis procedure of oil well is listed below.

1. Calculate the material balance time by

$$t_c = \frac{N_p}{q} \tag{4.98}$$

2. Calculate the normalized rate by

$$\frac{q}{\Delta p} = \frac{q}{p_i - p_{wf}} \tag{4.99}$$

3. Calculate the normalized rate integral by

$$\left(\frac{q}{\Delta p}\right)_i = \frac{1}{t_c}\int_0^{t_c} \frac{q}{p_i - p_{wf}} d\tau \tag{4.100}$$

4. Calculate the normalized rate integral derivative by

$$\left(\frac{q}{\Delta p}\right)_{id} = -\frac{d\left(\frac{q}{\Delta p}\right)_i}{d\ln t_c} = -t_c \frac{d\left(\frac{q}{\Delta p}\right)_i}{dt_c} \tag{4.101}$$

   The subscript i represents integration, and d represents derivation.

5. In the log–log coordinate system, plot the normalized rate, the normalized rate integral, and the rate integral derivative curves against the material balance time, respectively. $\frac{q}{\Delta p} \sim t_c$, $\left(\frac{q}{\Delta p}\right)_i \sim t_c$, $\left(\frac{q}{\Delta p}\right)_{id} \sim t_c$.

6. Choose any of the three curves or any combination thereof to match the actual points with the type curves (Figure 4.9), aiming to obtain the best matching of each curve.
7. According to the matching results, record the dimensionless drainage radius r_{eD}
8. Randomly select a matching point, and record the actual matching point $\left(t_c, q/\Delta p\right)_M$ and the corresponding theoretical matching point $(t_{cDd}, q_{Dd})_M$. If the reservoir thickness, total compressibility and wellbore radius are known, then we can calculate the reservoir permeability, skin factor, drainage area, and reserves, etc.
9. According to the rate matching point and Eq. (4.54), calculate the permeability (replacing ¾ in the equation with ½), we have

$$K = \frac{(q/\Delta p)_M}{\left(q_{Dd}\right)_M} \frac{\mu B}{2\pi h}\left(\ln r_{eD} - \frac{1}{2}\right) \tag{4.102}$$

10. On the basis of the time matching point and r_{eD} determined in step (7), calculate the effective wellbore radius r_{wa}

$$t_{cDd} = \frac{\dfrac{K}{\varphi C_t \mu} t_c}{\dfrac{1}{2} r_{wa}^2 \left(r_{eD}^2 - 1\right)\left(\ln r_{eD} - \dfrac{1}{2}\right)} \tag{4.103}$$

$$r_{wa} = \sqrt{\frac{2K/\varphi\mu C_t}{\left(r_{eD}^2 - 1\right)\left(\ln r_{eD} - \dfrac{1}{2}\right)}\left(\frac{t_c}{t_{cDd}}\right)_M} \tag{4.104}$$

11. Calculate the skin factor S

$$S = \ln \frac{r_w}{r_{wa}} \tag{4.105}$$

12. Calculate the drainage radius, drainage area, and reserves

$$r_e = r_{wa} r_{eD} \tag{4.106}$$

$$A = \pi r_e^2 \tag{4.107}$$

Or calculate drainage radius according to Eq. (3.103) through Eq. (3.105),

$$r_e = \sqrt{\frac{\dfrac{B}{C_t}\left(\dfrac{t_c}{t_{cDd}}\right)_M \left(\dfrac{q/\Delta p}{q_{Dd}}\right)_M}{\pi h \varphi}} \tag{4.108}$$

Then calculate the reserves according to Eq. (3.106)

$$N = \frac{1}{C_t}\left(\frac{t_c}{t_{Dd}}\right)_M \left(\frac{q/\Delta p}{q_{Dd}}\right)_M (1 - S_w) \tag{4.109}$$

4.3.2 Type Curves Matching Procedure—Gas Well

The procedure of Blasingame curve matching analysis for a gas well is similar to that for an oil well, which is given as follows:

1. Calculate the material balance pseudo-time

   Assuming reserves is G, calculate the material balance pseudo-time for each production data point.

$$t_{ca} = \frac{(\mu C_t)_i}{q} \int_0^t \frac{q}{\mu(\bar{p}) C_t(\bar{p})} dt = \frac{G C_{ti}}{q} (p_{p_i} - p_p) \tag{4.110}$$

   Please refer to the normalized pseudo-pressure p_p in Eq. (4.57), and the average reservoir pressure could be calculated according to Eq. (4.60).

2. Calculate the normalized rate

$$\frac{q}{\Delta p_p} = \frac{q}{p_{p_i} - p_{p_{wf}}} \tag{4.111}$$

3. Calculate the normalized rate integral

$$\left(\frac{q}{\Delta p_p}\right)_i = \frac{1}{t_{ca}} \int_0^{t_{ca}} \frac{q}{\Delta p_p} d\tau \tag{4.112}$$

   The subscript i represents integration.

4. Calculate the normalized rate integral derivative

$$\left(\frac{q}{\Delta p_p}\right)_{id} = -\frac{d\left(\frac{q}{\Delta p_p}\right)_i}{d \ln t_{ca}} = -t_{ca} \frac{d\left(\frac{q}{\Delta p_p}\right)_i}{dt_{ca}} \tag{4.113}$$

   The subscript i represents integration, and d represents derivation.

5. In the Cartesian coordinate system, plot the $\frac{\Delta p_p}{q} \sim t_{ca}$ curve and regress the data points linearly; then apply the slope of the line to Eq. (4.65), calculate the value of G, repeat steps (1)–(5), and make iterative computations until G meets the permissible error.

$$G = \frac{1}{Slope \times C_{ti}} \tag{4.114}$$

6. Prepare log–log plot of normalized rate, normalized rate integral, and normalized rate integral derivative changing with material balance pseudo-time, respectively, in the same coordinate system. In other words, we need to plot the $\frac{q}{\Delta p_\mathrm{p}} \sim t_\mathrm{ca}$, $\left(\frac{q}{\Delta p_\mathrm{p}}\right)_\mathrm{i} \sim t_\mathrm{ca}$, and $\left(\frac{q}{\Delta p_\mathrm{p}}\right)_\mathrm{id} \sim t_\mathrm{ca}$ curves.
7. Choose any of the three curves or any combination thereof to match the actual points with the type curves (Figure 4.9), aiming to obtain the best matching of each curve.
8. Record the dimensionless drainage radius r_eD according to the matching results.
9. Randomly choose a matching point, and record the actual matching point $(t_\mathrm{ca}, q/\Delta p_\mathrm{P})_\mathrm{M}$ and the corresponding theoretical matching point $(t_\mathrm{caDd}, q_\mathrm{Dd})_\mathrm{M}$. If the reservoir thickness, total compressibility, and wellbore radius are known, then we can calculate the reservoir permeability, skin factor, drainage area, and reserves, etc.
10. Calculate the permeability (replacing ¾ in the equation with ½) according to the rate matching point and Eq. (4.54), and we have

$$K = \frac{(q/\Delta p_\mathrm{p})_\mathrm{M}}{(q_\mathrm{Dd})_\mathrm{M}} \frac{\mu B}{2\pi h}\left(\ln r_\mathrm{eD} - \frac{1}{2}\right) \tag{4.115}$$

11. On the basis of the time matching point and r_eD determined in step (8), calculate the effective wellbore radius r_wa

$$r_\mathrm{wa} = \sqrt{\frac{2K/\varphi\mu C_\mathrm{t}}{(r_\mathrm{eD}^2 - 1)\left(\ln r_\mathrm{eD} - \frac{1}{2}\right)}\left(\frac{t_\mathrm{ca}}{t_\mathrm{caDd}}\right)_\mathrm{M}} \tag{4.116}$$

12. Calculate the skin factor S

$$S = \ln \frac{r_\mathrm{w}}{r_\mathrm{wa}} \tag{4.117}$$

13. Calculate the drainage radius, drainage area, and reserves

$$r_\mathrm{e} = r_\mathrm{wa} r_\mathrm{eD} \tag{4.118}$$

$$A = \pi r_\mathrm{e}^2 \tag{4.119}$$

Calculate drainage radius according to Eq. (4.108)

$$r_\mathrm{e} = \sqrt{\frac{\frac{B}{C_\mathrm{t}}\left(\frac{t_\mathrm{ca}}{t_\mathrm{caDd}}\right)_\mathrm{M}\left(\frac{q/\Delta p_\mathrm{p}}{q_\mathrm{Dd}}\right)_\mathrm{M}}{\pi h \varphi}} \tag{4.120}$$

Calculate the reserves according to Eq. (4.109)

$$G = \frac{1}{C_t}\left(\frac{t_{ca}}{t_{caDd}}\right)_M \left(\frac{q/\Delta p_p}{q_{Dd}}\right)_M (1 - S_w) \tag{4.121}$$

4.4 Case Study of Blasingame Decline Curves Analysis

A gas well centered in a closed circular reservoir is producing at a variable rate and variable BHFP. The initial reservoir pressure is 30 MPa, the reservoir temperature is 80°C, and the reservoir thickness is 10 m, the reservoir porosity is 0.1, the distance between the well and boundary is 500 m, and the gas reserves are 2.0×10^8 m^3. The relative density of gas phase is 0.6, the pseudo-critical temperature is 195.697 K, the pseudo-critical pressure is 4.66875 MPa, the reservoir permeability estimation from well test is 2.0 mD, the skin factor is –5.52, the production time is 800 days, the cumulative gas production is 11813×10^4 m^3, the fluid viscosity is 0.0228 mPa·s, the rock compressibility is 4.35113×10^{-4} MPa^{-1}, the gas compressibility under initial condition is 0.0246 MPa^{-1}, the gas formation volume factor under initial condition is 0.0039497, and the production data are listed in Table 4.1.

Neglecting the iterative computations for G and assuming $G = 2.0 \times 10^8$ m^3 immediately, the analysis process is as follows.

1–5. Calculate the material balance pseudo-time and the relative normalized rate parameters.

According to the gas properties, pressure and temperature, with the Standing empirical equation (1942), the relationship between apparent reservoir pressure p / Z and the actual pressure p can be obtained, as shown in Figure 4.10.

$$p_{p_i} - p_p = \left(\frac{\mu Z}{p}\right)_i \left[\int_0^{p_i} \frac{p}{\mu Z} dp - \int_0^{p} \frac{p}{\mu Z} dp\right] = \left(\frac{\mu Z}{p}\right)_i \int_p^{p_i} \frac{p}{\mu Z} dp \tag{4.122}$$

Calculate the average reservoir pressure, normalized pseudo-time, normalized rate, normalized rate integral, and normalized rate integral derivative according to Eq. (4.60), (4.110), (4.111), (4.112), and (4.113). The results are shown in Table 4.2.

6. Plot log–log relationships of the normalized rate, normalized rate integral, and normalized rate integral derivative with material balance pseudo-time respectively, that is $\frac{q}{\Delta p_p} \sim t_{ca}$, $\left(\frac{q}{\Delta p_p}\right)_i \sim t_{ca}$, and $\left(\frac{q}{\Delta p_p}\right)_{id} \sim t_{ca}$ curves, as shown in Figure 4.11.

7. Match type curve (Figure 4.9) with the actual production history curve, as shown in Figure 4.12.

8. Record the dimensionless drainage radius $r_{eD} = 20$.

9. Choose the actual matching point $(101, 47)_M$ and the corresponding theoretical matching point $(1, 0.1)_M$ to simplify the calculation.

Table 4.1 **Production history of well B**

Time	Rate	Flow pressure	Cumulative production	Time	Rate	Flow pressure	Cumulative production	Time	Rate	Flow pressure	Cumulative production
d	$10^3m^3/d$	MPa	10^4m^3	d	$10^3m^3/d$	MPa	10^4m^3	d	$10^3m^3/d$	MPa	10^4m^3
1	199.8	27.72	20	220	167.7	19.17	4033	520	131.9	12.57	8505
2	199.7	27.23	40	230	166.4	18.91	4200	530	130.9	12.38	8636
3	199.5	26.92	60	240	165.1	18.65	4366	540	129.8	12.19	8766
4	199.4	26.70	80	250	163.7	18.40	4530	550	128.8	12.00	8896
5	199.2	26.53	100	260	162.4	18.15	4693	560	127.8	11.82	9024
6	199.0	26.38	120	270	161.1	17.91	4855	570	126.8	11.63	9151
7	198.9	26.26	140	280	159.9	17.66	5015	580	125.8	11.45	9277
8	198.7	26.16	159	290	158.6	17.42	5174	590	124.8	11.27	9402
9	198.6	26.06	179	300	157.3	17.19	5332	600	123.8	11.09	9527
10	198.4	25.98	199	310	156.1	16.95	5489	610	122.8	10.91	9650
20	196.8	25.41	397	320	154.8	16.72	5644	620	121.8	10.73	9772
30	195.3	25.01	593	330	153.6	16.49	5798	630	120.8	10.55	9893
40	193.7	24.64	787	340	152.4	16.27	5951	640	119.9	10.37	10014
50	192.2	24.29	980	350	151.2	16.05	6103	650	118.9	10.20	10133
60	190.6	23.94	1171	360	150.0	15.83	6253	660	118.0	10.02	10251
70	189.1	23.60	1361	370	148.8	15.61	6403	670	117.0	9.84	10369
80	187.6	23.27	1549	380	147.6	15.39	6551	680	116.1	9.67	10485
90	186.1	22.94	1736	390	146.4	15.18	6698	690	115.2	9.50	10601
100	184.6	22.62	1921	400	145.2	14.97	6844	700	114.2	9.32	10715
110	183.2	22.31	2104	410	144.1	14.76	6988	710	113.3	9.15	10829
120	181.7	22.00	2287	420	142.9	14.55	7132	720	112.4	8.98	10942
130	180.2	21.69	2468	430	141.8	14.34	7274	730	111.5	8.80	11054
140	178.8	21.39	2648	440	140.7	14.14	7415	740	110.6	8.63	11165
150	177.4	21.10	2826	450	139.5	13.94	7555	750	109.8	8.46	11275
160	176.0	20.81	3002	460	138.4	13.74	7694	760	108.9	8.29	11384
170	174.6	20.53	3178	470	137.3	13.54	7832	770	108.0	8.12	11493
180	173.2	20.25	3351	480	136.2	13.34	7969	780	107.2	7.95	11600
190	171.8	19.97	3524	490	135.1	13.15	8104	790	106.3	7.77	11707
200	170.4	19.70	3695	500	134.1	12.95	8239	800	105.5	7.60	11813
210	169.1	19.43	3865	510	133.0	12.76	8372	—	—	—	—

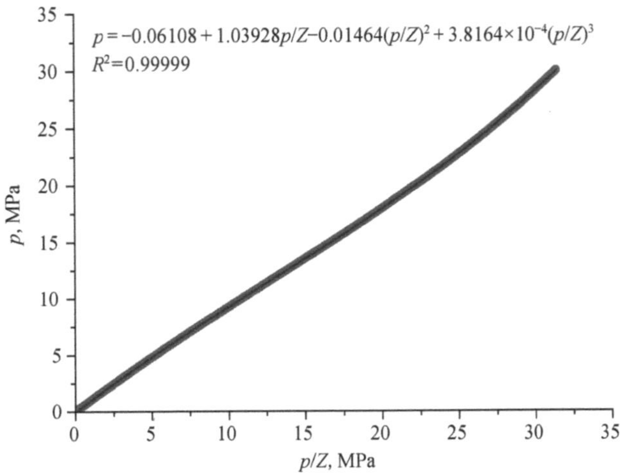

Figure 4.10 p vs. p/Z curve

Table 4.2 **Normalized rate parameters of well B**

t	q	p_{wf}	p_{avg}	$\Delta p_p(p_{avg})$	$\Delta p_p(p_{wf})$	t_{ca}	$q/\Delta p_p$	$(q/\Delta p_p)_i$	$(q/\Delta p_p)_{id}$
d	10^3 m^3/d	MPa				d	10^3 m^3/d/MPa		
1	199.8	27.72	30.00	0.01	2.26	0.12	88.41	144.21	
2	199.7	27.23	29.99	0.01	2.74	0.14	72.85	137.35	56.72
3	199.5	26.92	29.95	0.05	3.04	1.15	65.69	77.45	8.17
4	199.4	26.70	29.91	0.09	3.25	2.16	61.30	70.92	7.42
5	199.2	26.53	29.87	0.13	3.42	3.17	58.23	67.37	7.61
6	199.0	26.38	29.83	0.17	3.56	4.17	55.91	64.88	7.81
7	198.9	26.26	29.79	0.21	3.68	5.18	54.08	62.96	7.96
8	198.7	26.16	29.75	0.25	3.78	6.19	52.58	61.40	8.07
9	198.6	26.06	29.71	0.29	3.87	7.19	51.32	60.08	8.13
10	198.4	25.98	29.67	0.33	3.95	8.20	50.24	58.94	8.16
20	196.8	25.41	29.27	0.73	4.49	18.18	43.82	52.17	8.12
30	195.3	25.01	28.88	1.11	4.88	28.06	40.04	48.53	8.32
40	193.7	24.64	28.50	1.49	5.22	37.84	37.08	45.94	8.72
50	192.2	24.29	28.13	1.86	5.56	47.54	34.58	43.87	9.18
60	190.6	23.94	27.77	2.21	5.88	57.16	32.41	42.12	9.61
70	189.1	23.60	27.41	2.56	6.20	66.70	30.52	40.59	9.99
80	187.6	23.27	27.06	2.90	6.50	76.17	28.84	39.24	10.31
90	186.1	22.94	26.72	3.24	6.80	85.57	27.35	38.01	10.59
100	184.6	22.62	26.38	3.56	7.10	94.92	26.01	36.89	10.82
110	183.2	22.31	26.05	3.88	7.38	104.21	24.80	35.87	11.01
120	181.7	22.00	25.73	4.19	7.66	113.46	23.71	34.92	11.16
130	180.2	21.69	25.41	4.49	7.94	122.66	22.71	34.04	11.28
140	178.8	21.39	25.00	4.89	8.20	134.53	21.80	32.99	11.15

Table 4.2 **Normalized rate parameters of well B *(cont.)***

t	q	p_{wf}	p_{avg}	$\Delta p_p(p_{avg})$	$\Delta p_p(p_{wf})$	t_{ca}	$q/\Delta p_p$	$(q/\Delta p_p)_i$	$(q/\Delta p_p)_{id}$
d	10^3 m^3/d	MPa				d	10^3 m^3/d/MPa		
150	177.4	21.10	24.90	4.98	8.46	138.22	20.96	32.67	11.67
160	176.0	20.81	24.61	5.26	8.72	146.93	20.19	31.95	11.72
170	174.6	20.53	24.32	5.52	8.96	155.65	19.48	31.27	11.76
180	173.2	20.25	24.04	5.79	9.21	164.39	18.81	30.63	11.78
190	171.8	19.97	23.76	6.05	9.44	173.13	18.19	30.02	11.79
200	170.4	19.70	23.49	6.30	9.67	181.89	17.62	29.43	11.79
210	169.1	19.43	23.22	6.55	9.90	190.66	17.08	28.88	11.77
220	167.7	19.17	22.95	6.80	10.12	199.44	16.57	28.35	11.75
230	166.4	18.91	22.68	7.04	10.34	208.22	16.09	27.84	11.72
240	165.1	18.65	22.42	7.28	10.55	217.02	15.64	27.35	11.69
250	163.7	18.40	22.16	7.52	10.76	225.83	15.22	26.89	11.65
260	162.4	18.15	21.90	7.75	10.96	234.65	14.82	26.44	11.60
270	161.1	17.91	21.65	7.97	11.16	243.48	14.44	26.01	11.56
280	159.9	17.66	21.40	8.20	11.36	252.31	14.08	25.60	11.51
290	158.6	17.42	21.15	8.42	11.55	261.16	13.74	25.21	11.45
300	157.3	17.19	20.90	8.63	11.73	270.02	13.41	24.82	11.40
310	156.1	16.95	20.66	8.85	11.92	278.89	13.10	24.46	11.34
320	154.8	16.72	20.42	9.06	12.09	287.77	12.80	24.10	11.28
330	153.6	16.49	20.18	9.26	12.27	296.65	12.52	23.76	11.23
340	152.4	16.27	19.95	9.46	12.44	305.55	12.25	23.43	11.17
350	151.2	16.05	19.72	9.66	12.61	314.47	11.99	23.11	11.11
360	150.0	15.83	19.49	9.86	12.77	323.39	11.74	22.80	11.04
370	148.8	15.61	19.26	10.05	12.94	332.32	11.50	22.50	10.98
380	147.6	15.39	19.03	10.24	13.09	341.26	11.27	22.20	10.92
390	146.4	15.18	18.81	10.42	13.25	350.21	11.05	21.92	10.86
400	145.2	14.97	18.59	10.60	13.40	359.18	10.84	21.65	10.80
410	144.1	14.76	18.38	10.78	13.55	368.15	10.63	21.38	10.74
420	142.9	14.55	18.16	10.96	13.70	377.14	10.43	21.12	10.68
430	141.8	14.34	17.95	11.13	13.84	386.14	10.24	20.87	10.62
440	140.7	14.14	17.74	11.30	13.98	395.15	10.06	20.63	10.56
450	139.5	13.94	17.53	11.46	14.12	404.18	9.88	20.39	10.50
470	137.3	13.54	17.12	11.79	14.39	422.26	9.54	19.93	10.38
480	136.2	13.34	16.92	11.94	14.52	431.32	9.38	19.71	10.32
490	135.1	13.15	16.72	12.10	14.65	440.40	9.23	19.50	10.26
500	134.1	12.95	16.52	12.25	14.77	449.49	9.08	19.29	10.20
510	133.0	12.76	16.33	12.40	14.89	458.60	8.93	19.08	10.15
520	131.9	12.57	16.13	12.54	15.01	467.72	8.79	18.89	10.09
530	130.9	12.38	15.94	12.69	15.13	476.85	8.65	18.69	10.03
540	129.8	12.19	15.76	12.83	15.25	486.00	8.51	18.50	9.98
550	128.8	12.00	15.57	12.96	15.36	495.17	8.38	18.31	9.92
560	127.8	11.82	15.38	13.10	15.48	504.35	8.26	18.13	9.87
570	126.8	11.63	15.20	13.23	15.59	513.55	8.13	17.95	9.81
580	125.8	11.45	15.02	13.36	15.69	522.76	8.01	17.78	9.76

(Continued)

Table 4.2 Normalized rate parameters of well B *(cont.)*

t	q	p_{wf}	p_{avg}	$\Delta p_p(p_{avg})$	$\Delta p_p(p_{wf})$	t_{ca}	$q/\Delta p_p$	$(q/\Delta p_p)_i$	$(q/\Delta p_p)_{id}$
d	10^3 m^3/d	MPa				d	10^3 m^3/d/MPa		
590	124.8	11.27	14.84	13.49	15.80	531.99	7.90	17.61	9.71
600	123.8	11.09	14.67	13.61	15.90	541.25	7.78	17.44	9.65
610	122.8	10.91	14.49	13.74	16.00	550.51	7.67	17.28	9.60
620	121.8	10.73	14.32	13.86	16.10	559.80	7.56	17.12	9.55
630	120.8	10.55	14.15	13.98	16.20	569.11	7.46	16.96	9.50
640	119.9	10.37	13.98	14.09	16.30	578.43	7.35	16.81	9.45
650	118.9	10.20	13.81	14.21	16.39	587.79	7.25	16.66	9.40
660	118.0	10.02	13.64	14.32	16.49	597.15	7.16	16.51	9.35
670	117.0	9.84	13.48	14.43	16.58	606.54	7.06	16.36	9.30
680	116.1	9.67	13.32	14.53	16.67	615.95	6.97	16.22	9.25
690	115.2	9.50	13.16	14.64	16.75	625.39	6.87	16.08	9.20
700	114.2	9.32	13.00	14.74	16.84	634.84	6.78	15.94	9.15
710	113.3	9.15	12.84	14.84	16.92	644.33	6.70	15.81	9.10
720	112.4	8.98	12.69	14.94	17.01	653.83	6.61	15.67	9.06
730	111.5	8.80	12.53	15.04	17.09	663.36	6.53	15.54	9.01
740	110.6	8.63	12.38	15.13	17.17	672.91	6.44	15.41	8.96
750	109.8	8.46	12.23	15.23	17.25	682.50	6.36	15.29	8.92
760	108.9	8.29	12.08	15.32	17.32	692.10	6.29	15.16	8.87
770	108.0	8.12	11.93	15.41	17.40	701.74	6.21	15.04	8.83
780	107.2	7.95	11.79	15.49	17.47	711.40	6.13	14.92	8.78
790	106.3	7.77	11.64	15.58	17.55	721.09	6.06	14.80	8.74
800	105.5	7.60	11.50	15.66	17.62	730.81	5.99	14.68	8.69

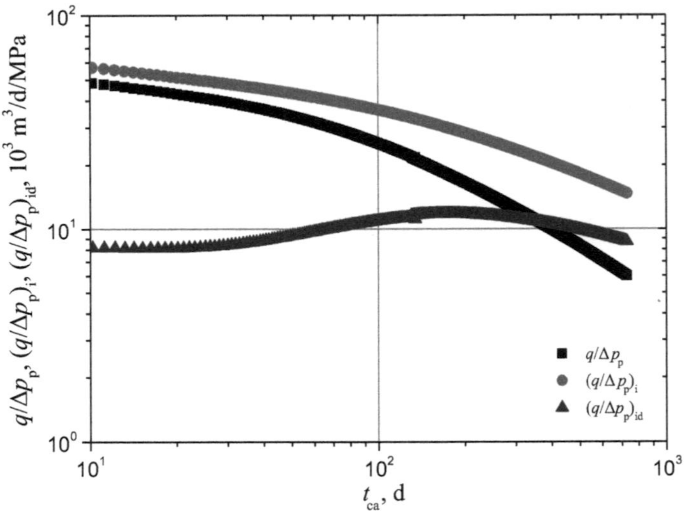

Figure 4.11 Rate functions for well B

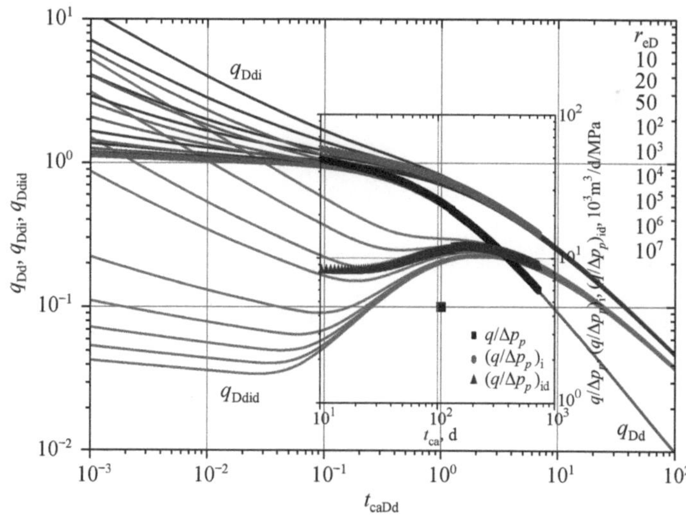

Figure 4.12 Matches of production data for well B

10. Calculate permeability according to the rate matching point and Eq. (4.115). The following expression can be obtained under the SI unit system,

$$\begin{aligned} K &= \frac{(q/\Delta p_p)_M}{(q_{Dd})_M} \frac{\mu B}{2\pi h}\left(\ln r_{eD} - \frac{1}{2}\right) \\ &= \frac{1.866 \times (47 \times 1000) \times 0.0228 \times 0.0039497 \times (\ln 20 - 0.5)}{0.1 \times 10} \\ &= 1.98\,\text{mD} \end{aligned} \tag{4.123}$$

11. Calculate the effective wellbore radius r_{wa}. On the basis of the time matching point and r_{eD} determined in step (8), the expression of r_{wa} under the SI unit system is as follows:

$$\begin{aligned} r_{wa} &= \sqrt{\frac{2K/\varphi\mu C_t}{(r_{eD}^2 - 1)\left(\ln r_{eD} - \frac{1}{2}\right)}\left(\frac{t_{ca}}{t_{caDd}}\right)_M} \\ &= \sqrt{\frac{3.6 \times 10^{-3} \times 2K/\varphi\mu C_t}{(r_{eD}^2 - 1)\left(\ln r_{eD} - \frac{1}{2}\right)}\left(\frac{24 \times t_{ca}}{t_{caDd}}\right)_M} \\ &= \sqrt{\frac{3.6 \times 10^{-3} \times 2 \times 1.98}{0.1 \times 0.0228 \times 0.0246(20^2 - 1)\left(\ln 20 - \frac{1}{2}\right)}\left(\frac{24 \times 101}{1}\right)} \\ &= 24.86\,\text{m} \end{aligned} \tag{4.124}$$

12. Calculate the skin factor according to Eq. (4.117)

$$S = \ln\frac{r_w}{r_{wa}} = \ln\frac{0.1}{24.86} = -5.52 \tag{4.125}$$

13. Calculate the drainage radius according to Eq. (4.118)

$$r_e = r_{wa} r_{eD} = 24.86 \times 20 = 497.2 \text{ m} \tag{4.126}$$

Calculate the drainage radius according to Eq. (4.120)

$$\begin{aligned} r_e &= \sqrt{\frac{\frac{B}{C_t}\left(\frac{t_{ca}}{t_{caDd}}\right)_M \left(\frac{q/\Delta p_p}{q_{Dd}}\right)_M}{\pi h \varphi}} \\ &= \sqrt{\frac{\frac{0.0039497}{0.0246}\left(\frac{24\times 101}{1}\right)_M \left(\frac{47\times 1000}{0.1}\right)_M}{3.14\times 10\times 0.1}} = 492.7 \text{ m} \end{aligned} \tag{4.127}$$

Calculate the reserves according to Eq. (4.121)

$$\begin{aligned} G &= \frac{1}{C_t}\left(\frac{t_{ca}}{t_{caDd}}\right)_M \left(\frac{q/\Delta p_p}{q_{Dd}}\right)_M (1 - S_w) \\ &= \frac{1}{0.0246}\left(\frac{101}{1}\right)_M \left(\frac{47\times 1000}{0.1}\right)_M \\ &= 1.93\times 10^8 \text{ m}^3 \end{aligned} \tag{4.128}$$

Agarwal–Gardner Decline Curves Analysis

5

The pseudo-pressure normalized flow rate ($q/\Delta p_p$) and material balance time function t_{ca} are introduced in the Blasingame method to establish type curves. The bottom hole flowing pressure variation in production and variation of gas pressure–volume–temperature (PVT) properties with formation pressure are also taken into account in this method. The pseudo-pressure normalized flow rate ($q/\Delta p_p$), material balance pseudo-time function t_{ca}, and the relationship between the dimensionless parameters in transient well test are used by Agarwal to establish the Agarwal–Gardner type curves. Because of the difference in the definitions of dimensionless variables, the early part of type curves is more dispersed than Blasingame type curves, which can reduce the ambiguity in the result of matching analysis. This chapter mainly presents the theoretical basis of the Agarwal–Gardner decline curves analysis method, which is suitable for the vertical well in closed circular homogeneous reservoirs, as well as the type curve plotting, and the curve matching method. Furthermore, several case studies using the previous method are given.

5.1 Agarwal–Gardner Type Curves Plotting

5.1.1 Agarwal–Gardner Dimensionless Flow Rate

Assuming that there is a vertical well centered in a closed circular reservoir producing at a constant rate q. The reservoir radius is r_e. The BHFP, reservoir thickness, initial reservoir pressure, wellbore radius, reservoir porosity, total compressibility, permeability, fluid viscosity, and formation volume factor are denoted by p_{wf}, h, p_i, r_w, φ, C_t, K, μ, B, respectively. The skin effect is not considered. Under the basic SI unit system, the following expression can be obtained as shown in Chapter 4.

$$\frac{1}{r_D}\frac{\partial}{\partial r_D}\left(r_D\frac{\partial p_D}{\partial r_D}\right)=\frac{\partial p_D}{\partial t_D} \tag{5.1}$$

$$p_D\left(r_D,0\right)=0 \tag{5.2}$$

$$\left(r_D\frac{\partial p_D}{\partial r_D}\right)_{r_D=1}=-1 \tag{5.3}$$

$$\left.\frac{\partial p_D}{\partial r_D}\right|_{r_D=r_{eD}}=0 \tag{5.4}$$

Advanced Production Decline Analysis and Application

The dimensionless variables are defined as follows

$$p_D = \frac{2\pi Kh(p_i - p_{wf})}{q\mu B} \quad t_D = \frac{Kt}{\varphi\mu C_t r_w^2} \quad r_D = \frac{r}{r_w} \quad r_{eD} = \frac{r_e}{r_w}$$

The dimensionless time introduced by Agarwal et al. (1998) based on drainage area is

$$t_{DA} = \frac{Kt}{\varphi\mu C_t r_w^2}\frac{r_w^2}{A} = t_D \frac{1}{\pi\left(r_{eD}^2 - 1\right)} = \beta t_D \tag{5.5}$$

Substituting Eq. (5.5) into Eq. (5.1), we have

$$\frac{1}{r_D}\frac{\partial}{\partial r_D}\left(r_D\frac{\partial p_D}{\partial r_D}\right) = \beta\frac{\partial p_D}{\partial t_{DA}} \tag{5.6}$$

The Laplace space solution for the definite solution problem of Eq. (5.6), (5.2), through (5.4) is

$$\bar{p}_D = \frac{1}{s\sqrt{\beta s}}\left[\frac{\dfrac{K_1\left(r_{eD}\sqrt{\beta s}\right)}{I_1\left(r_{eD}\sqrt{\beta s}\right)} + \dfrac{K_0\left(\sqrt{\beta s}\right)}{I_0\left(\sqrt{\beta s}\right)}}{\dfrac{K_1\left(\sqrt{\beta s}\right)}{I_0\left(\sqrt{\beta s}\right)} - \dfrac{I_1\left(\sqrt{\beta s}\right)}{I_0\left(\sqrt{\beta s}\right)}\dfrac{K_1\left(r_{eD}\sqrt{\beta s}\right)}{I_1\left(r_{eD}\sqrt{\beta s}\right)}}\right] \tag{5.7}$$

The definition of dimensionless flow rate is as follows

$$q_D = \frac{1}{L^{-1}\left[\bar{p}_D\right]} \tag{5.8}$$

Rearranging Eq. (5.8), we have

$$q_D = \frac{1}{p_D} = \frac{q\mu B}{2\pi Kh(p_i - p_{wf})} \tag{5.9}$$

5.1.2 Agarwal–Gardner Dimensionless Flow Rate Curves

The inverse of pressure solution at a constant flow rate $1/p_D$ and dimensionless time t_{DA} is plotted in Figure 5.1. The decline curves controlled by r_{eD} include a cluster

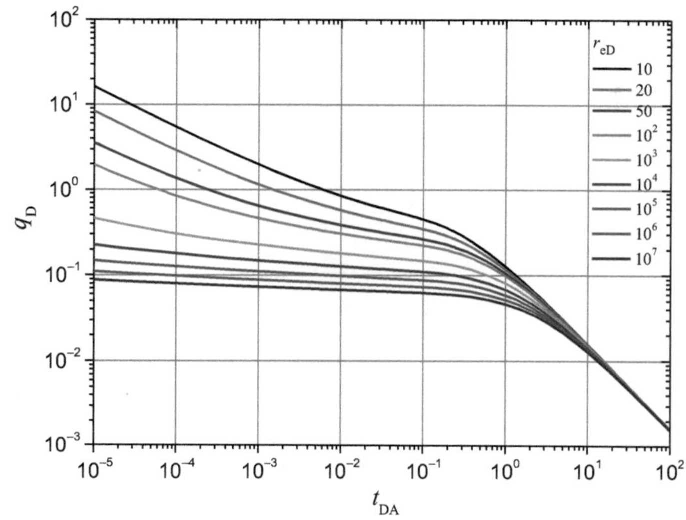

Figure 5.1 Agarwal–Gardner dimensionless flow rate curves

of curves in the transient flow stage. With the increase of r_{eD}, the decline curves shift downward. The curve cluster results in a straight line with the slope of −1 in the boundary-dominated flow stage. In the late stage of boundary-dominated flow (pseudo-steady state at constant flow rate production), we have

$$p_D \approx \frac{2t_D}{r_{eD}^2 - 1} + \ln r_{eD} - \frac{3}{4} \tag{5.10}$$

Substituting Eq. (5.5) into Eq. (5.10), we have

$$p_D = 2\pi t_{DA} + \ln r_{eD} - \frac{3}{4} \tag{5.11}$$

Substituting Eq. (5.11) into Eq. (5.10), we have

$$q_D = \frac{1}{p_D} = \frac{1}{2\pi t_{DA} + \ln r_{eD} - 3/4} \tag{5.12}$$

In the pseudo-steady state stage, when the dimensionless time t_{DA} is much larger than $\ln r_{eD}$, the decline curves result in a straight line with a slope of −1 in the log–log coordinate. Because of the difference in the definitions of dimensionless variables, the Agarwal–Gardner type curves are more dispersed than Palacio–Blasingame (Palacio and Blasingame, 1993) type curves (Figure 4.7), which can improve the matching analysis.

5.1.3 Agarwal–Gardner Dimensionless Inverse Normalized Pressure Derivative Curves

In order to improve the reliability of the analysis, Agarwal–Gardner introduced the inverse normalized pressure derivative function, that is, 1/DER, given as under

$$\frac{1}{\mathrm{DER}}=\frac{1}{\dfrac{\partial p_{\mathrm{D}}}{\partial \ln t_{\mathrm{DA}}}}=\frac{1}{t_{\mathrm{DA}}\dfrac{\partial p_{\mathrm{D}}}{\partial t_{\mathrm{DA}}}}=\frac{1}{t_{\mathrm{DA}}p_{\mathrm{D}}'} \tag{5.13}$$

In the Laplace space,

$$\bar{p}_{\mathrm{D}}'=\frac{1}{\sqrt{\beta s}}\left[\frac{\dfrac{K_1\left(r_{\mathrm{eD}}\sqrt{\beta s}\right)}{I_1\left(r_{\mathrm{eD}}\sqrt{\beta s}\right)}+\dfrac{K_0\left(\sqrt{\beta s}\right)}{I_0\left(\sqrt{\beta s}\right)}}{\dfrac{K_1\left(\sqrt{\beta s}\right)}{I_0\left(\sqrt{\beta s}\right)}-\dfrac{I_1\left(\sqrt{\beta s}\right)}{I_0\left(\sqrt{\beta s}\right)}\dfrac{K_1\left(r_{\mathrm{eD}}\sqrt{\beta s}\right)}{I_1\left(r_{\mathrm{eD}}\sqrt{\beta s}\right)}}\right] \tag{5.14}$$

$$\frac{1}{\mathrm{DER}}=\frac{1}{L^{-1}[\bar{p}_{\mathrm{D}}']t_{\mathrm{DA}}} \tag{5.15}$$

The Agarwal–Gardner inverse normalized pressure derivative curves are similar to the Fetkovich decline curves, and the type curves can be separated into two parts. In the left part, $t_{\mathrm{DA}} < 0.1$, which corresponds to the transient flow stage, which is mainly controlled by r_{eD}. In the right part, $t_{\mathrm{DA}} > 0.1$, which corresponds to the late pseudo-steady state flow. The type curves result in a straight line with the slope of –1 as shown in Figure 5.2.

In the Agarwal–Gardner decline curves as shown in Figure 5.3, the inverse normalized pressure derivative curve is equivalent to the pressure derivative curve and has a similar role to the well test pressure derivative, which could more easily recognize the different transient flow states. When $t_{\mathrm{DA}} = 0.1$, transient flow turns into boundary-dominated flow. The decline curve is a straight line with a slope of –1, and all curves have this feature. However, this parameter requires high-quality data, and the derivative curve analysis will be meaningless when the production data are distributed dispersedly.

5.1.4 Agarwal–Gardner Dimensionless Inverse Normalized Pressure Integral Derivative Curves

By introducing the inverse normalized pressure integral derivative function, the previous problems can be solved. It is defined as

$$p_{\mathrm{Di}}=\frac{1}{t_{\mathrm{DA}}}\int_0^{t_{\mathrm{DA}}} p_{\mathrm{D}}\, dt_{\mathrm{DA}} \tag{5.16}$$

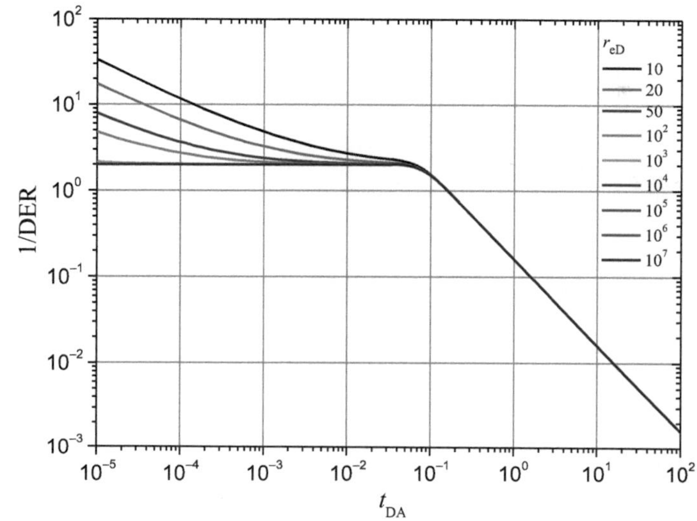

Figure 5.2 Agarwal–Gardner inverse normalized pressure derivative curves

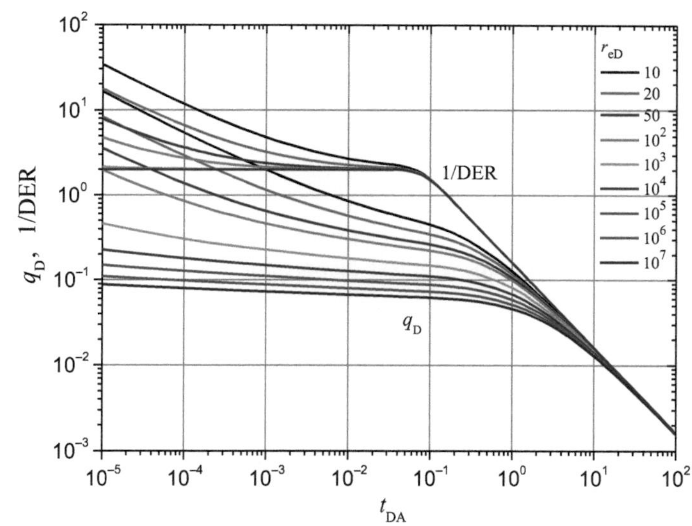

Figure 5.3 Agarwal–Gardner dimensionless flow rate q_D and inverse normalized pressure derivative 1/DER curves

Differentiating Eq. (5.16), we have

$$dp_{Di} = \frac{t_{DA}\, p_D dt_{DA} - \int_0^{t_{DA}} p_D \, dt_{DA}}{t_{DA}^2} \tag{5.17}$$

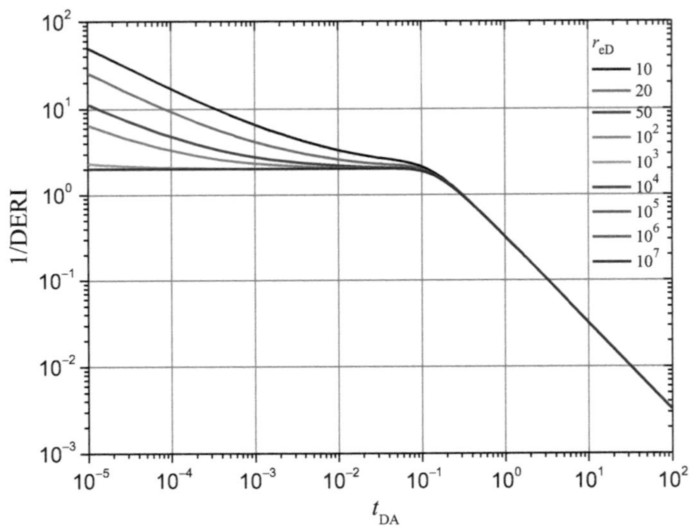

Figure 5.4 Agarwal–Gardner inverse normalized pressure integral derivative 1/DERI curves

Substituting Eq. (5.17) into Eq. (5.13), we have

$$\frac{1}{\text{DERI}} = \frac{1}{t_{DA}\dfrac{\partial p_{Di}}{\partial t_{DA}}} = \frac{1}{p_D - p_{Di}} \tag{5.18}$$

As shown in Figure 5.4, the derivative curve retains the majority of the characteristics of the "raw data" derivative, with less dispersion.

5.1.5 Agarwal–Gardner Decline Curve

The Agarwal–Gardner decline curves shown in Figure 5.5 are obtained by superposing the previous three curves. Because of the difference in the definitions of dimensionless variables, the Agarwal–Gardner decline curves are more dispersed than the Palacio and Blasingame (1993) decline curves (Figure 4.7) in the early stage, which can improve the matching analysis and achieve a lower ambiguity than the Blasingame method.

5.2 Decline Curves Analysis Using Agarwal–Gardner Style Type Curves

5.2.1 Type Curves Matching Procedure—Oil Well

Similar to the Blasingame matching analysis procedure, the Agarwal–Gardner decline curve matching analysis procedure of oil well is listed as under.

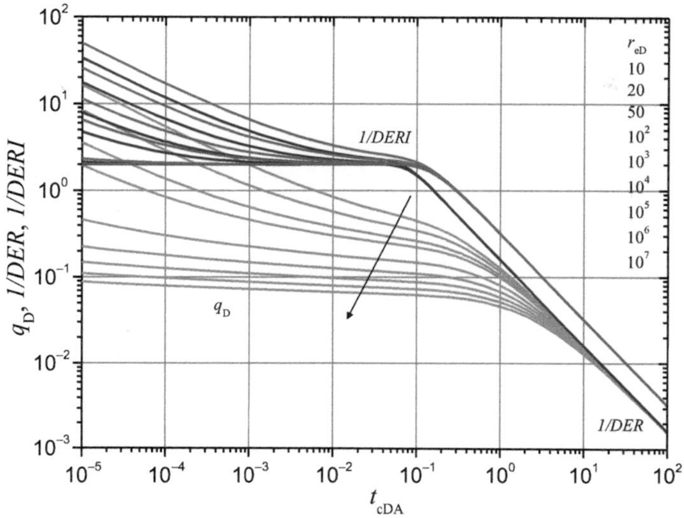

Figure 5.5 Agarwal–Gardner decline curves (after Agarwal and Gardner, 1998)

1. Calculate the material balance time

$$t_c = \frac{N_p}{q} \tag{5.19}$$

2. Calculate the normalized rate

$$\frac{q}{\Delta p} = \frac{q}{p_i - p_{wf}} \tag{5.20}$$

3. Calculate the normalized pressure integral

$$\left(\frac{\Delta p}{q}\right)_i = \frac{1}{t_c}\int_0^{t_c} \frac{p_i - p_{wf}}{q}\,d\tau \tag{5.21}$$

The subscript i represents integration.

4. Calculate the normalized pressure integral derivative

$$\left(\frac{\Delta p}{q}\right)_{id} = \frac{d\left(\frac{\Delta p}{q}\right)_i}{d\ln t_c} = t_c\frac{d\left(\frac{\Delta p}{q}\right)_i}{dt_c} \tag{5.22}$$

The subscript i represents integration, and d represents derivation.

5. Calculate the inverse normalized pressure derivative

$$\frac{1}{\mathrm{DER}}=\frac{1}{\dfrac{\partial\left(\dfrac{\Delta p}{q}\right)}{\partial \ln t_c}}=\frac{1}{t_c\dfrac{\partial\left(\dfrac{\Delta p}{q}\right)}{\partial t_c}} \tag{5.23}$$

6. Calculate the inverse normalized pressure integral derivative

$$\frac{1}{\mathrm{DERI}}=\frac{1}{\dfrac{\partial\left(\dfrac{\Delta p}{q}\right)}{\partial \ln t_c}}=\frac{1}{t_c\dfrac{d\left(\dfrac{\Delta p}{q}\right)_i}{dt_c}} \tag{5.24}$$

The subscript i represents integration, and d represents derivation.

7. In the log–log coordinate system, plot the normalized rate, inverse normalized pressure derivative, and inverse normalized pressure integral derivative curves versus material balance time respectively, $\dfrac{q}{\Delta p}\sim t_c$, $\dfrac{1}{DER}\sim t_c$, $\dfrac{1}{DERI}\sim t_c$.

8. Choose any of the three curves or any combination thereof to match the actual points with the type curves (Figure 5.5), aiming to acquire the best matching of each curve.

9. According to the matching results, record the dimensionless drainage radius r_{eD}.

10. Randomly select a matching point, and record the actual matching point $\left(t_c, q/\Delta p\right)_M$ and the corresponding theoretical matching point $(t_{cDA}, q_D)_M$. If the reservoir thickness, total compressibility, and wellbore radius are known, then we can calculate the reservoir permeability, skin factor, drainage area, reserves, etc.

11. According to the rate matching point and Eq. (5.9), calculate the permeability, and we have

$$K=\frac{(q/\Delta p)_M}{(q_D)_M}\frac{\mu B}{2\pi h} \tag{5.25}$$

12. On the basis of the time matching point and the permeability determined in step (11), calculate the drainage radius r_e

$$r_e=\sqrt{\frac{K}{\pi\varphi\mu C_t}\left(\frac{t_c}{t_{cDA}}\right)_M} \tag{5.26}$$

13. On the basis of r_{eD} determined in step (9), calculate the effective wellbore radius r_{wa}

$$r_{wa}=\frac{r_e}{r_{eD}} \tag{5.27}$$

14. Calculate the skin factor S

$$S = \ln\frac{r_{\mathrm{w}}}{r_{\mathrm{wa}}} \tag{5.28}$$

15. Calculate the original oil in place (OOIP) by the volumetric method

$$N = \frac{\pi r_e^2 \varphi h S_{\mathrm{o}}}{B_{\mathrm{o}}} \tag{5.29}$$

5.2.2 *Type Curves Matching Procedure—Gas Well*

The procedure of Agarwal–Gardner curve matching analysis for gas well is similar to that for oil well, as follows.

1. Calculate the material balance pseudo-time

Assuming reserves G, calculate the material balance pseudo-time for each production data point.

$$t_{\mathrm{ca}} = \frac{(\mu C_{\mathrm{t}})_{\mathrm{i}}}{q}\int_0^t \frac{q}{\mu(\bar{p})C_{\mathrm{t}}(\bar{p})}dt = \frac{GC_{\mathrm{ti}}}{q}\left(p_{p_{\mathrm{i}}} - p_{p}\right) \tag{5.30}$$

Please refer to the normalized pseudo-pressure p_{p} in Eq. (4.57), and the average reservoir pressure could be calculated according to Eq. (4.60).

2. Calculate the normalized rate

$$\frac{q}{\Delta p_{\mathrm{p}}} = \frac{q}{p_{p_{\mathrm{i}}} - p_{p_{\mathrm{wf}}}} \tag{5.31}$$

3. Calculate the normalized pressure integral

$$\left(\frac{\Delta p_{\mathrm{p}}}{q}\right)_{\mathrm{i}} = \frac{1}{t_{\mathrm{ca}}}\int_0^{t_{\mathrm{ca}}} \frac{\Delta p_{\mathrm{p}}}{q}d\tau \tag{5.32}$$

The subscript i represents integration.

4. Calculate the normalized pressure integral derivative

$$\left(\frac{\Delta p_{\mathrm{p}}}{q}\right)_{\mathrm{id}} = \frac{d\left(\frac{\Delta p_{\mathrm{p}}}{q}\right)_{\mathrm{i}}}{d\ln t_{\mathrm{ca}}} = t_{\mathrm{ca}}\frac{d\left(\frac{\Delta p_{\mathrm{p}}}{q}\right)_{\mathrm{i}}}{dt_{\mathrm{ca}}} \tag{5.33}$$

The subscript i represents integration, and d represents derivation.

5. Calculate the inverse normalized pressure derivative

$$\frac{1}{\text{DER}}=\frac{1}{\dfrac{\partial\left(\dfrac{\Delta p_{\text{p}}}{q}\right)}{\partial \ln t_{\text{ca}}}}=\frac{1}{t_{\text{ca}}\dfrac{\partial\left(\dfrac{\Delta p_{\text{p}}}{q}\right)}{\partial t_{\text{ca}}}} \tag{5.34}$$

6. Calculate the inverse normalized pressure integral derivative

$$\frac{1}{\text{DERI}}=\frac{1}{\dfrac{\partial\left(\dfrac{\Delta p_{\text{p}}}{q}\right)}{\partial \ln t_{\text{ca}}}}=\frac{1}{t_{\text{ca}}\dfrac{d\left(\dfrac{\Delta p_{\text{p}}}{q}\right)_{\text{i}}}{dt_{\text{ca}}}} \tag{5.35}$$

The subscript i represents integration, and d represents derivation.

7. In the Cartesian coordinate system, plot $\frac{\Delta p_{\text{p}}}{q} \sim t_{\text{ca}}$ curve and regress the data points linearly; then apply the slope of the line to Eq. (4.65), calculate the value of G, repeat steps (1)–(7) and make iterative computations until G meets the permissible error.

$$G=\frac{1}{\text{Slope}\times C_{\text{ti}}} \tag{5.36}$$

8. Prepare the log–log plot of the normalized rate, inverse normalized pressure derivative and inverse normalized pressure integral derivative changing with material balance pseudo-time, respectively, in the same coordinate system. In other words, we need to plot the $\frac{q}{\Delta p_{\text{p}}} \sim t_{\text{ca}}$, $\frac{1}{\text{DER}} \sim t_{\text{ca}}$, and $\frac{1}{\text{DERI}} \sim t_{\text{ca}}$ curves.

9. Choose any of the three curves or any combination thereof to match the actual points with the type curves (Figure 5.5), aiming to realize the best matching of each curve.

10. Record the dimensionless drainage radius r_{eD} according to the matching results.

11. Randomly choose a matching point, and record the actual matching point $\left(t_{\text{ca}}, q/\Delta p_{\text{p}}\right)_{\text{M}}$ and the corresponding theoretical matching point $(t_{\text{caDA}}, q_{\text{D}})_{\text{M}}$. If the reservoir thickness, total compressibility, and wellbore radius are known, then we can calculate the reservoir permeability, skin factor, drainage area, reserves, etc.

12. Calculate the permeability according to the rate matching point and Eq. (5.9), and we have

$$K=\frac{\left(q/\Delta p_{\text{p}}\right)_{\text{M}}}{\left(q_{\text{D}}\right)_{\text{M}}}\frac{\mu B}{2\pi h} \tag{5.37}$$

13. On the basis of the time matching point and the permeability determined in step (12), and Eq. (5.5), calculate the drainage radius r_e

$$r_e = \sqrt{\frac{K}{\pi\varphi\mu C_t}\left(\frac{t_{ca}}{t_{caDA}}\right)_M} \tag{5.38}$$

14. On the basis of r_{eD} determined in step (10), calculate the effective wellbore radius r_{wa}

$$r_{wa} = \frac{r_e}{r_{eD}} \tag{5.39}$$

15. Calculate the skin factor S

$$S = \ln\frac{r_w}{r_{wa}} \tag{5.40}$$

16. Calculate the original gas in place (OGIP) by the volumetric method

$$G = \frac{\pi r_e^2 \varphi h S_g}{B_{gi}} \tag{5.41}$$

5.3 Case Study of Agarwal–Gardner Decline Curves Analysis

We still take the case in Section 4.4 as an example. The main parameters are as follows. A gas well centered in a closed circular reservoir is producing at a variable rate and a variable BHFP. The initial reservoir pressure is 30 MPa, the reservoir temperature is 80°C, the reservoir thickness is 10 m, the reservoir porosity is 0.1, the distance between the well and boundary is 500 m, and the gas reserves are 2.0×10^8 m^3. The relative density of gas phase is 0.6, the pseudo-critical temperature is 195.697 K, the pseudo-critical pressure is 4.66875 MPa, the reservoir permeability estimation from well test is 2.0 mD, the skin factor is –5.52, the production time is 800 days, the cumulative gas production is 11813×10^4 m^3, the fluid viscosity is 0.0228 mPa·s, the rock compressibility is 4.35113×10^{-4} MPa^{-1}, the gas compressibility under initial condition is 0.0246 MPa^{-1}, the gas formation volume factor under initial condition is 0.0039497, and the production data have been listed in Table 4.1.

Neglecting the process of iterative computation for G, and assuming $G = 2.0 \times 10^8$ m^3 immediately, the analysis process is as follows.

1–5. Calculate the material balance pseudo-time and the relative normalized rate parameters.

Calculate the average reservoir pressure, normalized pseudo-pressure change, normalized pseudo-time, normalized rate, inverse normalized pressure derivative, and inverse normalized pressure integral derivative according to Eq. (4.60), (4.122), (5.30), (5.31), and (5.34), the results are shown in Table 5.1.

Table 5.1 Material balance pseudo-time and the relative normalized rate parameters of well B

t	q	p_{wf}	p_{avg}	Δp_p (p_{avg})	Δp_p (p_{wf})	t_{ca}	$q/\Delta p_p$	1/ DER	1/ DERI
d	10^3 m^3/d	MPa	MPa	MPa	MPa	d	10^3 m^3/ d/MPa	MPa/10^3 m^3/d	
1	199.8	27.72	30.00	0.01	2.26	0.12	88.41	158.46	—
2	199.7	27.23	29.99	0.01	2.74	0.14	72.85	44.64	256.88
3	199.5	26.92	29.95	0.05	3.04	1.15	65.69	588.56	1423.95
4	199.4	26.70	29.91	0.09	3.25	2.16	61.30	428.98	941.90
5	199.2	26.53	29.87	0.13	3.42	3.17	58.23	369.39	719.67
6	199.0	26.38	29.83	0.17	3.56	4.17	55.91	339.73	606.07
7	198.9	26.26	29.79	0.21	3.68	5.18	54.08	320.79	537.72
8	198.7	26.16	29.75	0.25	3.78	6.19	52.58	308.04	492.12
9	198.6	26.06	29.71	0.29	3.87	7.19	51.32	298.76	459.42
10	198.4	25.98	29.67	0.33	3.95	8.20	50.24	291.06	434.83
20	196.8	25.41	29.27	0.73	4.49	18.18	43.82	231.12	326.79
30	195.3	25.01	28.88	1.11	4.88	28.06	40.04	171.36	268.76
40	193.7	24.64	28.50	1.49	5.22	37.84	37.08	130.40	223.53
50	192.2	24.29	28.13	1.86	5.56	47.54	34.58	104.71	188.86
60	190.6	23.94	27.77	2.21	5.88	57.16	32.41	87.40	162.42
70	189.1	23.60	27.41	2.56	6.20	66.70	30.52	74.77	141.98
80	187.6	23.27	27.06	2.90	6.50	76.17	28.84	65.35	125.93
90	186.1	22.94	26.72	3.24	6.80	85.57	27.35	58.06	113.04
100	184.6	22.62	26.38	3.56	7.10	94.92	26.01	52.09	102.52
110	183.2	22.31	26.05	3.88	7.38	104.21	24.80	47.48	93.77
120	181.7	22.00	25.73	4.19	7.66	113.46	23.71	44.04	86.42
130	180.2	21.69	25.41	4.49	7.94	122.66	22.71	40.41	80.14
140	178.8	21.39	25.00	4.89	8.20	134.53	21.80	37.88	76.25
150	177.4	21.10	24.90	4.98	8.46	138.22	20.96	34.41	68.77
160	176.0	20.81	24.61	5.26	8.72	146.93	20.19	32.48	64.75
170	174.6	20.53	24.32	5.52	8.96	155.65	19.48	30.96	61.19
180	173.2	20.25	24.04	5.79	9.21	164.39	18.81	29.42	58.03
190	171.8	19.97	23.76	6.05	9.44	173.13	18.19	27.94	55.18
200	170.4	19.70	23.49	6.30	9.67	181.89	17.62	26.78	52.63
210	169.1	19.43	23.22	6.55	9.90	190.66	17.08	25.65	50.31
220	167.7	19.17	22.95	6.80	10.12	199.44	16.57	24.54	48.19
230	166.4	18.91	22.68	7.04	10.34	208.22	16.09	23.60	46.25
240	165.1	18.65	22.42	7.28	10.55	217.02	15.64	22.73	44.47
250	163.7	18.40	22.16	7.52	10.76	225.83	15.22	21.94	42.81
260	162.4	18.15	21.90	7.75	10.96	234.65	14.82	21.13	41.29
270	161.1	17.91	21.65	7.97	11.16	243.48	14.44	20.45	39.87
280	159.9	17.66	21.40	8.20	11.36	252.31	14.08	19.76	38.55
290	158.6	17.42	21.15	8.42	11.55	261.16	13.74	19.10	37.31
300	157.3	17.19	20.90	8.63	11.73	270.02	13.41	18.50	36.16
310	156.1	16.95	20.66	8.85	11.92	278.89	13.10	17.92	35.06
320	154.8	16.72	20.42	9.06	12.09	287.77	12.80	17.39	34.04
330	153.6	16.49	20.18	9.26	12.27	296.65	12.52	16.88	33.07
340	152.4	16.27	19.95	9.46	12.44	305.55	12.25	16.47	32.16

Table 5.1 **Material balance pseudo-time and the relative normalized rate parameters of well B *(cont.)***

t	q	p_{wf}	p_{avg}	Δp_p (p_{avg})	Δp_p (p_{wf})	t_{ca}	$q/\Delta p_p$	1/ DER	1/ DERI
350	151.2	16.05	19.72	9.66	12.61	314.47	11.99	16.08	31.29
360	150.0	15.83	19.49	9.86	12.77	323.39	11.74	15.60	30.47
370	148.8	15.61	19.26	10.05	12.94	332.32	11.50	15.19	29.69
380	147.6	15.39	19.03	10.24	13.09	341.26	11.27	14.81	28.95
390	146.4	15.18	18.81	10.42	13.25	350.21	11.05	14.45	28.24
400	145.2	14.97	18.59	10.60	13.40	359.18	10.84	14.05	27.57
410	144.1	14.76	18.38	10.78	13.55	368.15	10.63	13.72	26.92
420	142.9	14.55	18.16	10.96	13.70	377.14	10.43	13.41	26.30
430	141.8	14.34	17.95	11.13	13.84	386.14	10.24	13.11	25.71
440	140.7	14.14	17.74	11.30	13.98	395.15	10.06	12.77	25.15
450	139.5	13.94	17.53	11.46	14.12	404.18	9.88	12.50	24.60
460	138.4	13.74	17.32	11.63	14.25	413.21	9.71	12.29	24.08
470	137.3	13.54	17.12	11.79	14.39	422.26	9.54	11.94	23.58
480	136.2	13.34	16.92	11.94	14.52	431.32	9.38	11.66	23.10
490	135.1	13.15	16.72	12.10	14.65	440.40	9.23	11.47	22.64
500	134.1	12.95	16.52	12.25	14.77	449.49	9.08	11.21	22.19
510	133.0	12.76	16.33	12.40	14.89	458.60	8.93	11.00	21.76
520	131.9	12.57	16.13	12.54	15.01	467.72	8.79	10.75	21.34
530	130.9	12.38	15.94	12.69	15.13	476.85	8.65	10.51	20.94
540	129.8	12.19	15.76	12.83	15.25	486.00	8.51	10.23	20.55
550	128.8	12.00	15.57	12.96	15.36	495.17	8.38	10.10	20.17
560	127.8	11.82	15.38	13.10	15.48	504.35	8.26	9.92	19.80
570	126.8	11.63	15.20	13.23	15.59	513.55	8.13	9.75	19.45
580	125.8	11.45	15.02	13.36	15.69	522.76	8.01	9.59	19.11
590	124.8	11.27	14.84	13.49	15.80	531.99	7.90	9.31	18.78
600	123.8	11.09	14.67	13.61	15.90	541.25	7.78	9.20	18.46
610	122.8	10.91	14.49	13.74	16.00	550.51	7.67	9.09	18.14
620	121.8	10.73	14.32	13.86	16.10	559.80	7.56	8.91	17.84
630	120.8	10.55	14.15	13.98	16.20	569.11	7.46	8.69	17.55
640	119.9	10.37	13.98	14.09	16.30	578.43	7.35	8.56	17.26
650	118.9	10.20	13.81	14.21	16.39	587.79	7.25	8.46	16.98
660	118.0	10.02	13.64	14.32	16.49	597.15	7.16	8.30	16.71
670	117.0	9.84	13.48	14.43	16.58	606.54	7.06	8.18	16.45
680	116.1	9.67	13.32	14.53	16.67	615.95	6.97	8.02	16.19
690	115.2	9.50	13.16	14.64	16.75	625.39	6.87	7.91	15.95
700	114.2	9.32	13.00	14.74	16.84	634.84	6.78	7.76	15.70
710	113.3	9.15	12.84	14.84	16.92	644.33	6.70	7.65	15.46
720	112.4	8.98	12.69	14.94	17.01	653.83	6.61	7.51	15.23
730	111.5	8.80	12.53	15.04	17.09	663.36	6.53	7.40	15.01
740	110.6	8.63	12.38	15.13	17.17	672.91	6.44	7.27	14.79
750	109.8	8.46	12.23	15.23	17.25	682.50	6.36	7.17	14.57
760	108.9	8.29	12.08	15.32	17.32	692.10	6.29	7.07	14.36
770	108.0	8.12	11.93	15.41	17.40	701.74	6.21	6.95	14.16
780	107.2	7.95	11.79	15.49	17.47	711.40	6.13	6.82	13.96
790	106.3	7.77	11.64	15.58	17.55	721.09	6.06	6.74	13.76
800	105.5	7.60	11.50	15.66	17.62	730.81	5.99	6.62	13.57

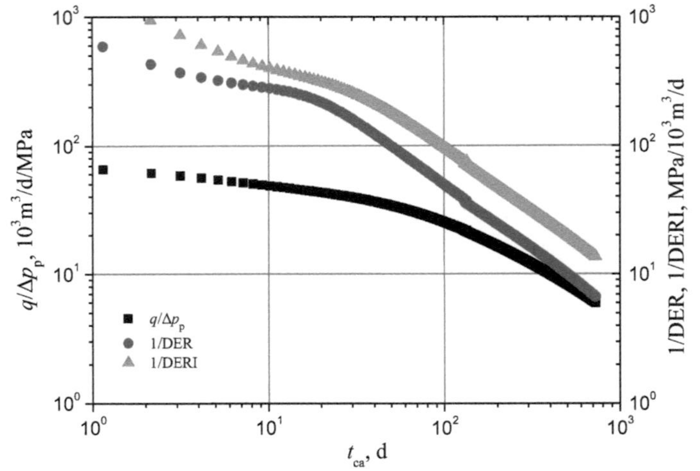

Figure 5.6 Agarwal–Gardner rate functions for well B

6. Plot log–log relationships of $\frac{q}{\Delta p_p} \sim t_{ca}$, $1/\text{DER} \sim t_{ca}$, and $1/\text{DERI} \sim t_{ca}$ curves, as shown in Figure 5.6.
7. Match type curve (Figure 5.6) with the actual production history curve, as shown in Figure 5.7.
8. Record the dimensionless drainage radius $r_{eD} = 20$.
9. Choose the actual matching point $(290, 108)_M$ and the corresponding theoretical matching point $(1, 1)_M$ to simplify the calculation.

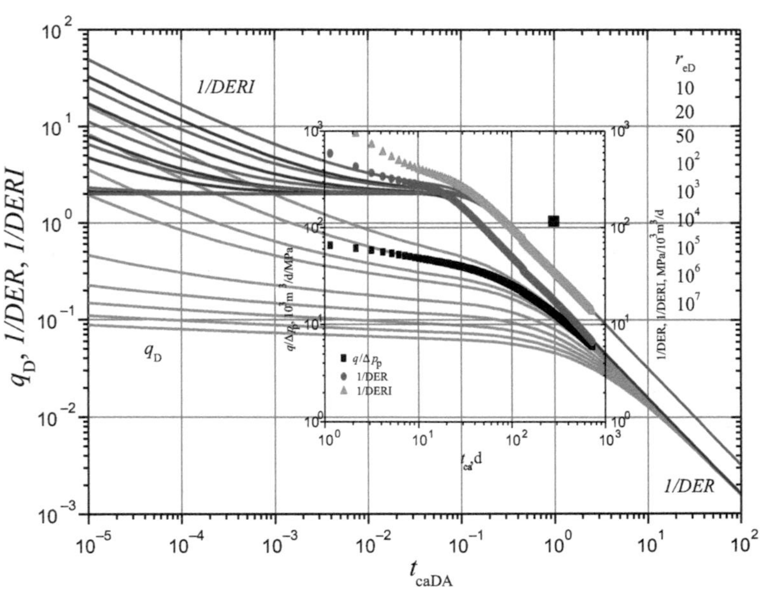

Figure 5.7 Matches of production data for well B: Agarwal–Gardner

10. Calculate permeability according to the rate matching point and Eq. (5.9). The following expression is obtained under the SI unit system

$$K = \frac{(q/\Delta p_{\mathrm{p}})_{\mathrm{M}}}{(q_{\mathrm{D}})_{\mathrm{M}}}\frac{\mu B}{2\pi h} = \frac{1.866\times(q/\Delta p_{\mathrm{p}})_{\mathrm{M}}}{(q_{\mathrm{D}})_{\mathrm{M}}}\frac{\mu B}{h}$$
$$= \frac{1.866\times(108\times1000)\times0.0228\times0.0039497}{1\times10}$$
$$= 1.81\,\mathrm{mD} \tag{5.42}$$

11. On the basis of the time matching point and the permeability determined in step (12), and Eq. (5.38), calculate the drainage radius r_{e}

$$r_{\mathrm{e}} = \sqrt{\frac{K}{\pi\varphi\mu C_{\mathrm{t}}}\left(\frac{t_{\mathrm{ca}}}{t_{\mathrm{caDA}}}\right)_{\mathrm{M}}} = \sqrt{\frac{3.6\times10^{-3}K}{\pi\varphi\mu C_{\mathrm{t}}}\left(\frac{t_{\mathrm{ca}}}{t_{\mathrm{caDA}}}\right)_{\mathrm{M}}}$$
$$= \sqrt{\frac{3.6\times10^{-3}\times1.81}{3.14159\times0.1\times0.0228\times0.0246}\left(\frac{290\times24}{1}\right)_{\mathrm{M}}} = 507.3\ \mathrm{m} \tag{5.43}$$

According to Eq. (5.38) and Eq. (5.42), the following expression is obtained under the SI unit system

$$r_{\mathrm{e}} = \sqrt{\frac{3.6\times10^{-3}B}{\pi\varphi hC_{\mathrm{t}}}\left(\frac{t_{\mathrm{ca}}}{t_{\mathrm{caDA}}}\right)_{\mathrm{M}}\frac{1.866\times(q/\Delta p_{\mathrm{p}})_{\mathrm{M}}}{(q_{\mathrm{D}})_{\mathrm{M}}}}$$
$$= \sqrt{\frac{2.1383\times10^{-3}B}{\varphi hC_{\mathrm{t}}}\left(\frac{t_{\mathrm{ca}}}{t_{\mathrm{caDA}}}\right)_{\mathrm{M}}\left(\frac{q/\Delta p_{\mathrm{p}}}{q_{\mathrm{D}}}\right)_{\mathrm{M}}} \tag{5.44}$$

$$r_{\mathrm{e}} = \sqrt{\frac{2.1383\times10^{-3}\times0.0039497}{0.1\times10\times0.0246}\left(\frac{290\times24}{1}\right)_{\mathrm{M}}\left(\frac{108\times1000}{1}\right)_{\mathrm{M}}}$$
$$= 508.0\ \mathrm{m} \tag{5.45}$$

12. Calculate the effective wellbore radius r_{wa}

$$r_{\mathrm{wa}} = \frac{r_{\mathrm{e}}}{r_{\mathrm{eD}}} = \frac{508}{20} = 25.4\mathrm{m} \tag{5.46}$$

13. Calculate the skin factor

$$S = \ln\frac{r_{\mathrm{w}}}{r_{\mathrm{wa}}} = \ln\frac{0.1}{25.4} = -5.54 \tag{5.47}$$

14. Calculate the OGIP by volumetric method

$$G = \frac{\pi r_e^2 \varphi h S_g}{B_{gi}} = \frac{3.14 \times 508^2 \times 0.1 \times 10 \times 1}{0.0039497} = 2.05 \times 10^8 \mathrm{m}^3 \tag{5.48}$$

Substituting Eq. (5.44) into Eq. (5.41), we have

$$\begin{aligned} G &= \frac{\pi r_e^2 \varphi h S_g}{B_{gi}} = \frac{\pi \varphi h S_g}{B_{gi}} \frac{2.1383 \times 10^{-3} B}{\varphi h C_t} \left(\frac{t_{ca}}{t_{caDA}} \right)_M \left(\frac{q/\Delta p_p}{q_D} \right)_M \\ &= \frac{6.7177 \times 10^{-3}}{C_t} \left(\frac{t_{ca}}{t_{caDA}} \right)_M \left(\frac{q/\Delta p_p}{q_D} \right)_M S_g \end{aligned} \tag{5.49}$$

We have

$$\begin{aligned} G &= \frac{6.7177 \times 10^{-3}}{0.0246} \left(\frac{290 \times 24}{1} \right)_M \left(\frac{108 \times 1000}{1} \right)_M \times 1.0 \\ &= 2.05 \times 10^8 \mathrm{m}^3 \end{aligned} \tag{5.50}$$

NPI Decline Curves Analysis

6

Both Blasingame and Agarwal-Gardner methods use pseudo-pressure normalized rate ($q/\Delta p_p$) and material balance pseudo-time function t_{ca} to plot type curves, whereas the normalized pressure integral (NPI) method uses rate normalized pressure integral to develop an analysis method that is relatively reliable and not affected by the scatter of data. In this chapter, we mainly introduce the theoretical basis of NPI decline curves analysis in a closed circular homogeneous reservoir, as well as the type curve plotting and the curve matching method. Furthermore, we also present several case studies using the previous method.

6.1 NPI Decline Curve Plotting

6.1.1 NPI Dimensionless Variables

Assuming that there is a vertical well centered in a closed circular reservoir producing at a constant rate of q. The reservoir radius is denoted by r_e. Besides, the bottom hole flowing pressure (BHFP), reservoir thickness, initial reservoir pressure, wellbore radius, reservoir porosity, total compressibility, permeability, fluid viscosity, and formation volume factor are denoted by p_{wf}, h, p_i, r_w, φ, C_t, K, μ, and B, respectively. In the meanwhile, the skin effect is not considered. Under the basic SI unit system, the following expression can be obtained, as shown in Eq. (5.7).

$$\bar{p}_D = \frac{1}{s\sqrt{\beta s}}\left[\frac{\dfrac{K_1\left(r_{eD}\sqrt{\beta s}\right)}{I_1\left(r_{eD}\sqrt{\beta s}\right)}+\dfrac{K_0\left(\sqrt{\beta s}\right)}{I_0\left(\sqrt{\beta s}\right)}}{\dfrac{K_1\left(\sqrt{\beta s}\right)}{I_0\left(\sqrt{\beta s}\right)}-\dfrac{I_1\left(\sqrt{\beta s}\right)}{I_0\left(\sqrt{\beta s}\right)}\dfrac{K_1\left(r_{eD}\sqrt{\beta s}\right)}{I_1\left(r_{eD}\sqrt{\beta s}\right)}}\right] \tag{6.1}$$

The dimensionless variables are defined as follows

$$p_D = \frac{2\pi Kh\left(p_i - p_{wf}\right)}{q\mu B} \tag{6.2}$$

$$t_{DA} = \frac{Kt}{\pi\varphi\mu C_t r_e^2} \tag{6.3}$$

Advanced Production Decline Analysis and Application

$$r_D = \frac{r}{r_w}, \quad r_{eD} = \frac{r_e}{r_w} \tag{6.4}$$

$$\beta = \frac{1}{\pi\left(r_{eD}^2 - 1\right)} \tag{6.5}$$

6.1.2 NPI Dimensionless Pressure Curves

The constant rate production pressure solution p_D and dimensionless time t_{DA} are plotted in a graph as shown by Figure 6.1. The decline curves controlled by r_{eD} include a cluster of curves in the transient flow stage. With the increase of r_{eD}, the decline curves shift upward. The curve cluster results in a straight line with a slope of 1 in the boundary-dominated flow stage. This results from the fact that during the period of boundary-dominated flow (namely, in the pseudo-steady state with a constant rate production case), we have

$$p_D = 2\pi t_{DA} + \ln r_{eD} - \frac{3}{4} \tag{6.6}$$

When dimensionless time t_{DA} is much larger than $\ln r_{eD}$, the curves assume a unit-slope straight line on the log–log plot.

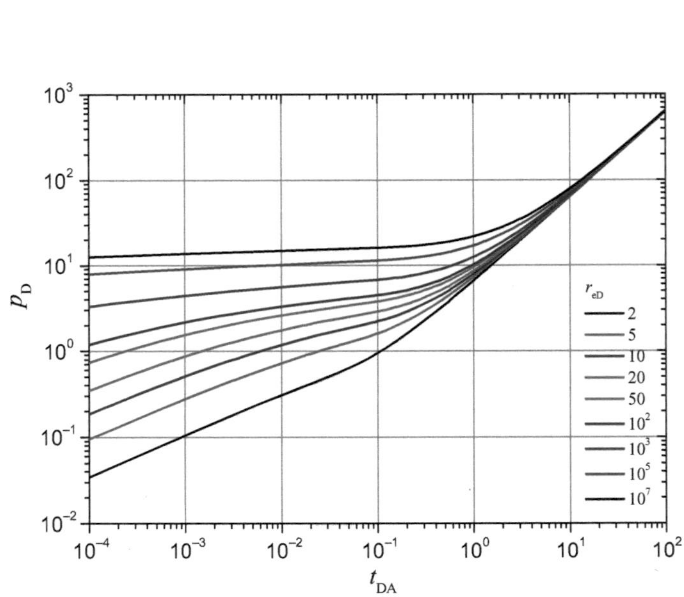

Figure 6.1 NPI dimensionless pressure curve

6.1.3 NPI Dimensionless Pressure Integral Curves

Pressure integral function is defined as

$$p_{\mathrm{Di}} = \frac{1}{t_{\mathrm{DA}}} \int_0^{t_{\mathrm{DA}}} p_{\mathrm{D}}\, dt_{\mathrm{DA}} \tag{6.7}$$

Eq. (6.7) can be expressed as

$$p_{\mathrm{Di}} = \frac{1}{t_{\mathrm{DA}}} L^{-1}\left[\frac{\bar{p}_{\mathrm{D}}}{s}\right] \tag{6.8}$$

The constant rate production pressure integral p_{Di} and dimensionless time t_{DA} are plotted in a coordinate system as shown in Figure 6.2. The curve is more open than the pressure curve and more favorable for reducing the ambiguity of matching.

6.1.4 NPI Dimensionless Pressure Integral Derivative Curves

Differentiating Eq. (6.7), we have

$$dp_{\mathrm{Di}} = \frac{t_{\mathrm{DA}} p_{\mathrm{D}} dt_{\mathrm{DA}} - \int_0^{t_{\mathrm{DA}}} p_{\mathrm{D}}\, dt_{\mathrm{DA}}}{t_{\mathrm{DA}}^2} \tag{6.9}$$

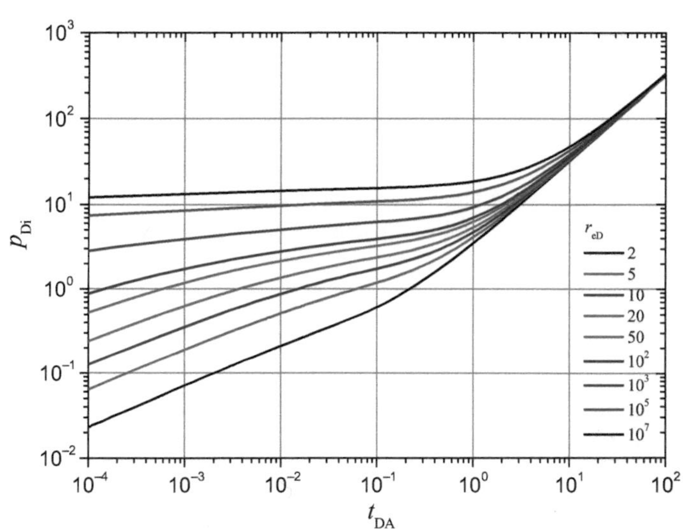

Figure 6.2 NPI dimensionless pressure integral curve

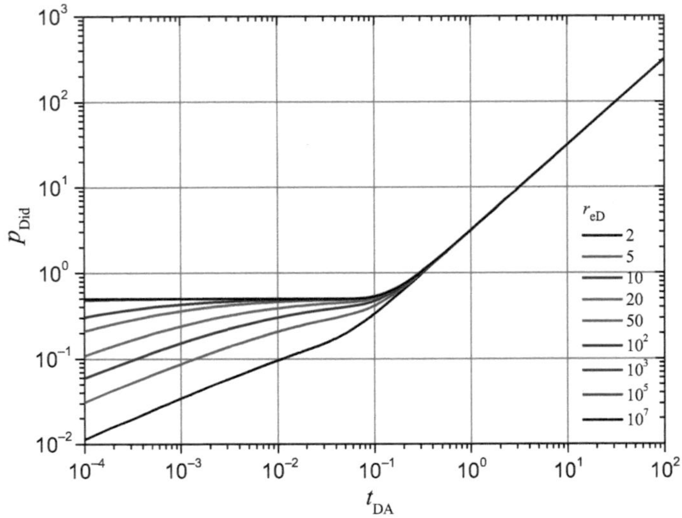

Figure 6.3 NPI dimensionless pressure integral derivative curve

The pressure integral derivative is defined as

$$p_{\mathrm{Did}} = \frac{dp_{\mathrm{Di}}}{d\ln t_{\mathrm{DA}}} = t_{\mathrm{DA}}\frac{dp_{\mathrm{Di}}}{dt_{\mathrm{DA}}} \tag{6.10}$$

Substituting Eq. (6.9) into Eq. (6.10), we have

$$p_{\mathrm{Did}} = p_{\mathrm{D}} - p_{\mathrm{Di}} \tag{6.11}$$

The constant rate production pressure integral derivative p_{Did} and dimensionless time t_{DA} are plotted in a coordinate system as shown in Figure 6.3. The curve is similar to the pressure derivative curve in well test analysis. When $t_{\mathrm{DA}} < 0.1$, the curves correspond to the period of transient flow, and the curves approach a horizontal line with a p_{Did} of 0.5 as the drainage radius increases. Whereas in the right part, when $t_{\mathrm{DA}} > 0.1$, the curves correspond to a pseudo-steady state period and become a unit-slope straight line.

6.1.5 NPI Decline Curves

The NPI decline curves shown in Figure 6.4 are obtained by superposing the previous three curves. The NPI decline curve is indeed an inverse of the Agarwal-Gardner decline curve, which can address the issues raised by a variable rate and variable flowing pressure because of the introduction of material balance pseudo-time. Similar to the Blasingame and Agarwal-Gardner methods, the NPI method can use the daily production data (time, flow rate, flowing pressure) to estimate the parameters

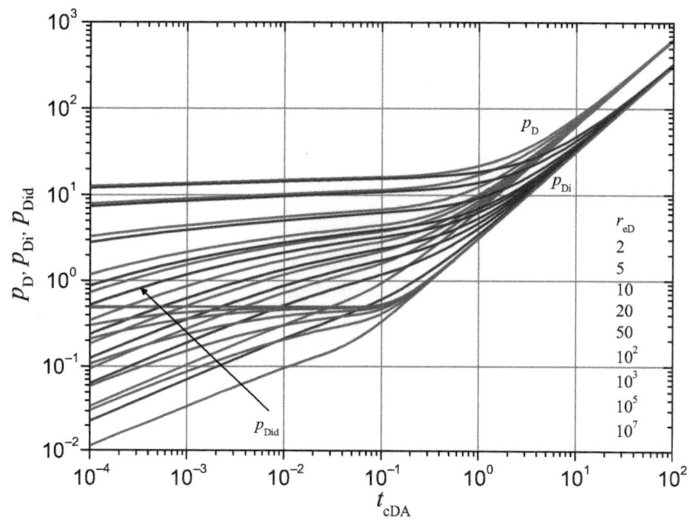

Figure 6.4 NPI decline curve

such as reservoir permeability, original oil in place (OOIP), skin factor, and drainage area.

6.2 Decline Curves Analysis Using NPI Style Type Curves

6.2.1 Type Curves Matching Procedure—Oil Well

The NPI decline curve matching analysis procedure of oil wells is listed as under.

1. Calculate the material balance time

$$t_c = \frac{N_p}{q} \tag{6.12}$$

2. Calculate the normalized pressure

$$\frac{\Delta p}{q} = \frac{p_i - p_{wf}}{q} \tag{6.13}$$

3. Calculate the normalized pressure integral

$$\left(\frac{\Delta p}{q}\right)_i = \frac{1}{t_c}\int_0^{t_c} \frac{p_i - p_{wf}}{q} d\tau \tag{6.14}$$

Where subscript i represents integration.

4. Calculate the normalized pressure integral derivative

$$\left(\frac{\Delta p}{q}\right)_{\mathrm{id}} = \frac{d\left(\frac{\Delta p}{q}\right)_{\mathrm{i}}}{d\ln t_{\mathrm{c}}} = t_{\mathrm{c}}\frac{d\left(\frac{\Delta p}{q}\right)_{\mathrm{i}}}{dt_{\mathrm{c}}} \tag{6.15}$$

Where subscript i represents integration, and d represents derivation.

5. In the log–log coordinate system, plot the normalized pressure, normalized pressure integral, and normalized pressure integral derivative curves versus material balance time, namely, $\frac{\Delta p}{q} \sim t_{\mathrm{c}}$, $\left(\frac{\Delta p}{q}\right)_{\mathrm{i}} \sim t_{\mathrm{c}}$, $\left(\frac{\Delta p}{q}\right)_{\mathrm{id}} \sim t_{\mathrm{c}}$, respectively.
6. Choose any of the three curves or any combination thereof to match the actual points with the type curves (Figure 6.4), aiming to acquire the best matching of each curve.
7. According to the matching results, record the dimensionless drainage radius r_{eD}.
8. Randomly select a matching point, and record the actual matching point $(t_{\mathrm{c}}, \Delta p/q)_{\mathrm{M}}$ and the corresponding theoretical matching point $(t_{\mathrm{cDA}}, p_{\mathrm{D}})_{\mathrm{M}}$. If the reservoir thickness, total compressibility, and wellbore radius are known, then we can calculate the reservoir permeability, skin factor, drainage area, reserves, etc.
9. According to the rate matching point and Eq. (6.2), calculate the permeability

$$K = \frac{\mu B}{2\pi h}\frac{(p_{\mathrm{D}})_{\mathrm{M}}}{(\Delta p/q)_{\mathrm{M}}} \tag{6.16}$$

10. On the basis of the time matching point and the permeability determined in step 9, calculate the drainage radius r_{e}

$$r_{\mathrm{e}} = \sqrt{\frac{K}{\pi\varphi\mu C_{\mathrm{t}}}\left(\frac{t_{\mathrm{c}}}{t_{\mathrm{cDA}}}\right)_{\mathrm{M}}} \tag{6.17}$$

11. On the basis of r_{eD} determined in step 7, calculate the effective wellbore radius

$$r_{\mathrm{wa}} = \frac{r_{\mathrm{e}}}{r_{\mathrm{eD}}} \tag{6.18}$$

12. Calculate the skin factor

$$S = \ln\frac{r_{\mathrm{w}}}{r_{\mathrm{wa}}} \tag{6.19}$$

13. Calculate the OOIP by the volumetric method

$$N = \frac{\pi r_{\mathrm{e}}^2\varphi h S_{\mathrm{o}}}{B_{\mathrm{o}}} \tag{6.20}$$

6.2.2 *Type Curves Matching Procedure—Gas Well*

The procedure of NPI curve analysis for gas well is similar to that for an oil well, as given below.

1. Calculate the material balance pseudo-time

   Assuming the reserves G, calculate the material balance pseudo-time for each production data point.

$$t_{\text{ca}} = \frac{(\mu C_{\text{t}})_{\text{i}}}{q} \int_0^t \frac{q}{\mu(\bar{p})C_{\text{t}}(\bar{p})} dt = \frac{GC_{\text{ti}}}{q}\left(p_{p_{\text{i}}} - p_p\right) \tag{6.21}$$

   Please refer to the normalized pseudo-pressure p_{p} in Eq. (4.57), and the average reservoir pressure could be calculated according to Eq. (4.60).

2. Calculate the normalized rate

$$\frac{q}{\Delta p_{\text{p}}} = \frac{q}{p_{p_{\text{i}}} - p_{p_{\text{wf}}}} \tag{6.22}$$

3. Calculate the normalized pressure

$$\frac{\Delta p_{\text{p}}}{q} = \frac{p_{p_{\text{i}}} - p_{p_{\text{wf}}}}{q} \tag{6.23}$$

4. Calculate the normalized pressure integral

$$\left(\frac{\Delta p_{\text{p}}}{q}\right)_{\text{i}} = \frac{1}{t_{\text{ca}}} \int_0^{t_{\text{ca}}} \frac{\Delta p_{\text{p}}}{q} d\tau \tag{6.24}$$

   where subscript i represents integration.

5. Calculate the normalized pressure integral derivative

$$\left(\frac{\Delta p_{\text{p}}}{q}\right)_{\text{id}} = \frac{d\left(\frac{\Delta p_{\text{p}}}{q}\right)_{\text{i}}}{d \ln t_{\text{ca}}} = t_{\text{ca}} \frac{d\left(\frac{\Delta p_{\text{p}}}{q}\right)_{\text{i}}}{dt_{\text{ca}}} \tag{6.25}$$

   where subscript i represents integration, and d represents derivation.

6. In the Cartesian coordinate system, plot the $\frac{\Delta p_{\text{p}}}{q} \sim t_{\text{ca}}$ curve and regress the data points linearly, then apply the slope of the line to Eq. (4.65), calculate the value of G, repeat steps 1–6 and make iterative computations until G meets the permissible error.

$$G = \frac{1}{Slope \times C_{\text{ti}}} \tag{6.26}$$

7. Prepare the log–log plot of the normalized pressure, normalized pressure integral and normalized pressure integral derivative changing with material balance pseudo-time, respectively, in the same coordinate system. In other words, we need to plot the $\frac{\Delta p_{\mathrm{p}}}{q} \sim t_{\mathrm{ca}}$, $\left(\frac{\Delta p_{\mathrm{p}}}{q}\right)_{\mathrm{i}} \sim t_{\mathrm{ca}}$, and $\left(\frac{\Delta p_{\mathrm{p}}}{q}\right)_{\mathrm{id}} \sim t_{\mathrm{ca}}$ curves.
8. Choose any of the three curves or any combination thereof to match the actual points with the type curves (Figure 6.4), aiming to realize the best matching of each curve.
9. Record the dimensionless drainage radius r_{eD} according to the matching results.
10. Randomly choose a matching point, and record the actual matching point $\left(t_{\mathrm{ca}}, \frac{\Delta p_{\mathrm{p}}}{q}\right)_{\mathrm{M}}$ and the corresponding theoretical matching point $(t_{\mathrm{caDA}}, p_{\mathrm{D}})_{\mathrm{M}}$. If the reservoir thickness, total compressibility, and wellbore radius are known, then we can calculate the reservoir permeability, skin factor, drainage area, reserves, etc.
11. Calculate the permeability according to the rate matching point and Eq. (6.2), and we have

$$K = \frac{\mu B}{2\pi h} \frac{(p_{\mathrm{D}})_{\mathrm{M}}}{\left(\Delta p_{\mathrm{p}}/q\right)_{\mathrm{M}}} \tag{6.27}$$

12. On the basis of the time matching point and the permeability determined in step 11, and Eq. (6.3), calculate the drainage radius r_{e}

$$r_{\mathrm{e}} = \sqrt{\frac{K}{\pi \varphi \mu C_{\mathrm{t}}} \left(\frac{t_{\mathrm{ca}}}{t_{\mathrm{caDA}}}\right)_{\mathrm{M}}} \tag{6.28}$$

13. On the basis of r_{eD} determined in step 9, calculate the effective wellbore radius

$$r_{\mathrm{wa}} = \frac{r_{\mathrm{e}}}{r_{\mathrm{eD}}} \tag{6.29}$$

14. Calculate the skin factor

$$S = \ln \frac{r_{\mathrm{w}}}{r_{\mathrm{wa}}} \tag{6.30}$$

15. Calculate the original gas in place (OGIP) by the volumetric method

$$G = \frac{\pi r_{\mathrm{e}}^2 \varphi h S_{\mathrm{g}}}{B_{\mathrm{gi}}} \tag{6.31}$$

6.3 Case Study of NPI Decline Curves Analysis

We still use the case in section 4.4 as the example, with the main parameters as follows. A gas well centered in a closed circular reservoir is producing at a variable rate and a variable BHFP. The initial reservoir pressure is 30 MPa, the reservoir temperature is 80 °C, the reservoir thickness is 10 m, the reservoir porosity is 0.1, the distance between the well and boundary is 500 m, the gas reserve is 2.0×10^8 m^3. The relative density of the gas phase is 0.6, the pseudo critical temperature is 195.697 K, the pseudo-critical pressure is 4.66875 MPa, the reservoir permeability estimation from well test is 2.0 mD, the skin factor is –5.52, the production time is 800 days, the cumulative gas production is 11813×10^4 m^3, the fluid viscosity is 0.0228 mPa·s, the rock compressibility is 4.35113×10^{-4} MPa^{-1}, the gas compressibility under the initial condition is 0.0246 MPa^{-1}, the gas formation volume factor under the initial condition is 0.0039497. The production data are listed in Table 4.1.

Ignoring the process of iterative computations for G, and assuming $G = 2.0 \times 10^8$ m^3 immediately, the analysis process is given as follows.

1–5. Calculate the material balance pseudo-time and the relative normalized rate parameters. Calculate the average reservoir pressure, normalized pseudo-pressure change, normalized rate, normalized pressure, normalized pressure integral, and the normalized pressure integral derivative according to Eq. (4.60), (4.122), (6.22), (6.23), (6.24), and (6.25), whose results are shown in Table 6.1.

6. Plot the log–log relationship of $\frac{\Delta p_p}{q} \sim t_{ca}$, $\left(\frac{\Delta p_p}{q}\right)_i \sim t_{ca}$, and $\left(\frac{\Delta p_p}{q}\right)_{id} \sim t_{ca}$ curves, as shown in Figure 6.5.

7. Match type curve (Figure 6.4) with the actual production history curve, as shown in Figure 6.6.

8. Record the dimensionless drainage radius $r_{eD} = 20$.

9. Choose the actual matching point $(290, 0.00926)_M$ and the corresponding theoretical matching point $(1, 1)_M$ to simplify the calculation.

10. Calculate permeability according to the rate matching point and Eq. (5.9). The following expression is obtained under the SI unit system

$$\begin{aligned} K &= \frac{\mu B}{2\pi h}\frac{(p_D)_M}{(\Delta p_p/q)_M} = \frac{1.866\times(p_D)_M}{(\Delta p_p/q)_M}\frac{\mu B}{h} \\ &= \frac{1.866\times 1\times 0.0228\times 0.0039497}{\frac{0.00926}{1000}\times 10} = 1.81\,\text{mD} \end{aligned} \tag{6.32}$$

11. On the basis of the time matching point and the permeability determined in step 10, and Eq. (6.28), calculate the drainage radius r_e

$$\begin{aligned} r_e &= \sqrt{\frac{3.6\times 10^{-3}K}{\pi\varphi\mu C_t}\left(\frac{t_{ca}}{t_{caDA}}\right)_M} \\ &= \sqrt{\frac{3.6\times 10^{-3}\times 1.81}{3.14159\times 0.1\times 0.0228\times 0.0246}\left(\frac{290\times 24}{1}\right)_M} = 507.3\,\text{m} \end{aligned} \tag{6.33}$$

Table 6.1 **Material balance pseudo-time and the relative normalized rate parameters of well B-NPI**

t d	q 10^3 m^3/d	p_{wf} MPa	p_{avg} MPa	$\Delta p_p(p_{avg})$ MPa	t_{ca} d	$q/\Delta p_p$ 10^3 $m^3/d/MPa$	$\Delta p_p/q$ $MPa/10^3$ m^3/d	$(\Delta p_p/q)_i$ $MPa/10^3$ m^3	$(\Delta p_p/q)_{id}$
1	199.8	27.72	30.00	0.01	0.12	88.41	0.0113	0.0082	
2	199.7	27.23	29.99	0.01	0.14	72.85	0.0137	0.0086	0.0039
3	199.5	26.92	29.95	0.05	1.15	65.69	0.0152	0.0138	0.0007
4	199.4	26.70	29.91	0.09	2.16	61.30	0.0163	0.0147	0.0011
5	199.2	26.53	29.87	0.13	3.17	58.23	0.0172	0.0154	0.0014
6	199.0	26.38	29.83	0.17	4.17	55.91	0.0179	0.0159	0.0016
7	198.9	26.26	29.79	0.21	5.18	54.08	0.0185	0.0163	0.0019
8	198.7	26.16	29.75	0.25	6.19	52.58	0.0190	0.0167	0.0020
9	198.6	26.06	29.71	0.29	7.19	51.32	0.0195	0.0171	0.0022
10	198.4	25.98	29.67	0.33	8.20	50.24	0.0199	0.0174	0.0023
20	196.8	25.41	29.27	0.73	18.18	43.82	0.0228	0.0196	0.0031
30	195.3	25.01	28.88	1.11	28.06	40.04	0.0250	0.0211	0.0037
40	193.7	24.64	28.50	1.49	37.84	37.08	0.0270	0.0224	0.0045
50	192.2	24.29	28.13	1.86	47.54	34.58	0.0289	0.0235	0.0053
60	190.6	23.94	27.77	2.21	57.16	32.41	0.0309	0.0246	0.0062
70	189.1	23.60	27.41	2.56	66.70	30.52	0.0328	0.0256	0.0070
80	187.6	23.27	27.06	2.90	76.17	28.84	0.0347	0.0266	0.0079
90	186.1	22.94	26.72	3.24	85.57	27.35	0.0366	0.0276	0.0088
100	184.6	22.62	26.38	3.56	94.92	26.01	0.0384	0.0286	0.0098
110	183.2	22.31	26.05	3.88	104.21	24.80	0.0403	0.0296	0.0107
120	181.7	22.00	25.73	4.19	113.46	23.71	0.0422	0.0305	0.0116
130	180.2	21.69	25.41	4.49	122.66	22.71	0.0440	0.0315	0.0125
140	178.8	21.39	25.00	4.89	134.53	21.80	0.0459	0.0327	0.0131
150	177.4	21.10	24.90	4.98	138.22	20.96	0.0477	0.0331	0.0145
160	176.0	20.81	24.61	5.26	146.93	20.19	0.0495	0.0340	0.0154
170	174.6	20.53	24.32	5.52	155.65	19.48	0.0513	0.0349	0.0163

180	173.2	20.25	24.04	5.79	164.39	18.81	0.0532	0.0358	0.0172
190	171.8	19.97	23.76	6.05	173.13	18.19	0.0550	0.0368	0.0181
200	170.4	19.70	23.49	6.30	181.89	17.62	0.0568	0.0377	0.0190
210	169.1	19.43	23.22	6.55	190.66	17.08	0.0586	0.0386	0.0199
220	167.7	19.17	22.95	6.80	199.44	16.57	0.0604	0.0395	0.0208
230	166.4	18.91	22.68	7.04	208.22	16.09	0.0621	0.0404	0.0216
240	165.1	18.65	22.42	7.28	217.02	15.64	0.0639	0.0413	0.0225
250	163.7	18.40	22.16	7.52	225.83	15.22	0.0657	0.0423	0.0234
260	162.4	18.15	21.90	7.75	234.65	14.82	0.0675	0.0432	0.0242
270	161.1	17.91	21.65	7.97	243.48	14.44	0.0693	0.0441	0.0251
280	159.9	17.66	21.40	8.20	252.31	14.08	0.0710	0.0450	0.0259
290	158.6	17.42	21.15	8.42	261.16	13.74	0.0728	0.0459	0.0268
300	157.3	17.19	20.90	8.63	270.02	13.41	0.0746	0.0468	0.0277
350	151.2	16.05	19.72	9.66	314.47	11.99	0.0834	0.0514	0.0320
400	145.2	14.97	18.59	10.60	359.18	10.84	0.0923	0.0559	0.0363
450	139.5	13.94	17.53	11.46	404.18	9.88	0.1012	0.0605	0.0406
500	134.1	12.95	16.52	12.25	449.49	9.08	0.1102	0.0650	0.0451
550	128.8	12.00	15.57	12.96	495.17	8.38	0.1193	0.0696	0.0496
600	123.8	11.09	14.67	13.61	541.25	7.78	0.1285	0.0742	0.0542
650	118.9	10.20	13.81	14.21	587.79	7.25	0.1379	0.0789	0.0589
700	114.2	9.32	13.00	14.74	634.84	6.78	0.1474	0.0836	0.0637
750	109.8	8.46	12.23	15.23	682.50	6.36	0.1571	0.0884	0.0686
800	105.5	7.60	11.50	15.66	730.81	5.99	0.1670	0.0933	0.0737

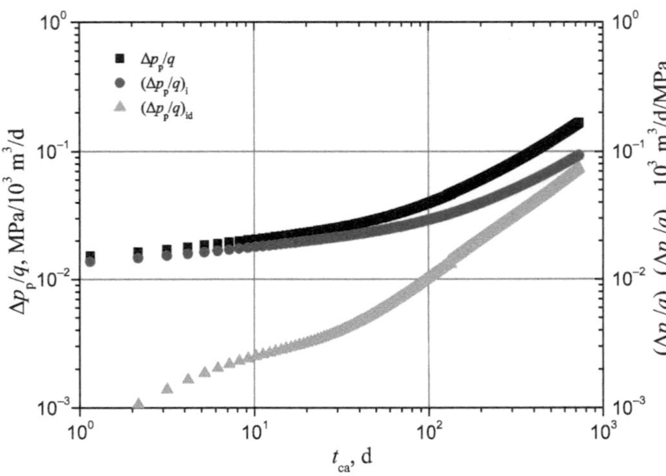

Figure 6.5 NPI pressure functions for well B

According to Eq. (6.32) and (6.33), the following expression is obtained under the SI unit system

$$
\begin{aligned}
r_{\mathrm{e}} &= \sqrt{\frac{3.6\times10^{-3}B}{\pi\varphi hC_{\mathrm{t}}}\left(\frac{t_{\mathrm{ca}}}{t_{\mathrm{caDA}}}\right)_{\mathrm{M}}\frac{1.866\times\left(p_{\mathrm{D}}\right)_{\mathrm{M}}}{\left(\Delta p_{\mathrm{p}}/q\right)_{\mathrm{M}}}\frac{\mu B}{h}} \\
&= \sqrt{\frac{2.1383\times10^{-3}B}{\varphi hC_{\mathrm{t}}}\left(\frac{t_{\mathrm{ca}}}{t_{\mathrm{caDA}}}\right)_{\mathrm{M}}\left(\frac{p_{\mathrm{D}}}{\Delta p_{\mathrm{p}}/q}\right)_{\mathrm{M}}}
\end{aligned}
\qquad (6.34)
$$

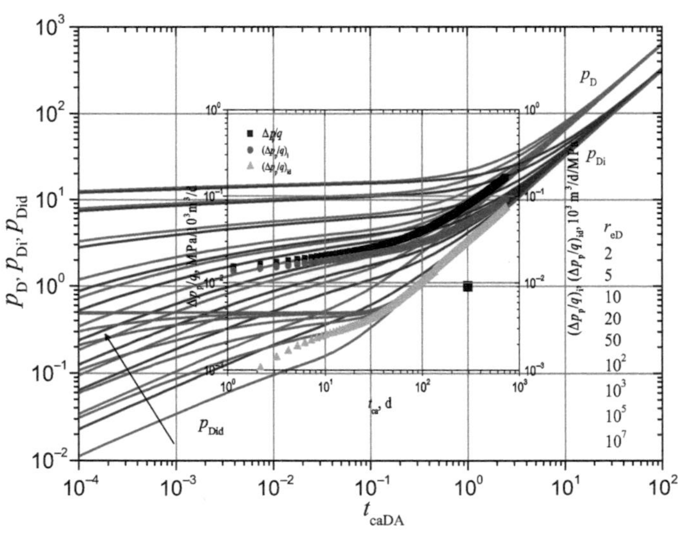

Figure 6.6 Matches of production data for well B-NPI

$$r_e = \sqrt{\frac{2.1383\times10^{-3}\times0.0039497}{0.1\times10\times0.0246}\left(\frac{290\times24}{1}\right)_M\left(\frac{1}{\frac{0.00926}{1000}}\right)_M} = 508.0\,\mathrm{m} \tag{6.35}$$

12. Calculate the effective wellbore radius

$$r_{wa} = \frac{r_e}{r_{eD}} = \frac{508}{20} = 25.4\mathrm{m} \tag{6.36}$$

13. Calculate the skin factor

$$S = \ln\frac{r_w}{r_{wa}} = \ln\frac{0.1}{25.4} = -5.54 \tag{6.37}$$

14. Calculate the OGIP by volumetric method

$$G = \frac{\pi r_e^2 \varphi h S_g}{B_{gi}} = \frac{3.14\times508^2\times0.1\times10\times1}{0.0039497} = 2.05\times10^8\mathrm{m}^3 \tag{6.38}$$

Substituting Eq. (6.34) into Eq. (6.38), we have

$$\begin{aligned} G &= \frac{\pi r_e^2 \varphi h S_g}{B_{gi}} = \frac{\pi\varphi h S_g}{B_{gi}}\frac{2.1383\times10^{-3}B}{\varphi h C_t}\left(\frac{t_{ca}}{t_{caDA}}\right)_M\left(\frac{p_D}{\Delta p_p/q}\right)_M \\ &= \frac{6.7177\times10^{-3}}{C_t}\left(\frac{t_{ca}}{t_{caDA}}\right)_M\left(\frac{p_D}{\Delta p_p/q}\right)_M S_g \end{aligned} \tag{6.39}$$

We have

$$G = \frac{6.7177\times10^{-3}}{0.0246}\left(\frac{290\times24}{1}\right)_M\left(\frac{1}{\frac{0.00926}{1000}}\right)_M\times1.0 = 2.05\times10^8\mathrm{m}^3 \tag{6.40}$$

Transient Decline Curves Analysis

7

Blasingame, Agarwal-Gardner, and normalized pressure integral (NPI) methods are all based on the boundary-dominated flow, and they can use the dimensionless time and flow rate relationship (transient method) to match the data in the period of transient flow, so as to reduce the ambiguity in the interpretation result. In this chapter, we mainly introduce the theoretical basis of the transient decline curves analysis method for data obtained in the period of transient flow in a closed circular homogeneous reservoir, as well as the type curve plotting and the curve matching method. Furthermore, we also present several case studies using the previous method.

7.1 Transient Decline Curve Plotting

7.1.1 Pressure Distribution in a Closed Circular Reservoir: Constant Rate Production

Assume that there is a vertical well centered in a closed circular reservoir producing at a constant rate of q. The reservoir radius is r_e. The bottom hole flowing pressure (BHFP), reservoir thickness, initial reservoir pressure, wellbore radius, reservoir porosity, total compressibility, permeability, fluid viscosity, and formation volume factor are represented by p_{wf}, h, p_i, r_w, φ, C_t, K, μ, and B, respectively. Besides, the skin effect is not considered. Under the basic SI unit system, the following expression can be obtained from Eq. (4.17).

$$\bar{p}_D = \frac{1}{s\sqrt{s}}\left[\frac{\dfrac{K_1\left(r_{eD}\sqrt{s}\right)}{I_1\left(r_{eD}\sqrt{s}\right)} + \dfrac{K_0\left(\sqrt{s}\right)}{I_0\left(\sqrt{s}\right)}}{\dfrac{K_1\left(\sqrt{s}\right)}{I_0\left(\sqrt{s}\right)} - \dfrac{I_1\left(\sqrt{s}\right)}{I_0\left(\sqrt{s}\right)}\dfrac{K_1\left(r_{eD}\sqrt{s}\right)}{I_1\left(r_{eD}\sqrt{s}\right)}}\right] \tag{7.1}$$

The dimensionless variables are defined as follows

$$p_D = \frac{2\pi Kh\left(p_i - p_{wf}\right)}{q\mu B} \tag{7.2}$$

$$t_D = \frac{Kt}{\varphi\mu C_t r_{wa}^2} \tag{7.3}$$

Advanced Production Decline Analysis and Application

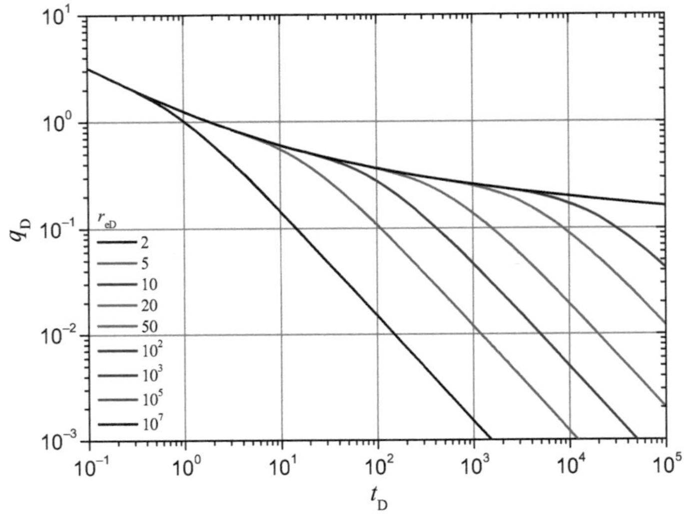

Figure 7.1 Transient dimensionless decline rate curves

$$r_{eD} = \frac{r_e}{r_{wa}} \tag{7.4}$$

$$q_D = \frac{q\mu B}{2\pi Kh\left(p_i - p_{wf}\right)} \tag{7.5}$$

7.1.2 Dimensionless Decline Flow Rate Curves

The inverse of the constant rate production pressure solution p_D and dimensionless time t_D are plotted in a coordinate system as shown in Figure 7.1. The curves are superposed in the period of the early transient flow, and drift apart as r_{eD} increases. This method can reduce the ambiguity of interpretation result because of the limited early data or low-permeability data (prior to the boundary-dominated flow period).

7.1.3 Dimensionless Decline Inverse Pressure Integral Curves

The dimensionless pressure integral function is defined as

$$p_{Di} = \frac{1}{t_D}\int_0^{t_{DA}} p_D \, dt_D \tag{7.6}$$

Eq. (7.6) can be expressed as

$$p_{Di} = \frac{1}{t_D} L^{-1}\left[\frac{\overline{p}_D}{s}\right] \tag{7.7}$$

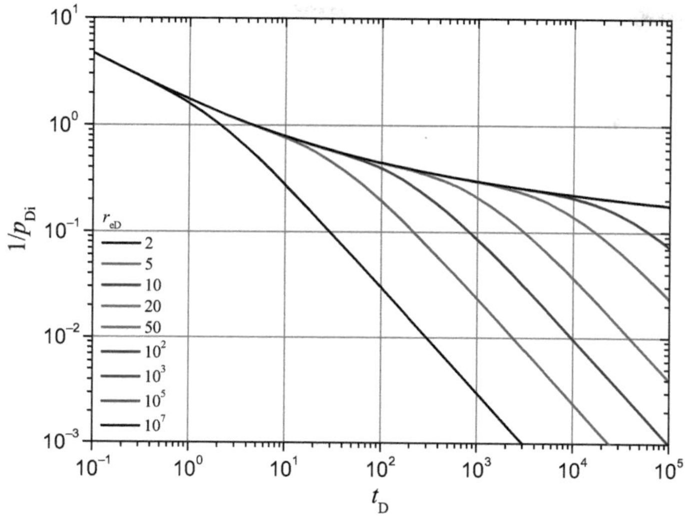

Figure 7.2 Dimensionless decline inverse pressure integral curves

The inverse of the constant rate production pressure integral solution p_{Di} and dimensionless time t_D are plotted in a coordinate system as shown in Figure 7.2.

7.1.4 Dimensionless Decline Inverse Pressure Integral Derivative Curves

Differentiating Eq. (7.6) yields

$$dp_{Di} = \frac{t_D p_D dt_D - \int_0^{t_D} p_D \, dt_D}{t_D^2} \tag{7.8}$$

The dimensionless pressure integral derivative is defined as

$$p_{Did} = \frac{dp_{Di}}{d\ln t_D} = t_D \frac{dp_{Di}}{dt_D} \tag{7.9}$$

Substituting Eq. (7.8) into Eq. (7.9) yields

$$\frac{1}{p_{Did}} = \frac{1}{p_D - p_{Di}} \tag{7.10}$$

The constant rate production pressure integral derivative reciprocal $1/p_{Did}$ and the dimensionless time t_D are plotted in a coordinate system as shown in Figure 7.3. The curve is similar to the pressure derivative curve in well test analysis.

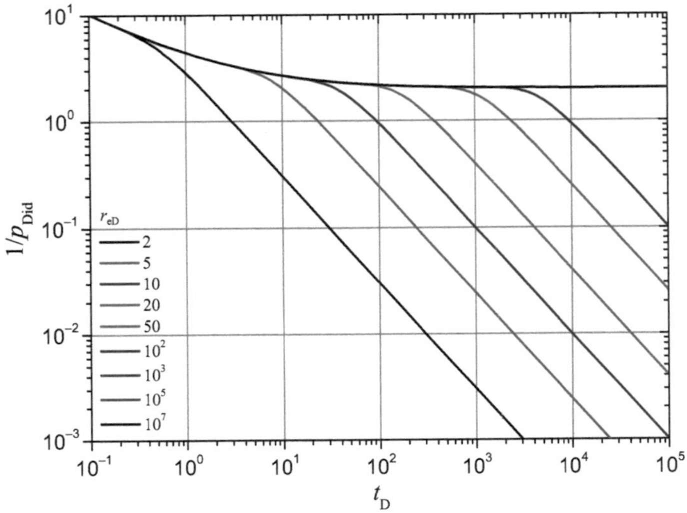

Figure 7.3 Transient dimensionless decline inverse pressure integral derivative curves

7.1.5 Transient Decline Curves

The transient decline curves shown in Figure 7.4 are obtained by superposing the curves of Figures 7.1, 7.2, and 7.3. The introduction of the material balance pseudo-time enables this method to address the issues of variable rate and variable flowing pressure. The Transient method can use the daily production data (time, flow rate, flowing pressure) obtained in the period of early transient flow to estimate the parameters such as reservoir permeability, original oil in place (OOIP), skin factor, and drainage area.

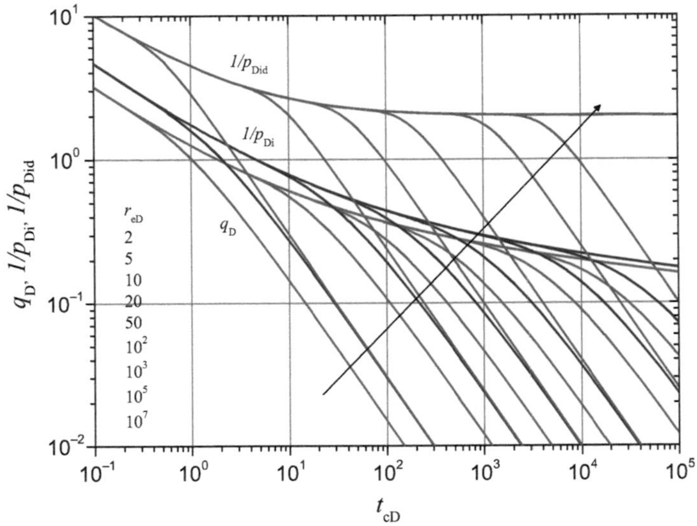

Figure 7.4 Transient decline curve

7.2 Decline Curves Analysis Using Transient Style Type Curves

7.2.1 Type Curves Matching Procedure—Oil Well

The Transient decline curve matching analysis procedure of oil well is given below.

1. Calculate the material balance time

$$t_c = \frac{N_p}{q} \tag{7.11}$$

2. Calculate the normalized rate

$$\frac{q}{\Delta p} = \frac{q}{p_i - p_{wf}} \tag{7.12}$$

3. Calculate the normalized pressure integral

$$\left(\frac{\Delta p}{q}\right)_i = \frac{1}{t_c}\int_0^{t_c} \frac{p_i - p_{wf}}{q}\, d\tau \tag{7.13}$$

   where subscript i represents integration.

4. Calculate the normalized pressure integral derivative

$$\left(\frac{\Delta p}{q}\right)_{id} = \frac{d\left(\frac{\Delta p}{q}\right)_i}{d \ln t_c} = t_c \frac{d\left(\frac{\Delta p}{q}\right)_i}{dt_c} \tag{7.14}$$

   where subscript i represents integration and d represents derivation.

5. Calculate the inverse normalized pressure

$$\frac{1}{p_{di}} = \frac{1}{\left(\frac{\Delta p}{q}\right)_i} \tag{7.15}$$

6. Calculate the inverse normalized pressure integral derivative

$$\frac{1}{p_{did}} = \frac{1}{\left(\frac{\Delta p}{q}\right)_{id}} \tag{7.16}$$

   where subscript i represents integration, and d represents derivation.

7. In the log–log coordinate system, plot the normalized rate, inverse normalized pressure derivative, and inverse normalized pressure integral derivative curves versus material balance time, namely $\frac{q}{\Delta p} \sim t_c$, $\frac{1}{p_{di}} \sim t_c$, and $\frac{1}{p_{did}} \sim t_c$, respectively.
8. Choose any of the three curves or any combination thereof to match the actual points with the type curves (Figure 7.4), aiming to acquire the best matching of each curve.
9. According to the matching results, record the dimensionless drainage radius r_{eD}.
10. Randomly select a matching point, and record the actual matching point $(t_c, q/\Delta p)_M$ and the corresponding theoretical matching point $(t_{cD}, q_D)_M$. If the reservoir thickness, total compressibility, and wellbore radius are known, then we can calculate the reservoir permeability, skin factor, drainage area, and reserves, etc.
11. According to the rate matching point and Eq. (7.5), calculate the permeability, and we have

$$K = \frac{(q/\Delta p)_M}{(q_D)_M} \frac{\mu B}{2\pi h} \tag{7.17}$$

12. On the basis of the time matching point and the permeability determined in step 11, calculate the effective wellbore radius r_{wa}

$$r_{wa} = \sqrt{\frac{K}{\varphi \mu C_t} \left(\frac{t_c}{t_{cD}} \right)_M} \tag{7.18}$$

13. On the basis of r_{eD} determined in step 9, calculate the drainage radius r_e

$$r_e = r_{wa} r_{eD} \tag{7.19}$$

14. Calculate the skin factor S

$$S = \ln \frac{r_w}{r_{wa}} \tag{7.20}$$

15. Calculate the OOIP by volumetric method

$$N = \frac{\pi r_e^2 \varphi h S_o}{B_o} \tag{7.21}$$

7.2.2 Type Curves Matching Procedure—Gas Well

The procedure of Transient curve analysis for gas well is similar to that for oil well, which is given as follows,

1. Calculate the material balance pseudo-time

   Assuming the reserve is G, calculate the material balance pseudo-time for each production data point according to Eq. (7.22)

$$t_{ca} = \frac{(\mu C_t)_i}{q} \int_0^t \frac{q}{\mu(\bar{p})C_t(\bar{p})} dt = \frac{GC_{ti}}{q}\left(p_{p_i} - p_p\right) \tag{7.22}$$

   Please refer to Eq. (4.57) for the normalized pseudo-pressure p_p, and the average reservoir pressure could be calculated according to Eq. (4.60).
2. Calculate the normalized rate

$$\frac{q}{\Delta p_p} = \frac{q}{p_{p_i} - p_{p_{wf}}} \tag{7.23}$$

3. Calculate the normalized pressure integral

$$\left(\frac{\Delta p_p}{q}\right)_i = \frac{1}{t_{ca}} \int_0^{t_{ca}} \frac{\Delta p_p}{q} d\tau \tag{7.24}$$

   where subscript i represents integration.
4. Calculate the normalized pressure integral derivative

$$\left(\frac{\Delta p_p}{q}\right)_{id} = \frac{d\left(\frac{\Delta p_p}{q}\right)_i}{d \ln t_{ca}} = t_{ca} \frac{d\left(\frac{\Delta p_p}{q}\right)_i}{dt_{ca}} \tag{7.25}$$

   where subscript i represents integration, and d represents derivation.
5. Calculate the inverse normalized pressure

$$\frac{1}{p_{di}} = \frac{1}{\left(\frac{\Delta p_p}{q}\right)_i} \tag{7.26}$$

6. Calculate the inverse normalized pressure integral derivative

$$\frac{1}{p_{did}} = \frac{1}{\left(\frac{\Delta p_p}{q}\right)_{id}} \tag{7.27}$$

7. In the Cartesian coordinate system, plot $\frac{\Delta p_p}{q} \sim t_{ca}$ curve and regress the data points linearly, apply the slope of the line to Eq. (4.65), calculate the value of G, repeat steps 1–7, and make iterative computations until G meets the permissible error.

$$G = \frac{1}{\text{Slope} \times C_{ti}} \tag{7.28}$$

8. Prepare log–log curves of the normalized rate, inverse normalized pressure derivative, and inverse normalized pressure integral derivative changing with material balance pseudo-time, respectively, in the same coordinate system. In other words, we need to plot $\frac{q}{\Delta p_p} \sim t_{ca}$, $\frac{1}{p_{di}} \sim t_{ca}$, and $\frac{1}{p_{did}} \sim t_{ca}$ curves.
9. Choose any of the three curves or any combination thereof to match the actual points with the type curves (Figure 7.4), aiming to obtain the best matching of each curve.
10. Record the dimensionless drainage radius r_{eD} according to the matching results.
11. Randomly choose a matching point, and record the actual matching point $(t_{ca}, q/\Delta p_p)_M$ and the corresponding theoretical matching point $(t_{caD}, q_D)_M$. If the reservoir thickness, total compressibility, and wellbore radius are known, then we can calculate the reservoir permeability, skin factor, drainage area, and reserves, etc.
12. Calculate the permeability according to the rate matching point and Eq. (7.5), we have

$$K = \frac{(q/\Delta p_p)_M}{(q_D)_M} \frac{\mu B}{2\pi h} \tag{7.29}$$

13. On the basis of the time matching point and the permeability determined in step 12, together with Eq. (7.3), calculate the effective wellbore radius r_{wa}

$$r_{wa} = \sqrt{\frac{K}{\varphi \mu C_t} \left(\frac{t_{ca}}{t_{caD}} \right)_M} \tag{7.30}$$

14. On the basis of r_{eD} determined in step 10, calculate the drainage radius r_e

$$r_e = r_{wa} r_{eD} \tag{7.31}$$

15. Calculate the skin factor

$$S = \ln \frac{r_w}{r_{wa}} \tag{7.32}$$

16. Calculate the original gas in place (OGIP) by the volumetric method

$$G = \frac{\pi r_e^2 \varphi h S_g}{B_{gi}} \tag{7.33}$$

7.3 Case Study of Transient Decline Curves Analysis

We still use the case in Section 4.4 as the example, where the main parameters are given as follows. A gas well centered in a closed circular reservoir is producing at a variable rate and a variable BHFP. The initial reservoir pressure is 30 MPa, the reservoir temperature is 80 °C, the reservoir thickness is 10 m, the reservoir porosity is 0.1, the distance between the well and boundary is 500 m, and the gas reserve is 2.0×10^8 m^3. The relative density of gas phase is 0.6, the pseudo-critical temperature is 195.697 K, the pseudo-critical pressure is 4.66875 MPa, the reservoir permeability estimation from well test is 2.0 mD, the skin factor is –5.52, the production time is 800 days, the cumulative gas production is 11813×10^4 m^3, the fluid viscosity is 0.0228 mPa·s, the rock compressibility is 4.35113×10^{-4} MPa^{-1}, the gas compressibility under initial conditions is 0.0246 MPa^{-1}, and the gas formation volume factor under initial conditions is 0.0039497. The production data are listed in Table 4.1.

Ignore the process of iterative computations for G, and assume $G = 2.0 \times 10^8$ m^3. The analysis process is given as follows.

1–5. Calculate the material balance pseudo-time and the relative normalized rate parameters.

Calculate the average reservoir pressure, normalized pseudo-pressure change, normalized rate, inverse normalized pressure, and inverse normalized pressure integral derivative according to Eq. (4.60), (4.122), (7.23), (7.26), and (7.27), and the results are shown in Table 7.1.

6. Plot the log–log curves of $\frac{q}{\Delta p_p} \sim t_{ca}$, $\frac{1}{p_{pi}} \sim t_{ca}$, and $\frac{1}{p_{pid}} \sim t_{ca}$ curves, as shown in Figure 7.5.

7. Match type curve (Figure 7.4) with the actual production history curve, as shown in Figure 7.6.

8. Record the dimensionless drainage radius $r_{eD} = 20$.

9. Choose the actual matching point $(20, 11)_M$ and the corresponding theoretical matching point $(100, 0.1)_M$ to simplify the calculation.

10. Calculate the permeability according to the rate matching point and Eq. (7.5). The following expression is obtained under the SI unit system

$$\begin{aligned} K &= \frac{(q/\Delta p_p)_M}{(q_D)_M} \frac{\mu B}{2\pi h} = \frac{1.866 \times (q/\Delta p_p)_M}{(q_D)_M} \frac{\mu B}{h} \\ &= \frac{1.866 \times (11 \times 1000) \times 0.0228 \times 0.0039497}{0.1 \times 10} \\ &= 1.85\,\text{mD} \end{aligned} \tag{7.34}$$

Table 7.1 Material balance pseudo-time and the relative normalized rate parameters of well B

t	q	p_{wf}	p_{avg}	$\Delta p_p(p_{avg})$	$\Delta p_p(p_{wf})$	t_{ca}	$q/\Delta p_p$	$1/p_{pi}$	$1/p_{pid}$
d	10^3 m^3/d	MPa	MPa	MPa	MPa	d	10^3 $m^3/d/$ MPa	$MPa/10^3$ m^3/d	
1	199.8	27.72	30.00	0.01	2.26	0.12	88.41	122.62	
2	199.7	27.23	29.99	0.01	2.74	0.14	72.85	115.93	256.88
3	199.5	26.92	29.95	0.05	3.04	1.15	65.69	72.61	1423.95
4	199.4	26.70	29.91	0.09	3.25	2.16	61.30	68.00	941.90
5	199.2	26.53	29.87	0.13	3.42	3.17	58.23	65.13	719.67
6	199.0	26.38	29.83	0.17	3.56	4.17	55.91	62.98	606.07
7	198.9	26.26	29.79	0.21	3.68	5.18	54.08	61.24	537.72
8	198.7	26.16	29.75	0.25	3.78	6.19	52.58	59.80	492.12
9	198.6	26.06	29.71	0.29	3.87	7.19	51.32	58.56	459.42
10	198.4	25.98	29.67	0.33	3.95	8.20	50.24	57.48	434.83
20	196.8	25.41	29.27	0.73	4.49	18.18	43.82	50.91	326.79
30	195.3	25.01	28.88	1.11	4.88	28.06	40.04	47.28	268.76
40	193.7	24.64	28.50	1.49	5.22	37.84	37.08	44.65	223.53
50	192.2	24.29	28.13	1.86	5.56	47.54	34.58	42.50	188.86
60	190.6	23.94	27.77	2.21	5.88	57.16	32.41	40.65	162.42
70	189.1	23.60	27.41	2.56	6.20	66.70	30.52	39.02	141.98
80	187.6	23.27	27.06	2.90	6.50	76.17	28.84	37.54	125.93
90	186.1	22.94	26.72	3.24	6.80	85.57	27.35	36.20	113.04
100	184.6	22.62	26.38	3.56	7.10	94.92	26.01	34.97	102.52
150	177.4	21.10	24.90	4.98	8.46	138.22	20.96	30.24	68.77
200	170.4	19.70	23.49	6.30	9.67	181.89	17.62	26.54	52.63
250	163.7	18.40	22.16	7.52	10.76	225.83	15.22	23.66	42.81
300	157.3	17.19	20.90	8.63	11.73	270.02	13.41	21.36	36.16
350	151.2	16.05	19.72	9.66	12.61	314.47	11.99	19.47	31.29
400	145.2	14.97	18.59	10.60	13.40	359.18	10.84	17.88	27.57
450	139.5	13.94	17.53	11.46	14.12	404.18	9.88	16.54	24.60
550	128.8	12.00	15.57	12.96	15.36	495.17	8.38	14.37	20.17
600	123.8	11.09	14.67	13.61	15.90	541.25	7.78	13.47	18.46
650	118.9	10.20	13.81	14.21	16.39	587.79	7.25	12.68	16.98
700	114.2	9.32	13.00	14.74	16.84	634.84	6.78	11.96	15.70
750	109.8	8.46	12.23	15.23	17.25	682.50	6.36	11.31	14.57
800	105.5	7.60	11.50	15.66	17.62	730.81	5.99	10.72	13.57

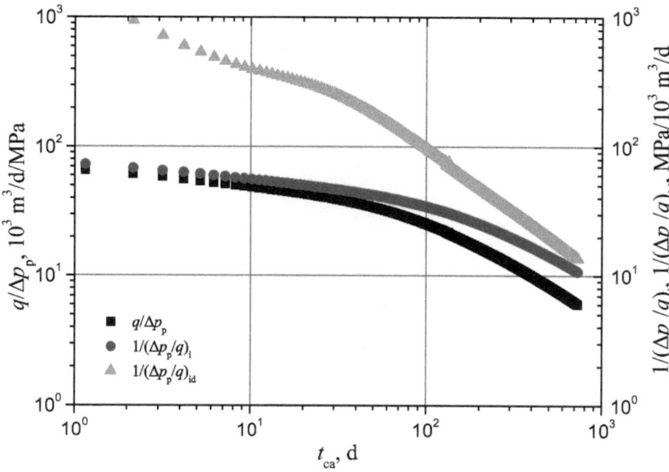

Figure 7.5 Transient rate functions for well B

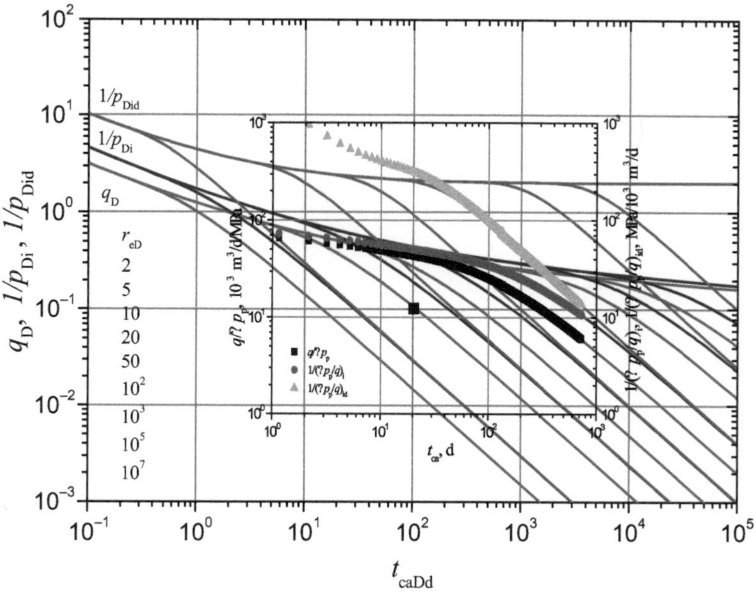

Figure 7.6 Matches of production data for well B-Transient

11. On the basis of the time matching point and the permeability determined in step 10, together with Eq. (7.30), calculate the effective wellbore radius

$$r_{\mathrm{wa}} = \sqrt{\frac{K}{\varphi\mu C_{\mathrm{t}}}\left(\frac{t_{\mathrm{ca}}}{t_{\mathrm{caD}}}\right)_{\mathrm{M}}} = \sqrt{\frac{3.6\times10^{-3}K}{\varphi\mu C_{\mathrm{t}}}\left(\frac{t_{\mathrm{ca}}}{t_{\mathrm{caD}}}\right)_{\mathrm{M}}} = \sqrt{\frac{3.6\times10^{-3}\times1.85}{0.1\times0.0228\times0.0246}\left(\frac{20\times24}{100}\right)_{\mathrm{M}}} = 23.9\,\mathrm{m} \tag{7.35}$$

According to Eq. (7.34) and Eq. (7.35), the following expression is obtained under the SI unit system

$$r_{\mathrm{wa}} = \sqrt{\frac{3.6\times10^{-3}B}{\pi\varphi hC_{\mathrm{t}}}\left(\frac{t_{\mathrm{ca}}}{t_{\mathrm{caD}}}\right)_{\mathrm{M}}\frac{1.866\times\left(q/\Delta p_{\mathrm{p}}\right)_{\mathrm{M}}}{\left(q_{\mathrm{D}}\right)_{\mathrm{M}}}} = \sqrt{\frac{6.7177\times10^{-3}B}{\varphi hC_{\mathrm{t}}}\left(\frac{t_{\mathrm{ca}}}{t_{\mathrm{caD}}}\right)_{\mathrm{M}}\left(\frac{q/\Delta p_{\mathrm{p}}}{q_{\mathrm{D}}}\right)_{\mathrm{M}}} \tag{7.36}$$

$$r_{\mathrm{wa}} = \sqrt{\frac{6.7177\times10^{-3}\times0.0039497}{0.1\times10\times0.0246}\left(\frac{20\times24}{100}\right)_{\mathrm{M}}\left(\frac{11\times1000}{0.1}\right)_{\mathrm{M}}} = 23.9\,\mathrm{m} \tag{7.37}$$

12. Calculate the drainage radius

$$r_{\mathrm{e}} = r_{\mathrm{wa}}r_{\mathrm{eD}} = 23.9\times20 = 478.0\,\mathrm{m} \tag{7.38}$$

13. Calculate the skin factor

$$S = \ln\frac{r_{\mathrm{w}}}{r_{\mathrm{wa}}} = \ln\frac{0.1}{23.9} = -5.48 \tag{7.39}$$

14. Calculate the OGIP by the volumetric method

$$G = \frac{\pi r_{\mathrm{e}}^2\varphi hS_{\mathrm{g}}}{B_{\mathrm{gi}}} = \frac{3.14\times478^2\times0.1\times10\times1}{0.0039497} = 1.82\times10^8\,\mathrm{m}^3 \tag{7.40}$$

Substituting Eq. (7.36) into Eq. (7.40), we have

$$G = \frac{\pi r_e^2 \varphi h S_g}{B_{gi}} = \frac{\pi \varphi h S_g}{B_{gi}} \frac{6.7177 \times 10^{-3} r_{eD}^2 B_{gi}}{\varphi h C_t} \left(\frac{t_{ca}}{t_{caD}} \right)_M \left(\frac{q/\Delta p_p}{q_D} \right)_M$$
$$= \frac{2.1104 \times 10^{-2}}{C_t} \left(\frac{t_{ca}}{t_{caD}} \right)_M \left(\frac{q/\Delta p_p}{q_D} \right)_M r_{eD}^2 S_g \tag{7.41}$$

We have

$$G = \frac{2.1104 \times 10^{-2}}{0.0246} \left(\frac{20 \times 24}{100} \right)_M \left(\frac{11 \times 1000}{0.1} \right)_M \times 1.0 \times 20^2$$
$$= 1.82 \times 10^8 \text{ m}^3 \tag{7.42}$$

Decline Curves Analysis of Long Linear Flow

8

The term *unconventional* describes production of reservoirs with very low permeability, such as due to shale or very tight sands. Linear flow is the most well-known flow regime in such reservoirs. It may last several years and is often the only flow regime observed in analysis. In this chapter, we mainly introduce the theoretical basis for decline curves analysis of long linear flow of a vertically fractured well in the rectangular homogeneous reservoir, dual-porosity reservoir, or closed circular homogeneous reservoir, as well as the type curve plotting and the curve matching method. Furthermore, we also present several case studies using the above method.

8.1 Vertically Fractured Well at the Center of a Rectangular Homogeneous Reservoir

8.1.1 Pressure Distribution of Vertically Fractured Well in a Rectangular Reservoir: Constant Rate Production

We assume the flow of a slightly compressible fluid of constant viscosity in a rectangular reservoir in which the outer boundaries $2x_e$ and $2y_e$ are closed. The well is located at the center of the reservoir and fluid is produced at a constant rate of q. The fracture penetrates the boundaries, $x_f = x_e$, and the distance from the well to the boundaries is y_w, as shown in Figure 8.1. Initially, the pressure is uniform throughout the reservoir. The skin and wellbore storage effect is not considered. The bottom hole flowing pressure (BHFP), reservoir thickness, initial reservoir pressure, wellbore radius, porosity, total compressibility, permeability, fluid viscosity, and volume factor are represented by p_{wf}, h, p_i, r_w, φ, C_t, K, μ, and B, respectively.

The basic partial differential equations for the system under investigation are given as,

$$\frac{\partial^2 p}{\partial y^2} = \frac{\varphi \mu C_t}{K} \frac{\partial p}{\partial t} \tag{8.1}$$

$$p(y,0) = p_i \tag{8.2}$$

$$\left(\frac{\partial p}{\partial y}\right)_{y=0} = \frac{q\mu B}{2x_f Kh} \tag{8.3}$$

Advanced Production Decline Analysis and Application

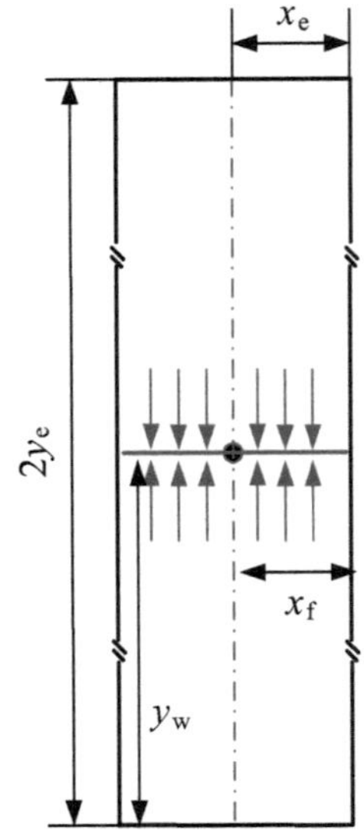

Figure 8.1 Vertically fractured well in a rectangular reservoir

$$\left(\frac{\partial p}{\partial y}\right)_{y=y_e}=0 \tag{8.4}$$

Introducing the variable $\Psi(y, t)$,

$$\Psi(y,t)=p_i-\frac{q\mu B}{2x_f Kh}\frac{(y_e-y)^2}{2y_e}+\eta t,\eta=\frac{K}{\varphi\mu C_t}$$

Substituting $\Psi(y, t)$ into Eq. (8.1) through (8.4) yields

$$\frac{\partial^2\Psi}{\partial y^2}=\frac{1}{\eta}\frac{\partial\Psi}{\partial t} \tag{8.5}$$

$$\Psi(y,0)=p_i-\frac{q\mu B}{2x_f Kh}\frac{(y_e-y)^2}{2y_e} \tag{8.6}$$

$$\left(\frac{\partial\Psi}{\partial y}\right)_{y=0}=\frac{q\mu B}{2x_f Kh} \tag{8.7}$$

$$\left(\frac{\partial \Psi}{\partial y}\right)_{y=y_e} = 0 \tag{8.8}$$

The solution is solved by Miller (1962) as

$$p(y,t) = p_i - \frac{q\mu B}{2x_f Kh}\left\{\left[\frac{(y_e - y)^2}{2y_e} - \frac{y_e}{6} + \frac{\eta t}{y_e}\right] - \frac{2y_e}{\pi^2}\sum_{n=1}^{\infty}\left(\frac{1}{n^2}\right)e^{-\eta\left(\frac{n\pi}{y_e}\right)^2 t}\cos\left(\frac{n\pi y}{y_e}\right)\right\} \tag{8.9}$$

When $y = 0$, we have

$$p(0,t) = p_i - \frac{q\mu B}{2x_f Kh}\left[\left(\frac{y_e}{3} + \frac{\eta t}{y_e}\right) - \frac{2y_e}{\pi^2}\sum_{n=1}^{\infty}\left(\frac{1}{n^2}\right)e^{-\eta\left(\frac{n\pi}{y_e}\right)^2 t}\right] \tag{8.10}$$

In dimensionless variables,

$$p_D = \frac{\pi}{2}\left[\left(\frac{y_e}{3x_f} + \frac{x_f}{y_e}t_{Dx_f}\right) - \frac{2y_e}{\pi^2 x_f}\sum_{n=1}^{\infty}\left(\frac{1}{n^2}\right)e^{-n^2\pi^2\left(\frac{x_f}{y_e}\right)^2 t_{Dxf}}\right] \tag{8.11}$$

where

$$p_D = \frac{2\pi Kh\Delta p}{q\mu B} \tag{8.12}$$

$$t_{Dx_f} = \frac{K}{\varphi\mu C_t x_f^2}t \tag{8.13}$$

Rearranging Eq. (8.11), we have

$$p_D = \frac{\pi}{2}\left[\left(\frac{y_e}{3x_f} + \frac{x_f}{y_e}t_{Dx_f}\right) - \frac{2y_e}{\pi^2 x_f}\sum_{n=1}^{\infty}\left(\frac{1}{n^2}\right)e^{-n^2\pi^2\left(\frac{x_f}{y_e}\right)^2 t_{Dxf}}\right] \tag{8.14}$$

$$\frac{x_f}{y_e}p_D = \frac{\pi}{2}\left\{\left[\frac{1}{3} + \left(\frac{x_f}{y_e}\right)^2 t_{Dx_f}\right] - \frac{2}{\pi^2}\sum_{n=1}^{\infty}\left(\frac{1}{n^2}\right)e^{-n^2\pi^2\left(\frac{x_f}{y_e}\right)^2 t_{Dxf}}\right\} \tag{8.15}$$

Let

$$p_{\mathrm{Dd}} = \frac{x_{\mathrm{f}}}{y_{\mathrm{e}}} p_{\mathrm{D}} \tag{8.16}$$

$$t_{\mathrm{Dd}} = \left(\frac{x_{\mathrm{f}}}{y_{\mathrm{e}}}\right)^2 t_{\mathrm{D}x_{\mathrm{f}}} \tag{8.17}$$

Rearranging Eq. (8.15), we have

$$p_{\mathrm{Dd}} = \frac{\pi}{2}\left[\left(\frac{1}{3} + t_{\mathrm{Dd}}\right) - \frac{2}{\pi^2}\sum_{n=1}^{\infty} \frac{e^{-n^2\pi^2 t_{\mathrm{Dd}}}}{n^2}\right] \tag{8.18}$$

For constant pressure production cases,

$$\frac{1}{q_{\mathrm{Dd}}} = \frac{\pi}{\sum_{n_{\mathrm{odd}}}^{\infty} \exp\left(\frac{-n^2\pi^2}{4} t_{\mathrm{Dd}}\right)} \tag{8.19}$$

8.1.2 Wattenbarger Type Curves

Wattenbarger and El-Banbi (1998) plotted the inverse of constant rate production pressure p_{Dd}, the inverse of pressure derivative and the dimensionless time t_{Dd} in a coordinate system, as shown in Figure 8.2. The linear flow is a straight line with a slope of –0.5 in the period of early transient flow, and becomes a unit-slope straight line during the later period. At early times, Eq. (8.14) could be reduced to

$$p_{\mathrm{D}} = \sqrt{\pi t_{\mathrm{D}x_{\mathrm{f}}}} \tag{8.20}$$

or

$$p_{\mathrm{Dd}} = \sqrt{\pi t_{\mathrm{Dd}}} \tag{8.21}$$

At late times,

$$p_{\mathrm{D}} = \frac{\pi}{2}\left(\frac{y_{\mathrm{e}}}{3x_{\mathrm{f}}} + \frac{x_{\mathrm{f}}}{y_{\mathrm{e}}} t_{\mathrm{D}x_{\mathrm{f}}}\right) \tag{8.22}$$

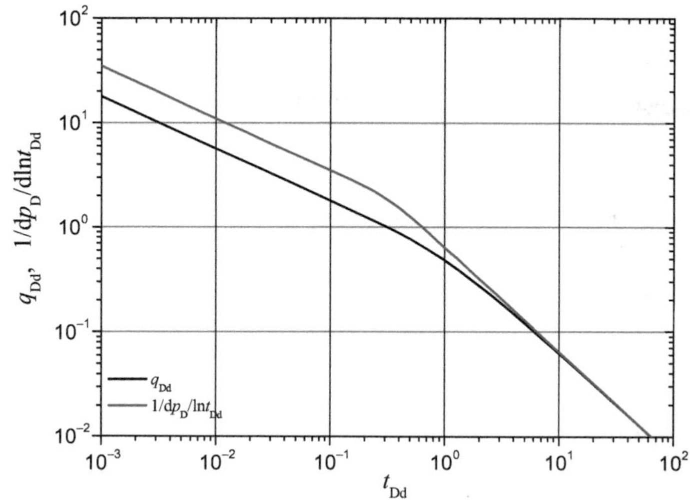

Figure 8.2 Wattenbarger type curves (after Wattenbarger and El-Banbi, 1998)

or

$$p_{Dd} = \frac{\pi}{2}\left(\frac{1}{3} + t_{Dd}\right) \tag{8.23}$$

8.1.3 Wattenbarger Type Curves Matching Procedure—Oil Well

1. Calculate the material balance time

$$t_c = \frac{N_p}{q} \tag{8.24}$$

2. Calculate the normalized rate

$$\frac{q}{\Delta p} = \frac{q}{p_i - p_{wf}} \tag{8.25}$$

3. Calculate the inverse of pressure derivative

$$\frac{1}{p_{di}} = \frac{1}{\dfrac{\partial\left(\dfrac{\Delta p}{q}\right)}{\partial \ln t_c}} \tag{8.26}$$

4. In the log–log coordinate system, plot the normalized rate and inverse of pressure derivative curves versus material balance time, $\frac{q}{\Delta p} \sim t_c$ and $\frac{1}{p_{di}} \sim t_c$, respectively.
5. Choose any or both of the two curves to match the actual points with the type curves (Figure 8.1), aiming to obtain the best matching of each curve.
6. Randomly select a matching point, and record the actual matching point $(t_c, q/\Delta p)_M$ and the corresponding theoretical matching point $(t_{cDd}, q_{Dd})_M$. If the reservoir thickness, total compressibility, and wellbore radius are known, then we can calculate the reservoir permeability, skin factor, drainage area, reserves, etc.
7. According to the rate matching point, calculate

$$K\frac{x_f}{y_e} = \frac{\mu B}{2\pi h}\frac{(q/\Delta p)_M}{(q_{Dd})_M} \tag{8.27}$$

8. According to the rate time point, calculate

$$\frac{y_e}{\sqrt{K}} = \sqrt{\frac{1}{\varphi \mu C_t}\left(\frac{t_c}{t_{cDd}}\right)_M} \tag{8.28}$$

Eq. (8.27) multiplies Eq. (8.28)

$$\sqrt{K}x_f = \frac{\mu B}{2\pi h}\frac{(q/\Delta p)_M}{(q_{Dd})_M}\sqrt{\frac{1}{\varphi \mu C_t}\left(\frac{t_c}{t_{cDd}}\right)_M} \tag{8.29}$$

9. Calculate the drainage area

$$A_e = 4x_e y_e = 4\left(\sqrt{K}x_f\right)\frac{y_e}{\sqrt{K}} = \frac{B}{2\pi\varphi C_t h}\frac{(q/\Delta p)_M}{(q_{Dd})_M}\left(\frac{t_c}{t_{cDd}}\right)_M \tag{8.30}$$

10. Calculate the original oil in place (OOIP) by the volumetric method

$$N = \frac{A\varphi h S_o}{B_o} \tag{8.31}$$

8.1.4 Wattenbarger Type Curves Matching Procedure—Gas Well

The procedure of Wattenbarger curve analysis for a gas well is similar to that for an oil well, as follows

1. Calculate the material balance pseudo-time

   Assuming the reserves is G, calculate the material balance pseudo-time for each production data point according to Eq. (8.32)

$$t_{ca} = \frac{(\mu C_t)_i}{q} \int_0^t \frac{q}{\mu(\bar{p}) C_t(\bar{p})} dt = \frac{G C_{ti}}{q}\left(p_{p_i} - p_p\right) \tag{8.32}$$

Please refer to the normalized pseudo-pressure p_p in Eq. (4.57), and the average reservoir pressure could be calculated according to Eq. (4.60).

2. Calculate the normalized rate

$$\frac{q}{\Delta p_p} = \frac{q}{p_{p_i} - p_{p_{wf}}} \tag{8.33}$$

3. Calculate the inverse of pressure derivative

$$\frac{1}{p_{di}} = \frac{1}{\dfrac{\partial\left(\dfrac{\Delta p_p}{q}\right)}{\partial \ln t_{ca}}} \tag{8.34}$$

4. In the log–log coordinate system, plot the normalized rate and inverse of pressure derivative curves versus material balance pseudo-time, $\frac{q}{\Delta p_p} \sim t_{ca}$ and $\frac{1}{p_{di}} \sim t_{ca}$, respectively.

5. Choose any or both of the two curves to match the actual points with the type curves (Figure 8.1), aiming to obtain the best matching of each curve.

6. Randomly select a matching point, and record the actual matching point $(t_{ca}, q/\Delta p_p)_M$ and the corresponding theoretical matching point $(t_{caDd}, q_{Dd})_M$. If the reservoir thickness, total compressibility, and wellbore radius are known, then we can calculate the reservoir permeability, skin factor, drainage area, reserves, etc.

7. According to the rate matching point, calculate

$$K\frac{x_f}{y_e} = \frac{\mu B}{2\pi h} \frac{(q/\Delta p_p)_M}{(q_{Dd})_M} \tag{8.35}$$

8. According to the rate time point, calculate

$$\frac{y_e}{\sqrt{K}} = \sqrt{\frac{1}{\varphi \mu C_t}\left(\frac{t_{ca}}{t_{caDd}}\right)_M} \tag{8.36}$$

Eq. (8.35) multiplies Eq. (8.36)

$$\sqrt{K} x_f = \frac{\mu B}{2\pi h} \frac{(q/\Delta p_p)_M}{(q_{caDd})_M} \sqrt{\frac{1}{\varphi \mu C_t}\left(\frac{t_{ca}}{t_{caDd}}\right)_M} \tag{8.37}$$

9. Calculate the drainage area

$$A_e = 4x_e y_e = 4\left(\sqrt{K}x_f\right)\frac{y_e}{\sqrt{K}} = \frac{B}{2\pi\varphi C_t h}\frac{\left(q/\Delta p_p\right)_M}{\left(q_{Dd}\right)_M}\left(\frac{t_{ca}}{t_{caDd}}\right)_M \tag{8.38}$$

10. Calculate the original gas in place (OGIP) by the volumetric method

$$G = \frac{A\varphi h S_g}{B_{gi}} \tag{8.39}$$

8.1.5 Case Study of Wattenbarger Decline Curves Analysis

Assume that there is a gas well in a bounded rectangular reservoir, producing at a variable rate and a variable BHFP. The initial reservoir pressure is 30 MPa, reservoir temperature is 80°C, net pay thickness is 10 m, porosity is 0.1, fracture half-length is 100 m, original well-control gas reserves are 2.0×10^8 m^3, relative density of gas is 0.6, pseudo-critical temperature is 195.697 K, pseudo-critical pressure is 4.66875 MPa, permeability estimation from well test analysis is 2.0 mD, production time is 200 days, cumulative gas production is $3{,}695 \times 10^4$ m^3, fluid viscosity is 0.0228 mPa·s, rock compressibility is 4.35113×10^{-4} MPa^{-1}, gas compressibility is 0.0246 MPa^{-1} under the original condition, and gas volume coefficient is 0.0039497 under the original condition. Detailed production data are shown in Table 8.1.

1. Calculate the material balance pseudo-time and the relative normalized rate parameters.
2. Calculate the average reservoir pressure, normalized pseudo-pressure difference, normalized pseudo-time, normalized rate, and inverse of pressure derivative; the results are shown in Table 8.2.
3. Plot log–log curves of the normalized rate and inverse of pressure derivative with material balance pseudo-time, that is, $\frac{q}{\Delta p_p} \sim t_{ca}$ and $\frac{1}{p_{pi}} \sim t_{ca}$ curves, as shown in Figure 8.3.
4. Match type curve (Figure 8.2) with the actual production history curve, as shown in Figure 8.4.
5. Choose the actual matching point $(67, 8.5)_M$ and the corresponding theoretical matching point $(0.1, 1)_M$ to simplify the calculation.
6. According to the rate matching point, calculate

$$\begin{aligned} K\frac{x_f}{y_e} &= \frac{\mu B}{2\pi h}\frac{\left(q/\Delta p_p\right)_M}{\left(q_{Dd}\right)_M} = \frac{1.866\times\mu B}{h}\frac{\left(q/\Delta p_p\right)_M}{\left(q_{Dd}\right)_M} \\ &= \frac{1.866\times 0.0228\times 0.0039497}{10}\frac{(8.5\times 1000)_M}{(1.0)_M} \\ &= 0.1428\,\text{mD} \end{aligned} \tag{8.40}$$

Table 8.1 Production history of well C

Time	Rate	Cumulative production	Time	Rate	Cumulative production	Time	Rate	Cumulative production	Time	Rate	Cumulative production
d	10^3 m^3/d	10^4 m^3	d	10^3 m^3/d	10^4 m^3	d	10^3 m^3/d	10^4 m^3	d	10^3 m^3/d	10^4 m^3
1	199.8	20	51	192.0	999	101	184.5	1940	151	177.2	2844
2	199.7	40	52	191.9	1018	102	184.3	1958	152	177.1	2861
3	199.5	60	53	191.7	1037	103	184.2	1977	153	177.0	2879
4	199.4	80	54	191.5	1057	104	184.0	1995	154	176.8	2897
5	199.2	100	55	191.4	1076	105	183.9	2013	155	176.7	2914
6	199.0	120	56	191.2	1095	106	183.7	2032	156	176.5	2932
7	198.9	140	57	191.1	1114	107	183.6	2050	157	176.4	2950
8	198.7	159	58	190.9	1133	108	183.4	2068	158	176.3	2967
9	198.6	179	59	190.8	1152	109	183.3	2087	159	176.1	2985
10	198.4	199	60	190.6	1171	110	183.2	2105	160	176.0	3002
11	198.2	219	61	190.5	1190	111	183.0	2123	161	175.8	3020
12	198.1	239	62	190.3	1209	112	182.9	2142	162	175.7	3038
13	197.9	259	63	190.2	1228	113	182.7	2160	163	175.5	3055
14	197.8	278	64	190.0	1247	114	182.6	2178	164	175.4	3073
15	197.6	298	65	189.9	1266	115	182.4	2196	165	175.3	3090
16	197.5	318	66	189.7	1285	116	182.3	2215	166	175.1	3108
17	197.3	338	67	189.6	1304	117	182.1	2233	167	175.0	3125
18	197.1	357	68	189.4	1323	118	182.0	2251	168	174.8	3143
19	197.0	377	69	189.3	1342	119	181.8	2269	169	174.7	3160
20	196.8	397	70	189.1	1361	120	181.7	2287	170	174.6	3178
21	196.7	416	71	189.0	1380	121	181.5	2306	171	174.4	3195
22	196.5	436	72	188.8	1399	122	181.4	2324	172	174.3	3213
23	196.4	456	73	188.7	1418	123	181.3	2342	173	174.2	3230
24	196.2	475	74	188.5	1436	124	181.1	2360	174	174.0	3247

(Continued)

Table 8.1 Production history of well C *(cont.)*

Time	Rate	Cumulative production	Time	Rate	Cumulative production	Time	Rate	Cumulative production	Time	Rate	Cumulative production
d	10^3 m^3/d	10^4 m^3	d	10^3 m^3/d	10^4 m^3	d	10^3 m^3/d	10^4 m^3	d	10^3 m^3/d	10^4 m^3
25	196.0	495	75	188.4	1455	125	181.0	2378	175	173.9	3265
26	195.9	514	76	188.2	1474	126	180.8	2396	176	173.7	3282
27	195.7	534	77	188.1	1493	127	180.7	2414	177	173.6	3299
28	195.6	554	78	187.9	1512	128	180.5	2432	178	173.5	3317
29	195.4	573	79	187.8	1530	129	180.4	2450	179	173.3	3334
30	195.3	593	80	187.6	1549	130	180.2	2468	180	173.2	3351
31	195.1	612	81	187.5	1568	131	180.1	2486	181	173.0	3369
32	194.9	632	82	187.3	1587	132	180.0	2504	182	172.9	3386
33	194.8	651	83	187.2	1605	133	179.8	2522	183	172.8	3403
34	194.6	671	84	187.0	1624	134	179.7	2540	184	172.6	3421
35	194.5	690	85	186.9	1643	135	179.5	2558	185	172.5	3438
36	194.3	709	86	186.7	1662	136	179.4	2576	186	172.3	3455
37	194.2	729	87	186.6	1680	137	179.2	2594	187	172.2	3472
38	194.0	748	88	186.4	1699	138	179.1	2612	188	172.1	3490
39	193.9	768	89	186.3	1717	139	179.0	2630	189	171.9	3507
40	193.7	787	90	186.1	1736	140	178.8	2648	190	171.8	3524
41	193.5	806	91	186.0	1755	141	178.7	2666	191	171.7	3541
42	193.4	826	92	185.8	1773	142	178.5	2684	192	171.5	3558
43	193.2	845	93	185.7	1792	143	178.4	2701	193	171.4	3575
44	193.1	864	94	185.5	1810	144	178.2	2719	194	171.2	3592
45	192.9	884	95	185.4	1829	145	178.1	2737	195	171.1	3610
46	192.8	903	96	185.2	1847	146	178.0	2755	196	171.0	3627
47	192.6	922	97	185.1	1866	147	177.8	2773	197	170.8	3644
48	192.5	941	98	184.9	1884	148	177.7	2790	198	170.7	3661
49	192.3	961	99	184.8	1903	149	177.5	2808	199	170.6	3678
50	192.2	980	100	184.6	1921	150	177.4	2826	200	170.4	3695

Table 8.2 Material balance pseudo-time and the relative normalized rate parameters of well C

t	q	p_{wf}	p_{avg}	$\Delta p_p(p_{avg})$	$\Delta p_p(p_{wf})$	t_{ca}	$q/\Delta p_p$	$1/p_{pi}$
d	10^3 m^3/d	MPa	MPa	MPa	MPa	d	10^3 m^3/d/ MPa	MPa/10^3 m^3/d
1	199.8	28.69	29.96	0.04	1.31	0.98	152.85	648.41
2	199.7	28.17	29.92	0.08	1.81	1.95	110.11	196.57
3	199.5	27.73	29.88	0.12	2.25	2.92	88.87	153.27
4	199.4	27.34	29.84	0.16	2.63	3.90	75.84	129.31
5	199.2	26.99	29.80	0.20	2.98	4.87	66.94	113.93
6	199.0	26.66	29.76	0.24	3.30	5.84	60.40	102.79
7	198.9	26.35	29.72	0.28	3.59	6.81	55.37	94.61
8	198.7	26.06	29.69	0.31	3.87	7.78	51.36	88.42
9	198.6	25.79	29.65	0.35	4.13	8.75	48.07	82.94
10	198.4	25.53	29.61	0.39	4.38	9.72	45.31	78.68
20	196.8	23.40	29.22	0.77	6.38	19.36	30.84	55.03
30	195.3	21.74	28.85	1.15	7.90	28.92	24.73	44.91
40	193.7	20.32	28.48	1.51	9.14	38.41	21.18	38.68
50	192.2	19.06	28.12	1.87	10.22	47.82	18.81	34.52
60	190.6	17.91	27.76	2.22	11.16	57.17	17.08	31.63
70	189.1	16.84	27.42	2.55	12.00	66.45	15.75	29.21
80	187.6	15.83	27.08	2.89	12.77	75.68	14.69	27.41
90	186.1	14.87	26.75	3.21	13.47	84.84	13.82	25.75
100	184.6	13.95	26.42	3.53	14.11	93.95	13.08	24.31
110	183.2	13.05	26.10	3.83	14.71	103.02	12.45	23.15
120	181.7	12.17	25.78	4.14	15.26	112.03	11.91	22.16
130	180.2	11.30	25.48	4.43	15.78	121.00	11.42	21.29
140	178.8	10.44	25.17	4.72	16.26	129.93	11.00	20.40
150	177.4	9.57	24.87	5.01	16.72	138.82	10.61	19.63
160	176.0	8.68	24.58	5.28	17.15	147.67	10.26	18.94
170	174.6	7.76	24.29	5.55	17.55	156.50	9.95	18.31
180	173.2	6.78	24.01	5.82	17.94	165.28	9.65	17.68
190	171.8	5.72	23.73	6.08	18.30	174.05	9.39	17.13
200	170.4	4.50	23.46	6.33	18.65	182.78	9.14	16.56

7. According to the rate time point, calculate

$$\begin{aligned}\frac{y_e}{\sqrt{K}} &= \sqrt{\frac{1}{\varphi\mu C_t}\left(\frac{t_{ca}}{t_{caDd}}\right)_M} = \sqrt{\frac{3.6\times10^{-3}}{\varphi\mu C_t}\left(\frac{t_{ca}}{t_{caDd}}\right)_M} \\ &= \sqrt{\frac{3.6\times10^{-3}}{0.1\times0.0228\times0.0246}\left(\frac{67\times24}{0.1}\right)_M} \\ &= 1015.9\ \text{m}/\text{mD}^{0.5}\end{aligned} \tag{8.41}$$

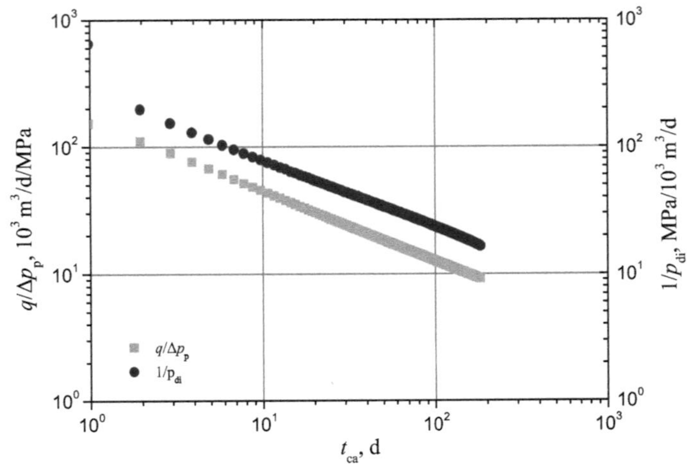

Figure 8.3 Wattenbarger rate functions for Well C

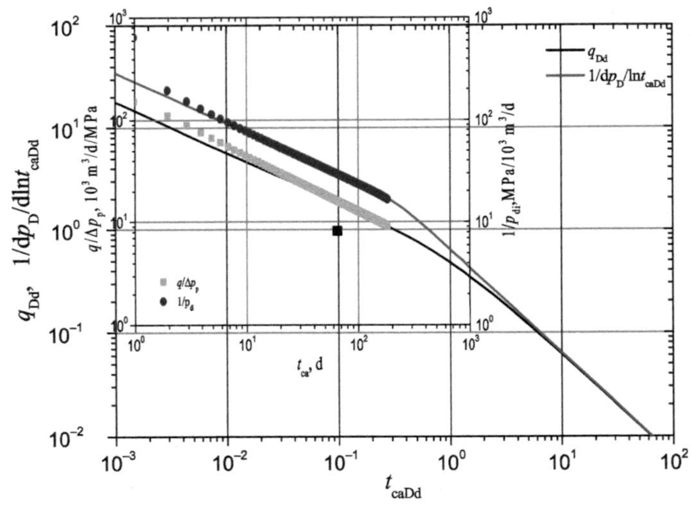

Figure 8.4 Matches of production data for Well C–Wattenbarger

Eq. (8.40) multiplies Eq. (8.41)

$$\sqrt{K}x_f = \frac{1.866 \times \mu B}{h} \frac{(q/\Delta p_p)_M}{(q_{caDd})_M} \sqrt{\frac{3.6 \times 10^{-3}}{\varphi \mu C_t}\left(\frac{t_{ca}}{t_{caDd}}\right)_M} \tag{8.42}$$

$$= 145.07\,\text{m.mD}^{0.5}$$

8. Calculate the drainage area

$$A_e = 4x_e y_e = 4\left(\sqrt{K}x_f\right)\frac{y_e}{\sqrt{K}} = 4\times145.07\times1015.9 = 0.5895\times10^6\text{m}^2 \tag{8.43}$$

9. Calculate the OGIP by the volumetric method

$$G = \frac{A\varphi hS_g}{B_{gi}} = \frac{0.5895\times0.1\times10\times1}{0.0039497} = 1.5\times10^8\text{m}^3 \tag{8.44}$$

8.2 Vertically Fractured Well at the Centre of a Rectangular Dual-Porosity Reservoir

8.2.1 Physical Model

We consider the flow of a slightly compressible fluid of constant viscosity in a rectangular reservoir in which the outer boundaries $2x_e$ and $2y_e$ are closed. The well is located at the center of the reservoir and fluid is produced at a constant rate q (or constant BHFP p_{wf}). The reservoir thickness is h, the fracture half-length is $x_f \le x_e$, the distance from the well to the boundaries is $y_w = y_e$, and the flow area is $A_c = 4x_f h$, as shown in Figure 8.5. Initially, the pressure is uniform throughout the reservoir. The skin and wellbore storage effect is not considered. The initial reservoir pressure, wellbore radius, porosity, total compressibility, permeability, fluid viscosity, and volume factor are represented by p_i, r_w, φ, C_t, K, μ, and B respectively.

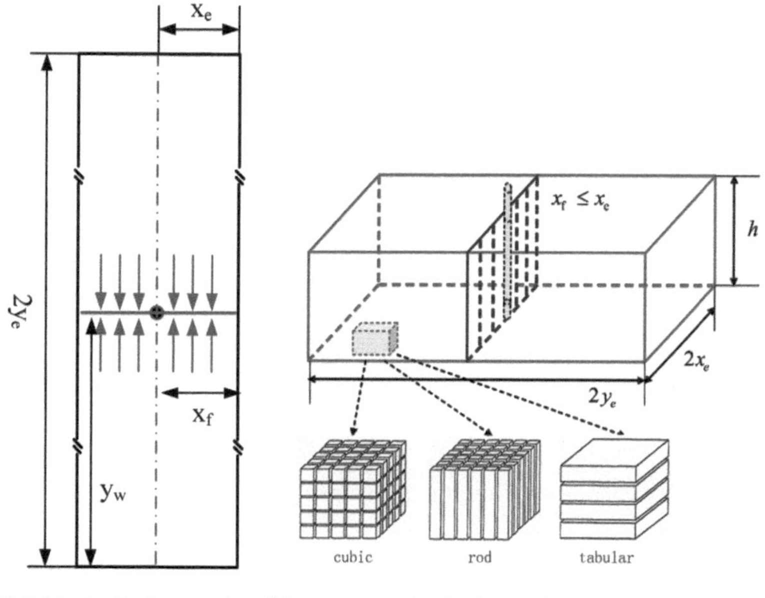

Figure 8.5 Vertically fractured well in a rectangular dual-porosity reservoir

8.2.2 Linear Pseudo-Steady Mathematical Model

El-Banbi and Wattenbarger (1998) introduced a mathematical model which describes the linear pseudo-steady crossflow in a dual-porosity reservoir, where the matrix and fracture control equation is

$$\frac{\partial^2 p_{\mathrm{Df}}}{\partial y_{\mathrm{D}}^2}=\omega\frac{\partial p_{\mathrm{Df}}}{\partial t_{\mathrm{DA_c}}}+(1-\omega)\frac{\partial p_{\mathrm{Dm}}}{\partial t_{\mathrm{DA_c}}} \tag{8.45}$$

$$\frac{\partial p_{\mathrm{Dm}}}{\partial t_{\mathrm{DA_c}}}=\frac{\lambda_{A_c}}{1-\omega}(p_{\mathrm{Df}}-p_{\mathrm{Dm}}) \tag{8.46}$$

where ω is the storativity ratio of the dual-porosity reservoir and λ_{A_c} is the interporosity flow coefficient

$$\omega=\frac{(\varphi VC_{\mathrm{t}})_{\mathrm{f}}}{(\varphi VC_{\mathrm{t}})_{\mathrm{f}}+(\varphi VC_{\mathrm{t}})_{\mathrm{m}}} \tag{8.47}$$

$$\lambda_{A_c}=\alpha\frac{K_{\mathrm{m}}}{K_{\mathrm{f}}}A_{\mathrm{c}} \tag{8.48}$$

with α being a shape-dependent constant of matrix block.

Initial condition, include

$$p_{\mathrm{Dm}}(y_{\mathrm{D}},0)=0 \tag{8.49}$$

$$p_{\mathrm{Df}}(y_{\mathrm{D}},0)=0 \tag{8.50}$$

Inner boundary condition of constant rate production is

$$\left(\frac{\partial p_{\mathrm{Df}}}{\partial y_{\mathrm{D}}}\right)_{y_{\mathrm{D}}=0}=-2\pi \tag{8.51}$$

Inner boundary condition of constant pressure production is

$$p_{\mathrm{Df}}\left(y_{\mathrm{D}}=0,t_{\mathrm{DA_c}}\right)=1 \tag{8.52}$$

Outer boundary condition—infinite reservoir is

$$p_{\mathrm{Df}}\left(y_{\mathrm{D}}\to\infty,t_{\mathrm{DA_c}}\right)=0 \tag{8.53}$$

Outer boundary condition—closed reservoir is

$$\left(\frac{\partial p_{\mathrm{Df}}}{\partial y_{\mathrm{D}}}\right)_{y_{\mathrm{D}}=\frac{L}{\sqrt{A_{\mathrm{c}}}}}=0 \tag{8.54}$$

Outer boundary condition—constant pressure is

$$p_{\mathrm{Df}}\left(y_{\mathrm{D}}=\frac{L}{\sqrt{A_{\mathrm{c}}}},t_{\mathrm{DA_c}}\right)=0 \tag{8.55}$$

The dimensionless variables are defined as

$$p_{\mathrm{D}}=\frac{2\pi K_{\mathrm{f}}\sqrt{A_{\mathrm{c}}}\,(p_{\mathrm{i}}-p)}{qB\mu} \quad \text{(constant rate)} \tag{8.56}$$

$$p_{\mathrm{D}}=\frac{p_{\mathrm{i}}-p}{p_{\mathrm{i}}-p_{\mathrm{wf}}} \quad \text{(constant pressure)} \tag{8.57}$$

$$\frac{1}{q_{\mathrm{D}}}=\frac{2\pi K_{\mathrm{f}}\sqrt{A_{\mathrm{c}}}\,(p_{\mathrm{i}}-p_{\mathrm{wf}})}{qB\mu} \quad \text{(constant pressure)} \tag{8.58}$$

$$t_{\mathrm{DA_c}}=\frac{K_{\mathrm{f}}}{\mu\left[(\varphi VC_{\mathrm{t}})_{\mathrm{f}}+(\varphi VC_{\mathrm{t}})_{\mathrm{m}}\right]A_{\mathrm{c}}}\cdot t \tag{8.59}$$

$$y_{\mathrm{D}}=\frac{y}{\sqrt{A_{\mathrm{c}}}} \tag{8.60}$$

8.2.3 Linear Transient Mathematical Model

Wattenbarger and El-Banbi (1998) introduced a mathematical model that describes the linear transient dual-porosity reservoir, where the matrix and fracture control equation is

$$\nabla^2 p_{\mathrm{Dm}}=\frac{3(1-\omega)}{\lambda_{A_{\mathrm{c}}}}\frac{\partial p_{\mathrm{Dm}}}{\partial t_{\mathrm{DA_c}}} \tag{8.61}$$

For the layered model, we have

$$\frac{\partial^2 p_{\mathrm{Dm}}}{\partial z_{\mathrm{D}}^2}=\frac{3(1-\omega)}{\lambda_{A_{\mathrm{c}}}}\frac{\partial p_{\mathrm{Dm}}}{\partial t_{\mathrm{DA_c}}} \tag{8.62}$$

$$\frac{\partial^2 p_{\mathrm{Df}}}{\partial y_{\mathrm{D}}^2} - \frac{\lambda_{A_c}}{3} \frac{\partial p_{\mathrm{Dm}}}{\partial z_{\mathrm{D}}}\bigg|_{z_{\mathrm{D}}=1} = \omega \frac{\partial p_{\mathrm{Df}}}{\partial t_{\mathrm{DA_c}}} \tag{8.63}$$

Subject to the following initial and boundary conditions

$$p_{\mathrm{Dm}}(z_{\mathrm{D}}, 0) = 0 \tag{8.64}$$

$$\frac{\partial p_{\mathrm{Dm}}\left(0, t_{\mathrm{DA_c}}\right)}{\partial z_{\mathrm{D}}} = 0 \tag{8.65}$$

$$p_{\mathrm{Dm}}\left(1, t_{\mathrm{DA_c}}\right) = p_{\mathrm{Df}}\left(1, t_{\mathrm{DA_c}}\right) \tag{8.66}$$

8.2.4 Laplace Space Solutions

The previous basic partial differential equation can be solved by the Laplace transformation. El-Banbi et al. (1998) gave the general solutions for the above two cases. As for the case with a constant rate case, the Laplace transform solution for constant BHFP for wells in reservoirs with different boundaries is

$$\bar{p}_{\mathrm{wD}} = \frac{2\pi}{s\sqrt{sf(s)}} \quad \text{(infinite reservoir)} \tag{8.67}$$

$$\bar{p}_{\mathrm{wD}} = \frac{2\pi}{s\sqrt{sf(s)}} \frac{\left\{1+\exp\left[-2\sqrt{sf(s)}\frac{L}{\sqrt{A_{\mathrm{c}}}}\right]\right\}}{\left\{1-\exp\left[-2\sqrt{sf(s)}\frac{L}{\sqrt{A_{\mathrm{c}}}}\right]\right\}} \quad \text{(closed reservoir)} \tag{8.68}$$

$$\bar{p}_{\mathrm{wD}} = \frac{2\pi}{s\sqrt{sf(s)}} \frac{\left\{1-\exp\left[-2\sqrt{sf(s)}\frac{L}{\sqrt{A_{\mathrm{c}}}}\right]\right\}}{\left\{1+\exp\left[-2\sqrt{sf(s)}\frac{L}{\sqrt{A_{\mathrm{c}}}}\right]\right\}} \quad \text{(constant pressure)} \tag{8.69}$$

where the Laplace function $f(s)$ expressions fitting for different reservoir models are shown in Tables 8.3 and 8.4.

As for a case with a constant pressure case, the flow rate solution in Laplace domain space under different boundary conditions is

Table 8.3 **$f(s)$ for different linear reservoir models (after El-Banbi et al., 1998)**

Model	$f(s)$
Homogenous	$f(s) = 1$
PSS dual-porosity	$f(s)=\dfrac{\omega(1-\omega)s+\lambda_{A_c}}{(1-\omega)s+\lambda_{A_c}}$
Transient dual-porosity (Slabs)	$f(s)=\omega+\dfrac{\lambda_{A_c}}{3s}\sqrt{\dfrac{3(1-\omega)s}{\lambda_{A_c}}}\tanh\left(\sqrt{\dfrac{3(1-\omega)s}{\lambda_{A_c}}}\right)$
Transient dual-porosity (Stick)	$f(s)=\omega+\dfrac{\lambda_{A_c}}{4s}\sqrt{\dfrac{8(1-\omega)s}{\lambda_{A_c}}}\dfrac{I_1\left(\sqrt{\dfrac{8(1-\omega)s}{\lambda_{A_c}}}\right)}{I_0\left(\sqrt{\dfrac{8(1-\omega)s}{\lambda_{A_c}}}\right)}$
Transient dual-porosity (Cubes)	$f(s)=\omega+\dfrac{\lambda_{A_c}}{5s}\sqrt{\dfrac{15(1-\omega)s}{\lambda_{A_c}}}\coth\left(\sqrt{\dfrac{15(1-\omega)s}{\lambda_{A_c}}}\right)$

Table 8.4 **Production history of well D (after Pratikno et al., 2003)**

Time	Pressure	Rate	Time	Pressure	Rate	Time	Pressure	Rate
d	MPa	10^4 m^3/d	d	MPa	10^4 m^3/d	d	MPa	10^4 m^3/d
1	8.32	31.04	105	6.70	9.54	209	6.54	3.82
2	11.07	25.75	106	6.62	9.04	210	3.95	3.31
3	10.94	27.82	107	6.39	8.67	211	4.20	3.84
4	10.21	25.02	108	6.34	8.43	212	5.31	5.05
5	9.85	23.02	109	6.19	8.19	213	5.07	4.57
6	8.87	21.53	110	6.03	7.94	214	4.78	3.93
7	8.73	20.38	111	5.76	7.52	215	5.36	5.16
8	8.50	19.39	112	5.82	7.11	216	4.76	3.89
9	7.83	18.62	113	5.67	6.88	217	4.86	4.13
10	7.26	17.86	114	5.65	6.80	218	5.89	5.18
11	7.14	17.16	115	5.68	6.85	219	5.91	5.23
12	7.02	16.51	116	5.79	6.88	220	5.78	4.94
13	6.82	15.94	117	5.88	7.02	221	6.11	4.58
14	6.70	15.42	118	6.18	6.90	222	5.97	3.99
15	6.50	14.93	119	6.10	6.75	223	5.24	0.79
16	6.35	14.49	120	6.05	6.66	224	16.80	5.65
17	6.29	14.13	121	6.02	6.62	225	16.80	5.67
18	6.18	13.80	122	6.02	6.58	226	6.58	5.16

(Continued)

Table 8.4 Production history of well D (after Pratikno et al., 2003) *(cont.)*

Time	Pressure	Rate	Time	Pressure	Rate	Time	Pressure	Rate
d	MPa	10^4 m^3/d	d	MPa	10^4 m^3/d	d	MPa	10^4 m^3/d
19	6.26	13.48	123	5.98	6.50	227	6.44	4.81
20	5.98	13.16	124	5.94	6.43	228	5.98	4.50
21	5.92	12.91	125	5.91	6.38	229	6.00	4.54
22	6.09	12.65	126	5.89	6.34	230	5.88	4.13
23	5.85	12.38	127	5.83	6.24	231	5.84	4.03
24	5.83	12.18	128	5.88	6.33	232	5.80	3.77
25	5.84	11.89	129	5.86	6.29	233	5.99	4.29
26	5.89	11.71	130	5.69	6.05	234	6.04	4.59
27	5.91	11.51	131	5.65	5.97	235	6.03	4.56
28	5.87	11.31	132	5.54	5.94	236	5.96	4.40
29	6.01	11.16	133	5.52	5.90	237	5.84	4.08
30	24.75	2.07	134	5.44	5.85	238	5.79	3.94
31	5.91	11.62	135	5.40	5.79	239	5.71	3.72
32	5.93	12.03	136	5.38	5.75	240	5.59	3.94
33	5.81	11.34	137	5.37	5.72	241	5.64	4.08
34	5.68	10.96	138	5.35	5.69	242	5.57	3.74
35	5.42	10.67	139	5.34	5.67	243	5.53	3.62
36	5.32	10.46	140	5.33	5.65	244	5.65	4.04
37	5.49	10.01	141	5.32	5.64	245	5.55	3.76
38	5.64	9.78	142	4.06	5.52	246	5.89	4.37
39	51.83	9.62	143	9.55	1.25	247	5.55	3.43
40	5.84	9.44	144	14.37	0.23	248	5.67	4.21
41	5.74	9.40	145	14.37	0.17	249	5.53	3.84
42	5.60	9.27	146	6.48	7.60	250	5.85	4.48
43	5.11	9.34	147	6.39	7.46	251	5.31	2.93
44	5.53	8.96	148	5.73	6.78	252	5.62	3.62
45	5.50	8.92	149	5.51	6.42	253	5.14	1.86
46	5.45	8.74	150	5.32	6.11	254	15.75	5.00
47	5.45	8.74	151	5.16	5.84	255	15.80	5.34
48	5.44	8.64	152	5.12	5.77	256	6.13	4.82
49	5.58	8.54	153	5.03	5.62	257	5.98	4.45
50	5.48	8.49	154	5.01	5.59	258	5.92	4.33
51	5.43	8.31	155	5.01	5.59	259	5.91	4.30
52	5.57	8.19	156	5.05	5.42	260	5.57	4.27
53	6.02	7.93	157	5.11	5.53	261	5.53	4.19
54	6.02	7.86	158	4.79	5.21	262	5.42	4.12
55	5.90	8.04	159	5.01	5.59	263	5.41	4.09
56	6.01	7.91	160	4.84	5.30	264	5.70	4.02
57	5.83	7.78	161	4.79	5.20	265	5.69	3.99
58	5.78	7.83	162	4.79	5.20	266	5.45	3.77
59	5.90	7.70	163	4.74	5.12	267	4.93	2.06
60	5.67	7.70	164	4.74	5.13	268	13.89	0.52
61	5.73	7.55	165	4.38	4.49	269	13.90	0.97

Table 8.4 **Production history of well D (after Pratikno et al., 2003) *(cont.)***

Time	Pressure	Rate	Time	Pressure	Rate	Time	Pressure	Rate
d	MPa	10^4 m^3/d	d	MPa	10^4 m^3/d	d	MPa	10^4 m^3/d
62	5.68	7.57	166	9.57	1.48	270	7.47	5.84
63	9.51	3.24	167	14.37	0.00	271	7.07	4.78
64	14.37	0.37	168	5.58	6.54	272	5.79	4.26
65	19.19	0.00	169	5.66	6.66	273	5.74	4.12
66	7.58	0.00	170	5.34	6.14	274	5.62	4.23
67	8.04	6.08	171	5.15	5.82	275	5.47	3.85
68	9.05	9.15	172	4.98	5.54	276	5.46	3.87
69	8.77	8.93	173	4.81	5.24	277	5.46	3.87
70	8.65	8.71	174	4.79	5.20	278	5.46	3.69
71	8.51	8.48	175	4.74	5.13	279	5.50	3.81
72	8.42	8.32	176	4.77	5.18	280	5.54	3.75
73	8.34	8.17	177	4.73	5.11	281	5.49	3.62
74	8.25	8.06	178	4.70	5.06	282	5.56	3.62
75	8.20	7.94	179	4.66	4.98	283	5.56	3.64
76	8.13	7.81	180	4.61	4.89	284	5.66	3.57
77	8.04	7.69	181	4.48	4.66	285	5.71	3.70
78	8.00	7.61	182	4.61	4.90	286	5.61	3.67
79	7.80	7.50	183	4.60	4.87	287	5.60	3.64
80	7.74	7.43	184	4.60	4.87	288	5.56	3.62
81	7.69	7.39	185	4.64	4.95	289	5.54	3.57
82	7.79	7.33	186	4.47	4.65	290	5.58	3.61
83	7.74	7.23	187	4.52	4.74	291	5.53	3.49
84	7.63	7.01	188	4.86	4.81	292	5.43	2.91
85	21.08	0.27	189	4.72	4.53	293	5.63	3.54
86	23.39	0.00	190	4.97	4.36	294	6.27	3.53
87	23.84	0.00	191	4.82	4.04	295	6.33	3.74
88	26.54	0.00	192	5.31	5.04	296	5.31	3.40
89	26.99	0.00	193	4.76	3.90	297	15.35	0.64
90	28.30	0.00	194	5.27	4.98	298	5.96	4.77
91	28.83	0.00	195	5.23	4.90	299	5.93	4.71
92	29.39	0.00	196	4.79	3.96	300	5.81	4.42
93	30.13	0.00	197	4.77	3.92	301	5.70	4.12
94	30.56	0.00	198	5.48	5.37	302	5.45	3.67
95	30.98	0.00	199	5.22	4.88	303	5.59	4.05
96	31.54	0.00	200	4.94	4.28	304	5.40	3.84
97	31.98	0.00	201	4.40	3.01	305	5.44	3.95
98	32.60	0.00	202	4.27	0.02	306	5.28	3.64
99	12.33	10.11	203	4.27	0.00	307	5.31	3.72
100	14.49	12.13	204	4.27	0.00	308	5.45	3.89
101	13.69	10.11	205	6.13	6.01	309	5.39	3.73
102	14.11	12.13	206	7.85	6.37	310	5.53	3.54
103	13.62	10.99	207	7.41	5.28	311	2.33	3.53
104	6.79	9.68	208	7.10	5.39	312	5.39	3.42

$$\frac{1}{\overline{q}_{\mathrm{D}}}=\frac{2\pi s}{\sqrt{sf(s)}} \quad \text{(infinite reservoir)} \tag{8.70}$$

$$\frac{1}{\overline{q}_{\mathrm{D}}}=\frac{2\pi s}{\sqrt{sf(s)}}\frac{\left\{1+\exp\left[-2\sqrt{sf(s)}\frac{L}{\sqrt{A_{\mathrm{c}}}}\right]\right\}}{\left\{1-\exp\left[-2\sqrt{sf(s)}\frac{L}{\sqrt{A_{\mathrm{c}}}}\right]\right\}} \quad \text{(closed reservoir)} \tag{8.71}$$

$$\frac{1}{\overline{q}_{\mathrm{D}}}=\frac{2\pi s}{\sqrt{sf(s)}}\frac{\left\{1-\exp\left[-2\sqrt{sf(s)}\frac{L}{\sqrt{A_{\mathrm{c}}}}\right]\right\}}{\left\{1+\exp\left[-2\sqrt{sf(s)}\frac{L}{\sqrt{A_{\mathrm{c}}}}\right]\right\}} \quad \text{(constant pressure)} \tag{8.72}$$

8.2.5 Curve Characteristics of El-Banbi

El-Banbi et al. (1998) used Eq. (8.68) to present the type curve characteristics of the case in which the outer boundary is closed. That is, the pressure curve and the pressure derivative reciprocal curve are parallel lines during the early period, become straight lines with a slope of –1 after the transition period, but are apt to be covered up and directly enter into the fracture characteristic period, and finally become straight lines with a slope of –1 (Figure 8.6).

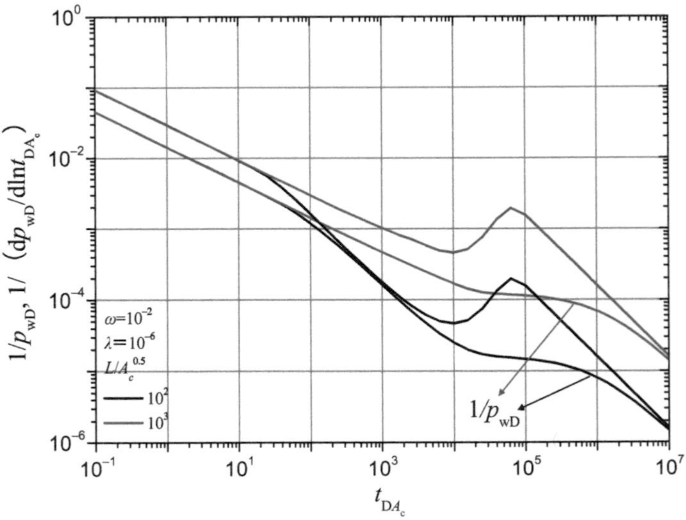

Figure 8.6 El-Banbi type curves (after El-Banbi et al., 1998)

8.3 Vertically Fractured Well at the Centre of a Closed Circular Reservoir

8.3.1 Type Curve Plotting

Pratikno et al. (2003) gave the pressure solution of a finite-conductivity vertically fractured well producing at a constant rate in a bounded circular reservoir as

$$\begin{aligned}
&\bar{p}_{\mathrm{D,ds}}\left(\left|x_{\mathrm{D}}\right| \le 1, y_{\mathrm{D}} = 0, r_{\mathrm{D}} \le r_{\mathrm{eD}}, s\right) \\
&= \frac{1}{2}\int_0^1 \bar{q}_{\mathrm{fD}}(x', s)\left[K_0\left(\left|x_{\mathrm{D}} - x'\right|\sqrt{s}\right) + K_0\left(\left|x_{\mathrm{D}} + x'\right|\sqrt{s}\right)\right]dx' \\
&- \frac{\pi}{F_{\mathrm{cD}}}\int_0^{x_{\mathrm{D}}}\int_0^{x'} \bar{q}_{\mathrm{fD}}(x'', s)\,dx''\,dx' + \frac{\pi}{sF_{\mathrm{CD}}}x_{\mathrm{D}} + S_{\mathrm{f}}\bar{q}_{\mathrm{fD}}(x_{\mathrm{D}}, s) \\
&+ \frac{1}{2s}\frac{1}{\sqrt{s}}\frac{K_1\left(\sqrt{s}r_{\mathrm{eD}}\right)}{I_1\left(\sqrt{s}r_{\mathrm{eD}}\right)}\left[\int_0^{\sqrt{s}(1 \mp x_{\mathrm{D}})} I_0(z)\,dz + \int_0^{\sqrt{s}(1 \pm x_{\mathrm{D}})} I_0(z)\,dz\right]
\end{aligned} \tag{8.73}$$

The dimensionless pressure in the period of pseudo-steady state is expressed as

$$p_{\mathrm{Dpss}}(t_{\mathrm{DA}}) = b_{\mathrm{Dpss}} + 2\pi t_{\mathrm{DA}} \tag{8.74}$$

As for the finite-conductivity vertically fractured well, b_{Dpss} is the function of r_{eD} and F_{CD}, independent of time. If the values of r_{eD} and F_{CD} are given, b_{Dpss} can be calculated using Eq. (8.75):

$$b_{\mathrm{Dpss}} = \ln r_{\mathrm{eD}} - 0.049298 + 0.43464 r_{\mathrm{eD}}^{-2} + \frac{a_1 + a_2 u + a_3 u^2 + a_4 u^3 + a_5 u^4}{1 + b_1 u + b_2 u^2 + b_3 u^3 + b_4 u^4} \tag{8.75}$$

where

$u = \ln F_{\mathrm{CD}}$
$a_1 = 0.93626800$, $a_2 = -1.00489000$, $a_3 = 0.31973300$, $a_4 = -0.04235320$, $a_5 = 0.00221799$
$b_1 = -0.38553900$, $b_2 = -0.06988650$, $b_3 = -0.04846530$, $b_4 = -0.00813558$

According to the decline curve variable presented by Fetkovich, q_{Dd} and t_{Dd} are, respectively, defined as

$$q_{\mathrm{Dd}} = q_{\mathrm{D}} b_{\mathrm{Dpss}} \tag{8.76}$$

$$t_{\mathrm{Dd}} = \frac{2\pi}{b_{\mathrm{Dpss}}} t_{\mathrm{DA}} \tag{8.77}$$

According to Eq. (8.73), given the dimensionless fracture conductivity F_{CD} and r_{eD} value, the type curve can be plotted. Similar to the homogeneous case, apart from the

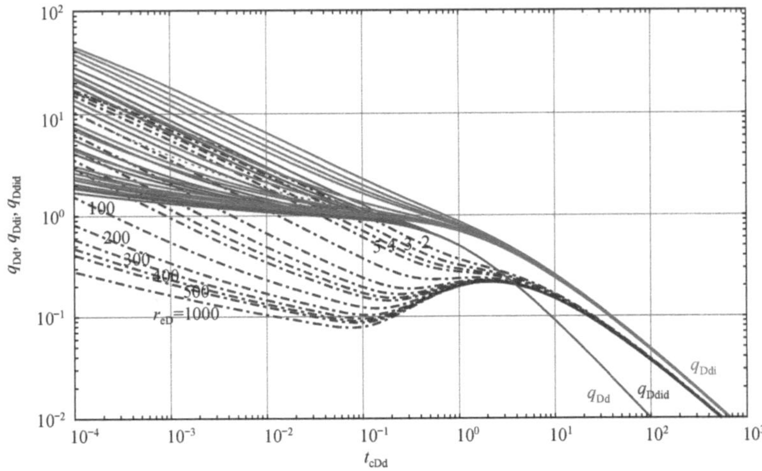

Figure 8.7 Blasingame decline curve for a well in a finite conductivity vertical fracture (F_{CD} = 200) (after Pratikno et al., 2003)

dimensionless decline rate curve, the type curve is also consistent with the rate integral function type curve and the rate integral derivative function type curve. The rate integral function q_{Ddi} is defined as

$$q_{Ddi} = \frac{N_{pDd}}{t_{Dd}} = \frac{1}{t_{Dd}} \int_0^{t_{Dd}} q_{Dd}(\tau)\mathrm{d}\tau \tag{8.78}$$

The rate integral derivative function q_{Ddid} is

$$q_{Ddid} = -\frac{\mathrm{d}q_{Ddi}}{\mathrm{d}\ln t_{Dd}} = -t_{Dd}\frac{\mathrm{d}q_{Ddi}}{\mathrm{d}t_{Dd}} = q_{Ddi} - q_{Dd} \tag{8.79}$$

In addition, the cumulative production function N_{pDd} is

$$N_{pDd} = \int_0^{t_{Dd}} q_{Dd}(\tau)\mathrm{d}\tau \tag{8.80}$$

The decline curves are shown in Figure 8.7.

8.3.2 Type Curves Matching Procedure

Similar to the decline curve analysis for a vertical well in a homogeneous reservoir, the decline curve matching for a vertically fractured well is conducted as follows,

1. Calculate the material balance time.

   As for an oil well, we have

$$t_c = \frac{N_p}{q} \tag{8.81}$$

As for a gas well, we have

$$t_{ca} = \frac{(\mu C_g)_i}{q} \int_0^t \frac{q}{\mu C_g} dt \tag{8.82}$$

2. Calculate the normalized rate, rate integral function, and rate integral derivative. As for an oil well,

$$q/\Delta p = q/(p_i - p_{wf}) \tag{8.83}$$

As for a gas well,

$$q/\Delta p_p = q/\left(p_{p_i} - p_{p_{wf}}\right) \tag{8.84}$$

As for an oil well,

$$(q/\Delta p)_i = \frac{1}{t_c} \int_0^t q/(p_i - p_{wf}) dt \tag{8.85}$$

As for a gas well,

$$\left(q/\Delta p_p\right)_i = \frac{1}{t_{ca}} \int_0^t q/\left(p_{p_i} - p_{p_{wf}}\right) dt \tag{8.86}$$

As for an oil well,

$$(q/\Delta p)_{id} = -\frac{1}{t_c} \frac{d}{dt}\left[(q/\Delta p)_i\right] \tag{8.87}$$

As for a gas well,

$$\left(q/\Delta p_p\right)_{id} = -\frac{1}{t_{ca}} \frac{d}{dt}\left[\left(q/\Delta p_p\right)_i\right] \tag{8.88}$$

3. Plot the normalized rate, normalized rate integral, and normalized rate integral derivative versus material balance time curves in a log–log coordinate system respectively, that is, $\frac{q}{\Delta p} \sim t_c$, $\left(\frac{q}{\Delta p}\right)_i \sim t_c$, $\left(\frac{q}{\Delta p}\right)_{id} \sim t_c$. As for the gas well, $\frac{q}{\Delta p_p} \sim t_{ca}$, plot $\left(\frac{q}{\Delta p_p}\right)_i \sim t_{ca}$, $\left(\frac{q}{\Delta p_p}\right)_{id} \sim t_{ca}$, respectively.

4. Choose any of the three curves or any combination thereof to match the theoretical type curves of certain F_{CD} value (Figure 8.7), aiming to obtain the best matching of each curve.
5. Based on the matching results, record the dimensionless reservoir drainage radius r_{eD}.
6. Randomly select a matching point and record the actual matching point $(t_c, q/\Delta p)_M$ and the corresponding theoretical matching point $(t_{cDd}, q_{Dd})_M$. If the reservoir thickness, total compressibility, and wellbore diameter are known, we can calculate the parameters such as reservoir permeability, skin factor, drainage area, and reserves.
7. On the basis of the value of F_{CD} and r_{eD}, determine the value of b_{Dpss} according to Eq. (8.75).
8. Based on the matching results, calculate the reservoir parameters.

$$N = \frac{1}{C_t}\left(\frac{t_c}{t_{cDd}}\right)_M \left(\frac{q/\Delta p}{q_{Dd}}\right)_M (1-S_w) \quad \text{(oil)} \tag{8.89}$$

$$G = \frac{1}{C_t}\left(\frac{t_{ca}}{t_{caDd}}\right)_M \left(\frac{q/\Delta p_p}{q_{Dd}}\right)_M (1-S_w) \quad \text{(gas)} \tag{8.90}$$

$$A = \frac{NB_o}{\varphi h(1-S_{wi})} \quad \text{(oil)} \tag{8.91}$$

$$A = \frac{GB_{gi}}{\varphi h(1-S_{wi})} \quad \text{(gas)} \tag{8.92}$$

$$r_e = \sqrt{A/\pi} \tag{8.93}$$

$$K = \frac{(q/\Delta p)_M}{(q_{Dd})_M}\frac{\mu B}{2\pi h} b_{Dpss} \quad \text{(oil)} \tag{8.94}$$

$$K = \frac{(q/\Delta p_p)_M}{(q_{Dd})_M}\frac{\mu B}{2\pi h} b_{Dpss} \quad \text{(gas)} \tag{8.95}$$

$$x_f = \frac{r_e}{r_{eD}} \tag{8.96}$$

$$S \approx \ln\frac{r_w}{x_f} + \frac{1.648546 - 3.002711\times10^{-1}u + 1.506532\times10^{-1}u^2}{1 + 2.136604\times10^{-1}u + 9.513761\times10^{-2}u^2 + 8.276998\times10^{-3}u^3} \tag{8.97}$$

8.3.3 Case Study

In this section, we take a gas well with low permeability in Texas, United States, as an example. The main parameters are given as follows. The wellbore radius is 0.1 m, pay thickness is 51.82 m, porosity is 8.8%, average residual water saturation is 13.1%, relative density of gas is 0.7, volume coefficient is 0.003087 m^3/m^3 under the original condition, viscosity is 0.0361 mPa·s under the original condition, gas compressibility is 0.0074 MPa^{-1} under the original condition, and the initial reservoir pressure is 64.33 MPa. Detailed production data are shown in Table 8.3 and 8.4.

Figure 8.8 shows the relation between the semi-log rate and Cartesian producing pressure history and the time. Overall, the data points have good quality and correlation. Plot the curve of the normalized rate $q/\Delta p_p$ versus cumulative gas production G_p, as shown in Figure 8.9. Extrapolate the straight-line portion to estimate the gas reserves. The result is 2785×10^4 m^3.

Then, plot the normalized rate, normalized rate integral, and normalized rate integral derivative versus material balance pseudo-time curves in a log–log coordinate system, respectively, that is, $\frac{q}{\Delta p_p} \sim t_{ca}$, $\left(\frac{q}{\Delta p_p}\right)_i \sim t_{ca}$, and $\left(\frac{q}{\Delta p_p}\right)_{id} \sim t_{ca}$. Match them with the theoretical type curves, for which results are shown in Figure 8.10.

We can figure out the OGIP, reservoir drainage area, effective permeability, fracture half-length, and skin factor, etc., using the r_{eD} and F_{CD} as well as the time and rate matching points. The type curve matching parameter $F_{CD} = 200$, the transient flow matching parameter $r_{eD} = 10.0$, the time matching point $\left(\frac{t_{ca}}{t_{caDd}}\right)_M = 54.95$ days, and the rate matching point $\left(\frac{q/\Delta p_p}{q_{Dd}}\right)_M = 3.66058 \times 10^3 \text{ m}^3/\text{d}/\text{MPa}$.

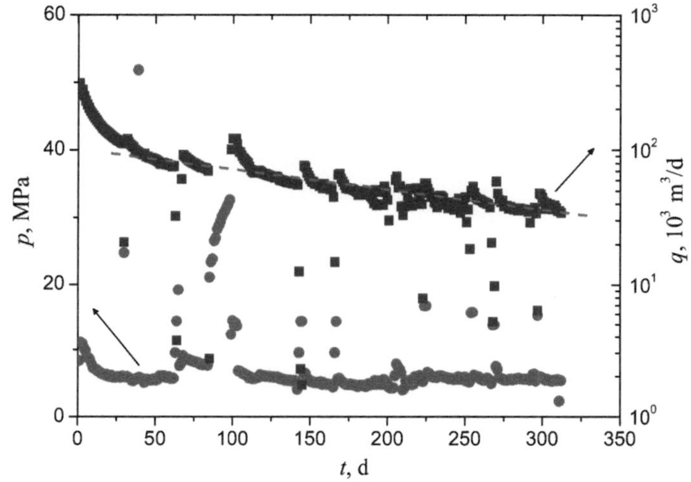

Figure 8.8 Pressure and rate data of Well D (after Pratikno et al., 2003)

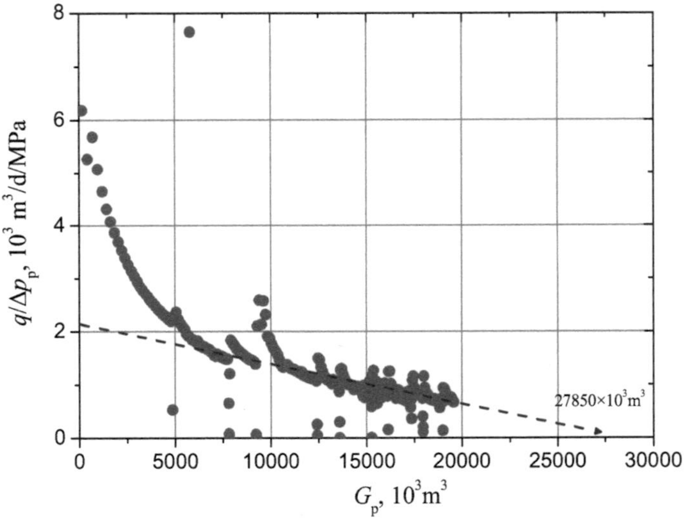

Figure 8.9 Normalized rate and cumulative gas production of Well D (after Pratikno et al., 2003)

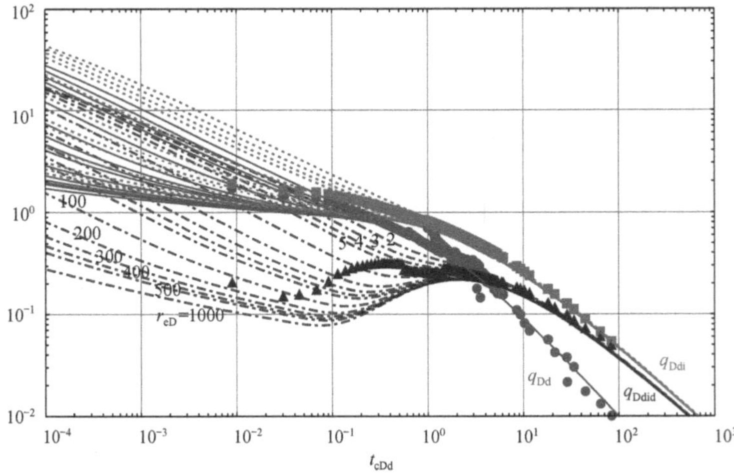

Figure 8.10 Matches of production data for Well D (F_{CD} = 200) (after Pratikno et al., 2003)

1. Calculate the pseudo-steady state flow constant b_{Dpss}

$$u = \ln F_{CD} = \ln 200 = 5.2983$$

$$b_{Dpss} = \ln r_{eD} - 0.049298 + 0.43464 r_{eD}^{-2} + \frac{a_1 + a_2 u + a_3 u^2 + a_4 u^3 + a_5 u^4}{1 + b_1 u + b_2 u^2 + b_3 u^3 + b_4 u^4} \tag{8.98}$$

$$= 2.2639478$$

2. Estimate the OGIP

$$G = \frac{1}{C_t}\left(\frac{t_{ca}}{t_{caDd}}\right)_M \left(\frac{q/\Delta p_p}{q_{Dd}}\right)_M = \frac{1}{0.0074}(54.95)_M (3.66058 \times 1000)_M = 2718 \times 10^4 \text{m}^3 \quad (8.99)$$

3. Calculate the drainage area and drainage radius

$$A = \frac{GB_{gi}}{\varphi h(1-S_{wi})} = \frac{2718 \times 10^4 \times 0.003087}{51.82 \times 0.088 \times (1-0.131)} = 2.117 \times 10^4 \text{ m}^2 \quad (8.100)$$

$$r_e = \sqrt{A/\pi} = \sqrt{2.117 \times 10^4/\pi} = 82 \text{ m} \quad (8.101)$$

4. Calculate the permeability

$$K = 1.866 \times \frac{(q/\Delta p_p)_M}{(q_{Dd})_M} \frac{\mu B}{h} b_{Dpss} = 1.866 \times (3.66058 \times 1000)_M \times \frac{0.0361 \times 0.003087}{51.82} \times 2.2639478 = 0.033 \text{ mD} \quad (8.102)$$

5. Calculate the fracture half-length

$$x_f = \frac{r_e}{r_{eD}} = \frac{82}{10} = 8.2 \text{ m} \quad (8.103)$$

6. Calculate the skin factor

$$S = \ln\frac{r_w}{x_f} + \frac{1.648546 - 3.002711 \times 10^{-1} u + 1.506532 \times 10^{-1} u^2}{1 + 2.136604 \times 10^{-1} u + 9.513761 \times 10^{-2} u^2 + 8.276998 \times 10^{-3} u^3} = \ln\frac{0.1}{8.2} + 0.71 = -3.70 \quad (8.104)$$

Dynamic Material Balance Method

9

In this chapter, we mainly introduce the fundamental principles of flow material balance and dynamic material balance methods presented by Mattar et al., as well as material balance methods presented by Agarwal et al. Furthermore, we also present several case studies using the previous methods.

9.1 Mattar Method

9.1.1 Flow Material Balance Method (Constant Rate)

The flow material balance method is applicable only when the flow reaches boundary-dominated flow period. In this case, the flow reaches a steady or pseudo-steady state, and the pressures at all points of the reservoir drops at an identical velocity, as shown in Figure 9.1. We can also see that the change in average pressure of reservoir equals to the change in bottom hole flowing pressure (BHFP). Thus, a pressure equation can be derived as

$$p_D = \frac{2t_D}{r_{eD}^2} + \left(\ln r_{eD} - \frac{3}{4} \right) \tag{9.1}$$

For an oil well, rearrange Eq. (9.1),

$$\frac{(p_i - p_{wf})}{q} = mt + b_{pss} \tag{9.2}$$

where

$$m = \frac{1}{NC_t} \tag{9.3}$$

$$b_{pss} = \frac{\mu B}{4\pi Kh} \ln\left(\frac{4A}{C_A e^{\gamma} r_w^2} \right) \tag{9.4}$$

For a closed circular reservoir,

$$b_{pss} = \frac{\mu B}{2\pi Kh} \left(\ln r_{eD} - \frac{3}{4} \right) \tag{9.5}$$

Advanced Production Decline Analysis and Application

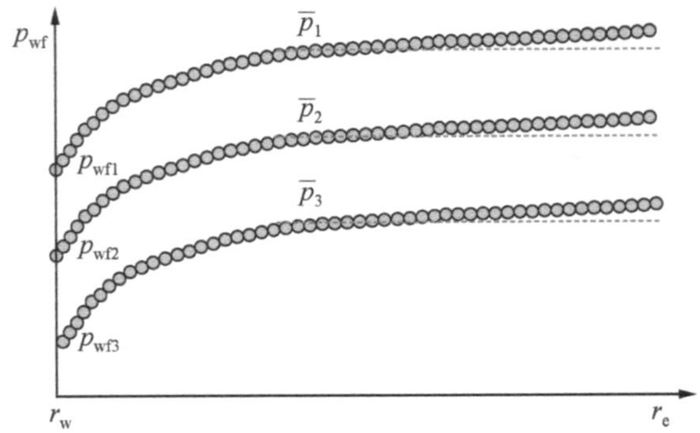

Figure 9.1 Pressure depletion during the boundary-dominated flow (after Mattar and McNeil, 1998)

Plot the $\frac{(p_i - p_{wf})}{q} \sim t$ relation curve using Eq. (9.2), and estimate the value of reserves N based on slope. If the formation flow capacity Kh is known, the reservoir drainage radius can be estimated based on the intercept, which is the well-known "reservoir limit test (RLT)" in well test. For a variable rate case, we only need to replace the time by the material balance time t_c.

For a gas well, draw the $\frac{p_{wf}}{Z} \sim G_p$ curve based on the material balance equation, regress the straight slope, draw the parallel line of $\frac{p_{wf}}{Z} \sim G_p$ crossing $\left(0, \frac{p_i}{Z_i}\right)$ point, and extrapolate it to $\frac{p}{Z} = 0$. The lateral axis is the original gas (oil) in place, as shown in Figure 9.2.

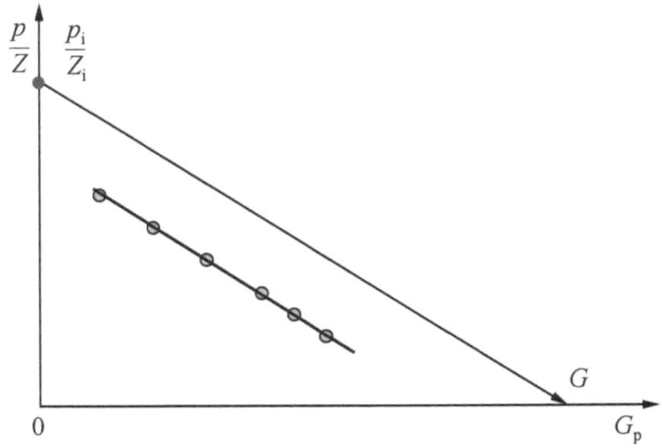

Figure 9.2 Flowing P/Z plot at constant rate production (after Mattar and McNeil, 1998)

9.1.2 Dynamic Material Balance Method (Variable Rate)

Regarding the variable rate case, Mattar (2006) modified his flow material balance method presented in 1998. For a gas well, Eq. (9.1) is rearranged as Eq. (4.67)

$$\frac{p_{p_i} - p_{p_{wf}}}{q} = \frac{t_{ca}}{GC_{ti}} + \frac{(\mu B)_i}{2\pi Kh}\left[\frac{1}{2}\ln\left(\frac{4A}{C_A e^{\gamma} r_w^2}\right)\right] \tag{9.6}$$

or

$$\frac{\Delta p_p}{q} = m_a t_{ca} + b_{a,pss} \tag{9.7}$$

where

$$\Delta p_p = p_{p_i} - p_{p_{wf}} \tag{9.8}$$

$$m_a = \frac{1}{GC_{ti}} \tag{9.9}$$

$$b_{a,pss} = \frac{(\mu B)_i}{2\pi Kh}\left[\frac{1}{2}\ln\left(\frac{4A}{C_A e^{\gamma} r_w^2}\right)\right] \tag{9.10}$$

Eq. (9.6) indicates that when a pseudo-steady flow occurs in a gas well, the total pressure drop (from the initial reservoir pressure to BHFP) consists of two parts, including the pressure drop resulting from depletion, that is, $\frac{t_{ca}}{GC_{ti}}$, and the pressure loss in the course of gas flowing from reservoir to the bottomhole, that is, $b_{a,pss} = \frac{(\mu B)_i}{2\pi Kh}\left[\frac{1}{2}\ln\left(\frac{4A}{C_A e^{\gamma} r_w^2}\right)\right]$.

It can be seen from Eq. (9.7) that $\frac{\Delta p_p}{q} \sim t_{ca}$ assumes a linear relation at the Cartesian coordinates. The dynamic material balance curve of a gas well with variable rate can be obtained through the following procedures:

1. Transform the pressure data into normalized pseudo-pressure Δp_p.
2. Given an original gas in place Δp_p, calculate the normalized pseudo-time t_{ca} using Eq. (4.64).
3. Plot the $\frac{\Delta p_p}{q} \sim t_{ca}$ relation curve, with the intercept of straight line as $b_{a,pss}$.
4. Calculate the average normalized pseudo-pressure of the reservoir using Eq. (4.66).
5. Transform the average normalized pseudo-pressure of the reservoir into an average reservoir pressure p_R.
6. Calculate p_R/Z, and plot the relation curve of p_R/Z versus the cumulative gas production G_p. It is similar to the drawing of a conventional material balance curve of a gas reservoir, where the intercept at x-axis is the original gas in place G.
7. Repeat steps 2–6 with a new value of G until G converges.

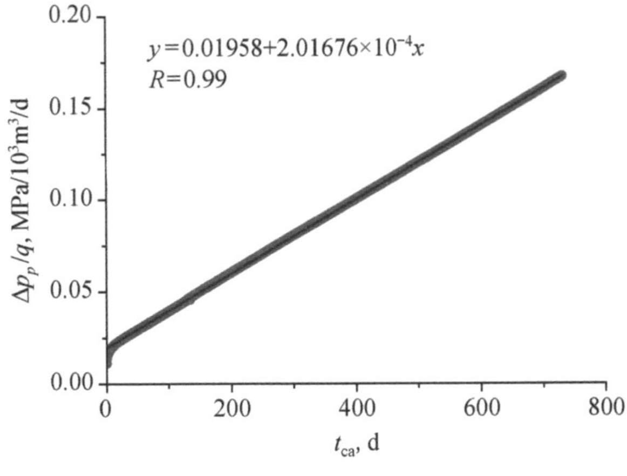

Figure 9.3 Determination of $b_{a,pss}$

The case in Section 4.4 is still used here. Ignore the iterative process of G, and assume $G = 2.0 \times 10^8$ m³. The calculating process is briefly described as follows:

1. Calculate the normalized pseudo-pressure and pseudo-time, and plot the $\frac{\Delta p_p}{q} \sim t_{ca}$ relation curve, as shown in Figure 9.3, with the intercept of straight line as $b_{a,pss} = 0.01958$ MPa/10³ m³/d.
2. According to Eq. (4.66), calculate the average normalized pseudo-pressure of the reservoir, and transform it into the average reservoir pressure p_R. Calculate p_R/Z, and plot the relation curve of p_R/Z vs. G_p. The intercept at x-axis is the original gas in place G, as shown in Figure 9.4.

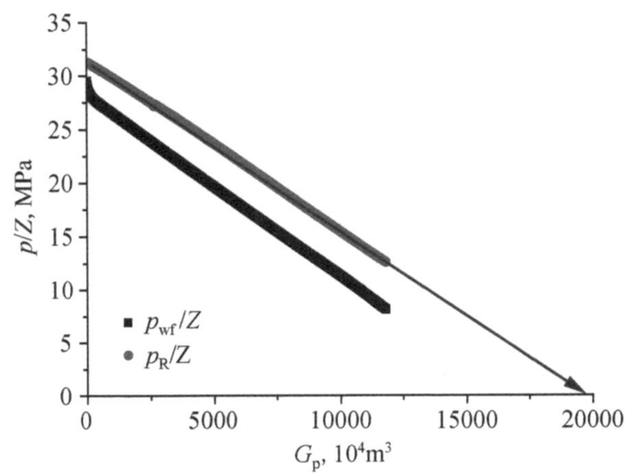

Figure 9.4 Dynamic material balance plot of Mattar method

9.2 Agarwal-Gardner Method

9.2.1 Agarwal-Gardner Method for Oil Well

For an oil well, in a variable rate case, replacing the time by material balance time, Eq. (9.2) can be expressed as

$$\frac{(p_i - p_{wf})}{q} = mt_c + b_{pss} \tag{9.11}$$

Multiplying $\frac{q}{(p_i - p_{wf})b_{pss}}$ at both sides of Eq. (9.11), we get

$$\frac{q}{\Delta p} = -\frac{N_p}{b_{pss} N \Delta p C_t} + \frac{1}{b_{pss}} \tag{9.12}$$

Plot the $\frac{q}{\Delta p} \sim \frac{N_p}{\Delta p C_t}$ curve and determine the reserves N based on the regression results.

9.2.2 Agarwal-Gardner Method for Gas Well

For a gas well, multiplying $\frac{q}{\Delta p_p b_{a,pss}}$ at both sides of Eq. (9.7) yields

$$\frac{1}{b_{a,pss}} = \frac{q}{\Delta p_p b_{a,pss}} m_a t_{ca} + \frac{q}{\Delta p_p} \tag{9.13}$$

Substituting Eq. (9.9) into Eq. (9.13), we have

$$\frac{q}{\Delta p_p} = \frac{1}{b_{a,pss}} - \frac{q}{\Delta p_p C_{ti}} \frac{1}{G b_{a,pss}} t_{ca} \tag{9.14}$$

Plot the $\frac{q}{\Delta p_p} \sim \frac{q}{\Delta p_p C_{ti}} t_{ca}$ curve, and the intercept of the straight line at x-axis is the reserve G.

The computational procedure of this method is similar to that of the Mattar method,

1. Predict the original gas in place (OGIP).
2. Calculate the normalized pseudo-time, normalized pseudo-pressure, and cumulative production, that is, $\frac{q}{\Delta p_p}, t_{ca}, \frac{q}{\Delta p_p C_{ti}} t_{ca}$.
3. Plot the $\frac{q}{\Delta p_p} \sim \frac{q}{\Delta p_p C_{ti}} t_{ca}$ curve, and the intercept of the straight line at x-axis is the reserve G.
4. Achieve the best matching, and extrapolate the reserve G.
5. Repeat above steps using a new value of G until G converges.

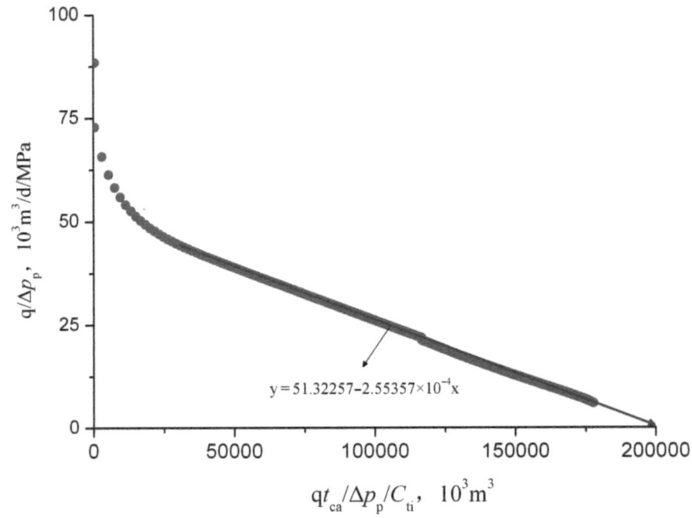

Figure 9.5 Dynamic material balance plot of the Agarwal-Gardner method

Figure 9.5 shows the $\frac{q}{\Delta p_p} \sim \frac{q}{\Delta p_p C_{ti}} t_{ca}$ relation curve, which is drawn on the basis of data in Section 4.1. It comes to a straight line at the late production period, and its point of intersection with the x-axis is $51.32257/(2.55357 \times 10^{-4}) = 2.01 \times 10^5\ 10^3\ m^3$, which is consistent with the original gas in place of $2.0 \times 10^8\ m^3$.

Decline Curves Analysis of Horizontal Well

10

This chapter mainly discusses the theoretical basis, type curve plotting, type curve matching method, and case studies of horizontal well decline curve in a rectangular closed homogeneous reservoir.

10.1 Constant-Rate Solution of Horizontal Well in Rectangular Closed Homogeneous Reservoir

Assume there is a rectangular reservoir with a closed outer boundary where the length and width are $2x_e$ and $2y_e$, and the reservoir thickness is h. A horizontal well with lateral section of L is producing at a constant rate of q in the central of reservoir, as shown in Figure 10.1. The bottom-hole flowing pressure is p_{wf}, the initial reservoir pressure is p_i, the wellbore radius is r_w, the reservoir porosity is φ, the total compressibility is C_t, the reservoir horizontal permeability is K_h, the vertical permeability is K_v, the fluid viscosity is μ, and the formation volume factor is B.

Shih (1994) introduced the line source solution expression of the previous model under the uniform flow rate condition,

$$
\begin{aligned}
\overline{p}_{wD} = {} & \frac{\pi}{x_{eD}}\left\{\frac{\cosh\left[\sqrt{s}\left(y_{eD}-\left|y_D-y_{wD}\right|\right)\right]+\cosh\left[\sqrt{s}\left(y_{eD}-y_D-y_{wD}\right)\right]}{\sqrt{s}\sinh\sqrt{s}\,y_{eD}}\right\} \\
& +\frac{2}{s}\sum_{m=1}^{\infty}\frac{1}{m}\sin\left(m\pi\frac{1}{x_{eD}}\right)\cos\left(m\pi\frac{x_{wD}}{x_{eD}}\right)\cos\left(m\pi\frac{x_D}{x_{eD}}\right) \\
& \times\left\{\frac{\cosh\left[\sqrt{s+\varepsilon_m}\left(y_{eD}-\left|y_D-y_{wD}\right|\right)\right]+\cosh\left[\sqrt{s+\varepsilon_m}\left(y_{eD}-y_D-y_{wD}\right)\right]}{\sqrt{s+\varepsilon_m}\sinh\sqrt{s+\varepsilon_m}\,y_{eD}}\right\} \\
& +\frac{2\pi}{x_{eD}s}\sum_{n=1}^{\infty}\cos\left(n\pi\hat{z}_D\right)\cos\left(n\pi\hat{z}_{wD}\right) \\
& \times\left\{\frac{\cosh\left[\sqrt{s+\varepsilon_n}\left(y_{eD}-\left|y_D-y_{wD}\right|\right)\right]+\cosh\left[\sqrt{s+\varepsilon_n}\left(y_{eD}-y_D-y_{wD}\right)\right]}{\sqrt{s+\varepsilon_n}\sinh\sqrt{s+\varepsilon_n}\,y_{eD}}\right\} \\
& +\frac{4}{s}\sum_{n=1}^{\infty}\cos\left(n\pi\hat{z}_D\right)\cos\left(n\pi\hat{z}_{wD}\right)\sum_{m=1}^{\infty}\frac{1}{m}\sin\left(m\pi\frac{1}{x_{eD}}\right)\cos\left(m\pi\frac{x_{wD}}{x_{eD}}\right)\cos\left(m\pi\frac{x_D}{x_{eD}}\right) \\
& \times\left\{\frac{\cosh\left[\sqrt{s+\varepsilon_{n,m}}\left(y_{eD}-\left|y_D-y_{wD}\right|\right)\right]+\cosh\left[\sqrt{s+\varepsilon_{n,m}}\left(y_{eD}-y_D-y_{wD}\right)\right]}{\sqrt{s+\varepsilon_{n,m}}\sinh\sqrt{s+\varepsilon_{n,m}}\,y_{eD}}\right\}
\end{aligned}
\tag{10.1}
$$

Advanced Production Decline Analysis and Application

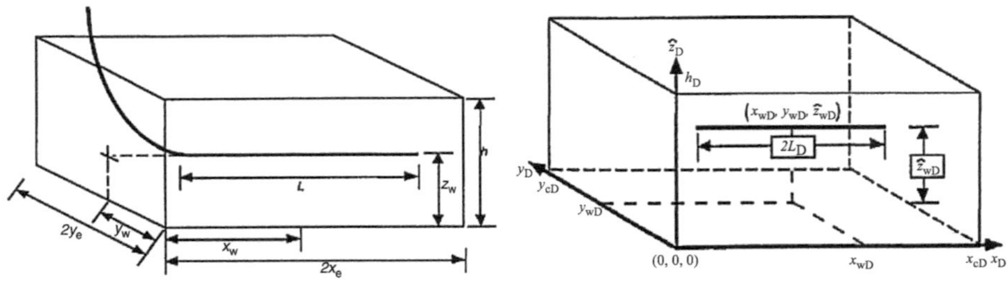

Figure 10.1 Diagrammatic sketch of the horizontal well model (after Shih, 1994)

The dimensionless variables in the expression are as follows

$$p_D = \frac{2\pi K_h h}{q\mu B}(p_i - p_{wf}) \quad t_D = \frac{K_h t}{\varphi\mu C_t B\left(\frac{L}{2}\right)^2}$$

$$x_D = \frac{2x}{L} \quad y_D = \frac{2y}{L} \quad \hat{z}_D = \frac{z}{h}$$

$$x_{wD} = \frac{2x_w}{L} \quad y_{wD} = \frac{2y_w}{L} \quad \hat{z}_{wD} = \frac{z_w}{h}$$

$$x_{eD} = \frac{4x_e}{L} \quad y_{eD} = \frac{4y_e}{L} \quad h_D = \frac{2h\beta}{L} \quad \beta = \sqrt{\frac{K_h}{K_v}} \quad L_D = \frac{L}{2h\beta}$$

$$r_{wD} = \frac{2r_w}{L} \quad r_{wDz} = L_D r_{wD}$$

$$\varepsilon_m = \frac{m^2\pi^2}{x_{eD}^2} \quad \varepsilon_n = n^2\pi^2 L_D^2 \quad \varepsilon_{n,m} = n^2\pi^2 L_D^2 + \frac{m^2\pi^2}{x_{eD}^2}$$

When $x_D = x_{wD}$, $y_D = y_{wD}$, and $z_D = z_{wD} + L_D r_{wD}$, the uniform flow rate solution can be obtained. When $x_D = x_{wD} + 0.732$, $y_D = y_{wD}$, and $z_D = z_{wD} + L_D r_{wD}$, the infinite conductivity solution can be obtained.

10.2 Blasingame Type Curves Plotting

10.2.1 Horizontal Well Shape Factor of Finite Conductivity

The asymptotic solution of Eq. (10.1) in the pseudo-steady state can be expressed as

$$p_{wD} = 2\pi t_{DA} + \frac{1}{2}\ln\frac{4A}{e^{\gamma}C_{AH}\left(\frac{L'}{2}\right)^2} \tag{10.2}$$

where

$$t_{\mathrm{DA}}=\frac{t_{\mathrm{D}}}{A_{\mathrm{D}}}=\frac{t_{\mathrm{D}}}{A}\left(\frac{L}{2}\right)^{2}=\frac{K_{\mathrm{h}}t}{\varphi\mu C_{\mathrm{t}}A} \tag{10.3}$$

$$L^{'}=Le^{-S} \tag{10.4}$$

Ozkan (1988) introduced the following formula to calculate the horizontal well shape factor C_{AH}

$$C_{\mathrm{AH}}=\exp\left\{\begin{array}{l}\ln\left(\dfrac{4}{e^{\gamma}A_{\mathrm{D}}}\right)-4\pi\dfrac{y_{\mathrm{eD}}}{x_{\mathrm{eD}}}\left(\dfrac{1}{3}-\dfrac{y_{\mathrm{D}}}{x_{\mathrm{D}}}+\dfrac{y_{\mathrm{D}}^{2}+y_{\mathrm{wD}}^{2}}{2y_{\mathrm{eD}}^{2}}\right)-\\ \dfrac{4x_{\mathrm{eD}}}{\pi}\displaystyle\sum_{k=1}^{\infty}\dfrac{1}{k^{2}}\sin\left(k\pi\dfrac{1}{x_{\mathrm{eD}}}\right)\cos\left(k\pi\dfrac{x_{\mathrm{wD}}}{x_{\mathrm{eD}}}\right)\cos\left(k\pi\dfrac{x_{\mathrm{D}}}{x_{\mathrm{eD}}}\right)\\ \times\left[\dfrac{\cosh\left[k\pi\left(\dfrac{y_{\mathrm{eD}}-\left|y_{\mathrm{D}}-y_{\mathrm{wD}}\right|}{x_{\mathrm{eD}}}\right)\right]+\cosh\left[\dfrac{k\pi\left(y_{\mathrm{eD}}-y_{\mathrm{D}}-y_{\mathrm{wD}}\right)}{x_{\mathrm{eD}}}\right]}{\sinh k\pi\dfrac{y_{\mathrm{eD}}}{x_{\mathrm{eD}}}}\right]-2F_{1}\end{array}\right\} \tag{10.5}$$

where

$$\begin{aligned}F_{1}=&\frac{2}{x_{\mathrm{eD}}L_{\mathrm{D}}}\sum_{n=1}^{\infty}\frac{1}{n}\cos\left(n\pi\hat{z}_{\mathrm{D}}\right)\cos\left(n\pi\hat{z}_{\mathrm{wD}}\right)\\&\times\left\{\frac{\cosh\left[n\pi L_{\mathrm{D}}\left(y_{\mathrm{eD}}-\left|y_{\mathrm{D}}-y_{\mathrm{wD}}\right|\right)\right]+\cosh\left[n\pi L_{\mathrm{D}}\left(y_{\mathrm{eD}}-y_{\mathrm{D}}-y_{\mathrm{wD}}\right)\right]}{\sinh n\pi L_{\mathrm{D}}y_{\mathrm{eD}}}\right\}\\&+\frac{4}{\pi}\sum_{n=1}^{\infty}\cos\left(n\pi\hat{z}_{\mathrm{D}}\right)\cos\left(n\pi\hat{z}_{\mathrm{wD}}\right)\sum_{m=1}^{\infty}\frac{1}{m}\sin\left(m\pi\frac{1}{x_{\mathrm{eD}}}\right)\cos\left(m\pi\frac{x_{\mathrm{wD}}}{x_{\mathrm{eD}}}\right)\cos\left(m\pi\frac{x_{\mathrm{D}}}{x_{\mathrm{eD}}}\right)\\&\times\left\{\frac{\cosh\left[\sqrt{\varepsilon_{n,m}}\left(y_{\mathrm{eD}}-\left|y_{\mathrm{D}}-y_{\mathrm{wD}}\right|\right)\right]+\cosh\left[\sqrt{\varepsilon_{n,m}}\left(y_{\mathrm{eD}}-y_{\mathrm{D}}-y_{\mathrm{wD}}\right)\right]}{\sqrt{\dfrac{\varepsilon_{n,m}}{\pi^{2}}}\sinh\sqrt{\varepsilon_{n,m}}\,y_{\mathrm{eD}}}\right\}\end{aligned} \tag{10.6}$$

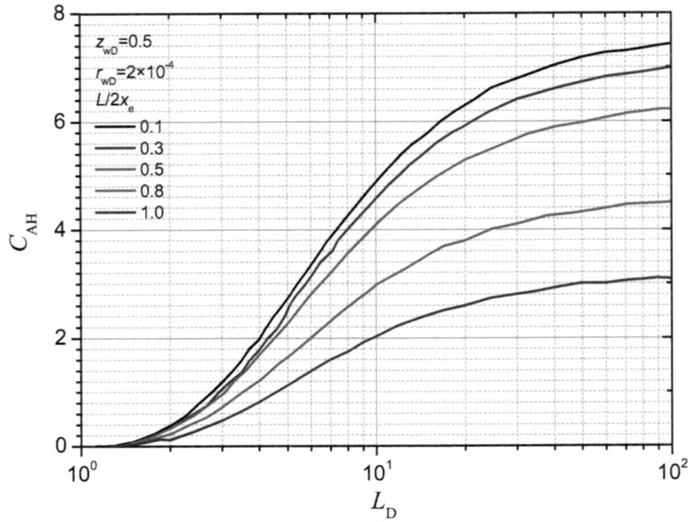

Figure 10.2 Finite conductivity horizontal well shape factor in square reservoir (after Shih, 1994)

Shih (1994) presented the shape factor of a horizontal well in a square reservoir, as shown in Figure 10.2. Because of the calculation complexity of Eq. (10.5), Figure 10.2 can be used to calculate the shape factor in the type curve matching process.

10.2.2 Blasingame Type Curve Plotting

Eq. (10.2) can be expressed as

$$p_{wD} = \alpha t_D + \beta_1 \tag{10.7}$$

where

$$\alpha = \frac{2\pi}{A_D} = \frac{2\pi}{A}\left(\frac{L}{2}\right)^2$$

$$\beta_1 = \frac{1}{2}\ln\frac{4A}{e^{\gamma} C_{AH}\left(\frac{L'}{2}\right)^2}$$

Taking the Laplace transformation of Eq. (10.7), we have

$$\bar{p}_{wD} = \frac{\alpha}{s^2} + \frac{\beta_1}{s} \tag{10.8}$$

According to the superposition principle, the relationship between constant-pressure and constant-rate solutions can be given by

$$\bar{p}_{\mathrm{wD}}\bar{q}_{\mathrm{D}} = \frac{1}{s^2} \tag{10.9}$$

Substituting Eq. (10.8) into Eq. (10.9), we have

$$\bar{q}_{\mathrm{D}} = \frac{1}{s^2\left(\dfrac{\alpha}{s^2}+\dfrac{\beta_1}{s}\right)} = \frac{1}{\beta_1\left(\dfrac{\alpha}{\beta_1}+s\right)} \tag{10.10}$$

Taking the Laplace inverse transformation of Eq. (10.10), we have

$$q_{\mathrm{D}} = \frac{1}{\beta_1}\exp\left(-\frac{\alpha}{\beta_1}t_{\mathrm{D}}\right) \tag{10.11}$$

According to the Fetkovich method, the definition of dimensionless decline time is

$$t_{\mathrm{Dd}} = \frac{\alpha}{\beta_1}t_{\mathrm{D}} \tag{10.12}$$

Dimensionless normalized rate is defined as

$$q_{\mathrm{Dd}} = \frac{\beta_1}{L^{-1}\left[\bar{p}_{\mathrm{D}}\right]} \tag{10.13}$$

Dimensionless normalized rate integral is defined as

$$q_{\mathrm{Ddi}} = \frac{N_{\mathrm{pDd}}}{t_{\mathrm{Dd}}} = \frac{1}{t_{\mathrm{Dd}}}\int_0^{t_{\mathrm{Dd}}} q_{\mathrm{Dd}}(\tau)d\tau \tag{10.14}$$

The dimensionless normalized rate integral derivative q_{Ddid} is defined as

$$q_{\mathrm{Ddid}} = -\frac{dq_{\mathrm{Ddi}}}{d\ln t_{\mathrm{Dd}}} = -t_{\mathrm{Dd}}\frac{dq_{\mathrm{Ddi}}}{dt_{\mathrm{Dd}}} = -t_{\mathrm{Dd}}\frac{d\left(N_{\mathrm{pDd}}/t_{\mathrm{Dd}}\right)}{dt_{\mathrm{Dd}}} \tag{10.15}$$

In the Blasingame chart, three type curves of relationship between the rate function and material balance time can be plotted, including normalized rate curve, normalized rate integral curve, and normalized rate integral derivative curve, as shown in Figure 10.3.

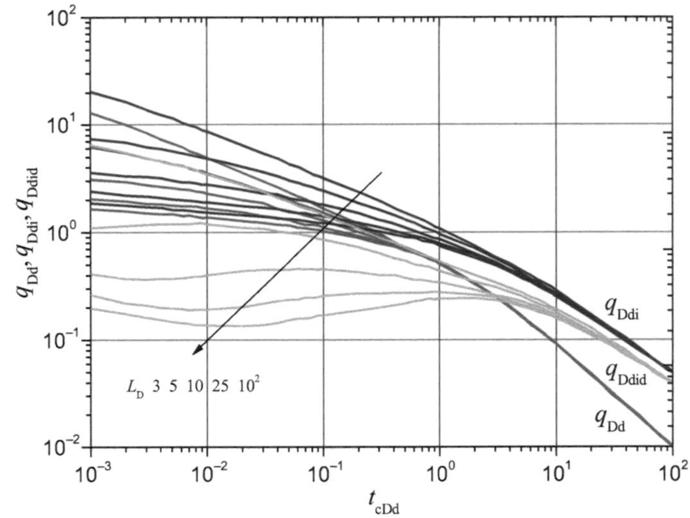

Figure 10.3 Blasingame decline curve of infinite conductivity horizontal well in square reservoir ($Z_{wD} = 0.5$) (after Shih, 1994)

10.3 Decline Curves Analysis Using Blasingame-Style Type Curves

10.3.1 Type Curves Matching Procedure—Oil Well

1. Calculate the material balance time

$$t_c = \frac{N_p}{q} \tag{10.16}$$

2. Calculate the normalized rate

$$\frac{q}{\Delta p} = \frac{q}{p_i - p_{wf}} \tag{10.17}$$

3. Calculate the normalized rate integral

$$\left(\frac{q}{\Delta p}\right)_i = \frac{1}{t_c}\int_0^{t_c} \frac{q}{p_i - p_{wf}} d\tau \tag{10.18}$$

where subscript i represents integration.

4. Calculate the normalized rate integral derivative

$$\left(\frac{q}{\Delta p}\right)_{id} = -\frac{d\left(\frac{q}{\Delta p}\right)_i}{d\ln t_c} = -t_c \frac{d\left(\frac{q}{\Delta p}\right)_i}{dt_c} \tag{10.19}$$

where subscript i represents integration and subscript d represents derivative.

5. Plot the relationships of normalized rate, normalized rate integral, normalized rate integral derivative with material balance time respectively in a log–log coordinate system. These curves are $\frac{q}{\Delta p} \sim t_c$, $\left(\frac{q}{\Delta p}\right)_i \sim t_c$, and $\left(\frac{q}{\Delta p}\right)_{id} \sim t_c$.
6. Choose any of the three curves or any combination thereof to match with the theoretical decline curve chart (Figure 10.3), aiming to obtain the best matching of each curve.
7. Record the value of $L/2x_e$, r_{wD}, and L_D according to the matching result.
8. Record actual matching point $(t_c, q/\Delta p)_M$ and theoretical matching point $(t_{cDd}, q_{Dd})_M$ by selecting any one of the matching points.
9. Calculate the skin factor according to the matching results of r_{wD}

$$S = \ln\left(\frac{r_w}{r_{wa}}\right) = \ln\left(\frac{2r_w}{r_{wD}L}\right) \tag{10.20}$$

10. Calculate the anisotropy coefficient β according to the dimensionless length L_D

$$\beta = \sqrt{\frac{K_h}{K_v}} = \frac{(L/2h)}{(L_D)_M} \tag{10.21}$$

11. Calculate the reserves according to time and rate matching point

$$N = \frac{1}{C_t}\left(\frac{t_c}{t_{Dd}}\right)_{MP}\left(\frac{q/\Delta p}{q_{Dd}}\right)_{MP}(1 - S_w) \tag{10.22}$$

12. Calculate the drainage area

$$A = 4x_e^2 = \frac{NB}{h\varphi(1 - S_{wi})} \tag{10.23}$$

13. Calculate the horizontal permeability according to the rate matching point

$$K_h = \frac{(q/\Delta p)_M}{(q_{Dd})_M}\frac{\mu B \beta_1}{2\pi h} \tag{10.24}$$

10.3.2 Type Curves Matching Procedure—Gas Well

1. Calculate the material balance pseudo-time
   Assuming the reserves is G, calculate the material balance pseudo-time for each production data point.

$$t_{ca} = \frac{(\mu C_t)_i}{q}\int_0^t \frac{q}{\mu(\overline{p})C_t(\overline{p})}dt = \frac{GC_{ti}}{q}\left(p_{p_i} - p_p\right) \tag{10.25}$$

Please refer to the normalized pseudo-pressure p_p in Eq. (4.57), and the average reservoir pressure could be calculated according to Eq. (4.60).

2. Calculate the normalized rate

$$\frac{q}{\Delta p_{\mathrm{p}}} = \frac{q}{p_{p_{\mathrm{i}}} - p_{p_{\mathrm{wf}}}} \tag{10.26}$$

3. Calculate the normalized rate integral

$$\left(\frac{q}{\Delta p_{\mathrm{p}}}\right)_{\mathrm{i}} = \frac{1}{t_{\mathrm{ca}}}\int_0^{t_{\mathrm{ca}}} \frac{q}{\Delta p_{\mathrm{p}}} d\tau \tag{10.27}$$

4. Calculate the normalized rate integral derivative

$$\left(\frac{q}{\Delta p_{\mathrm{p}}}\right)_{\mathrm{id}} = -\frac{d\left(\frac{q}{\Delta p_{\mathrm{p}}}\right)_{\mathrm{i}}}{d\ln t_{\mathrm{ca}}} = -t_{\mathrm{ca}}\frac{d\left(\frac{q}{\Delta p_{\mathrm{p}}}\right)_{\mathrm{i}}}{dt_{\mathrm{ca}}} \tag{10.28}$$

5. Plot $\frac{\Delta p_{\mathrm{p}}}{q} \sim t_{\mathrm{ca}}$ curve in the Cartesian system and calculate G by the regression line slope and Eq. (4.65)

$$G = \frac{1}{\mathrm{Slope} \times C_{\mathrm{ti}}} \tag{10.29}$$

Repeat steps 1–5 to perform iterative calculation until the result converges and meets the allowable error of G.

6. Plot the normalized rate, normalized rate integral, and normalized rate integral derivative changing with material balance pseudo-time, respectively, in a log–log coordinate system. In other words, we need to plot the $\frac{q}{\Delta p_{\mathrm{p}}} \sim t_{\mathrm{ca}}$, $\left(\frac{q}{\Delta p_{\mathrm{p}}}\right)_{\mathrm{i}} \sim t_{\mathrm{ca}}$, and $\left(\frac{q}{\Delta p_{\mathrm{p}}}\right)_{\mathrm{id}} \sim t_{\mathrm{ca}}$ curves.
7. Choose any of the three curves or any combination thereof to match the actual points with the type curves (Figure 10.3), aiming to obtain the best matching of each curve.
8. Record the value of $L/2x_{\mathrm{e}}$, r_{wD}, and L_{D} according to the matching result.
9. Record the actual matching point $(t_{\mathrm{ca}}, q/\Delta p_{\mathrm{p}})_{\mathrm{M}}$ and theoretical matching point $(t_{\mathrm{caDd}}, q_{\mathrm{Dd}})_{\mathrm{M}}$ by selecting any one of the matching points.
10. Calculate the skin factor according to the matching results of r_{wD}.

$$S = \ln\left(\frac{r_{\mathrm{w}}}{r_{\mathrm{wa}}}\right) = \ln\left(\frac{2r_{\mathrm{w}}}{r_{\mathrm{wD}}L}\right) \tag{10.30}$$

11. Calculate the anisotropy coefficient β according to the dimensionless length L_{D}

$$\beta = \sqrt{\frac{K_{\mathrm{h}}}{K_{\mathrm{v}}}} = \frac{(L/2h)}{(L_{\mathrm{D}})_{\mathrm{M}}} \tag{10.31}$$

12. Calculate the reserves according to time and rate matching point

$$G = \frac{1}{C_t}\left(\frac{t_c}{t_{Dd}}\right)_M \left(\frac{q/\Delta p}{q_{Dd}}\right)_M (1 - S_w) \tag{10.32}$$

13. Calculate the drainage area

$$A = 4x_e^2 = \frac{GB_{gi}}{h\varphi(1 - S_{wi})} \tag{10.33}$$

14. Calculate the horizontal permeability according to the rate matching point

$$K_h = \frac{(q/\Delta p)_M}{(q_{Dd})_M} \frac{\mu B_{gi} \beta_1}{2\pi h} \tag{10.34}$$

10.4 Case Study of Blasingame Decline Curves Analysis

Assume there is a gas well producing at a variable BHFP in a square reservoir. The initial reservoir pressure is 30 MPa, reservoir temperature is 80°C, pay thickness is 25 m, reservoir porosity is 0.1, closed boundary is 1000 m, OGIP is 6.35×10^8 m^3, gas relative density is 0.6, pseudo critical temperature is 195.697 K, pseudo critical pressure is 4.66875 MPa, horizontal permeability of well test interpretation is 1.0 mD, vertical permeability is 0.25 mD, skin factor is 0.0, production time is 1000 days, cumulative gas production is 3.0×10^8 m^3, fluid viscosity is 0.0228 mPa·s, rock compressibility is 4.35113×10^{-4} MPa^{-1}, initial gas compressibility is 0.0246 MPa^{-1}, initial gas formation volume factor is 0.0039497, and the production data are shown in Table 10.1.

Ignore the process of iterative computations for G, and assume $G = 6.35 \times 10^8$ m^3 immediately. And the analysis process is as follows.

1. Calculate the material balance pseudo-time and the relative normalized rate parameters.
2. Calculate the average reservoir pressure, normalized pseudo-time, normalized rate, normalized rate integral, and normalized rate integral derivative according to Eq. (4.60), (4.110), (4.111), (4.112), and (4.113). The results are shown in Table 10.2.
3. Plot log–log curves of the normalized rate, normalized rate integral, and normalized rate integral derivative, with material balance pseudo-time, respectively, that is $\frac{q}{\Delta p_p} \sim t_{ca}$, $\left(\frac{q}{\Delta p_p}\right)_i \sim t_{ca}$, and $\left(\frac{q}{\Delta p_p}\right)_{id} \sim t_{ca}$ curves, as shown in Figure 10.4.
4. Match the actual production curve (Figure 10.4) with the theoretical curve chart (Figure 10.3), and the matching result is shown in Figure 10.5.
5. According to the matching results, $L/2x_e$, r_{wD}, and L_D are 1.0, 2×10^{-4}, and 10.0, respectively.
6. The theoretical matching point is $(1, 0.1)_M$, and the corresponding actual matching point is $(78, 20)_M$.

Table 10.1 **Production data of well J**

Production time d	Gas production 10^3 m^3/d	Flowing pressure MPa	Cumulative production 10^4 m^3	Production time d	Gas production 10^3 m^3/d	Flowing pressure MPa	Cumulative production 10^4 m^3	Production time d	Gas production 10^3 m^3/d	Flowing pressure MPa	Cumulative production 10^4 m^3
1	474.4	28.97	47	335	339.7	21.51	13374	670	243.0	16.68	23174
5	472.5	28.47	237	340	338.0	21.43	13510	675	241.8	16.62	23295
10	470.1	28.14	472	345	336.3	21.34	13679	680	240.6	16.56	23416
15	467.8	27.91	707	350	334.6	21.26	13848	685	239.4	16.50	23536
20	465.4	27.73	940	355	332.9	21.17	14015	690	238.2	16.44	23655
25	463.1	27.58	1172	360	331.3	21.09	14182	695	237.0	16.39	23774
30	460.8	27.46	1403	365	329.6	21.00	14348	700	235.8	16.33	23892
35	458.5	27.34	1632	370	328.0	20.92	14513	705	234.6	16.27	24009
40	456.2	27.23	1861	375	326.4	20.84	14678	710	233.5	16.21	24126
45	453.9	27.12	2088	380	324.7	20.75	14841	715	232.3	16.16	24243
50	451.7	27.02	2315	385	323.1	20.67	15004	720	231.1	16.10	24359
55	449.4	26.92	2540	390	321.5	20.59	15166	725	230.0	16.04	24474
60	447.2	26.82	2764	395	319.9	20.51	15327	730	228.8	15.99	24588
65	445.0	26.72	2987	400	318.3	20.43	15487	735	227.7	15.93	24702
70	442.7	26.62	3209	405	316.7	20.35	15646	740	226.6	15.88	24816
75	440.5	26.52	3429	410	315.1	20.27	15805	745	225.4	15.82	24929
80	438.3	26.42	3649	415	313.6	20.19	15963	750	224.3	15.77	25041
85	436.1	26.32	3867	420	312.0	20.11	16120	755	223.2	15.71	25153
90	434.0	26.22	4085	425	310.4	20.03	16276	760	222.1	15.66	25264
95	431.8	26.12	4301	430	308.9	19.95	16432	765	221.0	15.60	25375
100	429.7	26.02	4516	435	307.3	19.88	16587	770	219.9	15.55	25485
105	427.5	25.92	4731	440	305.8	19.80	16741	775	218.8	15.50	25595

Production time d	Gas production 10^3 m^3/d	Flowing pressure MPa	Cumulative production 10^4 m^3	Production time d	Gas production 10^3 m^3/d	Flowing pressure MPa	Cumulative production 10^4 m^3	Production time d	Gas production 10^3 m^3/d	Flowing pressure MPa	Cumulative production 10^4 m^3
110	425.4	25.82	4944	445	304.3	19.72	16894	780	217.7	15.44	25704
115	423.3	25.72	5156	450	302.8	19.65	17046	785	216.6	15.39	25812
120	421.1	25.62	5367	455	301.3	19.57	17198	790	215.5	15.34	25920
125	419.0	25.52	5577	460	299.8	19.49	17349	795	214.4	15.29	26028
130	417.0	25.42	5786	465	298.3	19.42	17499	800	213.4	15.23	26135
135	414.9	25.32	5993	470	296.8	19.35	17649	805	212.3	15.18	26241
140	412.8	25.22	6200	475	295.3	19.27	17797	810	211.2	15.13	26347
145	410.7	25.12	6406	480	293.8	19.20	17945	815	210.2	15.08	26452
150	408.7	25.02	6611	485	292.4	19.12	18092	820	209.1	15.03	26557
155	406.7	24.92	6814	490	290.9	19.05	18239	825	208.1	14.98	26661
160	404.6	24.81	7017	495	289.4	18.98	18385	830	207.1	14.93	26765
165	402.6	24.71	7219	500	288.0	18.91	18530	835	206.0	14.88	26868
170	400.6	24.61	7420	505	286.6	18.83	18674	840	205.0	14.83	26971
175	398.6	24.51	7619	510	285.1	18.76	18817	845	204.0	14.78	27073
180	396.6	24.41	7818	515	283.7	18.69	18960	850	203.0	14.73	27175
185	394.6	24.31	8016	520	282.3	18.62	19102	855	201.9	14.68	27276
190	392.7	24.22	8212	525	280.9	18.55	19244	860	200.9	14.63	27377
195	390.7	24.12	8408	530	279.5	18.48	19385	865	199.9	14.58	27477
200	388.8	24.02	8603	535	278.1	18.41	19525	870	198.9	14.53	27576
205	386.8	23.92	8797	540	276.7	18.34	19664	875	197.9	14.49	27676
210	384.9	23.82	8990	545	275.3	18.27	19803	880	197.0	14.44	27774
215	383.0	23.73	9181	550	274.0	18.21	19941	885	196.0	14.39	27872
220	381.1	23.63	9372	555	272.6	18.14	20078	890	195.0	14.34	27970
225	379.2	23.53	9562	560	271.2	18.07	20214	895	194.0	14.29	28067

(Continued)

Table 10.1 **Production data of well J** ***(cont.)***

Production time d	Gas production $10^3\ m^3/d$	Flowing pressure MPa	Cumulative production $10^4\ m^3$	Production time d	Gas production $10^3\ m^3/d$	Flowing pressure MPa	Cumulative production $10^4\ m^3$	Production time d	Gas production $10^3\ m^3/d$	Flowing pressure MPa	Cumulative production $10^4\ m^3$
230	377.3	23.43	9751	565	269.9	18.01	20350	900	193.1	14.25	28164
235	375.4	23.34	9939	570	268.5	17.94	20485	905	192.1	14.20	28260
240	373.5	23.24	10127	575	267.2	17.87	20620	910	191.1	14.15	28356
245	371.7	23.15	10313	580	265.9	17.81	20754	915	190.2	14.11	28451
250	369.8	23.05	10498	585	264.5	17.74	20887	920	189.2	14.06	28546
255	368.0	22.96	10682	590	263.2	17.68	21020	925	188.3	14.02	28640
260	366.1	22.87	10866	595	261.9	17.61	21151	930	187.4	13.97	28734
265	364.3	22.77	11048	600	260.6	17.55	21283	935	186.4	13.93	28828
270	362.5	22.68	11230	605	259.3	17.48	21413	940	185.5	13.88	28921
275	360.7	22.59	11411	610	258.0	17.42	21543	945	184.6	13.84	29013
280	358.9	22.50	11590	615	256.7	17.36	21672	950	183.6	13.79	29105
285	357.1	22.40	11769	620	255.4	17.29	21801	955	182.7	13.75	29197
290	355.3	22.31	11947	625	254.2	17.23	21929	960	181.8	13.70	29288
295	353.5	22.22	12124	630	252.9	17.17	22056	965	180.9	13.66	29378
300	351.8	22.13	12301	635	251.6	17.10	22183	970	180.0	13.62	29468
305	350.0	22.04	12476	640	250.4	17.04	22309	975	179.1	13.57	29558
310	348.3	21.95	12651	645	249.1	16.98	22435	980	178.2	13.53	29647
315	346.5	21.86	12824	650	247.9	16.92	22559	985	177.3	13.49	29736
320	344.8	21.78	12997	655	246.7	16.86	22684	990	176.4	13.44	29825
325	343.1	21.69	13169	660	245.4	16.80	22807	995	175.6	13.40	29913
330	341.4	21.60	13340	665	244.2	16.74	23052	1000	174.7	13.37	30000

Table 10.2 **Normalized rate parameters of well J**

	q	p_{wf}	p_{avg}	$\Delta p_p(p_{avg})$	$\Delta p_p\ (p_{wf})$	t_{ca}	$q/\Delta p_p$	$(q/\Delta p_p)_i$	$(q/\Delta p_p)_{id}$
d	**10^3 m^3**	**MPa**	**MPa**	**MPa**	**MPa**	**d**	**10^3 m^3/MPa**	**10^3 m^3/MPa/d**	**10^3 m^3/MPa/d**
1	474.4	28.97	29.97	0.03	1.03	0.99	462.29	468.65	
10	470.1	28.14	29.70	0.30	1.85	9.91	254.47	332.57	74.16
20	465.4	27.73	29.41	0.59	2.25	19.79	207.16	280.06	71.29
30	460.8	27.46	29.12	0.87	2.52	29.65	182.98	251.49	67.53
40	456.2	27.23	28.84	1.15	2.74	39.49	166.40	232.25	65.12
50	451.7	27.02	28.57	1.43	2.95	49.31	153.37	217.79	63.82
60	447.2	26.82	28.30	1.69	3.14	59.12	142.40	206.16	63.25
70	442.7	26.62	28.03	1.95	3.33	68.92	132.84	196.40	63.11
80	438.3	26.42	27.77	2.21	3.52	78.71	124.37	187.96	63.18
90	434.0	26.22	27.52	2.46	3.72	88.50	116.77	180.51	63.37
100	429.7	26.02	27.27	2.70	3.91	98.28	109.93	173.81	63.56
110	425.4	25.82	27.02	2.94	4.10	108.06	103.72	167.74	63.73
120	421.1	25.62	26.78	3.18	4.29	117.85	98.08	162.19	63.84
130	417.0	25.42	26.54	3.41	4.49	127.63	92.93	157.08	63.89
140	412.8	25.22	26.31	3.63	4.68	137.43	88.23	152.33	63.88
150	408.7	25.02	26.08	3.85	4.87	147.24	83.92	147.92	63.79
160	404.6	24.81	25.86	4.07	5.06	157.05	79.96	143.79	63.65
170	400.6	24.61	25.64	4.28	5.25	166.88	76.31	139.92	63.44
180	396.6	24.41	25.42	4.49	5.44	176.73	72.93	136.28	63.19
190	392.7	24.22	25.21	4.69	5.62	186.59	69.82	132.85	62.89
200	388.8	24.02	25.00	4.89	5.81	196.47	66.93	129.61	62.54
210	384.9	23.82	24.79	5.09	5.99	206.38	64.24	126.53	62.16
220	381.1	23.63	24.59	5.28	6.17	216.31	61.74	123.62	61.76
230	377.3	23.43	24.39	5.46	6.35	226.26	59.40	120.84	61.33
240	373.5	23.24	24.19	5.65	6.53	236.24	57.21	118.20	60.88

(Continued)

Table 10.2 **Normalized rate parameters of well J *(cont.)***

	q	p_{wf}	p_{avg}	$\Delta p_p(p_{avg})$	$\Delta p_p\ (p_{wf})$	t_{ca}	$q/\Delta p_p$	$(q/\Delta p_p)_i$	$(q/\Delta p_p)_{id}$
d	$10^3\ m^3$	MPa	MPa	MPa	MPa	d	$10^3\ m^3/MPa$	$10^3\ m^3/MPa/d$	$10^3\ m^3/MPa/d$
250	369.8	23.05	24.00	5.83	6.70	246.26	55.17	115.68	60.41
260	366.1	22.87	23.81	6.01	6.88	256.31	53.25	113.27	59.92
270	362.5	22.68	23.62	6.18	7.05	266.38	51.45	110.96	59.42
280	358.9	22.50	23.43	6.35	7.21	276.50	49.75	108.75	58.92
290	355.3	22.31	23.25	6.52	7.38	286.66	48.15	106.63	58.40
300	351.8	22.13	23.07	6.68	7.54	296.84	46.64	104.60	57.89
310	348.3	21.95	22.90	6.85	7.70	307.08	45.21	102.64	57.36
320	344.8	21.78	22.72	7.01	7.86	317.36	43.86	100.76	56.84
330	341.4	21.60	22.55	7.16	8.02	327.68	42.58	98.95	56.31
340	338.0	21.43	22.38	7.31	8.17	338.04	41.36	97.20	55.79
350	334.6	21.26	22.22	7.46	8.32	348.45	40.20	95.52	55.26
360	331.3	21.09	22.05	7.61	8.47	358.92	39.10	93.89	54.74
370	328.0	20.92	21.89	7.76	8.62	369.44	38.05	92.31	54.22
380	324.7	20.75	21.73	7.90	8.77	380.00	37.04	90.79	53.70
390	321.5	20.59	21.58	8.04	8.91	390.62	36.09	89.32	53.18
400	318.3	20.43	21.42	8.18	9.05	401.30	35.17	87.89	52.67
410	315.1	20.27	21.27	8.31	9.19	412.03	34.30	86.50	52.16
420	312.0	20.11	21.12	8.44	9.32	422.82	33.46	85.16	51.66
430	308.9	19.95	20.97	8.58	9.46	433.68	32.66	83.85	51.16
440	305.8	19.80	20.83	8.70	9.59	444.59	31.89	82.59	50.67
450	302.8	19.65	20.68	8.83	9.72	455.57	31.15	81.36	50.18
460	299.8	19.49	20.54	8.95	9.85	466.62	30.43	80.16	49.69
470	296.8	19.35	20.40	9.08	9.98	477.73	29.75	79.00	49.21
480	293.8	19.20	20.26	9.20	10.10	488.90	29.09	77.86	48.74
490	290.9	19.05	20.12	9.31	10.22	500.14	28.46	76.76	48.27
500	288.0	18.91	19.99	9.43	10.34	511.47	27.84	75.68	47.81
510	285.1	18.76	19.85	9.54	10.46	522.85	27.26	74.64	47.35

	q	p_{wf}	p_{avg}	$\Delta p_p(p_{avg})$	$\Delta p_p\ (p_{wf})$	t_{ca}	$q/\Delta p_p$	$(q/\Delta p_p)_i$	$(q/\Delta p_p)_{id}$
d	**10^3 m^3**	**MPa**	**MPa**	**MPa**	**MPa**	**d**	**10^3 m^3/MPa**	**10^3 m^3/MPa/d**	**10^3 m^3/MPa/d**
520	282.3	18.62	19.72	9.66	10.58	534.32	26.69	73.61	46.90
530	279.5	18.48	19.59	9.77	10.69	545.86	26.14	72.61	46.45
540	276.7	18.34	19.46	9.88	10.81	557.48	25.61	71.64	46.01
550	274.0	18.21	19.34	9.98	10.92	569.17	25.09	70.69	45.57
560	271.2	18.07	19.21	10.09	11.03	580.95	24.60	69.76	45.14
570	268.5	17.94	19.09	10.19	11.14	592.80	24.12	68.85	44.71
580	265.9	17.81	18.97	10.29	11.24	604.74	23.65	67.96	44.29
590	263.2	17.68	18.85	10.39	11.35	616.77	23.20	67.09	43.87
600	260.6	17.55	18.73	10.49	11.45	628.88	22.76	66.25	43.46
610	258.0	17.42	18.61	10.59	11.55	641.07	22.34	65.41	43.06
620	255.4	17.29	18.49	10.68	11.65	653.36	21.92	64.60	42.65
630	252.9	17.17	18.38	10.78	11.75	665.73	21.53	63.80	42.26
640	250.4	17.04	18.27	10.87	11.85	678.18	21.14	63.02	41.87
650	247.9	16.92	18.15	10.96	11.94	690.75	20.76	62.26	41.48
660	245.4	16.80	18.04	11.05	12.04	703.38	20.39	61.51	41.10
670	243.0	16.68	17.93	11.14	12.13	716.12	20.04	60.77	40.72
680	240.6	16.56	17.83	11.23	12.22	728.96	19.69	60.05	40.35
690	238.2	16.44	17.72	11.31	12.31	741.89	19.35	59.35	39.98
700	235.8	16.33	17.61	11.40	12.40	754.94	19.02	58.65	39.62
710	233.5	16.21	17.51	11.48	12.48	768.09	18.70	57.97	39.26
720	231.1	16.10	17.41	11.56	12.57	781.33	18.39	57.30	38.90
730	228.8	15.99	17.30	11.64	12.65	794.68	18.08	56.65	38.55
740	226.6	15.88	17.20	11.72	12.74	808.14	17.79	56.00	38.20
750	224.3	15.77	17.10	11.80	12.82	821.69	17.50	55.37	37.86
760	222.1	15.66	17.00	11.88	12.90	835.38	17.22	54.75	37.52
770	219.9	15.55	16.91	11.95	12.98	849.16	16.94	54.14	37.18
780	217.7	15.44	16.81	12.03	13.06	863.03	16.67	53.54	36.85

(Continued)

Table 10.2 Normalized rate parameters of well J *(cont.)*

	q	p_{wf}	p_{avg}	$\Delta p_p(p_{avg})$	$\Delta p_p\ (p_{wf})$	t_{ca}	$q/\Delta p_p$	$(q/\Delta p_p)_i$	$(q/\Delta p_p)_{id}$
d	10^3 m^3	MPa	MPa	MPa	MPa	d	10^3 m^3/MPa	10^3 m^3/MPa/d	10^3 m^3/MPa/d
790	215.5	15.34	16.72	12.10	13.13	877.04	16.41	52.95	36.52
800	213.4	15.23	16.62	12.17	13.21	891.14	16.15	52.37	36.20
810	211.2	15.13	16.53	12.24	13.28	905.41	15.90	51.79	35.88
820	209.1	15.03	16.44	12.31	13.36	919.75	15.66	51.23	35.56
830	207.1	14.93	16.35	12.38	13.43	934.22	15.42	50.68	35.25
840	205.0	14.83	16.26	12.45	13.50	948.81	15.18	50.13	34.94
850	203.0	14.73	16.17	12.52	13.57	963.52	14.96	49.60	34.63
860	200.9	14.63	16.08	12.58	13.64	978.36	14.73	49.07	34.33
870	198.9	14.53	15.99	12.65	13.71	993.33	14.51	48.55	34.03
880	197.0	14.44	15.91	12.71	13.77	1008.42	14.30	48.04	33.73
890	195.0	14.34	15.82	12.78	13.84	1023.64	14.09	47.54	33.44
900	193.1	14.25	15.74	12.84	13.91	1038.98	13.88	47.04	33.15
910	191.1	14.15	15.65	12.90	13.97	1054.45	13.68	46.55	32.86
920	189.2	14.06	15.57	12.96	14.03	1070.06	13.48	46.07	32.58
930	187.4	13.97	15.49	13.02	14.10	1085.79	13.29	45.60	32.30
940	185.5	13.88	15.41	13.08	14.16	1101.71	13.10	45.13	32.02
950	183.6	13.79	15.33	13.14	14.22	1117.70	12.92	44.67	31.75
960	181.8	13.70	15.25	13.20	14.28	1133.87	12.73	44.22	31.47
970	180.0	13.62	15.17	13.25	14.34	1150.17	12.56	43.77	31.21
980	178.2	13.53	15.09	13.31	14.39	1166.62	12.38	43.33	30.94
990	176.4	13.44	15.02	13.36	14.45	1183.18	12.21	42.90	30.68
1000	174.7	13.37	14.94	13.42	14.50	1199.95	12.05	42.47	30.41

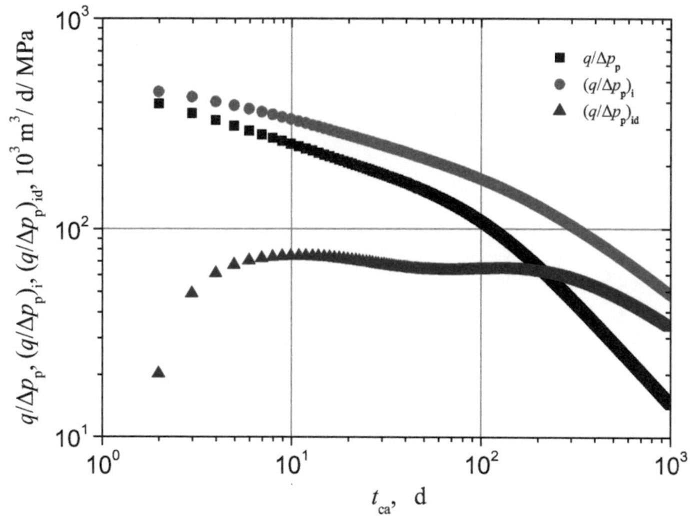

Figure 10.4 Rate functions for well J

7. Calculate the apparent wellbore radius according to the r_{wD} matching result.

$$r_{wa} = \frac{r_{wD}L}{2} = \frac{2\times10^{-4}\times1000}{2} = 0.1\,\text{m}$$

$$S = \ln\left(\frac{r_w}{r_{wa}}\right) = \ln\left(\frac{0.1}{0.1}\right) = 0.0$$

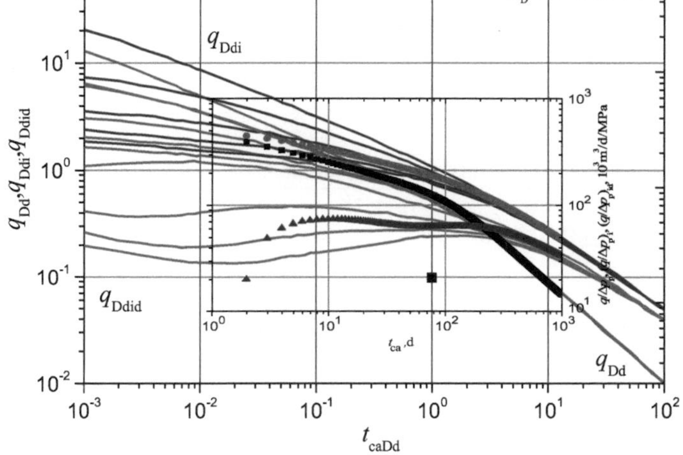

Figure 10.5 Matches of production data for well J

8. Calculate the heterogeneous coefficient according to L_D matching result.

$$\beta = \sqrt{\frac{K_h}{K_v}} = \frac{L/2h}{(L_D)_M} = \frac{1000/(2\times 25)}{10.0} = 2.0$$

9. Calculate the reserves according to time and rate matching points.

$$\begin{aligned} G &= \frac{1}{C_t}\left(\frac{t_{ca}}{t_{caDd}}\right)_M \left(\frac{q/\Delta p_p}{q_{Dd}}\right)_M (1-S_w) \\ &= \frac{1}{0.0246}\left(\frac{78.0}{1.0}\right)_M \left(\frac{20.0\times 1000}{0.1}\right)_M \\ &= 6.34\times 10^8 \, m^3 \end{aligned}$$

10. Calculate the drainage area and drainage radius.

$$A = \frac{GB_{gi}}{h\varphi(1-S_{wi})} = \frac{6.34\times 10^8 \times 0.0039497}{25\times 0.1} = 1001875 \, m^2$$

$$x_e = \sqrt{\frac{A}{4}} = 0.5\times\sqrt{1001875} = 500.5 \, m$$

11. Calculate the horizontal well shape factor according to Figure 10.2.

Substitute the corresponding shape factor $C_{AH} = 2.0$ when $L/2x_e = 1.0$, $L_D = 10.0$.

$$\beta_1 = \frac{1}{2}\ln\left[\frac{4A}{e^{\gamma} C_{Ah} (L'/2)^2}\right] = \frac{1}{2}\ln\left[\frac{4\times 1001875}{e^{0.5772}\times 2.0\times (1000/2)^2}\right] = 0.75207$$

Calculate horizontal permeability according to Eq. (10.34)

$$\begin{aligned} K_h &= \frac{(q/\Delta p_p)_M}{(q_{Dd})_M} \frac{\mu B \beta_1}{2\pi h} \\ &= \frac{1.866\times(20.0\times 1000)\times 0.0228\times 0.0039497\times 0.75207}{0.1\times 25} \\ &= 1.01 \, mD \end{aligned}$$

Calculate vertical permeability according to heterogeneous coefficient β.

$$K_v = \frac{K_h}{\beta^2} = \frac{1.01}{2^2} = 0.25 \, mD$$

The previous calculations for each parameter are consistent with the known parameters.

Decline Curves of Complex Reservoir

11

In this chapter, we mainly introduce the fundamental principles of decline curve analysis methods for vertical wells in the closed circular radial composite reservoir, the two-layered reservoir, the dual-porosity reservoir, and the triple-porosity reservoir. We also introduce the Blasingame type curve plotting and the type curve characteristics. Furthermore, we present the considerations in decline curve plotting. According to the solution method for closed circular homogeneous reservoirs, we can derive the Agarwal-Gardner, normalized pressure integral (NPI), and transient style type curves.

11.1 Decline Curve of Radial Composite Reservoir

11.1.1 Pressure Distribution in a Radial Composite Reservoir: Constant Rate Production

The physical model for a radial composite model reservoir with reservoir thickness of h is shown in Figure 11.1. To simplify the study, assume (1) the reservoir is uniform and isotropic, the radius of the inner and outer regions are r_{e1} and r_{e2} respectively, the outer boundary is closed, and the top and bottom boundaries are impermeable; (2) pores are filled with the single-phase slightly compressible fluid, of which the flow conforms to the Darcy's law; (3) the influence of gravity can be ignored; (4) skin effect and wellbore storage effect are not considered; (5) for both regions, the initial reservoir pressure is p_i, the volume coefficient is B, and the fluid viscosity is μ ($\mu_1 = \mu_2$), the total compressibility is C_{t1} and C_{t2}, respectively, the porosity is φ_1 and φ_2, respectively, and the permeability is K_1 and K_2, respectively; (6) the wellbore radius is r_w,

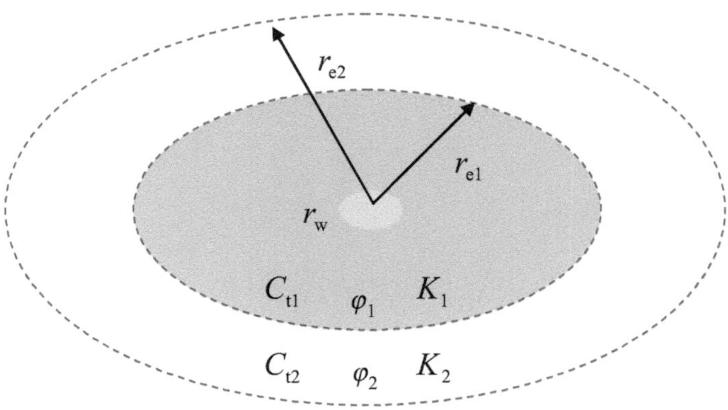

Figure 11.1 Radial composite reservoir

Advanced Production Decline Analysis and Application

the well produces at a constant flow rate of q, and the bottom hole flowing pressure (BHFP) is p_{wf}.

Under the basic SI unit system, the following expression can be obtained

$$\frac{\partial^2 p_1}{\partial r^2}+\frac{1}{r}\frac{\partial p_1}{\partial r}-\frac{1}{\eta_1}\frac{\partial p_1}{\partial t}=0 \quad (r_w<r<r_{e1}) \tag{11.1}$$

$$\frac{\partial^2 p_2}{\partial r^2}+\frac{1}{r}\frac{\partial p_2}{\partial r}-\frac{1}{\eta_2}\frac{\partial p_2}{\partial t}=0 \quad (r_{e1}<r\le r_{e2}) \tag{11.2}$$

The inner boundary condition is

$$\left(r\frac{\partial p_1}{\partial r}\right)_{r=r_w}=\frac{q\mu_1 B}{2\pi K_1 h} \tag{11.3}$$

The interface condition is

$$p_1(\mathrm{r}_{e1},t)=p_2(r_{e1},t) \tag{11.4}$$

$$\left(\frac{\partial p_1}{\partial r}\right)_{r=r_{e1}}=\frac{1}{M_C}\left(\frac{\partial p_2}{\partial r}\right)_{r=r_{e1}} \tag{11.5}$$

where

$$M_C=\frac{(K/\mu)_1}{(K/\mu)_2}$$

The outer boundary condition is

$$\left(\frac{\partial p_2}{\partial r}\right)_{r=r_{e2}}=0 \tag{11.6}$$

The initial condition is

$$p_1(\mathrm{r},0)=p_2(\mathrm{r},0)=0 \tag{11.7}$$

In dimensionless variables

$$\frac{\partial^2 p_{1D}}{\partial r_D^2}+\frac{1}{r_D}\frac{\partial p_{1D}}{\partial r_D}-\frac{\partial p_{1D}}{\partial t_D}=0 \tag{11.8}$$

$$\frac{\partial^2 p_{2D}}{\partial r_D^2} + \frac{1}{r_D}\frac{\partial p_{2D}}{\partial r_D} - \omega_C \frac{\partial p_{2D}}{\partial t_D} = 0 \tag{11.9}$$

where

$$\omega_C = \frac{(K/\varphi\mu C_t)_1}{(K/\varphi\mu C_t)_2}$$

The inner boundary condition is

$$\left(\frac{\partial p_{1D}}{\partial r_D}\right)_{r_D=1} = -1 \tag{11.10}$$

The interface condition is

$$p_{1D}(r_{e1D}, t_D) = p_{2D}(r_{e1D}, t_D) \tag{11.11}$$

$$\left(\frac{\partial p_{1D}}{\partial r_D}\right)_{r_D=r_{e1D}} = \frac{1}{M_C}\left(\frac{\partial p_{2D}}{\partial r_D}\right)_{r_D=r_{e1D}} \tag{11.12}$$

The outer boundary condition is

$$\left(\frac{\partial p_{2D}}{\partial r_D}\right)_{r=r_{e2D}} = 0 \tag{11.13}$$

The initial condition is

$$p_{1D}(r_D, 0) = p_{2D}(r_D, 0) = 0 \tag{11.14}$$

The dimensionless variables are defined as

$$p_{jD} = \frac{2\pi(Kh)_1(p_i - p_j)}{q\mu B} \quad j=1,2 \quad t_D = \left(\frac{K}{\varphi\mu C_t}\right)_1 \frac{t}{r_w^2} \qquad r_D = r/r_w$$

$$r_{e1D} = r_{e1}/r_w \quad r_{e2D} = r_{e2}/r_w$$

Taking the Laplace transformation of Eq. (11.8) through Eq. (11.14), we have

$$\frac{\partial^2 \bar{p}_{1D}}{\partial r_D^2} + \frac{1}{r_D}\frac{\partial \bar{p}_{1D}}{\partial r_D} - s\bar{p}_{1D} = 0 \tag{11.15}$$

$$\frac{\partial^2 \overline{p}_{2D}}{\partial r_D^2} + \frac{1}{r_D}\frac{\partial \overline{p}_{2D}}{\partial r_D} - s\omega_C \overline{p}_{2D} = 0 \tag{11.16}$$

$$\left(\frac{\partial \overline{p}_{1D}}{\partial r_D}\right)_{r_D=1} = -\frac{1}{s} \tag{11.17}$$

$$\overline{p}_{1D}\left(\mathrm{r}_{1D}, s\right) = \overline{p}_{2D}\left(r_{1D}, s\right) \tag{11.18}$$

$$\left(\frac{\partial \overline{p}_{1D}}{\partial r_D}\right)_{r_D=r_{e1D}} = \frac{1}{M_C}\left(\frac{\partial \overline{p}_{2D}}{\partial r_D}\right)_{r_D=r_{e1D}} \tag{11.19}$$

$$\left(\frac{\partial \overline{p}_{2D}}{\partial r_D}\right)_{r=r_{e2D}} = 0 \tag{11.20}$$

Let

$$\sigma_1 = s \quad \text{and} \quad \sigma_2 = s\omega_C \tag{11.21}$$

In the Laplace space, the solution of Eq. (11.15) and Eq. (11.16) are given as follows

$$\overline{p}_{1D}\left(r_D, s\right) = A_0 I_0\left(r_D\sqrt{\sigma_1}\right) + B_0 K_0\left(r_D\sqrt{\sigma_1}\right) \tag{11.22}$$

$$\overline{p}_{2D}\left(r_D, s\right) = C_0 I_0\left(r_D\sqrt{\sigma_2}\right) + D_0 K_0\left(r_D\sqrt{\sigma_2}\right) \tag{11.23}$$

According to the inner boundary condition Eq. (11.17), we have

$$A_0 - B_0\frac{K_1\left(\sqrt{\sigma_1}\right)}{I_1\left(\sqrt{\sigma_1}\right)} = -\frac{1}{s\sqrt{\sigma_1}I_1\left(\sqrt{\sigma_1}\right)} \tag{11.24}$$

According to the inner boundary condition Eq. (11.18), we have

$$A_0 + B_0\frac{K_0\left(r_{e1D}\sqrt{\sigma_1}\right)}{I_0\left(r_{e1D}\sqrt{\sigma_1}\right)} - C_0\frac{I_0\left(r_{e1D}\sqrt{\sigma_2}\right)}{I_0\left(r_{e1D}\sqrt{\sigma_1}\right)} - D_0\frac{K_0\left(r_{e1D}\sqrt{\sigma_2}\right)}{I_0\left(r_{e1D}\sqrt{\sigma_1}\right)} = 0 \tag{11.25}$$

According to the Interface condition Eq. (11.19), we have

$$A_0 - B_0 \frac{K_1\left(r_{e1D}\sqrt{\sigma_1}\right)}{I_1\left(r_{e1D}\sqrt{\sigma_1}\right)} - C_0 \frac{\sqrt{\sigma_2/\sigma_1}}{M_C} \frac{I_1\left(r_{e1D}\sqrt{\sigma_2}\right)}{I_1\left(r_{e1D}\sqrt{\sigma_1}\right)} + D_0 \frac{\sqrt{\sigma_2/\sigma_1}}{M_C} \frac{K_1\left(r_{e1D}\sqrt{\sigma_2}\right)}{I_1\left(r_{e1D}\sqrt{\sigma_1}\right)} = 0 \tag{11.26}$$

According to the outer boundary condition Eq. (11.20), we have

$$C_0 - D_0 \frac{K_1\left(r_{e2D}\sqrt{\sigma_2}\right)}{I_1\left(r_{e2D}\sqrt{\sigma_2}\right)} = 0 \tag{11.27}$$

For Eq. (11.24) to Eq. (11.27), which consider four variables, the solution by the determinant method is

$$\overline{p}_{1D}(1,z) = \frac{\Delta_A I_0\left(\sqrt{\sigma_1}\right) + \Delta_B K_0\left(\sqrt{\sigma_1}\right)}{\Delta} \tag{11.28}$$

where

$$a_{11} = 1 \quad a_{12} = -\frac{K_1\left(\sqrt{\sigma_1}\right)}{I_1\left(\sqrt{\sigma_1}\right)} \quad a_{13} = 0 \quad a_{14} = 0 \quad d_1 = -\frac{1}{s\sqrt{\sigma_1} I_1\left(\sqrt{\sigma_1}\right)}$$

$$a_{21} = 1 \quad a_{22} = -\frac{K_1\left(r_{e1D}\sqrt{\sigma_1}\right)}{I_1\left(r_{e1D}\sqrt{\sigma_1}\right)} \quad a_{23} = -\frac{\sqrt{\sigma_2/\sigma_1}}{M_C} \frac{I_1\left(r_{e1D}\sqrt{\sigma_2}\right)}{I_1\left(r_{e1D}\sqrt{\sigma_1}\right)}$$

$$a_{24} = \frac{\sqrt{\sigma_2/\sigma_1}}{M_C} \frac{K_1\left(r_{e1D}\sqrt{\sigma_2}\right)}{I_1\left(r_{e1D}\sqrt{\sigma_1}\right)} \quad d_2 = 0$$

$$a_{31} = 1 \quad a_{32} = \frac{K_0\left(r_{e1D}\sqrt{\sigma_1}\right)}{I_0\left(r_{e1D}\sqrt{\sigma_1}\right)} \quad a_{33} = -\frac{I_0\left(r_{e1D}\sqrt{\sigma_2}\right)}{I_0\left(r_{e1D}\sqrt{\sigma_1}\right)}$$

$$a_{34} = -\frac{K_0\left(r_{e1D}\sqrt{\sigma_2}\right)}{I_0\left(r_{e1D}\sqrt{\sigma_1}\right)} \quad d_3 = 0$$

$$a_{41} = 0 \quad a_{42} = 0 \quad a_{43} = 1 \quad a_{44} = -\frac{K_1\left(r_{e2D}\sqrt{\sigma_2}\right)}{I_1\left(r_{e2D}\sqrt{\sigma_2}\right)} \quad d_4 = 0$$

$$\Delta = \left(a_{33}a_{44} - a_{34}\right)\left(a_{11}a_{22} - a_{12}\right) + \left(a_{24} - a_{23}a_{44}\right)\left(a_{11}a_{32} - a_{12}\right)$$

$$\Delta_A = d_1\left[a_{22}\left(a_{33}a_{44} - a_{34}\right) + a_{32}\left(a_{24} - a_{23}a_{44}\right)\right]$$

$$\Delta_B = d_1\left[\left(a_{34} - a_{33}a_{44}\right) - \left(a_{24} - a_{23}a_{44}\right)\right]$$

11.1.2 Blasingame Type Curve Plotting

According to Eq. (A-5.27), the dimensionless time is expressed as

$$t_{\mathrm{Dd}}=\frac{\alpha_1}{\beta}t_{\mathrm{D}}=\alpha t_{\mathrm{D}} \tag{11.29}$$

Substituting Eq. (11.29) into Eq. (11.8) and Eq. (11.9), we have

$$\frac{\partial^2 p_{1\mathrm{D}}}{\partial r_{\mathrm{D}}^2}+\frac{1}{r_{\mathrm{D}}}\frac{\partial p_{1\mathrm{D}}}{\partial r_{\mathrm{D}}}-\alpha\frac{\partial p_{1\mathrm{D}}}{\partial t_{\mathrm{Dd}}}=0 \tag{11.30}$$

$$\frac{\partial^2 p_{2\mathrm{D}}}{\partial r_{\mathrm{D}}^2}+\frac{1}{r_{\mathrm{D}}}\frac{\partial p_{2\mathrm{D}}}{\partial r_{\mathrm{D}}}-\omega_{\mathrm{C}}\alpha\frac{\partial p_{2\mathrm{D}}}{\partial t_{\mathrm{Dd}}}=0 \tag{11.31}$$

In the Laplace space, the solution is given by

$$\bar{p}_{1\mathrm{D}}(1,z)=\frac{\Delta_{\mathrm{A}}I_0\left(\sqrt{\sigma_1}\right)+\Delta_{\mathrm{B}}K_0\left(\sqrt{\sigma_1}\right)}{\Delta}=\bar{p}_{\mathrm{wfD}} \tag{11.32}$$

where

$$\sigma_1=\alpha s \quad \sigma_2=\alpha\omega_{\mathrm{C}}s \tag{11.33}$$

The dimensionless decline flow rate is expressed as

$$q_{\mathrm{Dd}}=\frac{\beta}{L^{-1}\left[\bar{p}_{\mathrm{wfD}}\right]} \tag{11.34}$$

The dimensionless decline flow rate integral function q_{Ddi} is expressed as

$$q_{\mathrm{Ddi}}=\frac{N_{\mathrm{pDd}}}{t_{\mathrm{Dd}}}=\frac{1}{t_{\mathrm{Dd}}}\int_0^{t_{\mathrm{Dd}}}q_{\mathrm{Dd}}(\tau)d\tau \tag{11.35}$$

The dimensionless decline flow rate integral derivative q_{Ddid} is expressed as

$$q_{\mathrm{Ddid}}=-\frac{dq_{\mathrm{Ddi}}}{d\ln t_{\mathrm{Dd}}}=-t_{\mathrm{Dd}}\frac{dq_{\mathrm{Ddi}}}{dt_{\mathrm{Dd}}}=-t_{\mathrm{Dd}}\frac{d\left(N_{\mathrm{pDd}}/t_{\mathrm{Dd}}\right)}{dt_{\mathrm{Dd}}} \tag{11.36}$$

11.1.3 Blasingame Type Curve Characteristics

The Blasingame type curve characteristics of radial composite reservoir mainly depend on the composite radius ratio, mobility ratio, and storativity ratio of the inner and outer regions, as shown in Figure 11.2.

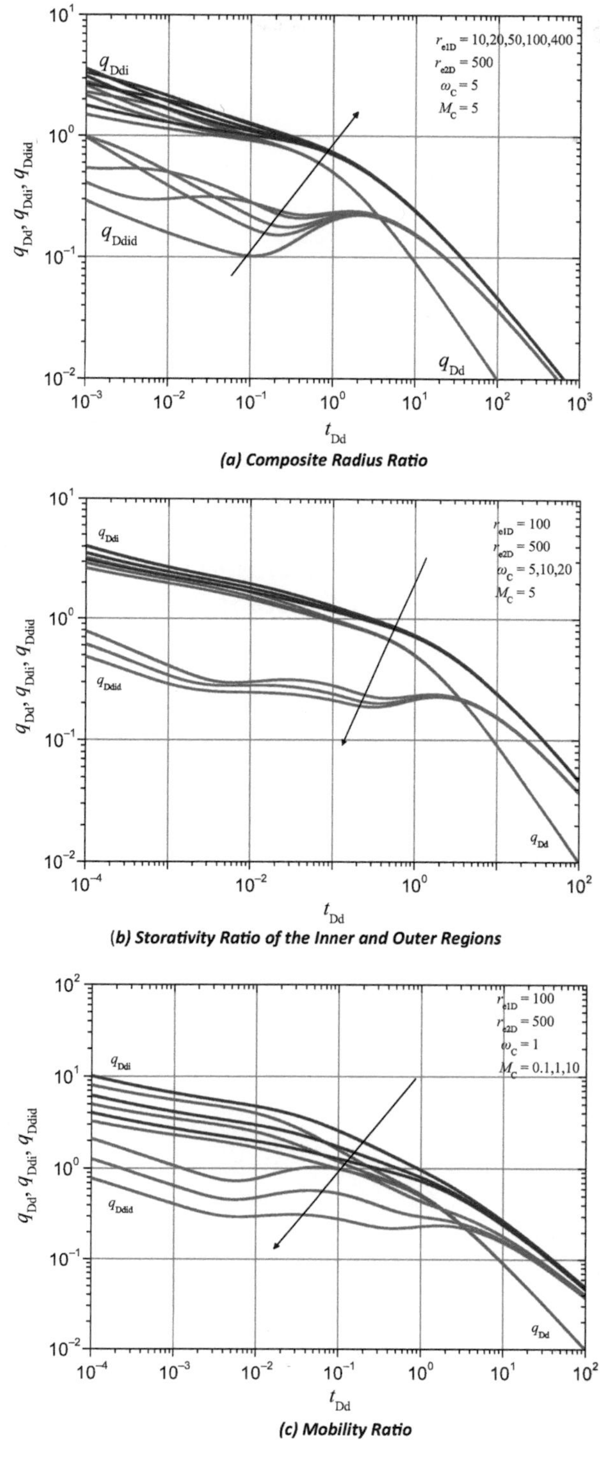

Figure 11.2 Blasingame type curves of radial composite reservoir. (a) Composite radius ratio; (b) storativity ratio of the inner and outer regions; (c) mobility ratio

With a similar method applied to the closed circular homogeneous reservoir, we can also derive the Agarwal-Gardner, NPI, and transient type curves of radial composite reservoir.

11.2 Decline Curve of Two-Layered Reservoir

Many reservoirs are composed of multiple layers as a result of sedimentation. These reservoirs can be classified into two types according to the presence or absence of crossflow, including the commingled reservoir, in which the layers are separated by the impermeable intercalation and crossflow does not exist, and the crossflow reservoir, in which there are communicating phenomena among the layers. In this section, we introduce the decline curve of the simplest multilayer reservoir, that is, the two-layered reservoir.

11.2.1 Pressure Distribution in a Closed Circular Two-Layered Commingled Reservoir with a Constant Rate Production

To simplify the study, assume (1) the reservoir is composed of two uniform and isotropic layers, the reservoir drainage radius is r_{e1} and r_{e2}, respectively, where $r_{e1} < r_{e2}$, the outer boundary is closed, and the top and bottom boundaries are impermeable; (2) pores are filled with single-phase slightly compressible fluid, of which the flow conforms to the Darcy's law; (3) the influence of gravity can be ignored; (4) there is no crossflow between the layers, and the skin effect and wellbore storage effect is not considered; (5) the initial reservoir pressure for both layers is p_i, the volume coefficient is B, the fluid viscosity is μ, the total compressibility is C_{t1} and C_{t2}, respectively, the reservoir thickness is h_1 and h_2, respectively, the porosity is φ_1 and φ_2, respectively, and the permeability is K_1 and K_2, respectively; (6) the two layers are drilled in simultaneously, the wellbore radius is r_w, the well produces at a constant flow rate of q, and the BHFP is p_{wf}, as shown in Figure 11.3.

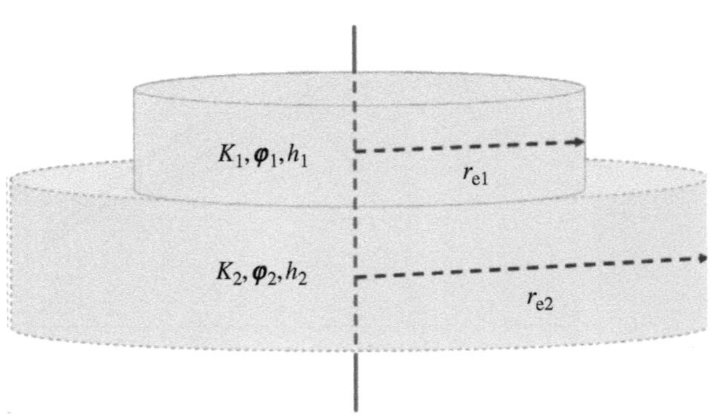

Figure 11.3 Two-layered commingled reservoir

Under the basic SI unit system, the following expressions can be obtained

$$\frac{1}{r}\frac{\partial}{\partial r}\left(r\frac{\partial p_1}{\partial r}\right)=\left(\frac{\varphi C_t}{K}\right)_1\mu\frac{\partial p_1}{\partial t} \tag{11.37}$$

$$\frac{1}{r}\frac{\partial}{\partial r}\left(r\frac{\partial p_2}{\partial r}\right)=\left(\frac{\varphi C_t}{K}\right)_2\mu\frac{\partial p_2}{\partial t} \tag{11.38}$$

$$p_1(r,0)=p_2(r,0)=p_i \tag{11.39}$$

$$p_1(r_w,t)=p_2(r_w,t)=p_{wf} \tag{11.40}$$

$$(Kh)_1\left(r\frac{\partial p_1}{\partial r}\right)_{r=r_w}+(Kh)_2\left(r\frac{\partial p_2}{\partial r}\right)_{r=r_w}=\frac{q\mu B}{2\pi} \tag{11.41}$$

$$\left.\frac{\partial p_1}{\partial r}\right|_{r=r_{e1}}=\left.\frac{\partial p_2}{\partial r}\right|_{r=r_{e2}}=0 \tag{11.42}$$

In dimensionless variables,

$$\frac{1}{r_D}\frac{\partial}{\partial r_D}\left(r_D\frac{\partial p_{1D}}{\partial r_D}\right)=\frac{\omega_1}{\kappa_1}\frac{\partial p_{1D}}{\partial t_D} \tag{11.43}$$

$$\frac{1}{r_D}\frac{\partial}{\partial r_D}\left(r_D\frac{\partial p_{1D}}{\partial r_D}\right)=\frac{\omega_2}{\kappa_2}\frac{\partial p_{1D}}{\partial t_D} \tag{11.44}$$

$$p_{1D}(r_D,0)=p_{2D}(r_D,0)=0 \tag{11.45}$$

$$p_{1D}(1,t_D)=p_{2D}(1,t_D)=p_{wfD} \tag{11.46}$$

$$\kappa_1\left(\frac{\partial p_{1D}}{\partial r_D}\right)_{r_D=1}+\kappa_2\left(\frac{\partial p_{2D}}{\partial r_D}\right)_{r_D=1}=-1 \tag{11.47}$$

$$\left.\frac{\partial p_{1D}}{\partial r_D}\right|_{r_D=r_{e1D}}=\left.\frac{\partial p_{2D}}{\partial r_D}\right|_{r_D=r_{e2D}}=0 \tag{11.48}$$

The dimensionless variables are defined as

$$p_{j\mathrm{D}}=\frac{2\pi(K_1h_1+K_2h_2)}{qB\mu}(p_\mathrm{i}-p_j)\quad p_{\mathrm{wfD}}=\frac{2\pi(K_1h_1+K_2h_2)}{qB\mu}(p_\mathrm{i}-p_{\mathrm{wf}})$$

$$t_\mathrm{D}=\frac{(K_1h_1+K_2h_2)}{(\varphi_1h_1+\varphi_2h_2)C_\mathrm{t}\mu r_\mathrm{w}^2}\cdot t\quad r_\mathrm{D}=\frac{r}{r_\mathrm{w}}\quad r_{\mathrm{e1D}}=\frac{r_{\mathrm{e1}}}{r_\mathrm{w}}\quad r_{\mathrm{e2D}}=\frac{r_{\mathrm{e2}}}{r_\mathrm{w}}$$

$$\omega_1=\frac{\varphi_1h_1C_{\mathrm{t1}}}{\varphi_1h_1C_{\mathrm{t1}}+\varphi_2h_2C_{\mathrm{t2}}}\quad \omega_2=1-\omega_1\quad \kappa_1=\frac{K_1h_1}{(K_1h_1+K_2h_2)}\quad \kappa_2=1-\kappa_1$$

Taking the Laplace transformation of Eq. (11.43) to Eq. (11.48), we have

$$\frac{d^2\bar{p}_{1\mathrm{D}}}{dr_\mathrm{D}^2}+\frac{1}{r_\mathrm{D}}\frac{d\bar{p}_{1\mathrm{D}}}{dr_\mathrm{D}}=\frac{\omega_1 s}{\kappa_1}\bar{p}_{1\mathrm{D}} \tag{11.49}$$

$$\frac{d^2\bar{p}_{2\mathrm{D}}}{dr_\mathrm{D}^2}+\frac{1}{r_\mathrm{D}}\frac{d\bar{p}_{2\mathrm{D}}}{dr_\mathrm{D}}=\frac{\omega_2 s}{\kappa_2}\bar{p}_{2\mathrm{D}} \tag{11.50}$$

$$\bar{p}_{1\mathrm{D}}(1,t_\mathrm{D})=\bar{p}_{2\mathrm{D}}(1,t_\mathrm{D})=\bar{p}_{\mathrm{wfD}} \tag{11.51}$$

$$\kappa_1\left(\frac{\partial\bar{p}_{1\mathrm{D}}}{\partial r_\mathrm{D}}\right)_{r_\mathrm{D}=1}+\kappa_2\left(\frac{\partial\bar{p}_{2\mathrm{D}}}{\partial r_\mathrm{D}}\right)_{r_\mathrm{D}=1}=-\frac{1}{s} \tag{11.52}$$

$$\left.\frac{d\bar{p}_{1\mathrm{D}}}{dr_\mathrm{D}}\right|_{r_\mathrm{D}=r_{\mathrm{e1D}}}=\left.\frac{d\bar{p}_{2\mathrm{D}}}{dr_\mathrm{D}}\right|_{r_\mathrm{D}=r_{\mathrm{e2D}}}=0 \tag{11.53}$$

In the Laplace space, the solution of Eq. (11.49) and Eq. (11.50) is given by

$$\bar{p}_{j\mathrm{D}}=A_jI_0(\sigma_jr_\mathrm{D})+B_jK_0(\sigma_jr_\mathrm{D})\qquad j=1,2 \tag{11.54}$$

Substituting Eq. (11.54) into Eq. (11.49) and Eq. (11.50), we have

$$\sigma_j=\sqrt{\frac{\omega_j s}{\kappa_j}} \tag{11.55}$$

According to the outer boundary condition

$$A_j=B_j\frac{K_1(r_{\mathrm{ejD}}\sigma_j)}{I_1(r_{\mathrm{ejD}}\sigma_j)}=B_jb_j \tag{11.56}$$

Substituting Eq. (11.56) into Eq. (11.54), we have

$$\overline{p}_{j\mathrm{D}} = B_j\left[b_j I_0\left(\sigma_j r_\mathrm{D}\right) + K_0\left(\sigma_j r_\mathrm{D}\right)\right] \tag{11.57}$$

Substituting Eq. (11.57) into Eq. (11.51), we have

$$B_j = \frac{\overline{p}_{\mathrm{wfD}}}{b_j I_0\left(\sigma_j r_\mathrm{D}\right) + K_0\left(\sigma_j r_\mathrm{D}\right)} \tag{11.58}$$

Substituting Eq. (11.58) into Eq. (11.52), we have

$$\overline{p}_{\mathrm{wfD}} = \frac{1}{s\sum_{j=1}^{2}\kappa_j\sigma_j \dfrac{\dfrac{K_1\left(\sigma_j\right)}{I_1\left(\sigma_j\right)} - b_j}{\dfrac{K_0\left(\sigma_j\right)}{I_1\left(\sigma_j\right)} + b_j\dfrac{I_0\left(\sigma_j\right)}{I_1\left(\sigma_j\right)}}} \tag{11.59}$$

11.2.2 Blasingame Type Curve Plotting

According to Eq. (A-6.12), dimensionless time is defined by

$$t_{\mathrm{Dd}} = t_\mathrm{D}\frac{1}{\dfrac{1}{2}\left(r_{\mathrm{e1D}}^2 - 1\right)\left(\ln r_{\mathrm{e1D}} - \dfrac{1}{2}\right)} = t_\mathrm{D}\alpha \tag{11.60}$$

Substituting Eq. (11.60) into Eq. (11.43) through Eq. (11.48), we have

$$\frac{1}{r_\mathrm{D}}\frac{\partial}{\partial r_\mathrm{D}}\left(r_\mathrm{D}\frac{\partial p_{1\mathrm{D}}}{\partial r_\mathrm{D}}\right) = \frac{\omega_1\alpha}{\kappa_1}\frac{\partial p_{1\mathrm{D}}}{\partial t_{\mathrm{Dd}}} \tag{11.61}$$

$$\frac{1}{r_\mathrm{D}}\frac{\partial}{\partial r_\mathrm{D}}\left(r_\mathrm{D}\frac{\partial p_{1\mathrm{D}}}{\partial r_\mathrm{D}}\right) = \frac{\omega_2\alpha}{\kappa_2}\frac{\partial p_{1\mathrm{D}}}{\partial t_\mathrm{D}} \tag{11.62}$$

$$p_{1\mathrm{D}}\left(r_\mathrm{D},0\right) = p_{2\mathrm{D}}\left(r_\mathrm{D},0\right) = 0 \tag{11.63}$$

$$p_{1\mathrm{D}}\left(1,t_\mathrm{D}\right) = p_{2\mathrm{D}}\left(1,t_\mathrm{D}\right) = p_{\mathrm{wfD}} \tag{11.64}$$

$$\sum_{j=1}^{2}\kappa_j\left(\frac{\partial p_{j\mathrm{D}}}{\partial r_\mathrm{D}}\right)_{r_\mathrm{D}=1} = -1 \tag{11.65}$$

$$\left.\frac{\partial p_{1\mathrm{D}}}{\partial r_{\mathrm{D}}}\right|_{r_{\mathrm{D}}=r_{\mathrm{e1D}}}=\left.\frac{\partial p_{2\mathrm{D}}}{\partial r_{\mathrm{D}}}\right|_{r_{\mathrm{D}}=r_{\mathrm{e2D}}}=0 \tag{11.66}$$

In the Laplace space, the solution of Eq. (11.61) to Eq. (11.66) is given by

$$\bar{p}_{\mathrm{wfD}}=\frac{1}{s\sum_{j=1}^{2}\kappa_j\sigma_j\dfrac{\dfrac{K_1(\sigma_j)}{I_1(\sigma_j)}-b_j}{\dfrac{K_0(\sigma_j)}{I_1(\sigma_j)}+b_j\dfrac{I_0(\sigma_j)}{I_1(\sigma_j)}}} \tag{11.67}$$

where

$$\sigma_j=\sqrt{\frac{\omega_j\alpha s}{\kappa_j}} \tag{11.68}$$

The dimensionless decline flow rate is defined by

$$q_{\mathrm{Dd}}=\frac{\beta}{L^{-1}\left[\bar{p}_{\mathrm{wfD}}\right]} \tag{11.69}$$

The dimensionless decline flow rate integral function q_{Ddi} is expressed as

$$q_{\mathrm{Dd\,i}}=\frac{N_{\mathrm{pDd}}}{t_{\mathrm{Dd}}}=\frac{1}{t_{\mathrm{Dd}}}\int_0^{t_{\mathrm{Dd}}} q_{\mathrm{Dd}}(\tau)d\tau \tag{11.70}$$

The dimensionless decline flow rate integral derivative q_{Ddid} is expressed as

$$q_{\mathrm{Dd\,id}}=-\frac{dq_{\mathrm{Dd\,i}}}{d\ln t_{\mathrm{Dd}}}=-t_{\mathrm{Dd}}\frac{d\left(N_{\mathrm{pDd}}/t_{\mathrm{Dd}}\right)}{dt_{\mathrm{Dd}}} \tag{11.71}$$

11.2.3 Blasingame Type Curve Characteristics

The Blasingame type curve characteristics of a two-layered commingled reservoir mainly depend on the drainage radius, the reservoir capacity ratio, and the formation capacity ratio of the two layers as shown in Figure 11.4.

With a similar method applied to the closed circular homogeneous reservoir, we can derive the Agarwal-Gardner, NPI, and transient type curves of the two-layered commingled reservoir.

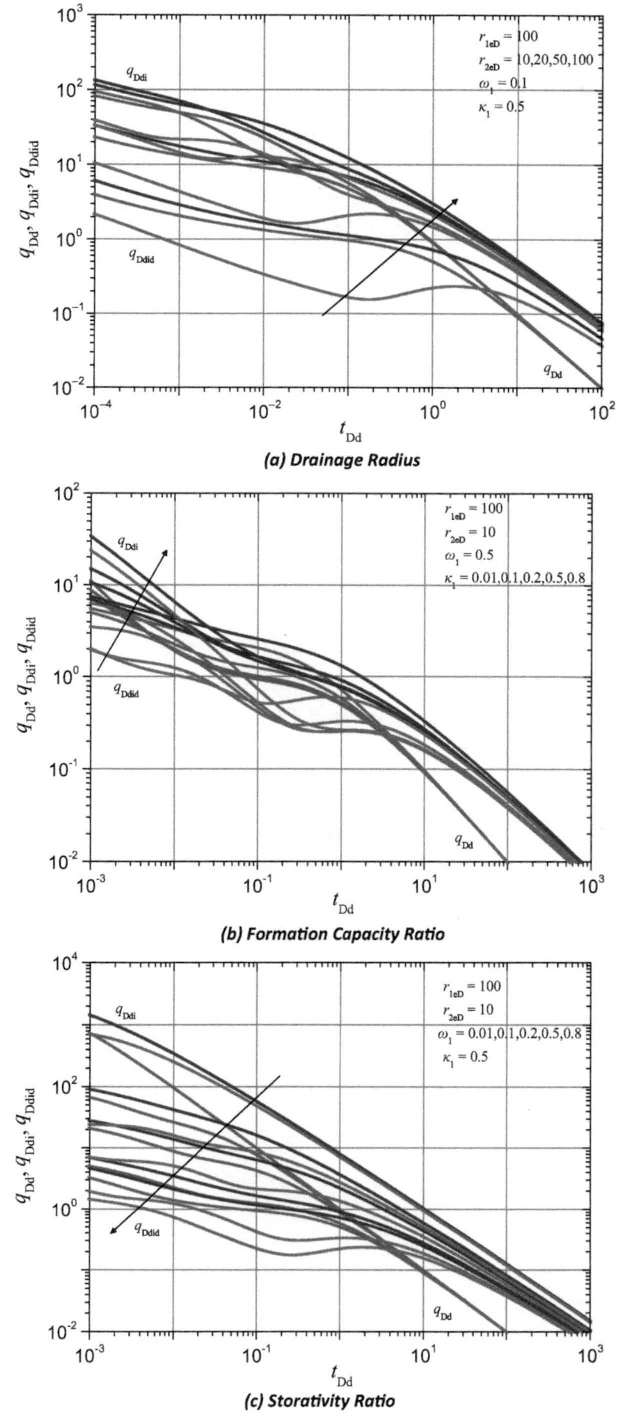

Figure 11.4 Blasingame type curves of two-layered commingled reservoir. (a) Drainage radius; (b) formation capacity ratio; (c) storativity ratio

11.2.4 Pressure Distribution in a Two-Layered Reservoir with Crossflow with Constant Rate Production

If there is crossflow between layers, we only need to reduce the triporosity and dual-permeability model introduced in Section 11.4. That is, when $\lambda_{\text{m1f}} = \lambda_{\text{m1m2}} = \omega_{\text{m1}} = 0$, the issue raised by triporosity and dual-permeability can be transformed into an issue of a two-layered crossflow reservoir, and the pressure solution at a constant rate in Laplace domain space is shown in Eq. (11.156). The solving process is omitted in this section, for which more information can be found in Section 11.4.

11.3 Decline Curve of Dual-Porosity Reservoir

The media with two or more than two pore textures simultaneously is referred to as multimedia. Naturally fractured reservoir can be divided into dual-porosity media and triple-porosity media. In the dual-porosity media filtration model, we assume that there are two systems like fracture and matrix block in the reservoir; the matrix is the major reservoir space, while the fracture is the major filtration channel. Over the past 50 years, people have extensively studied the well test problems related to this type of reservoir, and for the moment, the well test techniques are relatively mature and perfect. The flow behavior of this type of reservoir is shown in Figure 11.5.

11.3.1 Pressure Distribution in a Dual-Porosity Reservoir: Constant Rate Production

We consider the flow of a slightly compressible fluid of constant viscosity in a circular dual-porosity reservoir in which the outer boundary r_e is closed. The well is located at the center of the reservoir, and fluid is produced at a constant rate of q. Initially, the pressure is uniform throughout the reservoir. The skin and wellbore storage effects are not considered. The reservoir thickness, initial reservoir pressure, wellbore radius, porosity, total compressibility, permeability, fluid viscosity, and volume factor are represented by h, p_i, r_w, φ, C_t, K, μ, and B, respectively. The matrix is the reservoir space, the fracture is the filtration channel, and the flow from matrix to fracture is a

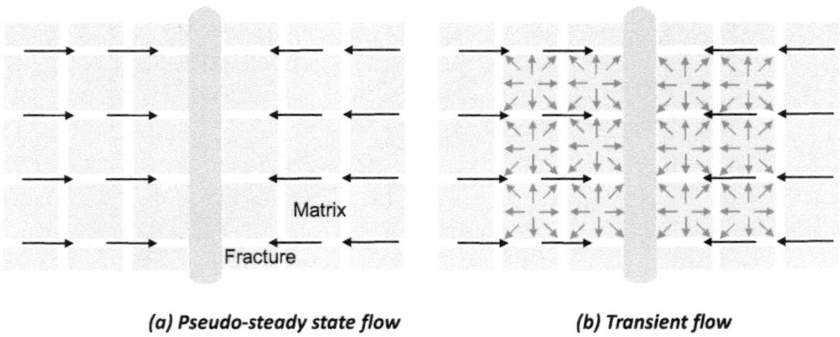

Figure 11.5 Dual-porosity media. (a) Pseudo-steady state flow; (b) transient flow

pseudo-steady state. The basic partial differential equation describing the flow of the above dual-porosity system is

$$\frac{\partial^2 p_{\mathrm{fD}}}{\partial r_{\mathrm{D}}^2}+\frac{1}{r_{\mathrm{D}}}\frac{\partial p_{\mathrm{fD}}}{\partial r_{\mathrm{D}}}=\omega\frac{\partial p_{\mathrm{fD}}}{\partial t_{\mathrm{D}}}+(1-\omega)\frac{\partial p_{\mathrm{mD}}}{\partial t_{\mathrm{D}}} \tag{11.72}$$

$$\frac{\partial p_{\mathrm{mD}}}{\partial t_{\mathrm{D}}}=\frac{\lambda}{1-\omega}\left(p_{\mathrm{fD}}-p_{\mathrm{mD}}\right) \tag{11.73}$$

The subscript f represents the fracture system, and m represents the matrix system, where

$$\omega=\frac{\left(\varphi VC_{\mathrm{t}}\right)_{\mathrm{f}}}{\left(\varphi VC_{\mathrm{t}}\right)_{\mathrm{f}}+\left(\varphi VC_{\mathrm{t}}\right)_{\mathrm{m}}}$$

$$\lambda=\alpha\frac{K_{\mathrm{m}}}{K_{\mathrm{f}}}r_{\mathrm{w}}^2$$

α is a shape-dependent constant of the matrix block.

Initial condition is

$$p_{\mathrm{fD}}\left(r_{\mathrm{D}},0\right)=p_{\mathrm{mD}}\left(r_{\mathrm{D}},0\right)=0 \tag{11.74}$$

Inner boundary condition is

$$\left(\frac{\partial p_{\mathrm{fD}}}{\partial r_{\mathrm{D}}}\right)_{r_{\mathrm{D}}=1}=-1 \tag{11.75}$$

Outer boundary condition is

$$\left(\frac{\partial p_{\mathrm{fD}}}{\partial r_{\mathrm{D}}}\right)_{r_{\mathrm{D}}=r_{\mathrm{eD}}}=0 \tag{11.76}$$

The dimensionless variables are defined as

$$p_{\mathrm{fD}}=\frac{2\pi K_{\mathrm{f}}h\left(p_{\mathrm{i}}-p_{\mathrm{f}}\right)}{qB\mu}\quad p_{\mathrm{mD}}=\frac{2\pi K_{\mathrm{m}}h\left(p_{\mathrm{i}}-p_{\mathrm{m}}\right)}{qB\mu};$$

$$t_{\mathrm{D}}=\frac{K_{\mathrm{f}}}{\mu\left[\left(\varphi VC_{\mathrm{t}}\right)_{\mathrm{f}}+\left(\varphi VC_{\mathrm{t}}\right)_{\mathrm{m}}\right]r_{\mathrm{w}}^2}\cdot t\quad r_{\mathrm{D}}=\frac{r}{r_{\mathrm{w}}}\quad r_{\mathrm{eD}}=\frac{r_{\mathrm{e}}}{r_{\mathrm{w}}}$$

Taking the Laplace transformation of Eq. (11.72) and Eq. (11.73), we have

$$\frac{d^2 \overline{p}_{\mathrm{fD}}}{dr_{\mathrm{D}}^2} + \frac{1}{r_{\mathrm{D}}} \frac{d\overline{p}_{\mathrm{fD}}}{dr_{\mathrm{D}}} = \omega s \overline{p}_{\mathrm{fD}} + (1-\omega) s \overline{p}_{\mathrm{mD}} \tag{11.77}$$

$$\overline{p}_{\mathrm{mD}} = \frac{\lambda}{[(1-\omega)s + \lambda]} \overline{p}_{\mathrm{fD}} \tag{11.78}$$

Substituting Eq. (11.78) into Eq. (11.77), we have

$$\frac{d^2 \overline{p}_{\mathrm{fD}}}{dr_{\mathrm{D}}^2} + \frac{1}{r_{\mathrm{D}}} \frac{d\overline{p}_{\mathrm{fD}}}{dr_{\mathrm{D}}} - s\left[\frac{\omega(1-\omega)s + \lambda}{(1-\omega)s + \lambda}\right] \overline{p}_{\mathrm{fD}} = 0 \tag{11.79}$$

In the Laplace space, the solution of Eq. (11.79) is given by

$$\overline{p}_{\mathrm{fD}}(r_{\mathrm{D}}, s) = A_0 I_0\left(r_{\mathrm{D}}\sqrt{\sigma_1}\right) + B_0 K_0\left(r_{\mathrm{D}}\sqrt{\sigma_1}\right) \tag{11.80}$$

where

$$\sigma = s\left[\frac{\omega(1-\omega)s + \lambda}{(1-\omega)s + \lambda}\right] \tag{11.81}$$

According to the inner boundary condition Eq. (11.75), we have

$$A_0 - B_0 \frac{K_1\left(\sqrt{\sigma}\right)}{I_1\left(\sqrt{\sigma}\right)} = -\frac{1}{s\sqrt{\sigma} I_1\left(\sqrt{\sigma}\right)} \tag{11.82}$$

According to the outer boundary condition Eq. (11.76), we have

$$A_0 - B_0 \frac{K_1\left(r_{\mathrm{eD}}\sqrt{\sigma}\right)}{I_1\left(r_{\mathrm{eD}}\sqrt{\sigma}\right)} = 0 \tag{11.83}$$

where

$$B_0 = \frac{1}{s\sqrt{\sigma} I_1\left(\sqrt{\sigma}\right)\left[\frac{K_1\left(\sqrt{\sigma}\right)}{I_1\left(\sqrt{\sigma}\right)} - \frac{K_1\left(r_{\mathrm{eD}}\sqrt{\sigma}\right)}{I_1\left(r_{\mathrm{eD}}\sqrt{\sigma}\right)}\right]}$$

$$A_0 = \frac{\dfrac{K_1\left(r_{eD}\sqrt{\sigma}\right)}{I_1\left(r_{eD}\sqrt{\sigma}\right)}}{s\sqrt{\sigma}I_1\left(\sqrt{\sigma}\right)\left[\dfrac{K_1\left(\sqrt{\sigma}\right)}{I_1\left(\sqrt{\sigma}\right)}-\dfrac{K_1\left(r_{eD}\sqrt{\sigma}\right)}{I_1\left(r_{eD}\sqrt{\sigma}\right)}\right]}$$

Substituting A_0 and B_0 into Eq. (11.80), we have

$$\overline{p}_{fD}(1,s) = \frac{\dfrac{K_1\left(r_{eD}\sqrt{\sigma}\right)}{I_1\left(r_{eD}\sqrt{\sigma}\right)}\dfrac{I_0\left(\sqrt{\sigma_1}\right)}{I_1\left(\sqrt{\sigma}\right)}+\dfrac{K_0\left(\sqrt{\sigma_1}\right)}{I_1\left(\sqrt{\sigma}\right)}}{s\sqrt{\sigma}\left[\dfrac{K_1\left(\sqrt{\sigma}\right)}{I_1\left(\sqrt{\sigma}\right)}-\dfrac{K_1\left(r_{eD}\sqrt{\sigma}\right)}{I_1\left(r_{eD}\sqrt{\sigma}\right)}\right]} = \overline{p}_{wfD} \tag{11.84}$$

11.3.2 Blasingame Type Curve Plotting

According to Da Prat et al. (1981), the dimensionless time is defined by

$$t_{Dd} = t_D \frac{1}{\dfrac{1}{2}\left(r_{eD}^2-1\right)\left(\ln r_{ed}-\dfrac{1}{2}\right)} = t_D\alpha \tag{11.85}$$

Substituting Eq. (11.85) into Eq. (11.72) and (11.73), we have

$$\frac{\partial^2 p_{fD}}{\partial r_D^2}+\frac{1}{r_D}\frac{\partial p_{fD}}{\partial r_D} = \omega\alpha\frac{\partial p_{fD}}{\partial t_{Dd}}+(1-\omega)\alpha\frac{\partial p_{mD}}{\partial t_{Dd}} \tag{11.86}$$

$$\frac{\partial p_{mD}}{\partial t_{Dd}} = \frac{\lambda}{\alpha(1-\omega)}\left(p_{fD}-p_{mD}\right) \tag{11.87}$$

Taking the Laplace transformation of Eq. (11.86) and (11.87), we have

$$\frac{d^2\overline{p}_{fD}}{dr_D^2}+\frac{1}{r_D}\frac{d\overline{p}_{fD}}{dr_D}-\alpha s\left[\frac{\omega\alpha(1-\omega)s+\lambda}{(1-\omega)\alpha s+\lambda}\right]\overline{p}_{fD} = 0 \tag{11.88}$$

The solution is,

$$\overline{p}_{wfD} = \frac{\dfrac{K_1\left(r_{eD}\sqrt{\sigma}\right)}{I_1\left(r_{eD}\sqrt{\sigma}\right)}\dfrac{I_0\left(\sqrt{\sigma}\right)}{I_1\left(\sqrt{\sigma}\right)}+\dfrac{K_0\left(\sqrt{\sigma}\right)}{I_1\left(\sqrt{\sigma}\right)}}{s\sqrt{\sigma}\left[\dfrac{K_1\left(\sqrt{\sigma}\right)}{I_1\left(\sqrt{\sigma}\right)}-\dfrac{K_1\left(r_{eD}\sqrt{\sigma}\right)}{I_1\left(r_{eD}\sqrt{\sigma}\right)}\right]} \tag{11.89}$$

where

$$\sigma = \alpha s \left[\frac{\omega\alpha(1-\omega)s+\lambda}{(1-\omega)\alpha s+\lambda} \right] \tag{11.90}$$

The dimensionless decline flow rate is defined as

$$q_{\mathrm{Dd}} = \frac{\ln r_{\mathrm{eD}} - \frac{1}{2}}{L^{-1}\left[\bar{p}_{\mathrm{wfD}}\right]} \tag{11.91}$$

The dimensionless decline flow rate integral function q_{Ddi} is expressed as

$$q_{\mathrm{Ddi}} = \frac{N_{\mathrm{pDd}}}{t_{\mathrm{Dd}}} = \frac{1}{t_{\mathrm{Dd}}} \int_0^{t_{\mathrm{Dd}}} q_{\mathrm{Dd}}(\tau)d\tau \tag{11.92}$$

The dimensionless decline flow rate integral derivative q_{Ddid} is expressed as

$$q_{\mathrm{Ddid}} = -\frac{dq_{\mathrm{Ddi}}}{d\ln t_{\mathrm{Dd}}} = -t_{\mathrm{Dd}} \frac{d\left(N_{\mathrm{pDd}}/t_{\mathrm{Dd}}\right)}{dt_{\mathrm{Dd}}} \tag{11.93}$$

11.3.3 Blasingame Type Curve Characteristics

The Blasingame type curve characteristics of dual-porosity reservoir mainly depend on the storativity ratio ω, interporosity flow coefficient λ, and drainage radius r_{eD}, as shown in Figure 11.6.

In the decline analysis, the solution form of the transient dual-porosity model, the transient columnar dual-porosity model and the transient spherical dual-porosity model is the same as that of the pseudo-steady state flow Eq. (11.89). The difference is that the σ value changes. However, we only need to replace the s in Table 8.3 by αs, as shown in Table 11.1.

With a similar method applied to the closed circular homogeneous reservoir, we can also derive the Agarwal-Gardner, NPI, and transient types of type curves of dual-porosity reservoir.

11.4 Decline Curve of Triple-Porosity Reservoir

11.4.1 Physical Model of Triple-Porosity Reservoir

Substantial carbonate reservoir development practices show that apart from matrix and fracture. There is also a cavity/vug system, that is, a reservoir with triple-porosity media.

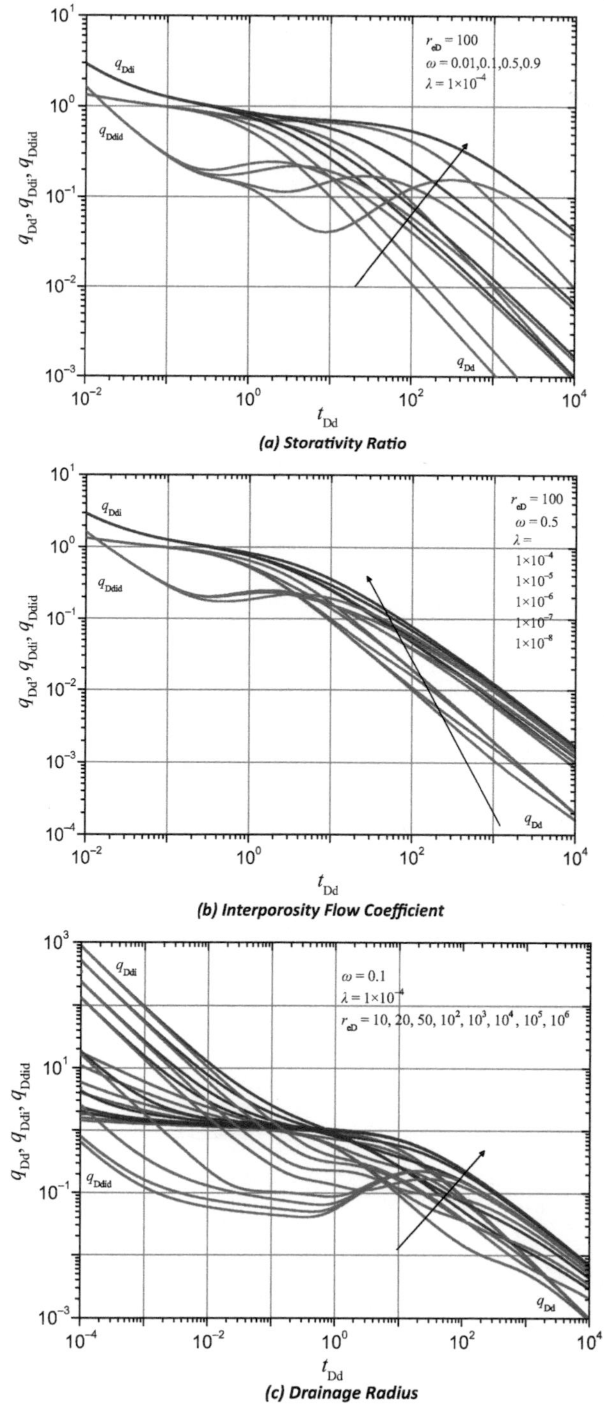

Figure 11.6 Blasingame type curves of dual-porosity reservoir. (a) Storativity ratio; (b) interporosity flow coefficient; (c) drainage radius

Table 11.1 σ **for different models in laplace space**

Model	σ
Pseudo-steady state	$\sigma = \alpha s \dfrac{\omega(1-\omega)\alpha s + \lambda}{(1-\omega)\alpha s + \lambda}$
Transient slab	$\sigma = \alpha s\left[\omega + \dfrac{\lambda}{3\alpha s}\sqrt{\dfrac{3(1-\omega)\alpha s}{\lambda}}\tanh\left(\sqrt{\dfrac{3(1-\omega)\alpha s}{\lambda}}\right)\right]$
Transient columnar	$\sigma = \alpha s\left[\omega + \dfrac{\lambda}{4\alpha s}\sqrt{\dfrac{8(1-\omega)\alpha s}{\lambda}}\dfrac{I_1\left(\sqrt{\dfrac{8(1-\omega)\alpha s}{\lambda}}\right)}{I_0\left(\sqrt{\dfrac{8(1-\omega)s}{\lambda\alpha}}\right)}\right]$
Transient spherical	$\sigma = \alpha s\left[\omega + \dfrac{\lambda}{5\alpha s}\sqrt{\dfrac{15(1-\omega)\alpha s}{\lambda}}\coth\left(\sqrt{\dfrac{15(1-\omega)\alpha s}{\lambda}}\right)\right]$

11.4.1.1 *Fracture + dual-matrix model*

Fracture connecting the wellbore is a filtration channel, whereas a matrix can be divided into two types according to the value of permeability. Connectivity with the fracture system of matrix 1 is better than matrix 2. Matrix can also be divided into two kinds, including layered and massive, as shown in Figure 11.7.

In a stratified model, the matrix system is cut horizontally by the fracture system, and the transient crossflow occurs between matrix and fracture. In contrast, in a massive model, the matrix is cut by a set of orthogonal fracture system, and the transient crossflow occurs between matrix and fracture.

11.4.1.2 *Matrix + dual-fracture model*

This model can be divided into three forms according to the possibility of nesting, including (1) fluid crossflows from matrix and microfracture toward macrofracture in a

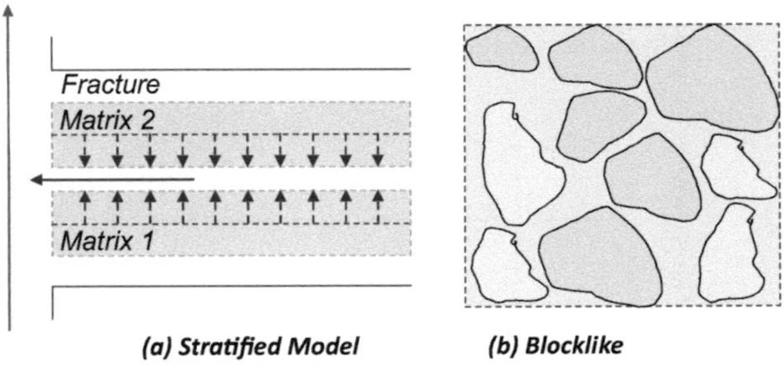

(a) Stratified Model ***(b) Blocklike***

Figure 11.7 Fracture + dual-matrix model (after Al-Ghamdi and Ershaghi, 1996).
(a) Stratified model; (b) block-like.

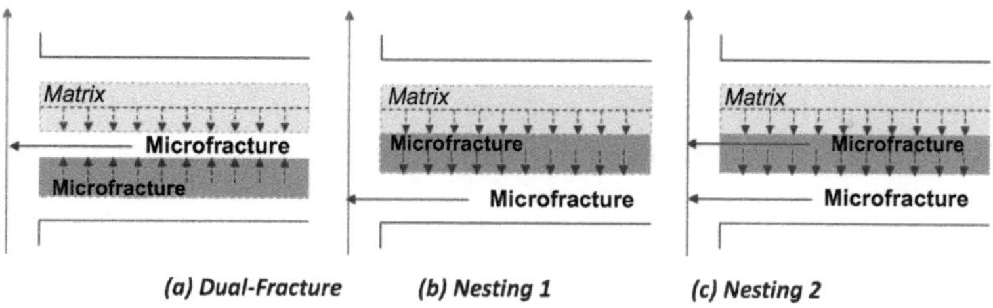

Figure 11.8 Matrix + dual-fracture model l (after Al-Ghamdi and Ershaghi, 1996). (a) Dual-fracture; (b) nesting 1; (c) nesting 2

pseudo-steady state, which is similar to the stratified model; (2) fluid crossflows from matrix toward microfracture in a pseudo-steady state, and then toward macrofracture still in a pseudo-steady state, and the macrofracture feeds fluid toward the wellbore; (3) fluid crossflow from matrix toward microfracture in a pseudo-steady state, and then toward macrofracture still in a pseudo-steady state, and both microfracture and macrofracture feed fluid toward the wellbore simultaneously, as shown in Figure 11.8.

11.4.1.3 Tricontinuum model

The tricontinuum model is mainly composed of fracture, cavity with different sizes (which directly or indirectly connects fracture via fine crack or matrix), and matrix (which possibly contains some vugs and connects fracture locally). Fracture is the filtration channel, and cavity and matrix are the reservoir space. Fluid exchange may occur between matrix and fracture, matrix and cavity, as well as fracture and cavity, as shown in Figure 11.9.

With a lot of efforts on the triple-porosity media model, researchers have got more and more information about the fracture, cavern, and matrix systems and better depiction of their relations.

11.4.2 Blasingame Decline Curves of Triporosity Monopermeability Model

Clossman (1975) presented the first triple-porosity media model. Abdassah and Ershaghi (1986), Al-Ghamdi and Ershaghi (1996), Camacho Velázquez et al. (2005), Rodriguez (2004), Héctor (2006), and Wu (2007) propelled the triple-porosity media research to another stage, successively presented, developing and perfecting the triporosity and

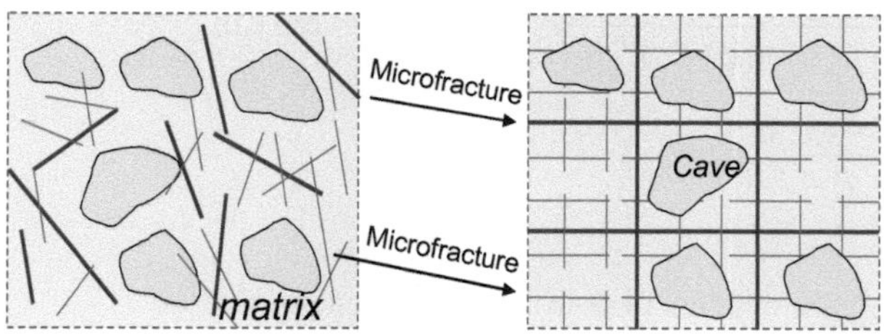

Figure 11.9 Tricontinuum model (after Wu, 2007)

monopermeability as well as the triporosity and dual-permeability (nesting form) well test models for single-phase (or multiphase) fluid and different well types (e.g., vertical well, incomplete-perforation vertical well, and horizontal well).

On the basis of the well test model, in this section, we introduce the decline analysis model and type curve characteristics of the simplest triporosity and the monopermeability model.

11.4.2.1 Pressure Distribution in a Triporosity and Monopermeability Reservoir with Constant Rate Production

We consider the flow of a slightly compressible fluid of constant viscosity in a circular triple-porosity reservoir in which the outer boundary r_e is closed. The well is located at the center of the reservoir and fluid is produced at a constant rate of q. Initially, the pressure is uniform throughout the reservoir. The skin and wellbore storage effect is not considered. The reservoir thickness, initial reservoir pressure, wellbore radius, porosity, total compressibility, permeability, fluid viscosity, and volume factor are represented by h, p_i, r_w, φ, C_t, K, μ, and B, respectively. Matrix 1 and matrix 2 are the reservoir space, the fracture is the filtration channel, and the flow from matrix to fracture is pseudo-steady state, as shown in Figure 11.8(a). The basic partial differential equation describing the flow of the previous triporosity and monopermeability system is

$$\frac{\partial^2 p_{fD}}{\partial r_D^2}+\frac{1}{r_D}\frac{\partial p_{fD}}{\partial r_D}=\omega_f\frac{\partial p_{fD}}{\partial t_D}+\omega_{m1}\frac{\partial p_{m1D}}{\partial t_D}+\omega_{m2}\frac{\partial p_{m2D}}{\partial t_D} \tag{11.94}$$

$$\frac{\partial p_{m1D}}{\partial t_D}=\frac{\lambda_{m1}}{\omega_{m1}}\left(p_{fD}-p_{m1D}\right) \tag{11.95}$$

$$\frac{\partial p_{m2D}}{\partial t_D}=\frac{\lambda_{m2}}{\omega_{m2}}\left(p_{fD}-p_{m2D}\right) \tag{11.96}$$

The subscript f represents the fracture system, while m1 and m2 represent the matrix system, where

$$\omega_j=\frac{(\varphi VC_t)_j}{(\varphi VC_t)_f+(\varphi VC_t)_{m1}+(\varphi VC_t)_{m2}} \quad j=\mathrm{f,m1,m2}$$

$$\lambda_j=\alpha_{mj}\frac{K_{mj}}{K_f}r_w^2 \quad j=1,2$$

and α_m is a shape-dependent constant of the matrix block.

Initial condition is

$$p_{fD}(r_D,0)=p_{m1D}(r_D,0)=p_{m2D}(r_D,0)=0 \tag{11.97}$$

Inner boundary condition is

$$\left(\frac{\partial p_{\mathrm{fD}}}{\partial r_{\mathrm{D}}}\right)_{r_{\mathrm{D}}=1} = -1 \tag{11.98}$$

Outer boundary condition is

$$\left(\frac{\partial p_{\mathrm{fD}}}{\partial r_{\mathrm{D}}}\right)_{r_{\mathrm{D}}=r_{\mathrm{eD}}} = 0 \tag{11.99}$$

The dimensionless variables are defined as

$$p_{\mathrm{fD}} = \frac{2\pi K_{\mathrm{f}} h(p_{\mathrm{i}} - p_{\mathrm{f}})}{qB\mu} \quad p_{\mathrm{m1D}} = \frac{2\pi K_{\mathrm{f}} h(p_{\mathrm{i}} - p_{\mathrm{m1}})}{qB\mu} \quad p_{\mathrm{m2D}} = \frac{2\pi K_{\mathrm{f}} h(p_{\mathrm{i}} - p_{\mathrm{m2}})}{qB\mu};$$

$$t_{\mathrm{D}} = \frac{K_{\mathrm{f}}}{\mu\left[(\varphi VC_{\mathrm{t}})_{\mathrm{f}} + (\varphi VC_{\mathrm{t}})_{\mathrm{m1}} + (\varphi VC_{\mathrm{t}})_{\mathrm{m2}}\right] r_{\mathrm{w}}^2} \cdot t \quad r_{\mathrm{D}} = \frac{r}{r_{\mathrm{w}}} \quad r_{\mathrm{eD}} = \frac{r_{\mathrm{e}}}{r_{\mathrm{w}}}$$

Taking the Laplace transformation of Eq. (11.94) to Eq. (11.96), we have

$$\frac{d^2 \overline{p}_{\mathrm{fD}}}{dr_{\mathrm{D}}^2} + \frac{1}{r_{\mathrm{D}}} \frac{d\overline{p}_{\mathrm{fD}}}{dr_{\mathrm{D}}} = \omega_{\mathrm{f}} s\overline{p}_{\mathrm{fD}} + \omega_{\mathrm{m1}} s\overline{p}_{\mathrm{m1D}} + \omega_{\mathrm{m2}} s\overline{p}_{\mathrm{m2D}} \tag{11.100}$$

$$\overline{p}_{\mathrm{m1D}} = \frac{\lambda_{\mathrm{m1}}}{(\omega_{\mathrm{m1}} s + \lambda_{\mathrm{m1}})} \overline{p}_{\mathrm{fD}} \tag{11.101}$$

$$\overline{p}_{\mathrm{m2D}} = \frac{\lambda_{\mathrm{m2}}}{(\omega_{\mathrm{m2}} s + \lambda_{\mathrm{m2}})} \overline{p}_{\mathrm{fD}} \tag{11.102}$$

Substituting Eq. (11.101) and (11.102) into Eq. (11.100), we have

$$\frac{d^2 \overline{p}_{\mathrm{fD}}}{dr_{\mathrm{D}}^2} + \frac{1}{r_{\mathrm{D}}} \frac{d\overline{p}_{\mathrm{fD}}}{dr_{\mathrm{D}}} - s\left[\omega_{\mathrm{f}} + \frac{\lambda_{\mathrm{m1}}\omega_{\mathrm{m1}}}{(\omega_{\mathrm{m1}} s + \lambda_{\mathrm{m1}})} + \frac{\lambda_{\mathrm{m2}}\omega_{\mathrm{m2}}}{(\omega_{\mathrm{m2}} s + \lambda_{\mathrm{m2}})}\right]\overline{p}_{\mathrm{fD}} = 0 \tag{11.103}$$

In the Laplace space, the solution of Eq. (11.103) is given by

$$\overline{p}_{\mathrm{fD}}(1,s) = \frac{\dfrac{K_1(r_{\mathrm{eD}}\sqrt{\sigma})}{I_1(r_{\mathrm{eD}}\sqrt{\sigma})}\dfrac{I_0(\sqrt{\sigma_1})}{I_1(\sqrt{\sigma})} + \dfrac{K_0(\sqrt{\sigma_1})}{I_1(\sqrt{\sigma})}}{s\sqrt{\sigma}\left[\dfrac{K_1(\sqrt{\sigma})}{I_1(\sqrt{\sigma})} - \dfrac{K_1(r_{\mathrm{eD}}\sqrt{\sigma})}{I_1(r_{\mathrm{eD}}\sqrt{\sigma})}\right]} = \overline{p}_{\mathrm{wfD}} \tag{11.104}$$

where

$$\sigma = s\left[\omega_{\mathrm{f}} + \frac{\lambda_{\mathrm{m1}}\omega_{\mathrm{m1}}}{\left(\omega_{\mathrm{m1}}s + \lambda_{\mathrm{m1}}\right)} + \frac{\lambda_{\mathrm{m2}}\omega_{\mathrm{m2}}}{\left(\omega_{\mathrm{m2}}s + \lambda_{\mathrm{m2}}\right)}\right] \tag{11.105}$$

11.4.2.2 Blasingame Type Curves Plotting

The dimensionless time is defined as

$$t_{\mathrm{Dd}} = t_{\mathrm{D}}\frac{1}{\frac{1}{2}\left(r_{\mathrm{eD}}^2 - 1\right)\left(\ln r_{\mathrm{eD}} - \frac{1}{2}\right)} = t_{\mathrm{D}}\alpha \tag{11.106}$$

Substituting Eq. (11.106) into Eq. (11.94) through Eq. (11.96), and taking the Laplace transformation of Eq. (11.94) through Eq. (11.96), we have

$$\frac{d^2\overline{p}_{\mathrm{fD}}}{dr_{\mathrm{D}}^2} + \frac{1}{r_{\mathrm{D}}}\frac{d\overline{p}_{\mathrm{fD}}}{dr_{\mathrm{D}}} - \alpha s\left[\omega_{\mathrm{f}} + \frac{\lambda_{\mathrm{m1}}\omega_{\mathrm{m1}}}{\left(\omega_{\mathrm{m1}}\alpha s + \lambda_{\mathrm{m1}}\right)} + \frac{\lambda_{\mathrm{m2}}\omega_{\mathrm{m2}}}{\left(\omega_{\mathrm{m2}}\alpha s + \lambda_{\mathrm{m2}}\right)}\right]\overline{p}_{\mathrm{fD}} = 0 \tag{11.107}$$

The solution is Eq. (11.104), where

$$\sigma = \alpha s\left[\omega_{\mathrm{f}} + \frac{\lambda_{\mathrm{m1}}\omega_{\mathrm{m1}}}{\left(\omega_{\mathrm{m1}}\alpha s + \lambda_{\mathrm{m1}}\right)} + \frac{\lambda_{\mathrm{m2}}\omega_{\mathrm{m2}}}{\left(\omega_{\mathrm{m2}}\alpha s + \lambda_{\mathrm{m2}}\right)}\right] \tag{11.108}$$

The dimensionless decline flow rate is defined as

$$q_{\mathrm{Dd}} = \frac{\ln r_{\mathrm{eD}} - \frac{1}{2}}{L^{-1}\left[\overline{p}_{\mathrm{wfD}}\right]} \tag{11.109}$$

The dimensionless decline flow rate integral function q_{Ddi} is expressed as

$$q_{\mathrm{Ddi}} = \frac{N_{\mathrm{pDd}}}{t_{\mathrm{Dd}}} = \frac{1}{t_{\mathrm{Dd}}}\int_0^{t_{\mathrm{Dd}}} q_{\mathrm{Dd}}(\tau)d\tau \tag{11.110}$$

The dimensionless decline flow rate integral derivative q_{Ddid} is expressed as

$$q_{\mathrm{Ddid}} = -\frac{dq_{\mathrm{Ddi}}}{d\ln t_{\mathrm{Dd}}} = -t_{\mathrm{Dd}}\frac{d\left(N_{\mathrm{pDd}}/t_{\mathrm{Dd}}\right)}{dt_{\mathrm{Dd}}} \tag{11.111}$$

11.4.2.3 Blasingame Type Curve Characteristics

Similar to the dual-porosity reservoir, the Blasingame type curve characteristics of triple-porosity reservoir depend on six parameters, including storativity ratio ω (3), interporosity flow coefficient λ (2) and drainage radius, as shown in Figure 11.10.

11.4.3 Blasingame Decline Curves of Triporosity Monopermeability Nesting Model

We consider the flow of a slightly compressible fluid of constant viscosity in a circular triple-porosity reservoir in which the outer boundary r_e is closed. The well is located at the center of the reservoir and fluid is produced at a constant rate of q. Matrix 1 and matrix 2 are the reservoir space, fluid crossflows from matrix 1 toward matrix 2 in a pseudo-steady state, then toward fracture still in a pseudo-steady state, the fracture is the filtration channel, as shown in Figure 11.8(b), and the other conditions are assumed to be the same as those for triporosity and monopermeability model. The basic partial differential equation of production decline analysis describing the flow of the above triporosity and monopermeability nesting system is

$$\frac{\partial^2 p_{\mathrm{fD}}}{\partial r_{\mathrm{D}}^2}+\frac{1}{r_{\mathrm{D}}}\frac{\partial p_{\mathrm{fD}}}{\partial r_{\mathrm{D}}}+\lambda_{\mathrm{m1}}\left(p_{\mathrm{m1D}}-p_{\mathrm{fD}}\right)=\omega_{\mathrm{f}}\alpha\frac{\partial p_{\mathrm{fD}}}{\partial t_{\mathrm{Dd}}} \tag{11.112}$$

$$\alpha\frac{\partial p_{\mathrm{m1D}}}{\partial t_{\mathrm{Dd}}}=\frac{\lambda_{\mathrm{m2}}}{\omega_{\mathrm{m1}}}\left(p_{\mathrm{m2D}}-p_{\mathrm{m1D}}\right)+\frac{\lambda_{\mathrm{m1}}}{\omega_{\mathrm{m1}}}\left(p_{\mathrm{m1D}}-p_{\mathrm{fD}}\right) \tag{11.113}$$

$$\alpha\frac{\partial p_{\mathrm{m2D}}}{\partial t_{\mathrm{Dd}}}=\frac{\lambda_{\mathrm{m2}}}{\omega_{\mathrm{m2}}}\left(p_{\mathrm{m1D}}-p_{\mathrm{m2D}}\right) \tag{11.114}$$

where the subscript f represents the fracture system, while m1 and m2 represent the matrix system.

Dimensionless variables and boundary conditions are the same as those in triporosity and monopermeability model.

Taking the Laplace transformation of Eq. (11.112) through Eq. (11.114), we have

$$\frac{\partial^2 \bar{p}_{\mathrm{fD}}}{\partial r_{\mathrm{D}}^2}+\frac{1}{r_{\mathrm{D}}}\frac{\partial \bar{p}_{\mathrm{fD}}}{\partial r_{\mathrm{D}}}+\lambda_{\mathrm{m1}}\left(\bar{p}_{\mathrm{m1D}}-\bar{p}_{\mathrm{fD}}\right)=\omega_{\mathrm{f}}\alpha s\bar{p}_{\mathrm{fD}} \tag{11.115}$$

$$s\alpha\bar{p}_{\mathrm{m1D}}=\frac{\lambda_{\mathrm{m2}}}{\omega_{\mathrm{m1}}}\left(\bar{p}_{\mathrm{m2D}}-\bar{p}_{\mathrm{m1D}}\right)-\frac{\lambda_{\mathrm{m1}}}{\omega_{\mathrm{m1}}}\left(\bar{p}_{\mathrm{m1D}}-\bar{p}_{\mathrm{fD}}\right) \tag{11.116}$$

$$\bar{p}_{\mathrm{m2D}}=\frac{\lambda_{\mathrm{m2}}}{\left(\omega_{\mathrm{m2}}\alpha s+\lambda_{\mathrm{m2}}\right)}\bar{p}_{\mathrm{m1D}} \tag{11.117}$$

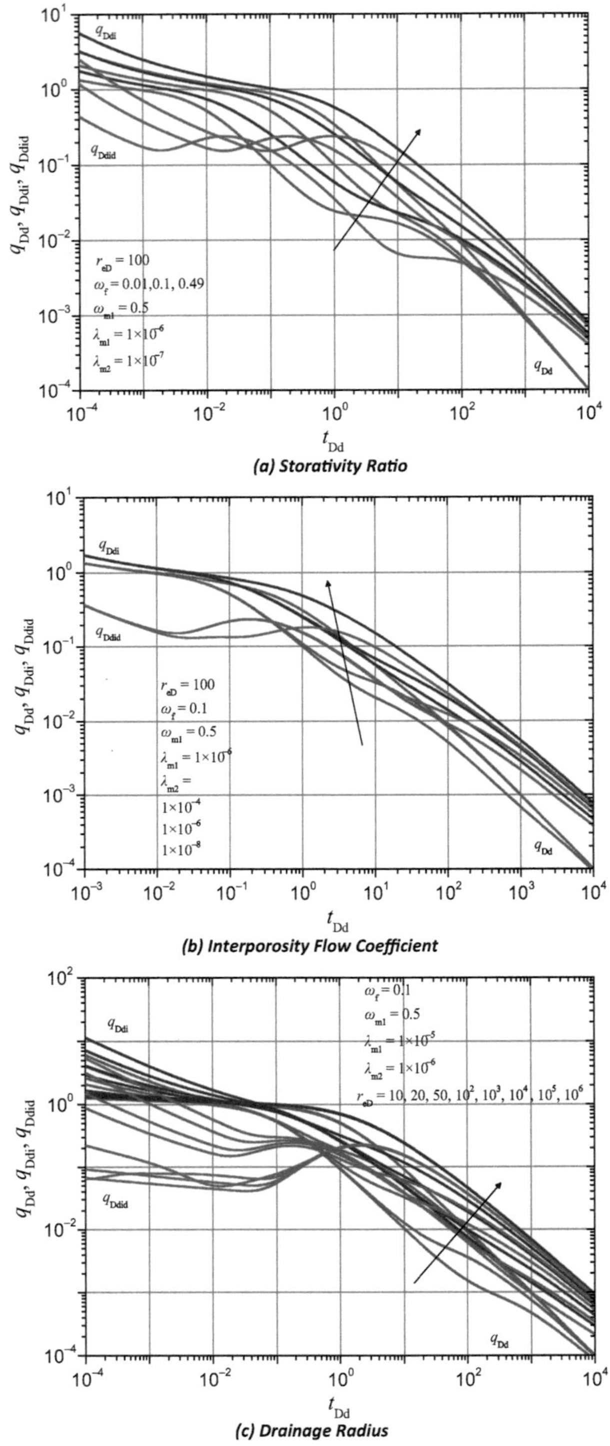

Figure 11.10 Blasingame type curves of triporosity and monopermeability model. (a) Storativity ratio; (b) interporosity flow coefficient; (c) drainage radius

Substituting Eq. (11.117) into Eq. (11.116), we have

$$\overline{p}_{\mathrm{m1D}} = \frac{\lambda_{\mathrm{m1}}}{\left[\omega_{\mathrm{m1}}\alpha s + \lambda_{\mathrm{m2}} + \lambda_{\mathrm{m1}} - \dfrac{\lambda_{\mathrm{m2}}^2}{\left(\omega_{\mathrm{m2}}\alpha s + \lambda_{\mathrm{m2}}\right)}\right]}\overline{p}_{\mathrm{fD}} \tag{11.118}$$

Substituting Eq. (11.117) into Eq. (11.115), we have

$$\frac{\partial^2 \overline{p}_{\mathrm{fD}}}{\partial r_{\mathrm{D}}^2} + \frac{1}{r_{\mathrm{D}}}\frac{\partial \overline{p}_{\mathrm{fD}}}{\partial r_{\mathrm{D}}} - \sigma \overline{p}_{\mathrm{fD}} = 0 \tag{11.119}$$

where

$$\sigma = \lambda_{\mathrm{m1}} + \omega_{\mathrm{f}}\alpha s - \frac{\lambda_{\mathrm{m1}}^2}{\left[\omega_{\mathrm{m1}}\alpha s + \lambda_{\mathrm{m2}} + \lambda_{\mathrm{m1}} - \dfrac{\lambda_{\mathrm{m2}}^2}{\left(\omega_{\mathrm{m2}}\alpha s + \lambda_{\mathrm{m2}}\right)}\right]} \tag{11.120}$$

Given that the inner boundary condition is constant flow rate and the outer boundary condition is closed boundary, the solution of Eq. (11.119) is

$$\overline{p}_{\mathrm{fD}}(1,s) = \overline{p}_{\mathrm{wfD}} = \frac{\dfrac{K_1\left(r_{\mathrm{eD}}\sqrt{\sigma}\right)}{I_1\left(r_{\mathrm{eD}}\sqrt{\sigma}\right)}\dfrac{I_0\left(\sqrt{\sigma}\right)}{I_1\left(\sqrt{\sigma}\right)} + \dfrac{K_0\left(\sqrt{\sigma}\right)}{I_1\left(\sqrt{\sigma}\right)}}{s\sqrt{\sigma}\left[\dfrac{K_1\left(\sqrt{\sigma}\right)}{I_1\left(\sqrt{\sigma}\right)} - \dfrac{K_1\left(r_{\mathrm{eD}}\sqrt{\sigma}\right)}{I_1\left(r_{\mathrm{eD}}\sqrt{\sigma}\right)}\right]} \tag{11.121}$$

The definitions of the dimensionless decline flow rate q_{Dd}, dimensionless decline rate integral q_{Ddi}, and dimensionless decline rate integral derivative q_{Ddid} of this model are the same as those for the triporosity and monopermeability model. The Blasingame decline type curves can be drawn using the previous three sets of parameters. The type curve characteristics mainly depend on the six parameters affecting the size of σ in Eq. (11.120). After the fluid having started to flow in a pseudo-steady state, the dimensionless decline flow rate q_{Dd}, dimensionless decline rate integral q_{Ddi}, and dimensionless decline rate integral derivative q_{Ddid} are slightly different from those of the triporosity and monopermeability model and the triporosity and monopermeability nesting model, as shown in Figure 11.11.

11.4.4 Blasingame Decline Curves of Triporosity Dual-Permeability Model

11.4.4.1 Building of mathematical model

We consider the flow of a slightly compressible fluid of constant viscosity in a circular triple-porosity reservoir in which the outer boundary r_{e} is closed. The well is located at the center of the reservoir and fluid is produced at a constant rate of q; Matrix 1 is

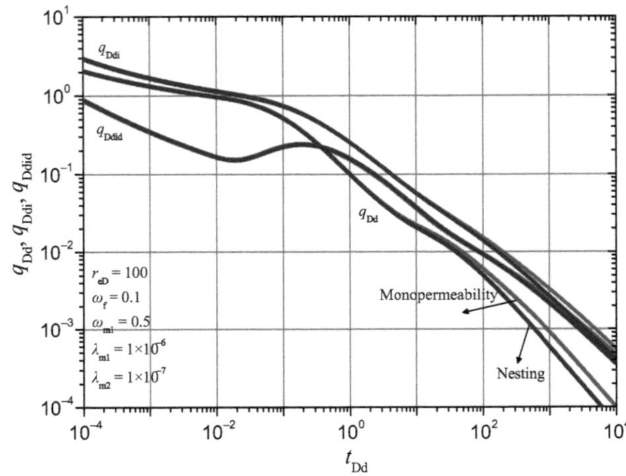

Figure 11.11 Blasingame type curves of triporosity and monopermeability nesting model and triporosity and monopermeability model

the reservoir space, matrix 2, and fracture are the filtration channel; fluid crossflows from matrix 1 to fracture and matrix 2 and from matrix 2 toward fracture in a pseudo-steady state as shown in Figure 11.8(c), and the other conditions are assumed to be the same as those for the triporosity and monopermeability model. The basic partial differential equations of decline analysis describing the flow of the above triporosity and dual-permeability nesting system include

$$\kappa_1 \nabla^2 p_{\mathrm{fD}} + \lambda_{\mathrm{m1f}} \left(p_{\mathrm{m1D}} - p_{\mathrm{fD}}\right) + \lambda_{\mathrm{m2f}} \left(p_{\mathrm{m2D}} - p_{\mathrm{fD}}\right) = \omega_{\mathrm{f}} \alpha \frac{\partial p_{\mathrm{fD}}}{\partial t_{\mathrm{Dd}}} \tag{11.122}$$

$$\kappa_2 \nabla^2 p_{\mathrm{m2D}} + \lambda_{\mathrm{m1m2}} \left(p_{\mathrm{m1D}} - p_{\mathrm{m2D}}\right) - \lambda_{\mathrm{m2f}} \left(p_{\mathrm{m2D}} - p_{\mathrm{fD}}\right) = \omega_{\mathrm{m2}} \alpha \frac{\partial p_{\mathrm{m2D}}}{\partial t_{\mathrm{Dd}}} \tag{11.123}$$

$$-\lambda_{\mathrm{m1f}} \left(p_{\mathrm{m1D}} - p_{\mathrm{fD}}\right) - \lambda_{\mathrm{m1m2}} \left(p_{\mathrm{m1D}} - p_{\mathrm{m2D}}\right) = \omega_{\mathrm{m1}} \alpha \frac{\partial p_{\mathrm{m1D}}}{\partial t_{\mathrm{Dd}}} \tag{11.124}$$

Initial condition is

$$p_{\mathrm{fD}} \left(r_{\mathrm{D}}, 0\right) = p_{\mathrm{m1D}} \left(r_{\mathrm{D}}, 0\right) = p_{\mathrm{m2D}} \left(r_{\mathrm{D}}, 0\right) = 0 \tag{11.125}$$

Inner boundary condition is

$$\kappa_1 \left(\frac{\partial p_{\mathrm{fD}}}{\partial r_{\mathrm{D}}}\right)_{r_{\mathrm{D}}=1} + \kappa_2 \left(\frac{\partial p_{\mathrm{m2D}}}{\partial r_{\mathrm{D}}}\right)_{r_{\mathrm{D}}=1} = -1 \tag{11.126}$$

Outer boundary condition is

$$\left(\frac{\partial p_{\mathrm{fD}}}{\partial r_{\mathrm{D}}}\right)_{r_{\mathrm{D}}=r_{\mathrm{eD}}}=\left(\frac{\partial p_{\mathrm{m2D}}}{\partial r_{\mathrm{D}}}\right)_{r_{\mathrm{D}}=r_{\mathrm{eD}}}=0 \tag{11.127}$$

The subscript f represents the fracture system, while m1 and m2 represent the matrix system, where

$$\begin{aligned}
\lambda_{\mathrm{m1f}} &= \alpha_{\mathrm{m1f}}\frac{K_{\mathrm{f}}}{K_{\mathrm{f}}+K_{\mathrm{m2}}}r_{\mathrm{w}}^2\\
\lambda_{\mathrm{m1m2}} &= \alpha_{\mathrm{m1m2}}\frac{K_{\mathrm{m1}}}{K_{\mathrm{f}}+K_{\mathrm{m2}}}r_{\mathrm{w}}^2\\
\lambda_{\mathrm{m2f}} &= \alpha_{\mathrm{m2f}}\frac{K_{\mathrm{m2}}}{K_{\mathrm{f}}+K_{\mathrm{m2}}}r_{\mathrm{w}}^2
\end{aligned}$$

The dimensionless variables are defined as

$$\begin{aligned}
p_{\mathrm{fD}} &= \frac{2\pi\left(K_{\mathrm{f}}+K_{\mathrm{m2}}\right)h\left(p_{\mathrm{i}}-p_{\mathrm{f}}\right)}{qB\mu}\\
p_{\mathrm{m1D}} &= \frac{2\pi\left(K_{\mathrm{f}}+K_{\mathrm{m2}}\right)h\left(p_{\mathrm{i}}-p_{\mathrm{m1}}\right)}{qB\mu}\\
p_{\mathrm{m2D}} &= \frac{2\pi\left(K_{\mathrm{f}}+K_{\mathrm{m2}}\right)h\left(p_{\mathrm{i}}-p_{\mathrm{m2}}\right)}{qB\mu}\\
t_{\mathrm{D}} &= \frac{\left(K_{\mathrm{f}}+K_{\mathrm{m2}}\right)}{\mu\left[\left(\varphi VC_{\mathrm{t}}\right)_{\mathrm{f}}+\left(\varphi VC_{\mathrm{t}}\right)_{\mathrm{m1}}+\left(\varphi VC_{\mathrm{t}}\right)_{\mathrm{m2}}\right]r_{\mathrm{w}}^2}\cdot t
\end{aligned}$$

11.4.4.2 Solving of mathematical model

Taking the Laplace transformation of Eq. (11.122) to Eq. (11.127), we have

$$\kappa_1\nabla^2\bar{p}_{\mathrm{fD}}+\lambda_{\mathrm{m1f}}\left(\bar{p}_{\mathrm{m1D}}-\bar{p}_{\mathrm{fD}}\right)+\lambda_{\mathrm{m2f}}\left(\bar{p}_{\mathrm{m2D}}-\bar{p}_{\mathrm{fD}}\right)=\omega_{\mathrm{f}}\alpha s\bar{p}_{\mathrm{fD}} \tag{11.128}$$

$$\kappa_2\nabla^2\bar{p}_{\mathrm{m2D}}+\lambda_{\mathrm{m1m2}}\left(\bar{p}_{\mathrm{m1D}}-\bar{p}_{\mathrm{m2D}}\right)-\lambda_{\mathrm{m2f}}\left(\bar{p}_{\mathrm{m2D}}-\bar{p}_{\mathrm{fD}}\right)=\omega_{\mathrm{m2}}\alpha s\bar{p}_{\mathrm{m2D}} \tag{11.129}$$

$$-\lambda_{\mathrm{m1f}}\left(\bar{p}_{\mathrm{m1D}}-\bar{p}_{\mathrm{fD}}\right)-\lambda_{\mathrm{m1m2}}\left(\bar{p}_{\mathrm{m1D}}-\bar{p}_{\mathrm{m2D}}\right)=\omega_{\mathrm{m1}}\alpha s\bar{p}_{\mathrm{m1D}} \tag{11.130}$$

$$\kappa_1\left(\frac{\partial \overline{p}_{\mathrm{fD}}}{\partial r_{\mathrm{D}}}\right)_{r_{\mathrm{D}}=1}+\kappa_2\left(\frac{\partial \overline{p}_{\mathrm{m2D}}}{\partial r_{\mathrm{D}}}\right)_{r_{\mathrm{D}}=1}=-\frac{1}{s} \tag{11.131}$$

$$\left(\frac{\partial \overline{p}_{\mathrm{fD}}}{\partial r_{\mathrm{D}}}\right)_{r_{\mathrm{D}}=r_{\mathrm{eD}}}=\left(\frac{\partial \overline{p}_{\mathrm{m2D}}}{\partial r_{\mathrm{D}}}\right)_{r_{\mathrm{D}}=r_{\mathrm{eD}}}=0 \tag{11.132}$$

According to Eq. (11.130), we have

$$\overline{p}_{\mathrm{m1D}}=\frac{\lambda_{\mathrm{m1f}}\overline{p}_{\mathrm{fD}}+\lambda_{\mathrm{m1m2}}\overline{p}_{\mathrm{m2D}}}{\omega_{\mathrm{m1}}\alpha s+\lambda_{\mathrm{m1f}}+\lambda_{\mathrm{m1m2}}} \tag{11.133}$$

Substituting Eq. (11.133) and Eq. (11.128) into Eq. (11.129), we have

$$\kappa_1\nabla^2\overline{p}_{\mathrm{fD}}+\Delta_1\overline{p}_{\mathrm{fD}}+\Delta_0\overline{p}_{\mathrm{m2D}}=0 \tag{11.134}$$

$$\kappa_2\nabla^2\overline{p}_{\mathrm{m2D}}+\Delta_2\overline{p}_{\mathrm{m2D}}+\Delta_0\overline{p}_{\mathrm{fD}}=0 \tag{11.135}$$

where

$$\Delta_1=\left(\frac{\lambda_{\mathrm{m1f}}^2}{\omega_{\mathrm{m1}}\alpha s+\lambda_{\mathrm{m1f}}+\lambda_{\mathrm{m1m2}}}\right)-\lambda_{\mathrm{m1f}}-\lambda_{\mathrm{m2f}}-\omega_{\mathrm{f}}\alpha s$$

$$\Delta_0=\frac{\lambda_{\mathrm{m1f}}\lambda_{\mathrm{m1m2}}}{\omega_{\mathrm{m1}}\alpha s+\lambda_{\mathrm{m1f}}+\lambda_{\mathrm{m1m2}}}+\lambda_{\mathrm{m2f}}$$

$$\Delta_2=\frac{\lambda_{\mathrm{m1m2}}^2}{\omega_{\mathrm{m1}}\alpha s+\lambda_{\mathrm{m1f}}+\lambda_{\mathrm{m1m2}}}-\lambda_{\mathrm{m1m2}}-\lambda_{\mathrm{m2f}}-\omega_{\mathrm{m2}}\alpha s$$

In the Laplace space, the solution of Eq. (11.134) and (11.135) is given by

$$\overline{p}_{\mathrm{fD}}(r_{\mathrm{D}},s)=A_1I_0\left(r_{\mathrm{D}}\sqrt{\sigma}\right)+B_1K_0\left(r_{\mathrm{D}}\sqrt{\sigma}\right) \tag{11.136}$$

$$\overline{p}_{\mathrm{m2D}}(r_{\mathrm{D}},s)=A_2I_0\left(r_{\mathrm{D}}\sqrt{\sigma}\right)+B_2K_0\left(r_{\mathrm{D}}\sqrt{\sigma}\right) \tag{11.137}$$

Substituting Eq. (11.136) and (11.137) into Eq. (11.134) and (11.135), we have

$$(\kappa_1\sigma+\Delta_1)\overline{p}_{\mathrm{fD}}+\Delta_0\overline{p}_{\mathrm{m2D}}=0 \tag{11.138}$$

$$\Delta_0\overline{p}_{\mathrm{fD}}+(\kappa_2\sigma+\Delta_2)\overline{p}_{\mathrm{m2D}}=0 \tag{11.139}$$

The nonzero solution condition of Eq. (11.138) and (11.139) is the zero determinant factor,

$$\begin{vmatrix} (\kappa_1\sigma+\Delta_1) & \Delta_0 \\ \Delta_0 & (\kappa_2\sigma+\Delta_2) \end{vmatrix} = 0 \tag{11.140}$$

According to Eq. (11.140), we have

$$\sigma_{1,2} = \frac{-(\kappa_1\Delta_2+\kappa_2\Delta_1)\pm\sqrt{(\kappa_1\Delta_2+\kappa_2\Delta_1)^2-4\kappa_1\kappa_2(\Delta_1\Delta_2-\Delta_0^2)}}{2\kappa_1\kappa_2} \tag{11.141}$$

Substituting $\sigma_{1,2}$ into Eq. (11.136) and (11.137), we have

$$\bar{p}_{\text{fD}} = A_1^1 I_0(r_{\text{D}}\sqrt{\sigma_1}) + B_1^1 K_0(r_{\text{D}}\sqrt{\sigma_1}) + A_1^2 I_0(r_{\text{D}}\sqrt{\sigma_2}) + B_1^2 K_0(r_{\text{D}}\sqrt{\sigma_2}) \tag{11.142}$$

$$\begin{aligned}\bar{p}_{\text{m2D}} &= a_1 A_1^1 I_0(r_{\text{D}}\sqrt{\sigma_1}) + a_1 B_1^1 K_0(r_{\text{D}}\sqrt{\sigma_1}) + a_2 A_1^2 I_0(r_{\text{D}}\sqrt{\sigma_2}) \\ &+ a_2 B_1^2 K_0(r_{\text{D}}\sqrt{\sigma_2})\end{aligned} \tag{11.143}$$

According to Eq. (11.138), we have

$$\bar{p}_{\text{fD}} = -\frac{\Delta_0}{(\kappa_1\sigma+\Delta_1)}\bar{p}_{\text{m2D}} \tag{11.144}$$

Then

$$a_{1,2} = -\frac{\kappa_1\sigma_{1,2}+\Delta_1}{\Delta_0} \tag{11.145}$$

According to the outer boundary condition Eq. (11.132), we have

$$A_j^k = b_k B_j^k, \quad j=1,2, \quad k=1,2 \tag{11.146}$$

where

$$b_k = \frac{K_1(\sqrt{\sigma_k}r_{\text{eD}})}{I_1(\sqrt{\sigma_k}r_{\text{eD}})} \tag{11.147}$$

Substituting Eq. (11.146) into Eq. (11.142) and (11.143), we have

$$\overline{p}_{\mathrm{fD}} = B_1^1 X_1 + B_1^2 X_2 \tag{11.148}$$

$$\overline{p}_{\mathrm{m2D}} = a_1 B_1^1 X_1 + a_2 B_1^2 X_2 \tag{11.149}$$

where

$$X_1 = b_1 I_0\left(r_{\mathrm{D}}\sqrt{\sigma_1}\right) + K_0\left(r_{\mathrm{D}}\sqrt{\sigma_1}\right)$$
$$X_2 = b_2 I_0\left(r_{\mathrm{D}}\sqrt{\sigma_2}\right) + K_0\left(r_{\mathrm{D}}\sqrt{\sigma_2}\right)$$

According to the inner boundary condition,

$$X_1\left(1-a_1\right)B_1^1 + X_2\left(1-a_2\right)B_1^2 = 0 \tag{11.150}$$

Then

$$B_1^2 = EB_1^1 \tag{11.151}$$

where

$$E = -\frac{\left(1-a_1\right)X_1}{\left(1-a_2\right)X_2} \tag{11.152}$$

According to Eq. (11.131), we have

$$\kappa_1\left(B_1^1 Y_1 + B_1^2 Y_2\right) + \kappa_2\left(a_1 B_1^1 Y_1 + a_2 B_1^2 Y_2\right) = \frac{1}{s} \tag{11.153}$$

Rearranging Eq.(11.153) yields

$$\left(\kappa_1 Y_1 + \kappa_2 a_1 Y_1\right)B_1^1 + \left(\kappa_1 Y_2 + \kappa_2 a_2 Y_2\right)B_1^2 = \frac{1}{s} \tag{11.154}$$

where

$$Y_1 = -\sqrt{\sigma_1}\left[b_1 I_1\left(\sqrt{\sigma_1}\right) - K_1\left(\sqrt{\sigma_1}\right)\right]$$
$$Y_2 = -\sqrt{\sigma_2}\left[b_2 I_1\left(\sqrt{\sigma_2}\right) - K_1\left(\sqrt{\sigma_2}\right)\right]$$

Substituting Eq. (11.151) into Eq. (11.154), we have

$$B_1^1 = \frac{1}{s\left[\left(\kappa_1 Y_1 + \kappa_2 a_1 Y_1\right) + E\left(\kappa_1 Y_2 + \kappa_2 a_2 Y_2\right)\right]} \tag{11.155}$$

Substituting Eq. (11.152) and (11.155) into Eq. (11.148), we have

$$\bar{p}_{\mathrm{wfD}} = \bar{p}_{\mathrm{fD}}(1,s) = \frac{X_1 + EX_2}{s\left[\left(\kappa_1 Y_1 + \kappa_2 a_1 Y_1\right) + E\left(\kappa_1 Y_2 + \kappa_2 a_2 Y_2\right)\right]} \tag{11.156}$$

This equation is the bottom hole flowing pressure (BHFP) solution of Blasingame decline analysis of the triporosity and dual-permeability model.

The definitions of dimensionless decline rate q_{Dd}, dimensionless decline rate integral q_{Ddi}, and dimensionless decline rate integral derivative q_{Ddid} of this model are the same as those for the triporosity and monopermeability model. The Blasingame decline type curves can be drawn using the previous three sets of parameters.

The type curve characteristics of triporosity and dual-permeability model are mainly controlled by nine parameters such as permeability (2), storativity ratio (3), interporosity flow coefficient (3), and drainage radius of three media. The type curve characteristics are complicated, as shown in Figure 11.12, whereas the real production data quality is poorer and does not well reflect the flow characteristics. Therefore, the triporosity and dual-permeability decline type curves are difficult to use in practice. In other words, they can only be analyzed theoretically.

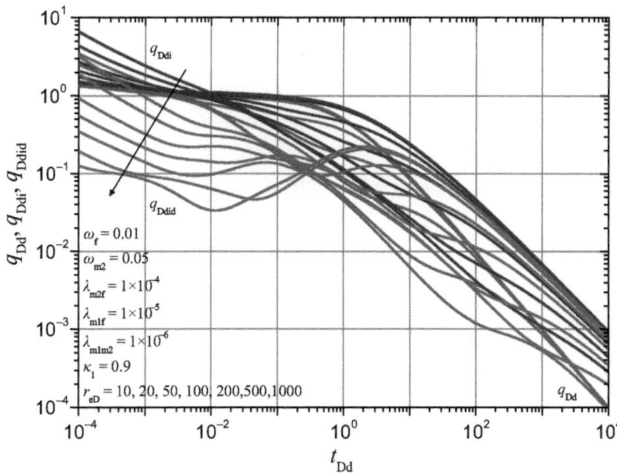

Figure 11.12 Blasingame type curves of triporosity and dual-permeability nesting model

11.4.5 Blasingame Decline Curves of a Tricontinuum Model

We consider the flow of a slightly compressible fluid of constant viscosity in a circular triple-porosity reservoir in which the outer boundary r_e is closed. The well is located at the center of the reservoir and fluid is produced at a constant rate of q; matrix 1 and matrix 2 are the reservoir space, the fracture is the filtration channel, fluid crossflows from matrix 1 toward fracture and matrix 2 in a pseudo-steady state, and fluid crossflows from matrix 2 toward fracture also in a pseudo-steady state, as shown in Figure 11.8, and the other conditions are assumed to be the same as those for the triporosity and monopermeability models. The basic partial differential equation of production decline analysis describing the flow of the previous tricontinuum system is

$$\nabla^2 p_{\mathrm{fD}} + \lambda_{\mathrm{m1f}}\left(p_{\mathrm{m1D}} - p_{\mathrm{fD}}\right) + \lambda_{\mathrm{m2f}}\left(p_{\mathrm{m2D}} - p_{\mathrm{fD}}\right) = \omega_{\mathrm{f}}\alpha\frac{\partial p_{\mathrm{fD}}}{\partial t_{\mathrm{Dd}}} \tag{11.157}$$

$$\lambda_{\mathrm{m1m2}}\left(p_{\mathrm{m1D}} - p_{\mathrm{m2D}}\right) - \lambda_{\mathrm{m2f}}\left(p_{\mathrm{m2D}} - p_{\mathrm{fD}}\right) = \omega_{\mathrm{m2}}\alpha\frac{\partial p_{\mathrm{m2D}}}{\partial t_{\mathrm{Dd}}} \tag{11.158}$$

$$-\lambda_{\mathrm{m1f}}\left(p_{\mathrm{m1D}} - p_{\mathrm{fD}}\right) - \lambda_{\mathrm{m1m2}}\left(p_{\mathrm{m1D}} - p_{\mathrm{m2D}}\right) = \omega_{\mathrm{m1}}\alpha\frac{\partial p_{\mathrm{m1D}}}{\partial t_{\mathrm{Dd}}} \tag{11.159}$$

Initial condition is

$$p_{\mathrm{fD}}\left(r_{\mathrm{D}},0\right) = p_{\mathrm{m1D}}\left(r_{\mathrm{D}},0\right) = p_{\mathrm{m2D}}\left(r_{\mathrm{D}},0\right) = 0 \tag{11.160}$$

Inner boundary condition is

$$\left(\frac{\partial p_{\mathrm{fD}}}{\partial r_{\mathrm{D}}}\right)_{r_{\mathrm{D}}=1} = -1 \tag{11.161}$$

Outer boundary condition is

$$\left(\frac{\partial p_{\mathrm{fD}}}{\partial r_{\mathrm{D}}}\right)_{r_{\mathrm{D}}=r_{\mathrm{eD}}} = 0 \tag{11.162}$$

The subscript f represents the fracture system, while m1 and m2 represent the matrix system.
where

$$\lambda_{\mathrm{m1f}} = \alpha_{\mathrm{m1f}}\frac{K_{\mathrm{f}}}{K_{\mathrm{f}} + K_{\mathrm{m2}}}r_{\mathrm{w}}^2$$

$$\lambda_{\mathrm{m1m2}} = \alpha_{\mathrm{m1m2}}\frac{K_{\mathrm{m1}}}{K_{\mathrm{f}} + K_{\mathrm{m2}}}r_{\mathrm{w}}^2$$

$$\lambda_{\mathrm{m2f}} = \alpha_{\mathrm{m2f}}\frac{K_{\mathrm{m2}}}{K_{\mathrm{f}} + K_{\mathrm{m2}}}r_{\mathrm{w}}^2$$

The dimensionless variables are defined as

$$p_{\text{fD}} = \frac{2K_{\text{f}}h(p_{\text{i}} - p_{\text{f}})}{qB\mu}$$

$$p_{\text{m1D}} = \frac{2K_{\text{f}}h(p_{\text{i}} - p_{\text{m1}})}{qB\mu}$$

$$p_{\text{m2D}} = \frac{2K_{\text{f}}h(p_{\text{i}} - p_{\text{m2}})}{qB\mu}$$

$$t_{\text{D}} = \frac{K_{\text{f}}}{\mu\left[(\varphi VC_{\text{t}})_{\text{f}} + (\varphi VC_{\text{t}})_{\text{m1}} + (\varphi VC_{\text{t}})_{\text{m2}}\right]r_{\text{w}}^2}\cdot t$$

Taking the Laplace transformation of Eq. (11.157) through Eq. (11.159), we have

$$\nabla^2\overline{p}_{\text{fD}} + \lambda_{\text{m1f}}(\overline{p}_{\text{m1D}} - \overline{p}_{\text{fD}}) + \lambda_{\text{m2f}}(\overline{p}_{\text{m2D}} - \overline{p}_{\text{fD}}) = \omega_{\text{f}}\alpha s\overline{p}_{\text{fD}} \tag{11.163}$$

$$\lambda_{\text{m1m2}}(\overline{p}_{\text{m1D}} - \overline{p}_{\text{m2D}}) - \lambda_{\text{m2f}}(\overline{p}_{\text{m2D}} - \overline{p}_{\text{fD}}) = \omega_{\text{m2}}\alpha s\overline{p}_{\text{m2D}} \tag{11.164}$$

$$-\lambda_{\text{m1f}}(\overline{p}_{\text{m1D}} - \overline{p}_{\text{fD}}) - \lambda_{\text{m1m2}}(\overline{p}_{\text{m1D}} - \overline{p}_{\text{m2D}}) = \omega_{\text{m1}}\alpha s\overline{p}_{\text{m1D}} \tag{11.165}$$

Rearranging Eq. (11.163) to Eq. (11.165), we have

$$\lambda_{\text{m1m2}}\overline{p}_{\text{m1D}} - (\lambda_{\text{m1m2}} + \lambda_{\text{m2f}} + \omega_{\text{m2}}\alpha s)\overline{p}_{\text{m2D}} = -\lambda_{\text{m2f}}\overline{p}_{\text{fD}} \tag{11.166}$$

$$(\lambda_{\text{m1f}} + \lambda_{\text{m1m2}} + \omega_{\text{m1}}\alpha s)\overline{p}_{\text{m1D}} - \lambda_{\text{m1m2}}\overline{p}_{\text{m2D}} = \lambda_{\text{m1f}}\overline{p}_{\text{fD}} \tag{11.167}$$

According to Eq. (11.166) and (11.167), we have

$$\overline{p}_{\text{m1D}} = \frac{\left[\lambda_{\text{m1m2}}\lambda_{\text{m2f}} + (\lambda_{\text{m1m2}} + \lambda_{\text{m2f}} + \omega_{\text{m2}}\alpha s)\lambda_{\text{m1f}}\right]\overline{p}_{\text{fD}}}{(\lambda_{\text{m1f}} + \lambda_{\text{m1m2}} + \omega_{\text{m1}}\alpha s)(\lambda_{\text{m2f}} + \lambda_{\text{m1m2}} + \omega_{\text{m2}}\alpha s) - \lambda_{\text{m1m2}}^2} = \Delta_1\overline{p}_{\text{fD}} \tag{11.168}$$

$$\overline{p}_{\text{m2D}} = \frac{\left[\lambda_{\text{m1m2}}\lambda_{\text{m1f}} + (\lambda_{\text{m1m2}} + \lambda_{\text{m1f}} + \omega_{\text{m1}}\alpha s)\lambda_{\text{m2f}}\right]\overline{p}_{\text{fD}}}{(\lambda_{\text{m1f}} + \lambda_{\text{m1m2}} + \omega_{\text{m1}}\alpha s)(\lambda_{\text{m2f}} + \lambda_{\text{m1m2}} + \omega_{\text{m2}}\alpha s) - \lambda_{\text{m1m2}}^2} = \Delta_2\overline{p}_{\text{fD}} \tag{11.169}$$

Substituting Eq. (11.168) and (11.169) into Eq. (11.163), we have

$$\nabla^2\overline{p}_{\text{fD}} - \sigma\overline{p}_{\text{fD}} = 0 \tag{11.170}$$

where

$$\sigma = \alpha s\left(\omega_{\mathrm{f}} + \omega_{\mathrm{m1}}\Delta_1 + \omega_{\mathrm{m2}}\Delta_2\right) \tag{11.171}$$

In the Laplace space, the solution of Eq. (11.170) is

$$\overline{p}_{\mathrm{fD}}(1,s) = \overline{p}_{\mathrm{wfD}} = \frac{\dfrac{K_1\left(r_{\mathrm{eD}}\sqrt{\sigma}\right)}{I_1\left(r_{\mathrm{eD}}\sqrt{\sigma}\right)}\dfrac{I_0\left(\sqrt{\sigma}\right)}{I_1\left(\sqrt{\sigma}\right)} + \dfrac{K_0\left(\sqrt{\sigma}\right)}{I_1\left(\sqrt{\sigma}\right)}}{s\sqrt{\sigma}\left[\dfrac{K_1\left(\sqrt{\sigma}\right)}{I_1\left(\sqrt{\sigma}\right)} - \dfrac{K_1\left(r_{\mathrm{eD}}\sqrt{\sigma}\right)}{I_1\left(r_{\mathrm{eD}}\sqrt{\sigma}\right)}\right]} \tag{11.172}$$

The definitions of dimensionless decline rate q_{Dd}, dimensionless decline rate integral q_{Ddi}, and dimensionless decline rate integral derivative q_{Ddid} of this model are the same as those for the triporosity and monopermeability model. The Blasingame decline type curves can be drawn using the previous three sets of parameters. The type curve characteristics depend mainly on seven parameters such as storativity ratio (3), interporosity flow coefficient (2), and reservoir drainage radius affecting the size of σ in Eq. (11.171), as shown in Figure 11.13. The tricontinuum model is the ultimate state of the triporosity and dual-permeability model when $\kappa_1 = 1$ and $\kappa_1 = 0$.

11.5 Considerations in Plotting Advanced Production Decline Type Curves

In this section, we present the Blasingame type curves of only four types of reservoirs. In the course of drawing type curves and determining analytical procedures, the vital role of material balance time can be fully understood.

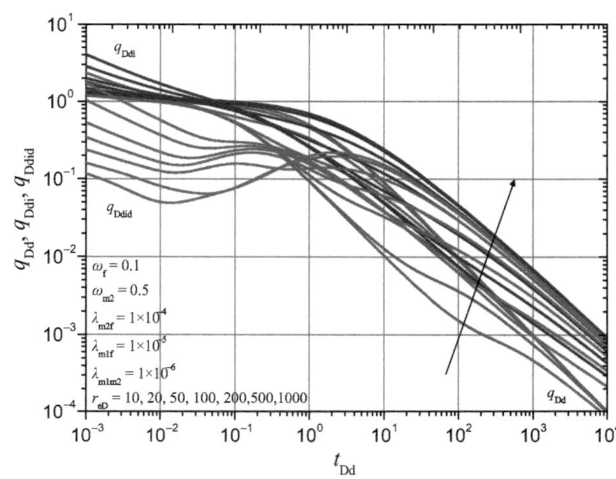

Figure 11.13 Blasingame type curves of tricontinuum model

11.5.1 Typical Errors in Plotting Blasingame Type Curves

As described in Chapter 1, Fetkovich (1980) introduced the transient flow equation in well test analysis to the decline analysis, and applied the Arps type curve to the transient flow period prior to boundary-dominated flow. Furthermore, Fetkovich combined the production decline curve and the Arps rate decline equation, to intuitively show the production decline regularity and the boundary effect. Finally, Fetkovich developed a set of relatively complete log–log production decline type curve matching methods similar to those for well test analysis. The greatest advantage of these methods is that they are able to reliably determine whether the production is in a transient flow period or a boundary-dominated flow period. Both Arps and Fetkovich methods assume a constant BHFP in production, and mainly analyze the flow rate data, without considering the change of pressure–volume–temperature (PVT) with pressure. Palacio and Blasingame (1993) introduced the pseudo-pressure normalized production and material balance pseudo-time to develop the production decline type curve, which considers the production at a variable BHFP and the change of PVT with reservoir pressure. This technique also introduces the rate integral and rate integral derivative curves as the auxiliary matching curve to reduce the ambiguity of interpretation results.

Although the advanced production decline analysis method represented by Blasingame has been developed for 20 years, some researchers get confused with the concepts of Blasingame material balance pseudo-time and Fetkovich dimensionless time, which results in type curve error. In this section, we take the closed circular homogeneous reservoir as an example, analyze the causes for errors, and present the Laplace domain space solutions and decline curves of production models for closed circular homogeneous reservoir, dual-porosity reservoir and triple-porosity reservoir.

By introducing the concept of material balance time, Blasingame sets up a bridge between a constant pressure production and a constant rate production. After the production entering a boundary-dominated flow period, whether producing at a constant pressure or a constant rate, the dimensionless decline rate curve q_{Dd} should be a harmonic decline curve, that is, a "–1"-slope straight line, as shown in Figure 11.14 (see Chapter 4 for details).

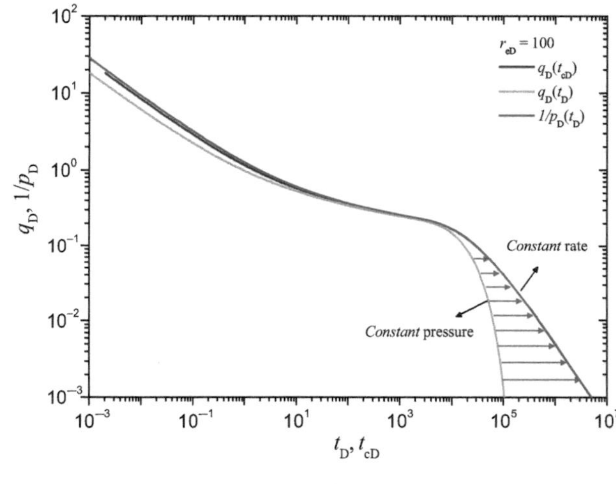

Figure 11.14 q_D, $1/p_D$ vs. t_D curve (after Palacio and Blasingame, 1993)

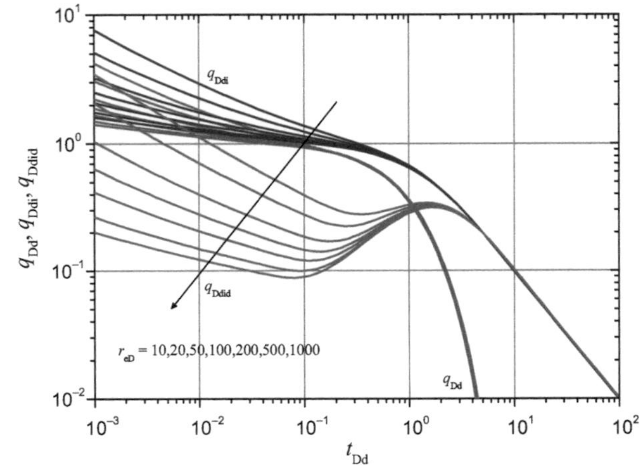

Figure 11.15 Error Blasingame type curves of closed circular homogeneous reservoir

For the published production decline curves, the most typical error is shown in Figure 11.15 (in this book, we take the closed circular homogeneous reservoir model as an example). After entering the boundary-dominated flow period, the declination of dimensionless rate type curve q_{Dd} aggravates, the curves of dimensionless rate integral q_{Ddi} and dimensionless rate integral derivative q_{Ddid} become a "–1"-slope straight line, and exhibit a characteristic of harmonic decline.

There are still many questions. For example, how is Figure 11.15 generated? Why are the slope of dimensionless rate integral q_{Ddi} and dimensionless rate integral derivative q_{Ddid} curves equal to "–1"?

The red line in Figure 11.15 represents for the relation curve of dimensionless decline rate q_{Dd} and dimensionless time t_{Dd}. When building the model, the BHFP is assumed as a constant, which essentially is the curve of Fetkovich dimensionless rate q_{Dd} of a well at the center of a closed circular homogeneous reservoir. The Laplace domain space solution of q_{Dd} (see Chapter 3 for details) is

$$\overline{q}_{Dd} = \frac{\beta\sqrt{\alpha}}{\sqrt{s}} \frac{\dfrac{K_1(\sqrt{\sigma})}{I_0(\sqrt{\sigma})} - \dfrac{K_1(r_{eD}\sqrt{\sigma})}{I_1(r_{eD}\sqrt{\sigma})}\dfrac{I_1(\sqrt{\sigma})}{I_0(\sqrt{\sigma})}}{\dfrac{K_0(\sqrt{\sigma})}{I_0(\sqrt{\sigma})} + \dfrac{K_1(r_{eD}\sqrt{\sigma})}{I_1(r_{eD}\sqrt{\sigma})}} \tag{11.173}$$

where

$$\alpha = \frac{1}{\dfrac{1}{2}\left(r_{eD}^2 - 1\right)\left(\ln r_{eD} - \dfrac{1}{2}\right)} \qquad \beta = \ln r_{eD} - \frac{1}{2} \qquad \sigma = \alpha s$$

As stated in Chapter 2, the flow in the boundary-dominated flow period of a constant pressure production conforms to exponential decline regularity. Therefore, the

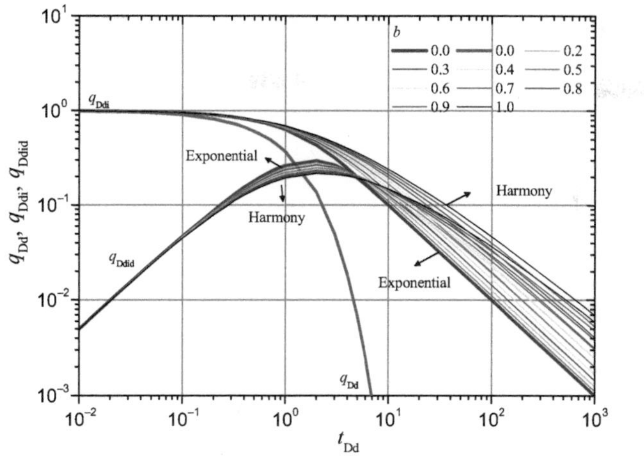

Figure 11.16 Palacio-Blasingame type curves of closed circular homogeneous reservoir-boundary-dominated flow period

red line in Figure 11.15 is indeed a cluster of exponential decline curves. This type curve essentially is a Palacio-Blasingame curve. Furthermore, after entering the boundary-dominated flow period, the flow conforms to exponential decline regularity, as shown in Figure 11.16.

In Figure 11.15, the dimensionless rate integral q_{Ddi} and the dimensionless rate integral derivative q_{Ddid} borrow the definitions of Blasingame method, where,

$$q_{\mathrm{Ddi}} = \frac{1}{t_{\mathrm{Dd}}} \int_0^{t_{\mathrm{Dd}}} q_{\mathrm{Dd}}(\tau) d\tau \tag{11.174}$$

$$q_{\mathrm{Ddid}} = -\frac{dq_{\mathrm{Ddi}}}{d \ln t_{\mathrm{Dd}}} = -t_{\mathrm{Dd}} \frac{dq_{\mathrm{Ddi}}}{dt_{\mathrm{Dd}}} \tag{11.175}$$

Therefore, Figure 11.15 is essentially a Fetkovich type curve. After entering the boundary-dominated flow period, it is forced to conform to exponential decline regularity. As a result, this type curve can only be used to analyze the constant-pressure production data.

As stated in Chapter 2, for exponential type decline, we have

$$q_{\mathrm{Ddi}} = \frac{1-e^{-t_{\mathrm{Dd}}}}{t_{\mathrm{Dd}}} = \frac{1-q_{\mathrm{Dd}}}{t_{\mathrm{Dd}}} \tag{11.176}$$

$$q_{\mathrm{Ddid}} = \frac{1-e^{-t_{\mathrm{Dd}}}}{t_{\mathrm{Dd}}} - q_{\mathrm{Dd}} = \frac{1-q_{\mathrm{Dd}}}{t_{\mathrm{Dd}}} - q_{\mathrm{Dd}} \tag{11.177}$$

Based on Figure 11.15, when $t_{Dd} > 4$ and $q_{Dd} < 0.01$, q_{Dd} is negligible. We have

$$q_{Ddi} \approx \frac{1}{t_{Dd}} \tag{11.178}$$

$$q_{Ddid} \approx \frac{1}{t_{Dd}} \tag{11.179}$$

Therefore, the dimensionless rate integral q_{Ddi} and the dimensionless rate integral derivative q_{Ddid} are a "–1" slope straight line on the log–log plot.

Importantly, we should never confuse the "–1" slope of dimensionless rate integral q_{Ddi} and dimensionless rate integral derivative q_{Ddid} curves in Figure 11.15 with the "–1" slope (harmonic decline) of Blasingame dimensionless rate q_{Dd} curve.

11.5.2 Analysis on Causes of Wrongly Plotting Blasingame Type Curves

From the previous analysis, it can be seen that when the Fetkovich constant pressure production solutions are used to generate the Blasingame decline type curves, the material balance processing is not carried out, but rather the Fetkovich dimensionless time t_{Dd} is taken as the material balance time t_{cDd}, which is the primary cause of errors. The root cause of the problem is the failure to fully comprehend the importance of material balance time (material balance pseudo-time). That is after having introduced the material balance time, the "flow rate" of transient pressure, transient flow rate, and gas well production all come down to a harmonic decline curve. Figure 11.15 is a Fetkovich type curve essentially, if the similar type curves as shown in Figure 11.15 are still used to analyze the transient pressure production data, which is widely off the mark!

11.5.3 Blasingame Type Curve Plotting

At the time of plotting the Blasingame type curves, we should use the constant rate production to calculate the dimensionless rate q_{Dd}, and the corresponding dimensionless time t_{Dd} at this time is the dimensionless material balance time t_{cDd} (gas well: t_{caDd}). Only in this case can we generate the correct Blasingame type curves.

As to the closed circular homogeneous reservoir, dual-porosity reservoir or triple-porosity reservoir, when vertical wells at their center produce at a constant rate, respectively, their dimensionless rate solution forms are the same. That is

$$q_{Dd} = \frac{\ln r_{eD} - \frac{1}{2}}{L^{-1}\left[\overline{p}_{wfD}\right]} \tag{11.180}$$

Table 11.2 σ **function in laplace space**

Model	σ
Homogeneous	$\sigma = \alpha s$
Dual-porosity	$\sigma = \alpha s \dfrac{\omega(1-\omega)\alpha s + \lambda}{(1-\omega)\alpha s + \lambda}$
Triporosity and monopermeability model	$\sigma = \alpha s \left[\omega_{\mathrm{f}} + \dfrac{\lambda_{\mathrm{m1}}\omega_{\mathrm{m1}}}{(\omega_{\mathrm{m1}}\alpha s + \lambda_{\mathrm{m1}})} + \dfrac{\lambda_{\mathrm{m2}}\omega_{\mathrm{m2}}}{(\omega_{\mathrm{m2}}\alpha s + \lambda_{\mathrm{m2}})} \right]$
Triporosity and monopermeability nesting model	$\sigma = \lambda_{\mathrm{m1}} + \omega_{\mathrm{f}}\alpha s - \dfrac{\lambda_{\mathrm{m1}}^2}{\left[\omega_{\mathrm{m1}}\alpha s + \lambda_{\mathrm{m2}} + \lambda_{\mathrm{m1}} - \dfrac{\lambda_{\mathrm{m2}}^2}{(\omega_{\mathrm{m2}}\alpha s + \lambda_{\mathrm{m2}})} \right]}$
Tricontinuum model	$\sigma = \alpha s(\omega_{\mathrm{f}} + \omega_{\mathrm{m1}}\Delta_1 + \omega_{\mathrm{m2}}\Delta_2)$ $\Delta_1 = \dfrac{\lambda_{\mathrm{m1m2}}\lambda_{\mathrm{m2f}} + (\lambda_{\mathrm{m1m2}} + \lambda_{\mathrm{m2f}} + \omega_{\mathrm{m2}}\alpha s)\lambda_{\mathrm{m1f}}}{(\lambda_{\mathrm{m1f}} + \lambda_{\mathrm{m1m2}} + \omega_{\mathrm{m1}}\alpha s)(\lambda_{\mathrm{m2f}} + \lambda_{\mathrm{m1m2}} + \omega_{\mathrm{m2}}\alpha s) - \lambda_{\mathrm{m1m2}}^2}$ $\Delta_2 = \dfrac{\lambda_{\mathrm{m1m2}}\lambda_{\mathrm{m1f}} + (\lambda_{\mathrm{m1m2}} + \lambda_{\mathrm{m1f}} + \omega_{\mathrm{m1}}\alpha s)\lambda_{\mathrm{m2f}}}{(\lambda_{\mathrm{m1f}} + \lambda_{\mathrm{m1m2}} + \omega_{\mathrm{m1}}\alpha s)(\lambda_{\mathrm{m2f}} + \lambda_{\mathrm{m1m2}} + \omega_{\mathrm{m2}}\alpha s) - \lambda_{\mathrm{m1m2}}^2}$

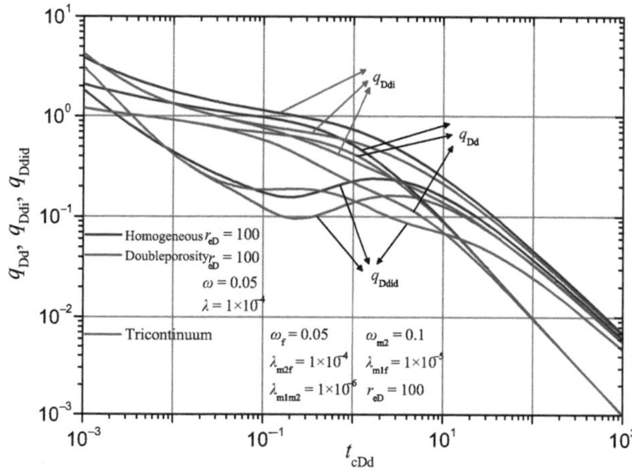

Figure 11.17 Blasingame type curves of different reservoir model

$$\bar{p}_{\mathrm{wfD}}=\frac{\dfrac{K_1\left(r_{\mathrm{eD}}\sqrt{\sigma}\right)I_0\left(\sqrt{\sigma}\right)}{I_1\left(r_{\mathrm{eD}}\sqrt{\sigma}\right)I_1\left(\sqrt{\sigma}\right)}+\dfrac{K_0\left(\sqrt{\sigma}\right)}{I_1\left(\sqrt{\sigma}\right)}}{s\sqrt{\sigma}\left[\dfrac{K_1\left(\sqrt{\sigma}\right)}{I_1\left(\sqrt{\sigma}\right)}-\dfrac{K_1\left(r_{\mathrm{eD}}\sqrt{\sigma}\right)}{I_1\left(r_{\mathrm{eD}}\sqrt{\sigma}\right)}\right]} \tag{11.181}$$

The only difference is that the expression of σ is different in different models, as shown in Table 11.2.

The dimensionless rate q_{Dd}, dimensionless rate integral q_{Ddi}, and dimensionless rate integral derivative q_{Ddid} type curves of some models are shown in Figure 11.17. After having entered the boundary-dominated flow period, no matter what the reservoir types are, the Blasingame dimensionless rate q_{Dd} type curves are all shown as a harmonic decline curve, where the production integral q_{Ddi} curve and production integral derivative q_{Ddid} curve converge gradually.

It is known from the Blasingame matching procedure that only boundary-dominated flow data are true. That is, the production data conform to harmonic decline, and thus the ambiguity in evaluation results of various parameters can be reduced.

Methodology and Cases for Decline Curve Analysis

12

In this chapter, we mainly introduce the process of dynamic analysis using advanced production decline analysis (APDA) methods. In addition, we also present several case studies.

12.1 Methodology of APDA

12.1.1 Data Preparation

During the period of data preparation, we mainly collect the data of well completion and daily production. Please refer to Table 12.1 for the data.

12.1.2 Data Diagnoses

In the production data analysis and interpretation, we often encounter several issues as listed in Table 12.2. Therefore, prior to the production data analysis, it is worthwhile to know these issues.

We can find apparent or potential discordance by a simple visual inspection on the production data report (flowing pressure, rate, and time), and the time–pressure–rate data should be checked in a way as follows.

If the rate suddenly increases or decreases, the flowing pressure should drop or rise accordingly. If not, the rate and pressure data could be unmatched. The unmatched rate and pressure data may result from the change in completion conditions, incorrect measuring position of flowing pressure, or liquid loading in wellbore, etc. The discordant distribution of tubing pressure and casing pressure may result from the liquid loading in wellbore or leaking on tubing.

As for pressure and rate issues, it is usually acceptable to take the mean value of rate, but it makes no sense to take the mean value of pressure. The rate distribution perhaps conforms to the flow rate on site or of pipe manifold, but it may be incapable of representing the single-well performance, especially when the rate is interfered in such cases as shut in or short-term nonuniform flow. If the rate and flowing pressure data are widely scattered, it indicates that the flow is unstable in the wellbore and/or the working conditions are unstable. If the measured value or split value of rate appears as step change for a long time, it usually indicates that the measuring frequency is low and the measurement is most likely inaccurate. This type of flow rate data would reduce the accuracy of bottom hole flowing pressure (BHFP) calculated value, even if the well head pressure (WHP) measurement quality is quite high. If the measured value of flowing pressure assumes step change for a long time, it usually indicates that the

Advanced Production Decline Analysis and Application

Table 12.1 Data preparation of decline curve analysis

Reservoir quality	Porosity, effective thickness, rock compressibility, total compressibility
Oil physical property	Volume factor, gas–oil ratio, viscosity, compressibility
Gas physical property	Volume factor, Z-factor, viscosity, compressibility, gas gravity, CO_2%, H_2S%, N_2%
Pressure convert	Middepth, initial formation pressure, tubing size, casing size, reservoir temperature, flowing well head temperature
Daily production	Tubing pressure, casing pressure, oil–gas–water flow rate data, flowing pressure gradient, hydrostatic gradient

measuring frequency is low. In such a case, we can compare the pressure data with the flow rate data, so as to ensure the pressure data meet the requirements.

12.1.3 Model Diagnoses

The production decline curves have two main functions, including (a) to determine the reservoir permeability parameter by matching the production decline curve with it, which belongs to quantitative analysis, and (b) to diagnose the production performance characteristics, that is, to qualitatively judge the flow state, the change of production index, the offset interference, and the extraneous energy supplement and so on.

Table 12.2 Challenges/Pitfalls of decline curves analysis (after Anderson and Mattar, 2007)

Issue		Influence
Pressure	No pressure measurement	High
	Incorrect initial pressure estimate	High
	Poor p_{tf}–p_{wf} conversion	Moderate
	Liquid loading: effect on p_{tf}–p_{wf}	Moderate
	Incorrect location of pressure measurement	Very high
Flow rate	Rate allocations (potential for errors)	Moderate
	Liquid loading: effect on gas flow rate	Moderate
Well Completion	Zone changes: new/old perforations	Very high
	Changes in the wellbore tubular	High
	Changes in surface equipment	Moderate/high
	Stimulation: hydraulic fracturing	High
General	Reservoir properties	Moderate
	Oil properties	Moderate
	Gas properties	Moderate
	Poor time–pressure–rate synchronization	Moderate/high
	Poor time–pressure–rate correlation	Very high

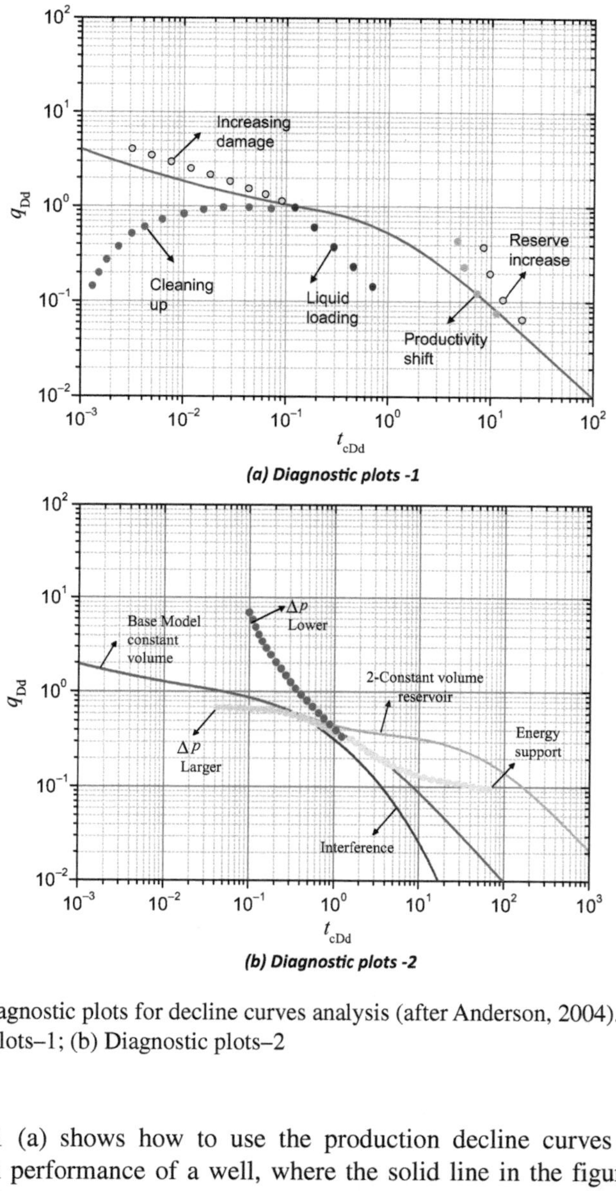

Figure 12.1 Diagnostic plots for decline curves analysis (after Anderson, 2004). (a) Diagnostic plots–1; (b) Diagnostic plots–2

Figure 12.1 (a) shows how to use the production decline curves to diagnose the production performance of a well, where the solid line in the figure represents a normal decline curve in the constant-volume reservoir. In the period of the early transient flow, if the real production curve gradually goes upward to approach the normal curve, it indicates that the skin factor of the well is dropping gradually, that is, a cleanup process is undergoing. On the contrary, it indicates that the well is being damaged more and more seriously. In the period of late boundary-dominated flow, if the real production data skews from the harmonic straight line portion and gradually moves leftwards, it indicates that there is interwell interference, which reduces the original gas (oil) in place of the well. On the contrary, if the real production data curve gradually shifts rightwards, it indicates that there is extraneous energy supply, as shown in Figure 12.1 (b).

12.1.4 Methodology of Decline Curve Analysis

Theoretically, the production data analysis is the same as the pressure transient analysis. To improve the production data analysis, we diagnose, analyze, and interpret the production data using the following process,

1. Convert flow rate. For a high gas–liquid ratio (GLR), convert the condensate oil into condensate gas, and conduct processing as per pseudo-single phase; if the flowing time is less than 24 hours, convert the flow rate into daily rate.
2. Convert flowing pressure. If there are no bottom hole flowing pressure (BHFP) data, use the related wellbore pipe flow software to convert the tubing pressure or casing pressure data into BHFP data under the constraint of the measured pressure gradient.
3. Review data. Review the data quality/correlativity of the production curve.
4. Remove/reduce data. Remove the anomalous data points in the log–log plot used by diagnoses.
5. Identify characteristic flow regime. Similar to the pressure derivative curve in the well test analysis (WTA), we can also use the pressure curve and rate integral derivative curve to identify the flow characteristics in the period of transient flow, for example, the normalized pressure integral (NPI) method and Blasingame method.
6. Match type curve. Match the measured data with the corresponding theoretical decline curves in the reservoir model to get relevant parameters.
7. Match history. Conduct ultimate "history matching" of model and original data of single-well performance (p_{wf} and q).
8. If the analytic solution cannot be matched very well, we can carry out numerical analyses based on it to derive relevant parameters.
9. Forcast production performance. On the basis of building the dynamic model from the above analysis, we can conduct performance prediction of constant pressure or constant rate based on the matching results, so as to direct the production.

As mentioned by Mattar and Anderson (2003), the production decline curve analysis technique is developed at a very high speed, where a variety of methods can be applied. However, there is no method that can always be capable of presenting the most reliable interpretation. In practice, various methods can be combined rationally, constraining each other and supplementing each other, so as to reduce the uncertainty in reserve estimate.

12.2 Example of Gas Well

12.2.1 Basic Parameters

The initial reservoir pressure is 9.20 MPa, the reservoir temperature is 48.9°C, the net thickness is 6.04 m, the porosity is 0.20, and the gas saturation is 0.80. The gas flow rate is far higher than the critical fluid carrying flow rate. The data in the example are taken from the continuous measurement values of gas flow rate and BHFP, as shown in Figure 12.2.

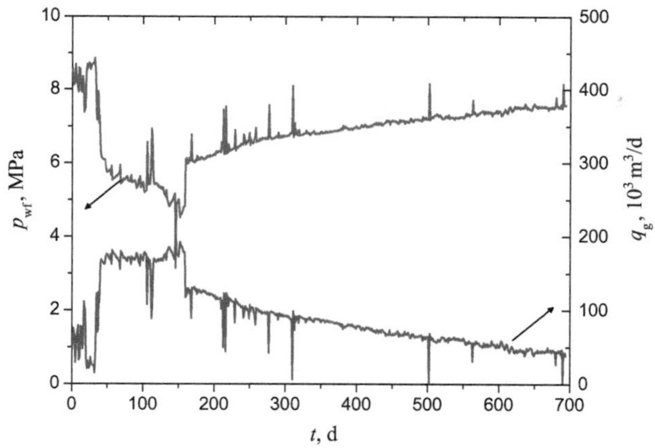

Figure 12.2 Production history curve of well E

12.2.2 Decline Curves Analysis

First, use the "production history curve" to review the data. In this example, the correlation between flow rate and pressure is very good, which indicates that the data quality is high and favorable for carrying out reliable analysis.

Second, conduct Arps production decline analysis, as shown in Figure 12.3. When the flow rate is zero, the cumulative production is predicted as only 0.82×10^8 m^3 according to the exponential decline.

Third, conduct type curve matching, which shows that the type curve matching results of the Blasingame, Agarwal-Gardner, and NPI methods are consistent, the production is in boundary-dominated flow period, and the original gas in place (OGIP) is close to 6.5×10^8 m^3 (see Figure 12.4).

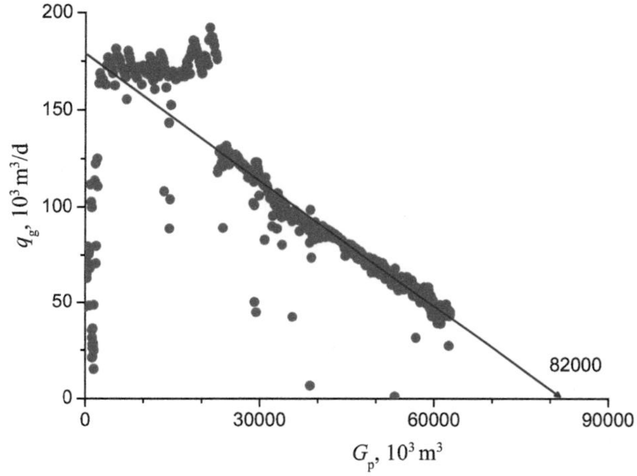

Figure 12.3 Arps decline curve analysis of well E

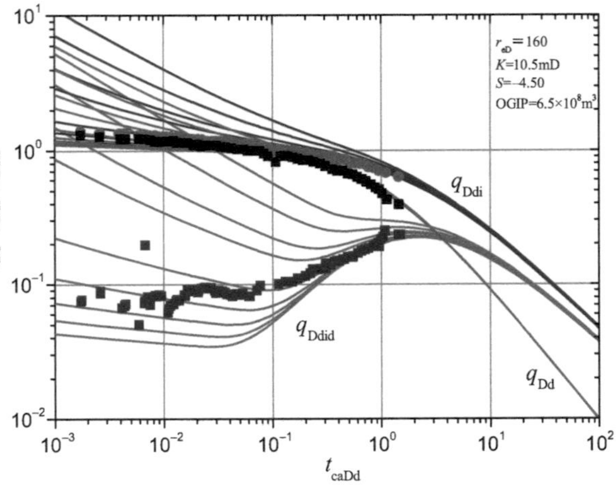

Figure 12.4 Blasingame decline curve analysis of well E

Fourth, draw the dynamic material balance curve, as shown in Figure 12.5, from which the OGIP is calculated as 6.3×10^8 m^3. It approaches the result of curve matching. Supposing the abandonment pressure is 2.1 MPa, the ultimate recoverable reserve is expected as 5.1×10^8 m^3.

At first sight, there is a great difference in the two analysis methods. Mattar and Anderson (2003) pointed out that if the impact of flowing pressure had been taken into consideration, it was very apt to interpret this difference. The reason is that the Arps decline analysis correlates with the production conditions, and the expected ultimate recoverable reserve depends on the value of pressure drop (which is negligible). Once the impact of flowing pressure (scale and trend) is taken into consideration (according to flow material balance [FMB]), it is observed that the actual OGIP is far higher

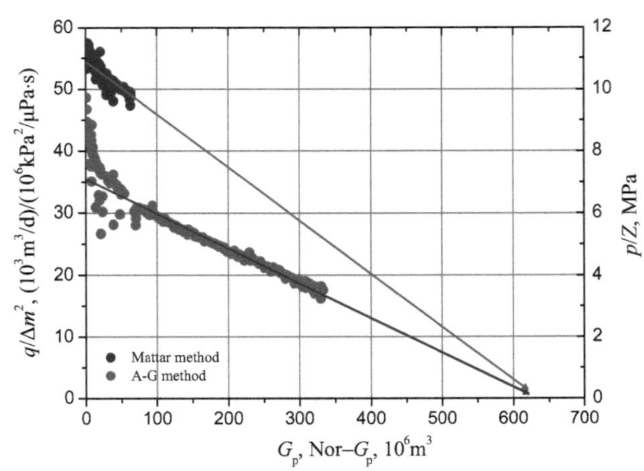

Figure 12.5 FMB analysis of well E

than the value obtained from the Arps method. Therefore, under current production conditions (i.e., if no stimulation is adopted), the recovery factor would be very low (0.13%). On the contrary, the expected ultimate recoverable reserve presented by the FMB method is not restricted by the lower pressure drop, and it gives much higher recoverable reserve. From the above, it can be seen that this well is an excellent candidate for pressurization.

12.3 Example of Oil Well

12.3.1 Basic Parameters

The initial reservoir pressure is 70.8 MPa, the reservoir temperature is 136.4°C, the net thickness is 26.0m, the porosity is 0.03, the gas saturation is 0.80, and there are data from six gradient tests. The production data is shown in Figure 12.6.

12.3.2 Decline Curves Analysis

Step 1: The flow rate has a very good correlation to pressure. Use the wellbore pipe flow software to convert the tubing pressure into flowing pressure at midpoint of pay zone, as shown in Figure 12.6. In conversion, the gradient test data points are used to carry out correction.

Step 2: Combine a variety of analytic methods for determining the reservoir parameters and the original oil in place (OOIP). Use Eq. (9.12) to draw the Agarwal-Gardner FMB curve, as shown in Figure 12.7, and the OOIP is calculated as 9.2×10^4 m^3.

The type curve matching results of Blasingame, Agarwal-Gardner, NPI, and transient methods show that the production is in boundary-dominated flow period, as shown in Figure 12.8. The matching results of various methods are basically the same, and the OOIP is close to 9.2×10^4 m^3, as shown in Table 12.3.

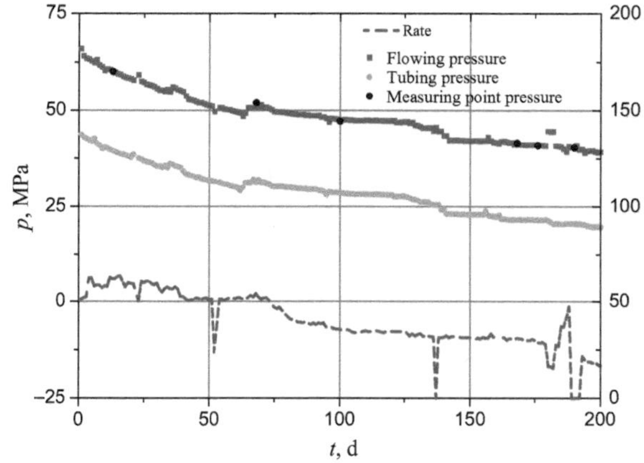

Figure 12.6 Production history curve of well F

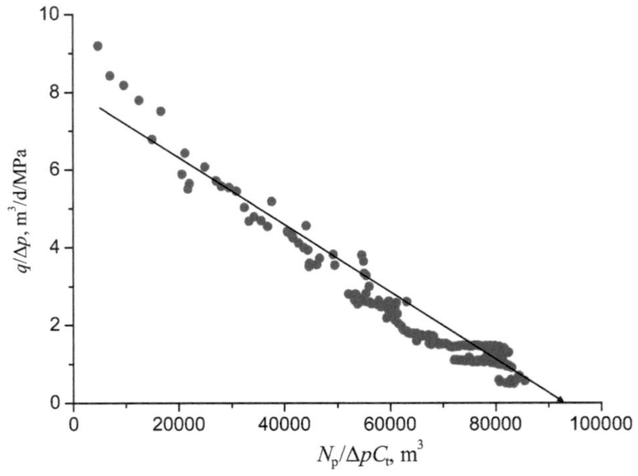

Figure 12.7 Agarwal-Gardner FMB analysis of well F

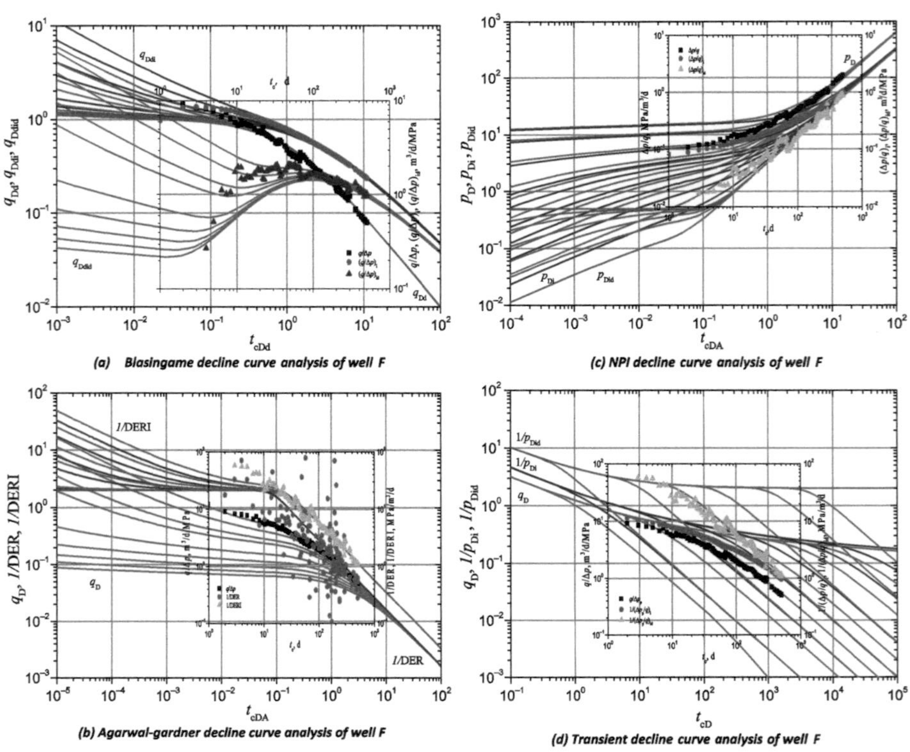

Figure 12.8 Decline curve analysis of well F. (a) Blasingame decline curve analysis of well F; (b) Agarwal-Gardner decline curve analysis of well F; (c) NPI decline curve analysis of well F; (d) Transient decline curve analysis of well F

Table 12.3 **Result of decline curve analysis of well F**

	Dimensionless radius at external drainage	OOIP	Drainage area	Permeability	Skin factor
	r_{eD}	N	A	K	S
Method		10^4 m^3	10^4 m^2	mD	
Blasingame	7	9.27	27.8	0.21	−5.26
Agarwal-Gardner	7	8.84	26.5	0.23	−5.96
NPI	7	9.70	26.8	0.25	−5.97
Transient	7	8.94	29.1	0.19	−6.01
Average		9.19	27.6	0.22	−5.80

A great number of analytic cases show that there is no independent production data analysis method that is capable of processing all types of data and reservoir, and flowing pressure has nearly the same important impact on the analysis quality as the flow rate. If the flowing pressure is ignored, there is a high possibility to get a wrong interpretation. Therefore, for a medium- and high-yielding well in which the flow is complicated, we must pay attention to the periodical monitoring of BHFP, so as to lay a foundation for the performance prediction and stimulation adjustment of the well. Based on the above process, we build a single-well performance model, which can be used to predict the performance of a single well, as shown in Figure 12.9.

12.4 Example for Integrating APDA and WTA

Tazhong area is located in the middle of the Central Uplift Zone, Tarim Basin, of China, where three 1000×10^8 m^3 large condensate gas fields were proved in Ordovician carbonate reservoirs in recent years. The reservoirs in the area are

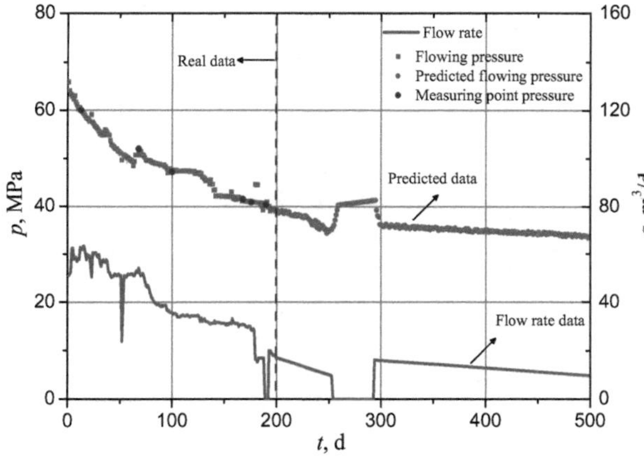

Figure 12.9 Performance forecasting analysis of well F

complicated in types, including cavern reservoir, fracture–vug reservoir and fracture reservoir, and dominantly matrix–pore type reservoir. The flow mechanism is also complicated, including seepage flow, slit flow, and cavern flow. Moreover, the fluid property is complicated, resulting in condensate gas reservoir containing low, medium, to high content of condensate oil and volatile oil reservoir. In practice, the controllability is poor, and the production appears at a variable pressure and a variable flow rate. The difference of reservoir pressure and dew point pressure is small, retrograde condensation exists in the reservoir, multiphase flow occurs in the wellbore, and the BHFP conversion is difficult. There are only a few hydrostatic gradient points, which do not possess monotone decreasing property. Therefore, the conventional material balance method is not applicable. For the field development personnel, it is a great challenge to estimate the dynamic reserves of a complicated carbonate condensate gas well (reservoir) in a complex production state (Sun and Shi, 2013).

Based on the geologic characteristics and availability of data, we developed a dynamic description technique for a heterogeneous gas reservoir. This technique combines the WTA and the APDA to mainly estimate the dynamic reserves. This technique, with the help of dynamic description of a single well or whole gas reservoir solves the key issues such as reservoir type classification, single-well productivity, single-well drainage area and dynamic reserve, as well as reservoir permeability. The calculation results of the case show that this technical process reduces the ambiguity of well test interpretation results and the uncertainty of dynamic reserves.

12.4.1 Adaptability Analysis of Classical Computational Method

The reservoir type and flow mechanism are somewhat complicated in the Tazhong-I gas field. It is essential to determine whether the filtration theory–based dynamic reserve estimate method is applicable. We conducted analysis on the single-well seismic reflection type, drilling venting, WTA, and production test characteristics. The analysis results show that the gas field is characterized by low production index and formation flow capacity, dominant seepage flow, and no "pipe flow" or "cavity flow;" the reservoir type is dominantly fracture–vug. Therefore, the filtration theory based dynamic reserve estimate method can be applied in Tazhong area, as shown in Table 12.4.

Based on the definition, it is known that "dynamic reserve" is a value between the measured petroleum initially in place and the technically recoverable reserves. Therefore, it can be calculated using the computational method of recoverable reserves (SY/T 6098-2010). In such a case, we only need to allow the "flow rate to be zero, and the pressure to be 1 normal atmospheric pressure." Restricted by the complex reservoir, limited static pressure test data, particular production mode at variable pressure and variable rate, and unpredictable production performance, the recoverable reserve extrapolation method is not suitable for the dynamic reserves estimate in the Tazhong area. Refer to Table 12.5 for more information.

Table 12.4 **Static state and dynamic state statistical list in tazhong area**

Well	Seismic reflection	Drilling venting	Well test analysis	Formation flow capacity	Testing					Production index	
					Choke	Tubing pressure	Daily oil production	Daily gas production	Daily water production	Gas well	Oil well
		m		mD·m	mm	MPa	m^3	m^3	m^3	$m^3/MPa^2/d$	$m^3/MPa/d$
TZG1	String of beads				5	45.7	74.4	130921	0	154.4	
TZG2	Flaky				3	24.5	22.9	29556	0	9.0	
TZG3	String of beads	2.3	Homo-geneous reservoir	100	6	43.4	15.3	169537	22.6	111.2	
TZG4					5	21.4	40.8	70260	1.4	20.2	
TZG5					5	40.4	37.7	146082	0	80.9	
TZG6					5	42.7	62.9	114366	1.9	91.5	
TZG7		4.7	Dual-porosity reservoir	44	5	30.0	52.5	95440	0	32.4	
TZG8					4	35.8	25.1	69485	/	40.6	
TZG9					5	41.4	38.1	134935	/	435.6	
TZG10					4	31.2	85.4	35890	/		30.5
TZG11					4	14.9	76.2	3276	/		7.8
TZG12	Flaky				5	9.0	42.0	15576	11.1		1.3
TZG13		0.4			4	27.1	32.4	45102	/	16.7	
TZG14			Dual-porosity reservoir	194	4	39.9	52.0	74006	/	234.4	
TZG15			Composite reservoir	338	5	23.8	79.9	45301	/	27.2	
TZG16	Clutter				5	5.7	9.2	36201	/	9.8	

Table 12.5 SY/T 6098-2010 applicability assessment on recoverable reserve extrapolation method

Computational method	Applicable conditions	Applicability		Practical situation	Conclusion
		Single well	Unit		
Material balance	At least 2 static pressure survey points	√	√	A few static pressure points, nonmonotone decreasing	×
Elastic two-phase	Constant-rate production + closed boundary	√		Variable flow rate + fracture + multiple reservoirs	×
Production decline	Constant-pressure production	√		Variable pressure production	×
Improved attenuation	Cumulative production–time relationship	√	√	Multisection, trend uncertainty is large	×
Forecast model	Annual output–time relationship	√	√	Multisection, trend uncertainty is large	×
Numerical simulation	Geologic model	√	√	Modeling is difficult	×
Analogy method	Applicable in new area and reservoir that are not put into production or developed only for a short time, as well as in old area short of performance data	√	√		×

12.4.2 Dynamic Reserves Estimate from Dynamic Description Method

In view of the issues like strong reservoir heterogeneity, complicated fluid property, multiphase flow in wellbore, variable pressure and variable rate production modes, and poor performance data quality, based on the geological characteristics of the gas reservoir, we use the WTA software and the APDA (PA) software to develop a dynamic reserves estimate and analysis technique for heterogeneous gas reservoirs. The technique combines the WTA with the APDA.

12.4.2.1 Transient Well Test Reserve Estimate Technique of Full Life Cycle

Considering the variety, ambiguity, and complexity in interpretation, a WTA technique by using the "entire history match" can greatly increase the utilization ratio of data and decrease the ambiguity of interpretation results (Sun 2012). Besides, it can be used to estimate the single-well dynamic reserves. The essence of this method is the material balance, that is, to estimate the drainage area of gas well by combining the short-term well test with the long-term production performance, and use the volumetric method to calculate the dynamic reserves of a single well.

The single-well drainage area can be determined using the following three methods, including

1. For a log–log plot with boundary response, the volumetric method can be directly used to calculate the reserves.
2. For a log–log plot without boundary response, the infinite approximating method can be used to match the long-term production history to determine the boundary range, and then the dynamic reserves within the equivalent scope spread by the current pressure wave are derived, as shown in Figure 12.10. The reservoir drainage radius of this well is 350 m.
3. For a well producing for a long time and having past pressure build-up test data, the deconvolution method can be used to determine the boundary, as shown in Figure 12.11. Deconvolution results show that the well has obvious closed boundary, where the interpretation boundary is 500 m.

When the analytic solution WTA method is insufficient to describe the reservoir, the distribution characteristics of it on the reservoir profile can be properly adjusted, so as to obtain the numerical well test description of the reservoir, as shown in Figure 12.12.

This well is characterized by dual-porosity multi-region composite model, as shown in Figure 12.13. It is a 66-m-radius dual-porosity reservoir in the region around the well, with a permeability of 5.2 mD and a storativity ratio ω of 0.223, which are consistent with the results obtained from the drilling and geological study. The pressure hopper of depletion has already swept into the gas reservoir boundary.

On the basis of single-well performance description by combining the short-term well test and the long-term production performance, we use the volumetric method to calculate the dynamic reserves of the single well. For this well, the drainage area is 2.95 km^2, and the dynamic reserves calculated by well test dynamic description method are 6.0×10^8 m^3; see Table 12.6 for details. This technique has greatly enhanced the functions of WTA. It can be used to evaluate the stimulation effect and identify the reservoir parameters, as well as to calculate the dynamic reserves, predict the production performance, optimize flow rate, and carry out multiwell connectivity analysis.

12.4.2.2 APDA Reserves Estimate

The advanced production decline analysis technique, based on the daily production data of a gas well and the "type curves matching" process, presents a new method to estimate the dynamic reserves. Represented by the Blasingame method, this technique can be used to process complex data like variable pressure and variable rate in Tazhong area, as shown in Table 12.7. The parameters estimation by this technique can be made as follows. Firstly, carry out BHFP conversion under the constraint of pressure gradient data. Then, make preliminary analysis by means of the type curves matching. Finally, match the production history using the single-well analytical or numerical model.

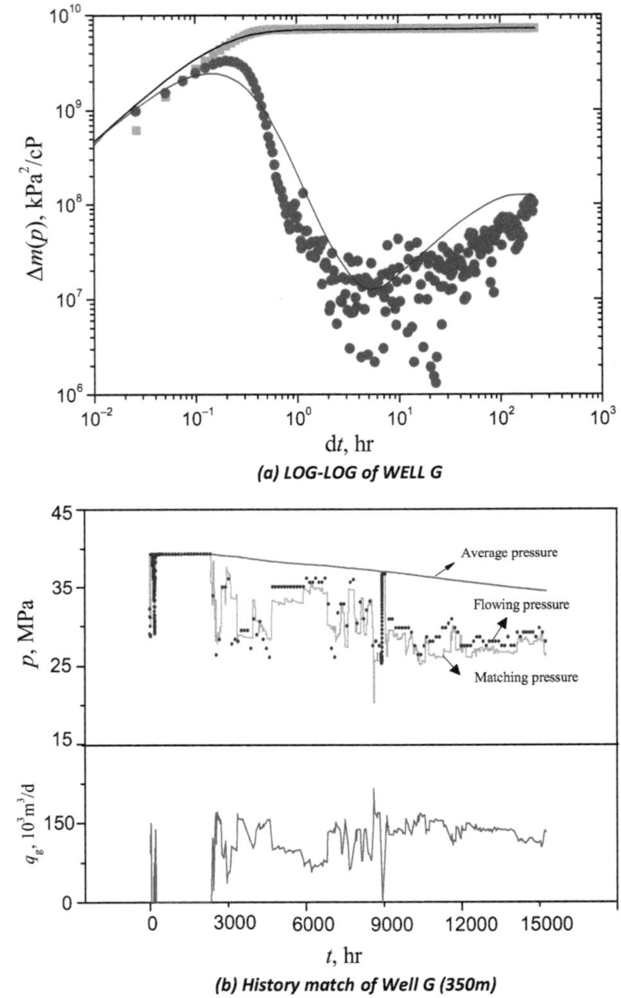

Figure 12.10 Infinite approximation method of well G. (a) Log–log of well G; (b) History match of well G (350 m)

12.4.2.2.1 Influential Factors in a Higher Gas–Oil Ratio (GOR) Case

For a case with higher GOR similar to dry gas, the accuracy of dynamic reserves estimate is affected by two factors, including (a) the error resulting from inaccurate BHFP conversion, and (b) the error due to the initial reservoir pressure, relevant fluid pressure–volume–temperature (PVT), and reservoir parameters.

The accuracy of BHFP is the key factor influencing the calculation result of drainage area. The calculation result shows that the error percentage of dynamic reserves is about twice that of flowing pressure conversion. If the flowing pressures from conversion and from flowing pressure gradient test are 5% lower, the dynamic reserves can be 10% lower; on the contrary, the dynamic reserves can be 10% higher.

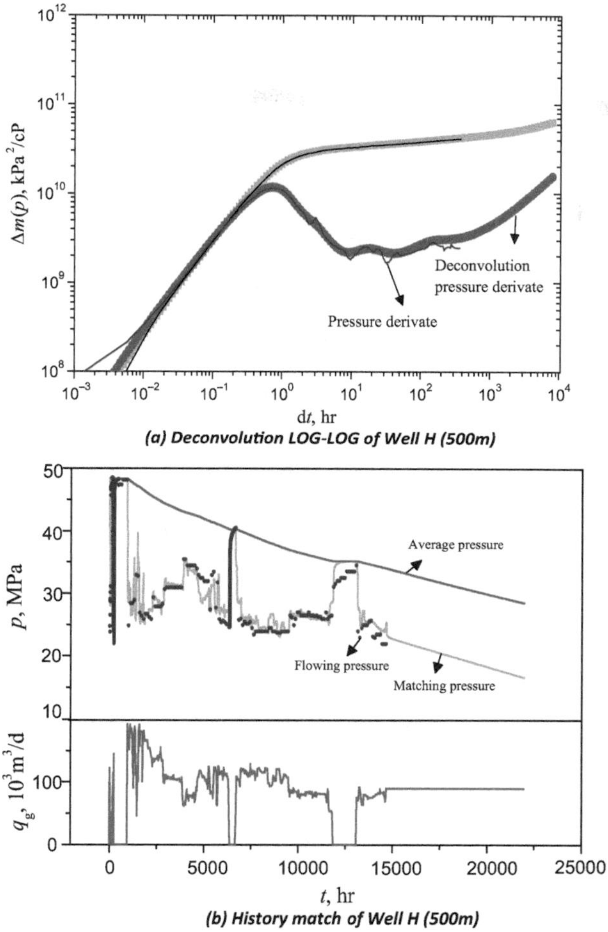

Figure 12.11 Deconvolution method of well G. (a) Deconvolution log–log of well H (500 m); (b) History match of well H (500 m)

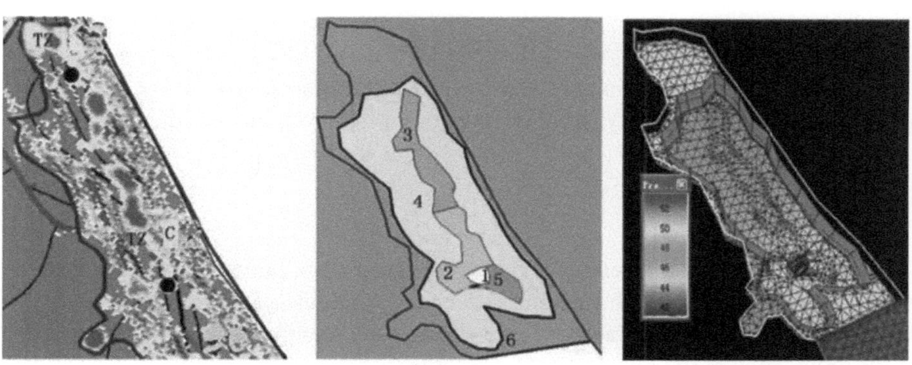

Figure 12.12 Reservoir prediction, blocking and pressure field of well I

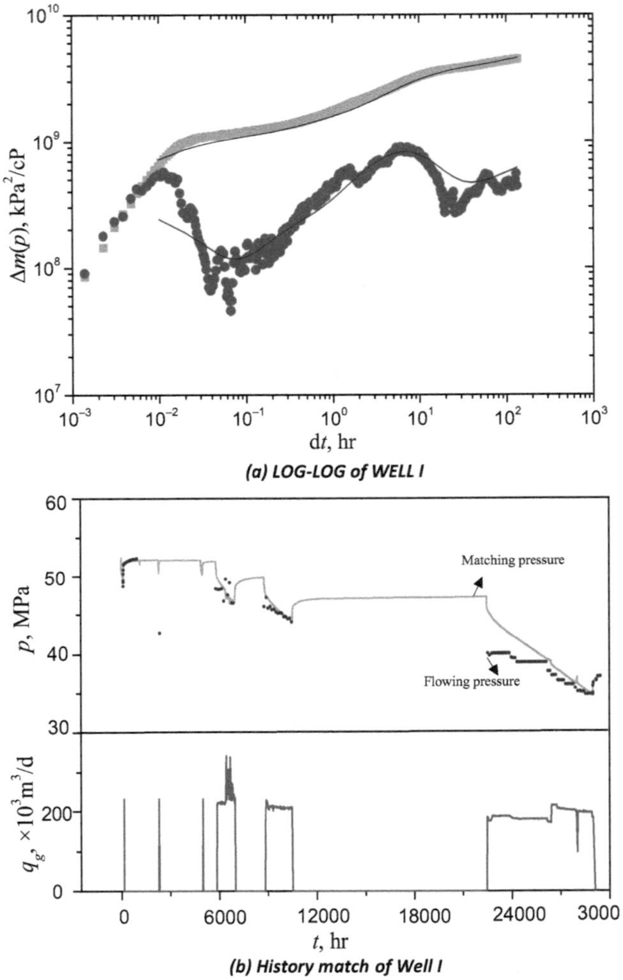

Figure 12.13 Numerical well test method of well I. (a) Log–log of well I; (b) History match of well I

In the advanced production decline analysis method, if the BHFP is expressed by pseudo-pressure $m(p)$, the dynamic reserves of gas reservoir can be calculated by

$$G = \alpha \frac{S_g p_i}{\mu_{gi} Z_i C_{ti}} f_M \tag{12.1}$$

Therefore, the initial reservoir pressure, gas saturation, porosity, and relative density of gas can all bring forth a certain influence on the calculation results. Simulation shows that the previous parameters may affect the dynamic reserves by about ±5%. If the input value of the initial reservoir pressure is higher than the true value, the calculated dynamic reserves are higher than the true dynamic reserves. On the contrary,

Table 12.6 Numerical well test result of well I

Region	C (m^3/MPa)	Kh (mD·m)	S	ω	λ	M	D	A (km^2)	OGIP (10^8 m^3)
1	0.18	221	−3.36	0.22	1.42E–05			0.014	0.040
2				0.01	8.20E–07	40	40	0.06	0.004
3				0.05	2.00E–06	2	0.2	0.17	2.476
4				0.10	2.00E–06	0.25	1	0.19	0.547
5						1.75	1.55	1.41	2.619
6						10	10	1.10	0.317
Sum								2.95	6.000

the calculated dynamic reserves are lower than the true dynamic reserves, as shown in Figure 12.14. Although porosity is a static parameter, it can also affect the calculation results, because it is related to the rock compressibility, as shown in Figure 12.15. Given that the other parameters are constant, the deviation resulting from porosity mainly shows in the total reservoir compressibility, which lies within 5%. If the compressibility is 5% higher, the dynamic reserves are 5% lower; if the compressibility is 5% lower, the dynamic reserves are about 5% higher.

Table 12.7 Applicability assessment on production decline analysis method

SN.	Development stage		Computational method	Applicable condition	Practical situation	Conclusion
1	Empirical method		Arps	Constant pressure production + boundary-dominated flow	Variable pressure and variable rate production	Not applicable
2	Curve matching method	Classical	Fetkovich	Constant rate production + closed boundary		Not applicable
3		Advanced	Blasingame	Boundary-dominated flow		Applicable
4			AG (Agarwal-Gardner)			Applicable
5			NPI			Applicable
6	Numerical method		Numerical	Refinement based on analytical matching		Applicable

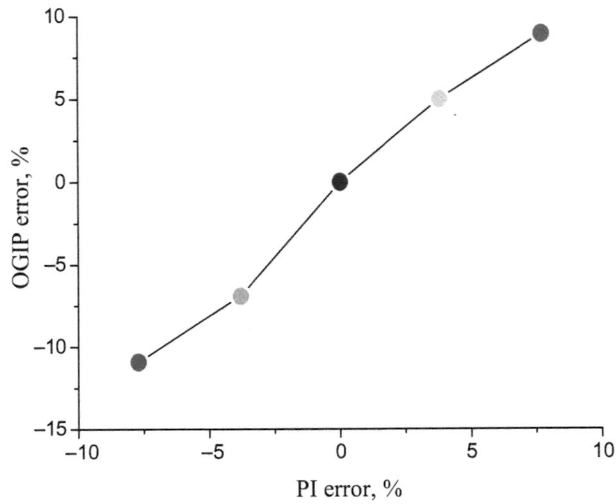

Figure 12.14 PI on the OGIP result

12.4.2.2.2 Influential Factors in a Lower GOR Case

For a case with lower GOR, the PVT data of condensate gas reservoir are the major factors influencing the calculation results of dynamic reserves. Supposing a well is located at the center of a closed circular reservoir in Tazhong area, we take the simulation parameters from the real data of the well, as shown in Table 12.8. In the model, the gas in place is $1.29 \times 10^8\ m^3$ and the condensate oil in place is 10.8×10^4 t. At the time of abandonment, the recovery percentages of gas and condensate oil are 62.01% and 32.2%, respectively.

Converting the flow rate of condensate oil and gas into the flow rate of condensate gas, and calculating the dynamic reserves using the pseudo-single-phase method, the

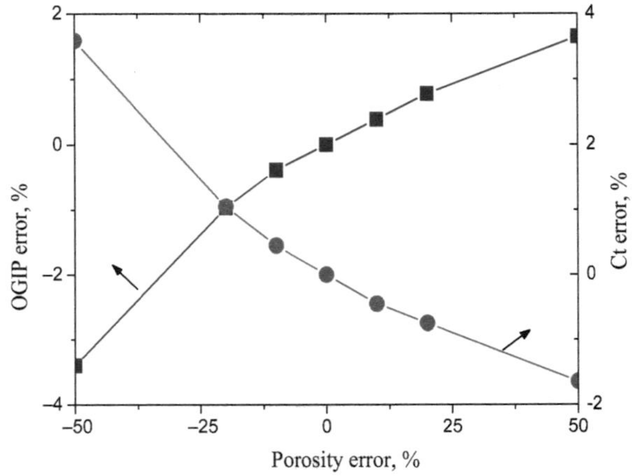

Figure 12.15 Porosity on the OGIP result

Table 12.8 **Simulation variable for low gas–oil ratio cases**

Parameter	Unit	Value	Parameter	Unit	Value
Top and bottom	m	6293–6308	Abandonment rate	m^3	3000
Net pay	m	15.0	Well bore diameter	mm	89.0
Porosity	%	1.85	Relative density of gas		0.621
Permeability	mD	11.0	Gas–oil ratio	m^3/m^3	1945.9
Saturation	%	75.0	Separator pressure	MPa	1.5
Compressibility factor	kPa^{-1}	1.50E–06	Separator temperature	°C	26.8
Initial reservoir pressure	MPa	68.11	Dew point pressure	MPa	51.8
Reservoir temperature	°C	151.86	Dew point temperature	°C	151.9
Reservoir boundary	m	1000.0	Relative density of oil		0.7681
Allocation	10^4 m^3	10.0	N_2 mol%	%	3.09
Abandonment pressure	MPa	18.0	CO_2 mol%	%	2.66

dynamic reserves can be obtained as 0.93×10^8 m^3, which is 28% lower than the actual value. Then, if we consider the impact of PVT, and calculate the dynamic reserves as 1.34×10^8 m^3 using the gas–liquid two-phase pseudo-pressure method, it is 4% higher than the actual value. Therefore, the error can be regarded as an operating error. From the above it can be seen that, for a condensate gas well, when the GOR is lower, the dynamic reserves cannot be calculated simply by using the condensate gas pseudo-single-phase method.

12.4.2.2.3 Other Influential Factors

Further analysis shows that when the basic parameters are fixed, the production time, produced water, and routine of work are the main factors that influence the accuracy of dynamic reserves estimate, as shown in Table 12.9. A short production time leads to the a lower OGIP estimate. For low permeability or fractured reservoirs, gas supply is not sufficient in the immediate vicinity of wellbore, so the BHFP decreases quickly. If data of the early stage is used, the dynamic reserves tends to be low; if the production time is short, outside tanks are not connected, which also leads to the low estimate.

12.4.2.3 Integrated Dynamic Reserve Estimate Technique of Full Life Cycle

In view of the geological characteristic and the availability of data, based on data diagnostics, model diagnosis, reservoir type judgment, damage analysis, and liquid loading, WTA and PA are combined to evaluate the reserves. The combination can decrease the uncertainty in estimation. The evaluation of the single-well dynamic reserves can be combined with geological achievements to further evaluate the producing reserves of block.

Table 12.9 Other factors influencing the uncertainty of dynamic reserves

Influential factor		Influence on dynamic reserves		Remarks
		Greater	Smaller	
Production time	Multitank		√	If the production time is short, the outside tank will not be production communicated.
	Low-permeability, fractured		√	The gas supply is insufficient in the immediate vicinity of well bore, the BHFP drops fast, but becomes slow at late stage.
Water production	Connecting large bottom water	√		Communicated with large bottom water, present an illusion of stable pressure and high yield in the early period of production.
	Communicated with fracture		√	Water energy is insufficient; after water breakthrough, flowing pressure drops rapidly.
Routine of work	Large differential pressure		√	Similar to low-permeability case.
	Retrograde condensation		√	Gas–liquid two-phase flow, flowing pressure drops rapidly.

Also, taking the well I (a circular area with a diameter of 880 m) as an example, the effective drainage area is 244.7×10^4 m^2 which is calculated by using the PA method. The dynamic reserves are 5.72×10^8 m^3, as shown in Figure 12.16. The OGIP is estimated to be 6.0×10^8 m^3 using the WTA method. The previous two estimations are approximations. The equivalent control area of PA is less than that of WTA. This is because the well shows a composite model (as indicated by the well test interpretation) while the PA method does not consider the heterogeneity of formations. The equivalent control area obtained from the PA method is smaller than that obtained from the WTA method. The reason can be that the well test interpretation results show that this well is indicative of a "good inside and poor outside" composite model characteristic, whereas the production decline analysis method does not take reservoir heterogeneity into account, as shown in Figure 12.17.

According to the WTA and PA theories, the relevant software is used to evaluate the dynamic reserves in the full life cycle. Well spacing and dynamic reserves of a complex carbonate gas condensate well can be determined under the complex production specification. Besides, several issues during the production, such as gas deliverability, rate maintenance capability, interwell communication, analysis of water producing, and determination of permeability, can also be addressed.

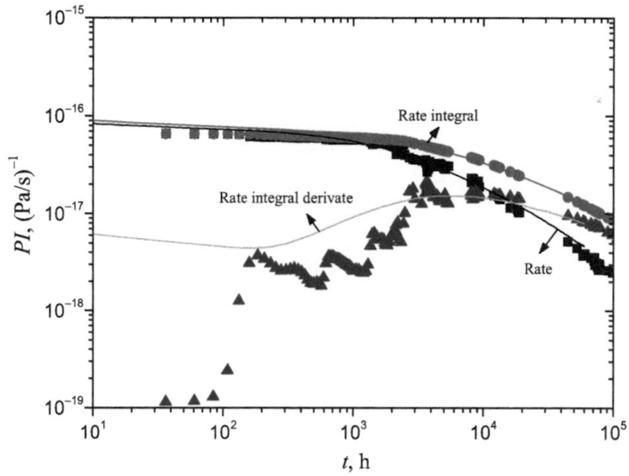

Figure 12.16 Blasingame match curve of well I

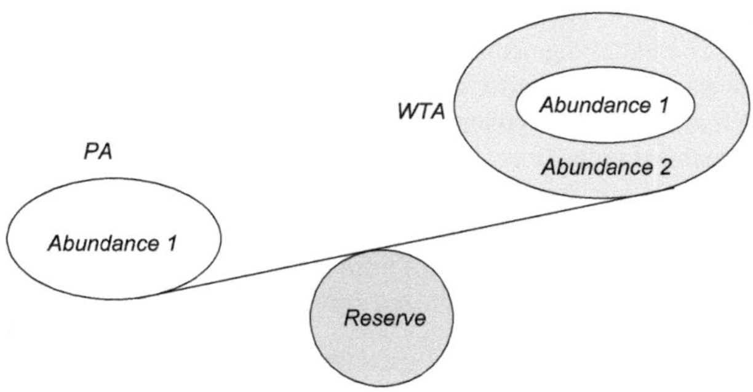

Figure 12.17 Well spacing of WTA and PA

12.5 Application of APDA Method in Oil and Gas Reservoir Engineering

The APDA and the entire life cycle WTA methods are principal in the dynamic reservoir characterization. By using daily production data, the APDA can quantitatively analyze the fluid flow in porous media. In addition, this method can be applied for water breakthrough forewarning analysis, multiwell and multitank reserves evaluation, and producing pressure drop estimation.

12.5.1 Water Breakthrough Forewarning Analysis

For the oil/gas well with edge/bottom aquifer, the decline analysis curve can be separated into four stages, which are listed as follows with the corresponding characteristics, as shown in Figure 12.18.

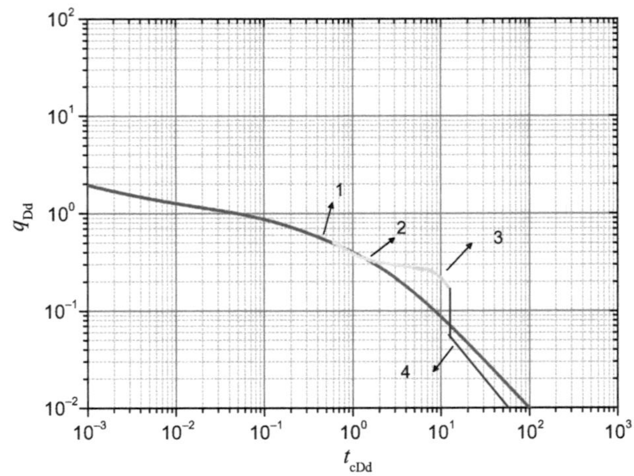

Figure 12.18 Blasingame decline curve for water invasion

1. Normal production stage. There is an overlap between the normalized rate curve and a certain type curve. The initial segment of the FMB curve is a straight line.
2. Energy supply stage due to aquifer coning. The normalized rate curve deviates from the boundary-dominated straight segment to the upper right. The FMB curve deviates from the initial straight segment to the upper right.
3. Production variation stage due to aquifer coning. The normalized rate curve deviates from the boundary-dominated straight segment to the lower left and does not intersect with the theoretical curve. The FMB curve deviates from the initial straight segment to the lower left.
4. Unloading production stage. The normalized curve falls below the theoretical curve.

The water invasion can be classified into two major categories, as shown in Table 12.10, including no water production and water breakthrough production by integrating individual well production, pressure, and rate decline characteristics. The water-free production can be classified into three minor types, including no water invasion stage, early water invasion stage, and middle and late water invasion stage. Those four water invasion types are different in production, pressure, and rate decline characteristics.

A water breakthrough gas well in the Tazhong gas reservoir was analyzed. The corresponding type curve illustrated that the data points deviate to the upper right first and then deviated to the lower left. The deviation to the upper right reflects the energy supply. The deviation to the lower left reflects the decrease of gas deliverability, which indicates water breakthrough, as shown in Figure 12.19.

Therefore, the water breakthrough forewarning analysis can be performed through rate decline analysis to extend the water-free production period as long as possible. Figure 12.20 is an actual field case of three wells. There is a water breakthrough trend during the production. The pressure-control production mode is applied and gains greater cumulative production.

The curve matching should be performed cautiously and the initial data should be selected as far as possible during the reserve evaluation. Otherwise, the reserve will be extremely overestimated, as shown in Figure 12.21. The OGIP is 1.1×10^8 m^3 when

Table 12.10 Classification of single-well water invasion type

No.	Type		Production characteristic	Decline curve characteristic
1	No water production	No water invasion stage	Production: steady Pressure: similar decreasing trend	Type curve: overlaps with a certain type curve FMB curve: straight line
2		Early water invasion stage	Production: steady or increasing Pressure: a relatively low decreasing rate and lower than the previous, an obvious inflection point	Type curve: data points deviate from the type curve to the upper right FMB curve: deviates from straight line and rises
3		Middle and late water invasion stage	Production steady decline The decline rate of pressure is larger than that of the initial stage	Type curve: data points deviate from the type curve to the lower left FMB curve: deviates from straight line and drops
4	Water break-through production	Unloading production stage	Production: decline gradually Pressure: decline	Type curve: data points deviate to the lower part FMB curve: deviates from a straight line to rise firstly and then drops

the initial data are applied in the reserve analysis. The OGIP is 1.4×10^8 m^3 when the later data are applied in the dynamic reserve analysis. The relative estimated reserve difference is 27%.

12.5.2 Multitank Reservoir Dynamic Reserve Evaluation

There is a strong reservoir heterogeneity in the fractured–vuggy carbonatite reservoir, and a multitank reservoir can be connected during production. The rate decline curve of this reservoir rises in different steps, as shown in Figure 12.22. The OOIP is

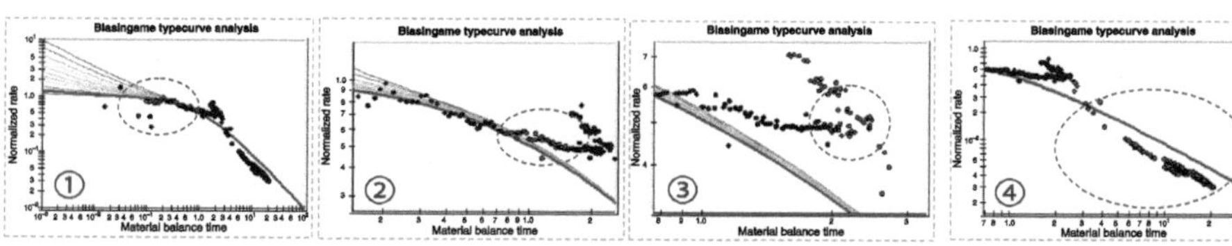

No Water production			Water breakthrough production
No water invasion stage: Satisfactory matching	Early water invasion stage: Data points deviate to the upper right	Middle and late water invasion stage: data points deviate to the lower	Unloading production stage: data points deviate to the left

Figure 12.19 TZ63-C well water breakthrough characteristic

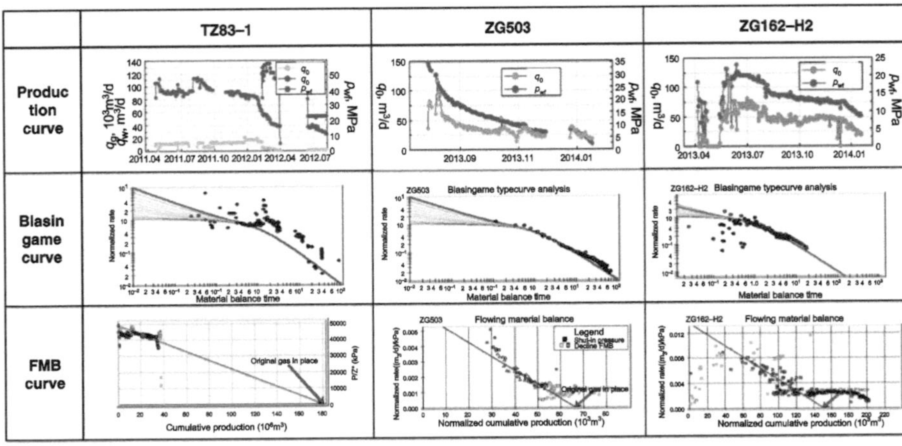

Figure 12.20 Water breakthrough forewarning analysis of three typical wells in the Tazhong gas field

2.66×10^4 m^3, 10.0×10^4 m^3, or 21.2×10^4 m^3, respectively, corresponding to when the first, second, or third section data is applied to match with the normalized rate curve.

The production shows obvious multitank reservoir feed flowing characteristic, and there is a visible production rate and pressure jump data point, as shown in Figure 12.23. This result is consistent with the seismic interpretation. Well drilling indicates that the well drills across the upper reservoir. The upper reservoir may be connected with the lower reservoir during the production, as shown in Figure 12.22.

However, the OOIP estimated with this method may be underestimated. Hower and Collins (1989) and Sun (2011) have performed theoretical analysis on the two-tank reservoir.

The reserve derived from the multisection $(p/Z)_1$ curve is lower than that of the conventional (p/Z) curve, and the reserve difference ΔG can be expressed as

$$\Delta G = \frac{\mu C_g W}{K^* Lh}\left(\frac{G_2^2}{G_1 + G_2}\right) q_g$$

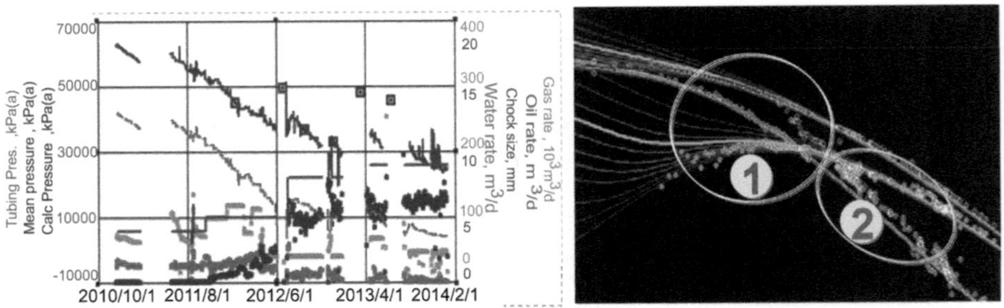

Figure 12.21 Production curve and Blasingame analysis of well ZG441

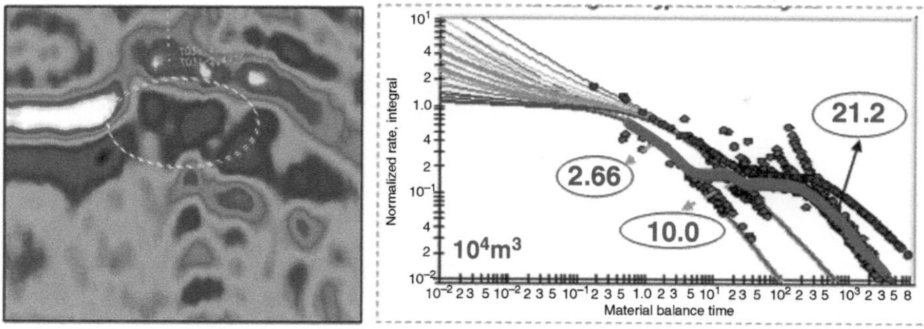

Figure 12.22 Blasingame decline analysis of well TZ622

The previous formula indicates that when the multiregion reserve remains constant, ΔG depends on the characteristic parameters of the gas reservoir as well as the initial production associated with the production area. The residual reserve decreases with the increase of gas reservoir permeability and the contact area between the blocks. The residual reserve also decreases with the decrease of the two intervals' stripe width. The residual increases with the increase of all the allocated production rate, which is consistent with the laboratory experiment results, as shown in Figure 12.24.

12.5.3 Dynamic Reserve Analysis of Connected Well Group

Anderson and Mattar (2004) performed a reserve estimation of multiwell interference through connected well group analysis. The subject example (Figure 12.25) shows that the $q/\Delta p$ data drop in the later periods, even though the production rate is larger than the critical liquid lift threshold. The data, in general, are smooth and well behaved. It is important to note that the remaining data lead to an interpretation that represents the dynamic reserve for the entire reservoir. Thus, the ultimate recovery of the subject well will be much lower because of the competitive drainage.

It is possible to correct the late-time data to make it match on the boundary-dominated stem of the type curve. This can be done by including the cumulative production from the offsets in the material balance time (MBT) function (Figure 12.26). This

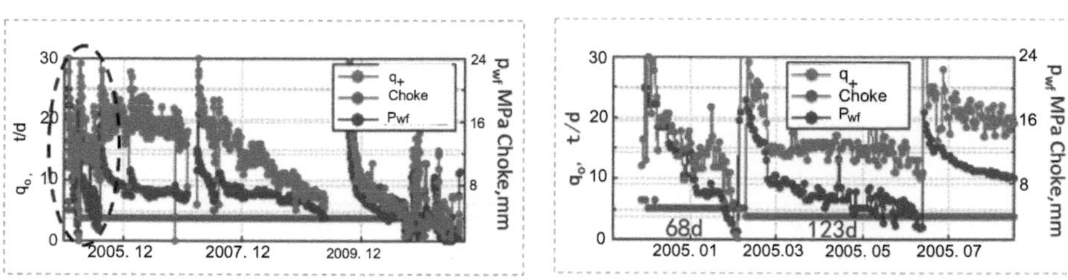

Figure 12.23 The production curve of well TZ622

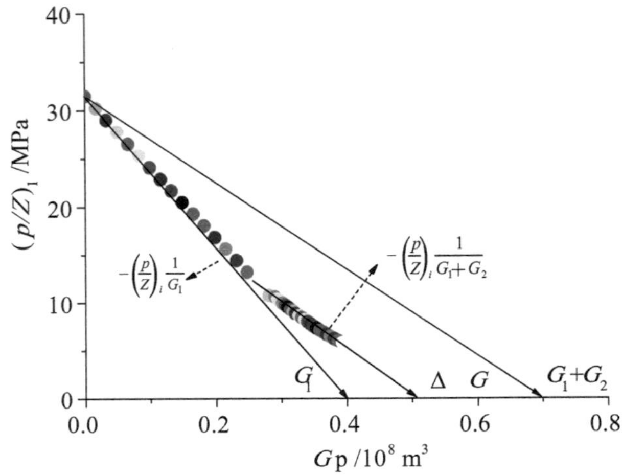

Figure 12.24 *p/Z* curve of gas material balance equation with a supplying region (after Sun, 2011)

method was proposed and documented by Blasingame in 2001. In most cases, the identities of the interfering offsets are unknown. Thus, the process of determining the communicating wells is generally one of trial and error. The objective is to obtain a type curve match where there is no fall-off or overshoot on the boundary-dominated stem by including the correct cumulative production versus time function, based on offset production. The interpreted dynamic reserve is identical to that of Figure 12.25, validating the simple "remove and ignore" approach.

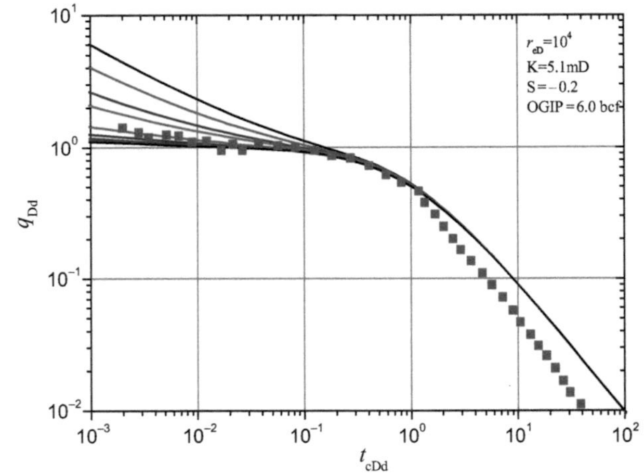

Figure 12.25 Multiwell interference curve analysis (after Anderson and Mattar, 2004)

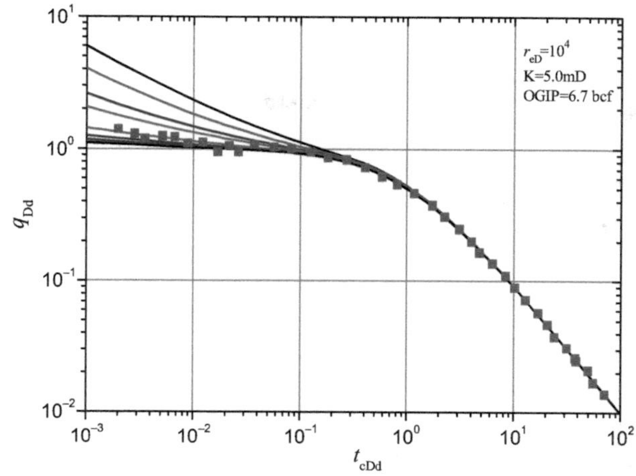

Figure 12.26 Type curve match, corrected using multiwell MBT method (after Anderson and Mattar, 2004)

12.5.4 Producing Pressure Drop Estimation

There are relatively less static pressure test data points due to various factors. The single-well decline curve analysis can be applied to estimate the average formation pressure and calculate the single well producing pressure drop; as shown in Figure 12.27, it is about 5.0 MPa.

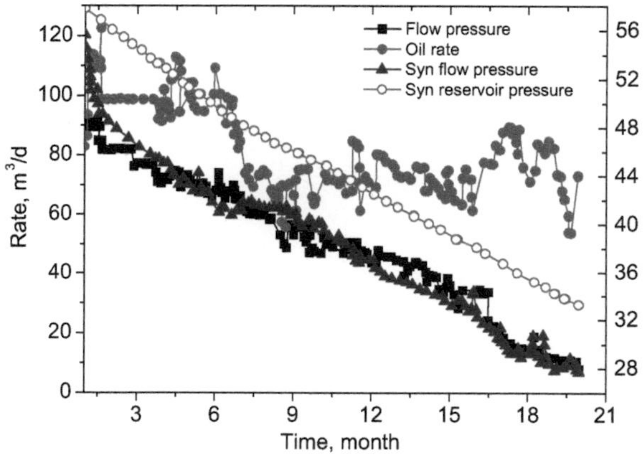

Figure 12.27 Producing pressure drop estimation of well ZG15-5H

Appendix 1

Nomenclature (with China Statutory Units CSU)

A	drainage area of well, m^2
A_c	El-Banbi cross-sectional area to flow, $A_c = 4x_f h$, m^2
A_g	area of gas reservoir, m^2, Eq. (2.91)
A_0	coefficient
A	coefficient of laminar flow term in LIT deliverability equation, $MPa^2/(10^4\ m^3/d)$, Eq. (2.90), (2.91), (2.95), (2.97), and (2.98)
b	Fetkovich/Arps decline curve exponent
b_{pss}	intercept of Eq. (4.48), constant in the pseudo-steady state equation for liquid flow, $b_{pss} = \frac{\mu B}{4\pi Kh}\ln\left(\frac{4A}{C_A e^{\gamma} r_w^2}\right)$, $(MPa\times d/m^3)^{-1}$
$b_{a,pss}$	intercept of Eq. (4.68), constant in the pseudo-steady state equation for gas flow, $b_{a,pss} = \frac{(\mu B)_i}{4\pi Kh}\ln\left(\frac{4A}{C_A e^{\gamma} r_w^2}\right)$, $(MPa\times d/m^3)^{-1}$
B	formation volume factor, m^3/m^3
B_i	formation volume factor evaluated at p_i, m^3/m^3
B	coefficient of turbulent flow term in LIT deliverability equation, $MPa^2/(10^4\ m^3/d)^2$, Eq. (2.90), (2.92), (2.95), (2.97), and (2.98)
B_0	coefficient
B_g, B_o, B_w	formation volume factors of gas, oil, and water respectively, m^3/m^3
B_{gi}	formation volume factor of gas evaluated at p_i, m^3/m^3
B_{oi}	formation volume factor of oil evaluated at p_i, m^3/m^3
Const	constant
C	wellbore storage coefficient, m^3/MPa
C_A	Dietz shape factor for single well drainage area, dimensionless
C_i	integral constant
C_g	gas compressibility, MPa^{-1}
C_o	oil compressibility, MPa^{-1}
C_r	rock compressibility, MPa^{-1}
C_f	effective rock compressibility, MPa^{-1}
C_w	water compressibility, MPa^{-1}
C_t	total compressibility, MPa^{-1}
C_{ti}	total compressibility evaluated at pi, MPa^{-1}
C_D	dimensionless wellbore storage coefficient
$C_D e^{2S}$	shape parameter group of log–log pressure and pressure derivative curves
D	non-Darcy flow coefficient, $(10^4 m^3/d)^{-1}$ or $(m^3/d)^{-1}$, Eq. (2.92)
D	continuous decline rate, d^{-1}
D_i	initial decline rate, d^{-1}

Advanced Production Decline Analysis and Application

D_{ji}	initial decline rate of *j*th layer, d^{-1}
e	2.71828
Ei	exponential integration function
EUR	cumulative production, $10^3 m^3$
f_M	Parameter group of curve match
F_{CD}	conductivity of fracture, $F_{CD} = K_{fD} \cdot W_{fD}$
G	original gas in place (OGIP), 10^8 m^3
G_p	cumulative gas production, 10^8 m^3
h	net-pay thickness, m
$I_0(x)$	modified Bessel function of the first kind, zero order
$I_1(x)$	modified Bessel function of the first kind, first order
J	productivity index, $m^3/(MPa \cdot d)$
J_o, J_w, J_L	oil, water, and liquid productivity index respectively, $m^3/(MPa \cdot d)$
$J_0(x)$	Bessel function of the first kind, zero order
$J_1(x)$	Bessel function of the first kind, first order
K	formation permeability, mD
K_a	absolute permeability, mD
K_f	permeability of fracture proppant, mD
K_{fD}	dimensionless permeability of fracture proppant, $K_{fD} = K_f / K$
K_h	horizontal formation permeability, mD
K_v	vertical formation permeability, mD
K_o	effective permeability to oil, mD
K_r	radial formation permeability, mD
K_{ro}	relative permeability to oil, mD
K_s	permeability of altered zone (skin effect), mD
Kh	formation flow capacity, mD·m
Kh/μ	formation flow coefficient, (mD·m)/(mPa·s)
$K_0(x)$	modified Bessel function of the second kind, zero order
$K_1(x)$	modified Bessel function of the second kind, first order
lg	common logarithm
ln	natural logarithm
m	real gas pseudo-pressure, $MPa^2/mPa \cdot s$
m	slope of Eq. (4.48), $m = \frac{1}{NC_t}$, MPa^2/m^3
m_a	slope of Eq. (4.68), $m_a = \frac{1}{GC_{ti}}$, MPa^2/m^3
m_s	slope of Mattar power function, d^{-2}
M	mobility ratio, dimensionless
M_c	mobility ratio of composite reservoir
M_g	molar mass of gas, kg/kmol
N	original oil in place (OOIP), 10^4 m^3 or 10^4 t
N_p	cumulative oil production, 10^4 m^3 or 10^4 t
N_{pDd}	dimensionless cumulative production, $N_{pDd} = \int_0^{t_{Dd}} q_{Dd}(\tau) d\tau$
N_p	normalized cumulative production, $N_p = \int_0^t \frac{q}{p_i - p_{wf}} d\tau$, m^3/MPa

p	pressure, MPa
p_D	dimensionless pressure
p_D	El-Banbi dimensionless pressure, $p_D = \dfrac{2\pi K_f \sqrt{A_c}(p_i - p)}{qB\mu}$
p_D'	derivative of dimensionless pressure p_D with respect to t_D/C_D
p_{Di}	dimensionless pressure integral, $p_{Di} = \dfrac{1}{t_{DA}} \int_0^{t_{DA}} p_D \, dt_{DA}$
p_{Did}	dimensionless pressure integral derivative, $p_{Did} = \dfrac{dp_{Di}}{d \ln t_{DA}} = t_{DA} \dfrac{dp_{Di}}{dt_{DA}}$
p_{Dd}	Wattenbarger dimensionless pressure, $p_{Dd} = \dfrac{x_f}{y_e} p_D$
p_p	normalized pseudo-pressure, MPa
$(p_D)_M$	matched value of dimensionless pressure, that is, the value of dimensionless pressure at the match point on type curve
p_e	pressure at external boundary, MPa
p_i	initial pressure, MPa
p_c	critical pressure of gas, MPa
p_{pc}	pseudo critical pressure of gas, MPa
p_r	reduced pressure of gas, dimensionless
p_R	reservoir pressure, MPa
p_{pi}	normalized pseudo-pressure at initial reservoir pressure, MPa
p_{pr}	pseudo reduced pressure of gas, dimensionless
p_{sc}	standard pressure condition, $p_{sc} = 0.101325$ MPa
p_{tf}	flowing surface tubing pressure, MPa
p_{wf}	flowing bottom hole pressure, MPa
$\overline{p}$	average pressure, MPa
Δp	pressure difference, MPa
Δp_p	normalized pseudo-pressure difference, MPa
Δp_{ideal}	pressure drop with zero skin, MPa
Δp_{skin}	additional pressure drop across altered zone, MPa
q	daily flow rate, m^3/d or t/d
q_{AOF}	absolute open flow potential (AOFP) of a gas well, $10^4 m^3/d$
q_f	forecast abandonment rate, m^3/d
q_i	initial flow rate, m^3/d
q_{ab}	abandonment rate, m^3/d
q_D	Agarwal-Gardner dimensionless flow rate, $q_D = \dfrac{1}{p_D} = \dfrac{q\mu B}{2\pi Kh(p_i - p_{wf})}$
q_{Dd}	Fetkovich dimensionless decline flow rate function, $q_{Dd} = \dfrac{q}{q_i} = q_D\left(\ln r_{eD} - \dfrac{1}{2}\right)$
q_{Dd}	Blasingame dimensionless decline flow rate function for oil well, $q_{Dd} = \dfrac{q}{p_i - p_{wf}} b_{pss}$

q_{Dd} — Blasingame dimensionless decline flow rate function for gas well,

$$q_{Dd} = \frac{b_{a,pss}}{\Delta p_p}$$

q_{Ddi} — dimensionless decline flow rate integral function,

$$q_{Ddi} = \frac{N_{pDd}}{t_{Dd}} = \frac{1}{t_{Dd}} \int_0^{t_{Dd}} q_{Dd}(\tau)d\tau$$

q_{Ddid} — dimensionless decline flow rate integral derivative function,

$$q_{Ddid} = -t_{Dd}\frac{d}{dt_{Dd}}\left[\frac{1}{t_{Dd}}\int_0^{t_{Dd}} q_{Dd}(\tau)d\tau\right]$$

q_g — gas flow rate, 10^4 m^3/d
$\hat{q}_i$ — instantaneous initial flow rate, 10^4 m^3/d or 10^3 m^3/d
q_{ji} — initial flow rate of *j*th layer, 10^4 m^3/d or 10^3 m^3/d
q_L — liquid flow rate, m^3/d
q_o — oil flow rate, m^3/d or t/d
q_w — water flow rate, m^3/d
q_{sc} — gas flow rate of a gas well under standard conditions, $10^4m^3/d$
q_n — flow rate in the *n*th period, respectively, during multirate testing, m^3/d
Q — cumulative gas production, 10^3 m^3
Q_f — forecast cumulative production, 10^3 m^3
r — radial distance from a well, m
r_{eD} — dimensionless radius at external drainage, $r_{eD} = r_e/r_w$
r_D — dimensionless radius, $r_D = r/r_w$
r_e — external drainage radius, m
r_i — radius of investigation, m
r_M — inner radius of oil bank in radial composite reservoir, m
r_s — radius of zone of altered permeability, m
r_w — wellbore radius, m
r_{wa} — apparent wellbore radius, m
R — universal gas law constant, $(MPa\cdot m^3)/(kmol\cdot K)$
s — Laplace space variable, dimensionless
S — skin factor, dimensionless
S_a — apparent skin factor, dimensionless
S_f — fracture skin factor of a fractured well, dimensionless
S_g, S_o, S_w — gas, oil, and water saturations respectively, %
S_{wc} — connate water saturation, %
Slope — slope of $\frac{q}{\Delta p_p} \sim t_{ca}$, 10^3 $m^3/d/MPa/d$
S_{gi}, S_{oi}, S_{wi} — initial gas, oil, and water saturations respectively, fraction
t — time, h
t_a — pseudo-time, h
t_c — material balance time for oil well, $t_c = N_p / q$, d
t_{cD} — dimensionless material balance time for oil well,

$$t_{cD} = \frac{K}{\varphi\mu C_t A}t_c$$

t_{cDd} Blasingame dimensionless material balance time for oil well, $t_{cDd}=\frac{m}{b_{pss}}t_c=t_{cD}\frac{1}{\frac{1}{2}\left(r_{eD}^2-1\right)\left(\ln r_{eD}-\frac{1}{2}\right)}$

t_{ca} material balance pseudo-time for gas well, $t_{ca}=\frac{(\mu C_t)_i}{q}\int_0^t\frac{q}{\mu C_t}dt$, d

t_{caDd} Blasingame dimensionless material balance pseudo-time for gas well, $t_{caDd}=\frac{m_a}{b_{a,pss}}t_{ca}$

t_{Dd} Fetkovich decline curve dimensionless time, $t_{Dd}=t_D\cdot\frac{1}{\frac{1}{2}\left(r_{eD}^2-1\right)\left(\ln r_{eD}-\frac{1}{2}\right)}$

t_{DA_c} El-Banbi dimensionless time based on cross-sectional area to flow in linear reservoirs, $t_{DA_c}=\frac{K_f}{\mu\left[(\varphi VC_t)_f+(\varphi VC_t)_m\right]A_c}\cdot t$

t_{Dd} Wattenbarger dimensionless time, $t_{Dd}=\left(\frac{x_f}{y_e}\right)^2 t_{Dx_f}$

t_D dimensionless time in terms of wellbore radius, $t_D=\frac{3.6\times10^{-3}Kt}{\mu\varphi C_t r_w^2}$

t_{DA} dimensionless time with drainage, $t_{DA}=\frac{3.6\times10^{-3}Kt}{\mu\varphi C_t A}$

t_{De} dimensionless time with r_e, $t_{De}=\frac{3.6\times10^{-3}Kt}{\mu\varphi C_t r_e^2}$

t_{Dxf} dimensionless time with x_f, $t_{Dx_f}=\frac{3.6\times10^{-3}Kt}{\mu\varphi C_t x_f^2}$

$(t_D)_{pss}$ dimensionless time when pseudo-steady state flow begins

t_f time at start of forecast date, h

t_{ab} abandonment time, h

t_{ps}, t_{pss} time when pseudo-steady state flow begins, h

t_{ss} time when steady state flow begins, h

T temperature, K

T_f temperature of reservoir, K

T_r reduced temperature, dimensionless

T_{pr} pseudo reduced temperature, dimensionless

T_c critical temperature, K

T_{pc} pseudo-critical temperature, K

T_{sc} standard temperature, K, 293.15 K

u variable

v variable

V volume, m^3

V_f	fissures volume per unit reservoir volume, dimensionless
V_m	matrix volume per unit reservoir volume, dimensionless
V_u	volume per unit length of wellbore, m^3/m
V_w	wellbore volume, m^3
W	width of fracture, m
W_{fD}	dimensionless width of fracture, $W_{fD} = W/x_f$
W_p	cumulative water production, m^3
x_e	distance from the well to the boundary in x-direction in a closed rectangular reservoir, m
x_f	half fracture length, m
y_e	distance from the well to the boundary in y-direction in closed rectangular reservoir, m
y_D	El-banbi dimensionless distance in y-direction, $y_D = \frac{y}{\sqrt{A_c}}$
$Y_0(x)$	Bessel function of the second kind, zero order
$Y_1(x)$	Bessel function of the second kind, first order
Z	real gas deviation factor, dimensionless
$\bar{Z}$	real gas deviation factor at average test pressure, dimensionless
$\Psi(y, t)$	variable function, $\Psi(y,t) = p_i - \frac{q\mu B}{2x_f Kh}\frac{(y_e - y)^2}{2y_e} + \eta t$, MPa
α	shape factor (used for calculating the interporosity flow coefficient in double medium system), dimensionless
γ	Euler constant, 0.5772
γ_g	relative density of gas, dimensionless
γ_o	relative density of stock tank oil, dimensionless
∇	divergence operator
η	diffusivity, $\eta = K/\mu\varphi C_t$
μ	viscosity, mPa·s
μ_g	reservoir gas viscosity, mPa·s
μ_o	reservoir oil viscosity, mPa·s
μ_w	formation water viscosity, mPa·s
$\bar{\mu}$	viscosity at average test pressure, mPa·s
ρ_g, ρ_o, ρ_w	densities of gas, oil, and water, respectively, kg/m^3
φ	porosity, %
φC_t	storability coefficient, MPa^{-1}
ω	storativity ratio of dual-porosity reservoir, dimensionless
ω_c	storativity ratio of composite formation, dimensionless
$\frac{1}{DER}$	inverse normalized pressure derivative for oil well, $\frac{1}{DER} = \frac{1}{t_c \frac{\partial\left(\frac{\Delta p}{q}\right)}{\partial t_c}}$, $m^3/d/MPa$
$\frac{1}{DER}$	inverse normalized pressure derivative for gas well, $\frac{1}{DER} = \frac{1}{\frac{\partial\left(\frac{\Delta p_p}{q}\right)}{\partial \ln t_{ca}}} = \frac{1}{t_{ca}\frac{\partial\left(\frac{\Delta p_p}{q}\right)}{\partial t_{ca}}}$, $m^3/d/MPa$

$\frac{1}{DERI}$ inverse normalized pressure integral derivative for oil well,

$$\frac{1}{DERI}=\frac{1}{t_c\frac{d\left(\frac{\Delta p}{q}\right)_i}{dt_c}}, \text{ m}^3\text{/d/MPa}$$

$\frac{1}{DERI}$ inverse normalized pressure integral derivative for gas well,

$$\frac{1}{DERI}=\frac{1}{t_{ca}\frac{d\left(\frac{\Delta p_p}{q}\right)_i}{dt_{ca}}}, \text{ m}^3\text{/d/MPa}$$

$\frac{q}{\Delta p}$ normalized rate for oil well, $\frac{q}{\Delta p}=\frac{q}{p_i-p_{wf}}$, m^3/d/MPa

$\frac{q}{\Delta p_p}$ normalized rate for gas well, $\frac{q}{\Delta p_p}=\frac{q}{p_{p_i}-p_{p_{wf}}}$, m^3/d/MPa

$\left(\frac{q}{\Delta p}\right)_i$ normalized rate integral for oil well, $\left(\frac{q}{\Delta p}\right)_i=\frac{1}{t_c}\int_0^{t_c}\frac{q}{p_i-p_{wf}}d\tau$, m^3/d/MPa

$\left(\frac{q}{\Delta p_p}\right)_i$ normalized rate integral for gas well, $\left(\frac{q}{\Delta p_p}\right)_i=\frac{1}{t_{ca}}\int_0^{t_{ca}}\frac{q}{\Delta p_p}d\tau$, m^3/d/MPa

$\left(\frac{q}{\Delta p}\right)_{id}$ normalized rate integral derivative for oil well, $\left(\frac{q}{\Delta p}\right)_{id}=-t_c\frac{d\left(\frac{q}{\Delta p}\right)_i}{dt_c}$, m^3/d/MPa

$\left(\frac{q}{\Delta p_p}\right)_{id}$ normalized rate integral derivative for gas well,

$$\left(\frac{q}{\Delta p_p}\right)_{id}=-t_{ca}\frac{d\left(\frac{q}{\Delta p_p}\right)_i}{dt_{ca}}, \text{ m}^3\text{/d/MPa}$$

$\frac{\Delta p}{q}$ normalized pressure for oil well, $\frac{\Delta p}{q}=\frac{p_i-p_{wf}}{q}$, MPa/m^3/d

$\frac{\Delta p_p}{q}$ normalized pressure for gas well, $\frac{\Delta p_p}{q}=\frac{p_{p_i}-p_{p_{wf}}}{q}$, MPa/m^3/d

$\left(\frac{\Delta p}{q}\right)_i$ normalized pressure integral for oil well, $\left(\frac{\Delta p}{q}\right)_i=\frac{1}{t_c}\int_0^{t_c}\frac{p_i-p_{wf}}{q}d\tau$, MPa/m^3/d

$\left(\frac{\Delta p_p}{q}\right)_i$ normalized pressure integral for gas well, $\left(\frac{\Delta p_p}{q}\right)_i=\frac{1}{t_{ca}}\int_0^{t_{ca}}\frac{\Delta p_p}{q}d\tau$, MPa/m^3/d

$\left(\frac{\Delta p}{q}\right)_{id}$ normalized pressure integral derivative for oil well,

$$\left(\frac{\Delta p}{q}\right)_{id}=t_c\frac{d\left(\frac{\Delta p}{q}\right)_i}{dt_c}, \text{ MPa/m}^3\text{/d}$$

$\left(\frac{\Delta p_p}{q}\right)_{id}$ normalized pressure integral derivative for gas well,

$$\left(\frac{\Delta p_p}{q}\right)_{id} = t_{ca}\frac{d\left(\frac{\Delta p_p}{q}\right)_i}{dt_{ca}}, \text{ MPa/m}^3\text{/d}$$

$\frac{1}{p_{di}}$ inverse normalized pressure integral for oil well, $\frac{1}{p_{di}} = \frac{1}{\left(\frac{\Delta p}{q}\right)_i}$, m^3/d/MPa

$\frac{1}{p_{did}}$ inverse normalized pressure integral derivative for oil well,

$$\frac{1}{p_{did}} = \frac{1}{\left(\frac{\Delta p}{q}\right)_{id}}, \text{ m}^3\text{/d/MPa}$$

$\frac{1}{p_{di}}$ inverse normalized pressure integral for gas well, $\frac{1}{p_{di}} = \frac{1}{\left(\frac{\Delta p_p}{q}\right)_i}$, m^3/d/MPa

$\frac{1}{p_{did}}$ inverse normalized pressure integral derivative for gas well,

$$\frac{1}{p_{did}} = \frac{1}{\left(\frac{\Delta p_p}{q}\right)_{id}}, \text{ m}^3\text{/d/MPa}$$

A_D	Dimensionless drainage area
C_{AH}	Shape factor for a horizontal well
F_1	Pseudo-skin function for horizontal well
h_D	Dimensionless formation thickness
L	horizontal well length, m
L_D	Dimensionless horizontal well half-length
r_{wD}	Dimensionless wellbore radius (based on the horizontal well half-length)
r_{wDz}	Dimensionless wellbore radius (based on the formation thickness)
x_D	Dimensionless distance in the x-direction
x_e	Reservoir length in the x-direction, m
x_{eD}	Dimensionless distance in the x-direction
x_w	Well location in the x-direction, m
x_{wD}	Dimensionless well location in the x-direction
y_D	Dimensionless distance in the y-direction
y_e	Reservoir length in the y-direction, m
y_{eD}	Dimensionless distance in the y-direction
y_w	Well location in the y-direction, m
y_{wD}	Dimensionless well location in the y-direction
z	Distance in the z-direction, m
z_D	Dimensionless distance in the z-direction
z_w	Well location in the z-direction, m
z_{wD}	Dimensionless well location in the z-direction
$\hat{z}_D$	Dimensionless distance in the z-direction (based on the formation thickness)
$\hat{z}_{wD}$	Dimensionless well location in the z-direction(based on the formation thickness)
β	Anisotropy factor

Subscripts

avg	average
b	pressure buildup
d	pressure drawdown
D	dimensionless
e	external boundary
f	formation
f	fracture
g	gas
h	horizontal
i	initial
j	the number of sequence
L	liquid
m	matrix
max	maximum
min	minimum
M	type curve match point
o	oil
R	reservoir
S	skin altered zone
t	total
v	vertical
w	well bottom
wf	wellbore and flowing

Appendix 2

Commonly Used Units in Different Unit Systems

Parameters	Basic units of SI	China statutory units	Darcy units	Imperial field units
Pressure, p, p_e, p_i, p_o, p_R, p_{sc}, p_w, p_{ws}, p_{wf}, Δp, Δp_s, Δp_M	Pa or kPa	MPa	atm	psi
Pseudo gas pressure, m	$Pa^2/(Pa \cdot s)$	$MPa^2/(mPa \cdot s)$	atm^2/cP	psi^2/cP
Normalized pseudo gas pressure, p_p	Pa or kPa	MPa	atm	psi
Gas flow rate, q, q_{AOF}, q_{sc}	m^3/s or m^3/d	10^4 m^3/d	cm^3/s	MMscfd
Oil or other liquid flow rate, q_o, q_w	m^3/s or m^3/d	m^3/d	cm^3/s	bbl/D
Cumulative oil production, N_p	m^3	10^4 m^3	cm^3	bbl
Cumulative gas production, G_p	m^3	10^4 m^3	cm^3	MMscf
OOIP, N	m^3	10^8 m^3	cm^3	bbl
OGIP, G	m^3	10^8 m^3	cm^3	MMscf
Time, t, t_M, t_p, t_{ps}, t_{ss}, t_c, t_{ca}, Δt, Δt_M, Δt_P, Δt_C	s or h	h	s	h
Thickness, h	m	m	cm	ft
Distance, L, r, r_e, r_i, r_f, r_M, r_w, r_{we}	m	m	cm	ft
Area, A	m^2	m^2 or km^2	cm^2	ft^2
Permeability, K, K_h, K_v, K_r, K_f, K_g, K_o	m^2	mD	D	md
Porosity, φ	Fraction or %	Fraction or %	Fraction or %	Fraction or %
Saturation, S_o, S_w, S_g, S_{oi}, S_{gi}, S_{wi}	Fraction or %	Fraction or %	Fraction or %	Fraction or %
Compressibility, C_o, C_w, C_g, C_f, C_t	Pa^{-1}	MPa^{-1}	atm^{-1}	psi^{-1}
Wellbore storage coefficient, C	$m^3/$ Pa	$m^3/$ MPa	$cm^3/$ atm	bbl/ psi
Viscosity, μ_o, μ_w, μ_g	Pa·s	mPa·s	cP	cP
Temperature, T, T_f, T_c, T_{pc}, T_{sc}	K	K or °C	°C	°F or °R
Formation volume factor, B, B_o, B_w, B_g	m^3/m^3	m^3/m^3	cm^3/cm^3	RB/STB (o, w) RB/scf (g)

Advanced Production Decline Analysis and Application

Parameters	Basic units of SI	China statutory units	Darcy units	Imperial field units
Universal gas law constant, R	$Pa \cdot m^3 (kmol \cdot K)^{-1}$	$MPa \cdot m^3 (kmol \cdot K)^{-1}$	$atm \cdot cm^3 (mol \cdot K)^{-1}$	$psi \cdot ft^3 (lbmol \cdot {}^\circ R)^{-1}$
Coefficient of non-Darcy gas flow, D	$(m^3/s)^{-1}$	$(10^4\ m^3/d)^{-1}$	$(cm^3/s)^{-1}$	$(MMscfd)^{-1}$
Density, ρ, ρ_g, ρ_o, ρ_w	kg/m^3	g/cm^3 or kg/m^3	g/cm^3	Lb/ft^3

Appendix 3

Unit Conversions from China Statutory Unit to Other Unit Systems

1. **Length**

	km	m	cm	mile	ft	in.
1 × km	1	1000	10^5	0.6214	3280.84	39370.08
1 × m	0.001	1	100	6.214×10^{-4}	3.28084	39.37008
1 × cm	10^{-5}	0.01	1	6.214×10^{-6}	0.0328084	0.393701
1 × mile	1.60934	1609.34	1.60934×10^5	1	5280	63360
1 × ft	3.48×10^{-4}	0.3048	30.48	1.839×10^{-4}	1	12
1 × in	2.54×10^{-5}	0.0254	2.54	1.5783×10^{-5}	0.08333	1

2. **Area**

	m^2	cm^2	ft^2	$in.^2$
1 × m^2	1	10^4	10.7639	1550
1 × cm^2	10^{-4}	1	1.07639×10^{-3}	0.1550
1 × ft^2	0.092903	929.03	1	144
1 × in^2	6.4516×10^{-4}	6.4516	6.9444×10^{-3}	1

3. **Volume**

	m^3	cm^3	ft^3	bbl	L
1 × m^3	1	10^6	35.3147	6.28978	1000
1 × cm^3	10^{-6}	1	3.53147×10^{-5}	6.28978×10^{-6}	10^{-3}
1 × ft^3	0.0283168	2.83168×10^4	1	0.17811	28.3168
1 × bbl	0.158988	1.58988×10^5	5.6146	1	158.99
1 × L	10^{-3}	10^3	3.53147×10^{-2}	6.28978×10^{-3}	1

4. **Pressure**

	MPa	kPa	atm	bar	kg/cm^2	psi
1 × MPa	1	1000	9.86923	10	10.1972	145.038
1 × kPa	0.001	1	9.86923×10^{-3}	0.01	0.0101972	0.145038
1 × atm	0.101325	101.325	1	1.01325	1.03323	14.6959
1 × bar	0.1	100	0.986923	1	1.01972	14.5038
1 × kg/cm^2	0.0980665	98.0665	0.967841	0.980665	1	14.2233
1 × psi	0.00689476	6.89476	0.068406	0.0689476	0.070307	1

Advanced Production Decline Analysis and Application

5. Temperature

	°C	K	°F	°R
1 × °C	t	$t + 273.15$	$1.8t + 32$	$1.8t + 491.67$
1× K	$T - 273.15$	T	$1.8T - 459.67$	$1.8T$
1 × °F	$5(f - 32)/9$	$5(f + 459.67)/9$	f	$f + 459.67$
1 × °R	$5r/9 - 273.15$	$5r/9$	$r - 459.67$	r

6. Oil Production

	m^3/d	cm^3/s	bbl/D
1 × m^3/d	1	$10^4/864$	6.28978
1 × cm^3/s	0.0864	1	0.543437
1 × bbl/d	0.158988	1.84014	1

7. Gas production

	$10^4m^3/d$	cm^3/s	Mscf/D	MMscf/D
1 × $10^4m^3/d$	1	$10^8/864$	353.147	0.353147
1 × cm^3/s	864×10^{-8}	1	3.05119×10^{-3}	3.05119×10^{-6}
1 × $10^3ft^3/d$	2.83168×10^{-3}	327.741	1	0.001
1 × $10^6ft^3/d$	2.83168	327741	1000	1

8. Compressibility

	1/MPa	1/atm	1/(kg/cm^2)	1/psi
1/MPa	1	0.101325	0.0980665	0.00689476
1/atm	9.86923	1	0.967841	0.068406
1/(kg/cm^2)	10.1972	1.03323	1	0.070307
1/psi	145.038	14.6959	14.2233	1

9. Permeability

	μm^2	cm^2	m^2	mD	D
1 × μm^2	1	10^{-8}	10^{-12}	1.0133×10^3	1.0133
1 × mD	0.98692×10^{-3}	9.8692×10^{-12}	9.8692×10^{-16}	1	10^{-3}

10. Viscosity

$$1\,\text{mPa}\cdot\text{s} = 10^{-3}\,\text{Pa}\cdot\text{s} = 10^3\,\mu\text{Pa}\cdot\text{s} = 1\text{cP}$$

11. Relative density of stock tank oil, (go) and °API

$$°\text{API} = \frac{141.5}{\gamma_0} - 131.5$$

$$\gamma_0 = 141.5/(131.5 + °\text{API})$$

12. Gas–oil ratio

$$1\,m^3/m^3 = 5.615\,scf/STB$$

$$1\,scf/STB = 0.1781\,m^3/m^3$$

Appendix 4

Formulae Commonly Used in Main Text (with China Statutory Units CSU)

Chapter 3 Fetkovich Decline Curves Analysis

q_i	$q_i = \left(\frac{q}{q_{Dd}}\right)_M$	/
D_i	$D_i = \left(\frac{t_{Dd}}{t}\right)_M$	(3.108)
K	$K = \frac{12.7355\times 10^4 \times (\mu Z)_{avg} T\left(\ln r_{eD} - \frac{1}{2}\right)}{h(p_i^2 - p_{wf}^2)}\left(\frac{q}{q_{Dd}}\right)_M$	(3.109)
r_{wa}	$r_{wa} = \sqrt{\frac{3.6\times 10^{-3}\times 2K/\varphi\mu C_t}{\left(r_{eD}^2 - 1\right)\left(\ln r_{eD} - \frac{1}{2}\right)}\left(\frac{24t}{t_{Dd}}\right)_M}$	(3.110)
S	$S = \ln\frac{r_w}{r_{wa}}$	(3.111)
x_f	$x_f = 2r_w e^{-S}$	(3.112)
V_p	$V_p = \frac{0.6912\times 10^3 \times ZT}{C_t(p_i^2 - p_{wf}^2)} \times \left(\frac{t}{t_{Dd}}\right)_M \left(\frac{q}{q_{Dd}}\right)_M$	(3.117)
r_e	$r_e = \sqrt{\frac{V_p}{\pi h\varphi}}$	(3.119)
OGIP (OOIP)	$N = \frac{V_p(1 - S_w)}{B_i}$	(3.85)
	$N \approx \frac{q_i}{(1-b)D_i}$	(3.86)

Chapter 4 Blasingame Decline Curves Analysis

K	$K = \frac{1.866\times\left(q/\Delta p_p\right)_M}{\left(q_{Dd}\right)_M}\frac{\mu B}{h}\left(\ln r_{eD} - \frac{1}{2}\right)$	(4.123)

Advanced Production Decline Analysis and Application

r_{wa}	$r_{wa}=\sqrt{\dfrac{3.6\times10^{-3}\times2K/\varphi\mu C_t}{\left(r_{eD}^2-1\right)\left(\ln r_{eD}-\dfrac{1}{2}\right)}\left(\dfrac{24\times t_{ca}}{t_{caDd}}\right)_M}$	(4.124)
S	$S=\ln\dfrac{r_w}{r_{wa}}$	(4.125)
r_e	$r_e=r_{wa}r_{eD}$	(4.126)
	$r_e=\sqrt{\dfrac{\dfrac{B}{C_t}\left(\dfrac{24t_{ca}}{t_{caDd}}\right)_M\left(\dfrac{q/\Delta p_p}{q_{Dd}}\right)_M}{\pi h\varphi}}$	(4.127)
OGIP (OOIP)	$G=\dfrac{1}{C_t}\left(\dfrac{t_{ca}}{t_{caDd}}\right)_M\left(\dfrac{q/\Delta p_p}{q_{Dd}}\right)_M\left(1-S_w\right)$	(4.128)

Chapter 5 Agarwal-Gardner Decline Curves Analysis

K	$K=\dfrac{1.866\times\left(q/\Delta p_p\right)_M}{\left(q_D\right)_M}\dfrac{\mu B}{h}$	(5.42)
r_e	$r_e=\sqrt{\dfrac{3.6\times10^{-3}K}{\pi\varphi\mu C_t}\left(\dfrac{24t_{ca}}{t_{caDA}}\right)_M}$	(5.43)
	$r_e=\sqrt{\dfrac{2.1383\times10^{-3}B}{\varphi h C_t}\left(\dfrac{24t_{ca}}{t_{caDA}}\right)_M\left(\dfrac{q/\Delta p_p}{q_D}\right)_M}$	(5.44)
r_{wa}	$r_{wa}=\dfrac{r_e}{r_{eD}}$	(5.46)
S	$S=\ln\dfrac{r_w}{r_{wa}}$	(5.47)
OGIP (OOIP)	$G=\dfrac{\pi r_e^2\varphi h S_g}{B_{gi}}=\dfrac{6.7177\times10^{-3}}{C_t}\left(\dfrac{24t_{ca}}{t_{caDA}}\right)_M\left(\dfrac{q/\Delta p_p}{q_D}\right)_M S_g$	(5.49)

Chapter 6 NPI Decline Curves Analysis

K	$K=\dfrac{1.866\times\left(p_D\right)_M}{\left(\Delta p_p/q\right)_M}\dfrac{\mu B}{h}$	(6.32)

r_e	$r_e=\sqrt{\dfrac{3.6\times10^{-3}K}{\pi\varphi\mu C_t}\left(\dfrac{24t_{ca}}{t_{caDA}}\right)_M}$	(6.33)
	$r_e=\sqrt{\dfrac{2.1383\times10^{-3}B}{\varphi hC_t}\left(\dfrac{24t_{ca}}{t_{caDA}}\right)_M\left(\dfrac{p_D}{\Delta p_p/q}\right)_M}$	(6.34)
r_{wa}	$r_{wa}=\dfrac{r_e}{r_{eD}}$	(6.36)
S	$S=\ln\dfrac{r_w}{r_{wa}}$	(6.37)
OGIP (OOIP)	$G=\dfrac{\pi r_e^2\varphi hS_g}{B_{gi}}=\dfrac{6.7177\times10^{-3}}{C_t}\left(\dfrac{24\times t_{ca}}{t_{caDA}}\right)_M\left(\dfrac{p_D}{\Delta p_p/q}\right)_M S_g$	(6.39)

Chapter 7 Transient Decline Curves Analysis

K	$K=\dfrac{1.866\times(q/\Delta p_p)_M}{(q_D)_M}\dfrac{\mu B}{h}$	(7.34)
r_{wa}	$r_{wa}=\sqrt{\dfrac{3.6\times10^{-3}K}{\varphi\mu C_t}\left(\dfrac{24t_{ca}}{t_{caD}}\right)_M}$	(7.35)
	$r_{wa}=\sqrt{\dfrac{6.7177\times10^{-3}}{\varphi hC_t}\left(\dfrac{24t_{ca}}{t_{caD}}\right)_M\left(\dfrac{q/\Delta p_p}{q_D}\right)_M}$	(7.36)
r_e	$r_e=r_{wa}r_{eD}$	(7.38)
S	$S=\ln\dfrac{r_w}{r_{wa}}$	(7.39)
OGIP (OOIP)	$G=\dfrac{\pi r_e^2\varphi hS_g}{B_{gi}}=\dfrac{2.1104\times10^{-2}}{C_t}\left(\dfrac{24t_{ca}}{t_{caD}}\right)_M\left(\dfrac{q/\Delta p_p}{q_D}\right)_M r_{eD}^2S_g$	(7.41)

Chapter 8 Decline Curves Analysis of Long Linear Flow

Wattenbarger

Kx_f/y_e	$K\dfrac{x_f}{y_e}=\dfrac{1.866\times\mu B}{h}\dfrac{(q/\Delta p_p)_M}{(q_{Dd})_M}$	(8.40)
$y_e/\sqrt{K}$	$\dfrac{y_e}{\sqrt{K}}=\sqrt{\dfrac{3.6\times10^{-3}}{\varphi\mu C_t}\left(\dfrac{24t_{ca}}{t_{caDd}}\right)_M}$	(8.41)

$\sqrt{K}x_f$	$\sqrt{K}x_f = \frac{1.866\times\mu B}{h}\frac{(q/\Delta p_p)_M}{(q_{caDd})_M}\sqrt{\frac{3.6\times10^{-3}}{\varphi\mu C_t}\left(\frac{24t_{ca}}{t_{caDd}}\right)_M}$	(8.42)
A_e	$A_e = 4x_e y_e = 4(\sqrt{K}x_f)\frac{y_e}{\sqrt{K}}$	(8.43)
OGIP (OOIP)	$G = \frac{A\varphi hS_g}{B_{gi}}$	(8.44)

Vertical Fractured Well

b_{Dpss}	$b_{Dpss} = \ln r_{eD} - 0.049298 + 0.43464 r_{eD}^{-2} + \frac{a_1 + a_2u + a_3u^2 + a_4u^3 + a_5u^4}{1 + b_1u + b_2u^2 + b_3u^3 + b_4u^4}$	(8.98)
OGIP (OOIP)	$G = \frac{1}{C_t}\left(\frac{t_{ca}}{t_{caDd}}\right)_M\left(\frac{q/\Delta p_p}{q_{Dd}}\right)_M$	(8.99)
A	$A = \frac{GB_{gi}}{\varphi h(1-S_{wi})}$	(8.100)
r_e	$r_e = \sqrt{A/\pi}$	(8.101)
K	$K = 1.866\times\frac{(q/\Delta p_p)_M}{(q_{Dd})_M}\frac{\mu B}{h}b_{Dpss}$	(8.102)
x_f	$x_f = \frac{r_e}{r_{eD}}$	(8.103)
S	$S = \ln\frac{r_w}{x_f} + \frac{1.648546 - 3.002711\times10^{-1}u + 1.506532\times10^{-1}u^2}{1 + 2.136604\times10^{-1}u + 9.513761\times10^{-2}u^2 + 8.276998\times10^{-3}u^3}$	(8.104)

Others

Dimensionless Pressure	$p_D = \frac{2.714\times10^{-1}KhT_{sc}\Delta m}{q_g T_f p_{sc}}$
	$p_D = \frac{2.714\times10^{-1}KhT_{sc}\Delta(p^2)}{q_g\bar{\mu}_g\bar{Z}T_f p_{sc}}$
	$p_D = \frac{0.5428Kh\Delta p}{q\mu B}$
Dimensionless time	$t_D = \frac{3.6\times10^{-3}K}{\varphi\mu C_t r_w^2}\cdot\Delta(24t)$

Production in steady flow stage

$$q_g = \frac{2.714\times10^{-1}KhT_{sc}\left[m(p_R)-m(p_{wf})\right]}{p_{sc}T_f\left(\ln\frac{r_e}{r_w}+S_a\right)}$$

$$q_g = \frac{2.714\times10^{-1}KhT_{sc}\left(p_R^2-p_{wf}^2\right)}{p_{sc}\bar{\mu}_g\bar{Z}T_f\left(\ln\frac{r_e}{r_w}+S_a\right)}$$

Production in pseudo-steady flow stage

$$q_g = \frac{2.714\times10^{-1}KhT_{sc}\left[m(p_R)-m(p_{wf})\right]}{p_{sc}T_f\left(\ln\frac{r_e}{r_w}+S_a-\frac{3}{4}\right)}$$

$$q_g = \frac{2.714\times10^{-1}KhT_{sc}\left(p_R^2-p_{wf}^2\right)}{p_{sc}\bar{\mu}_g\bar{Z}T_f\left(\ln\frac{r_e}{r_w}+S_a-\frac{3}{4}\right)}$$

$$q_o = \frac{0.5358Kh\left(p_R-p_{wf}\right)}{B\mu\left(\ln\frac{r_e}{r_w}+S\right)}$$

$$q_o = \frac{0.5358Kh\left(p_R-p_{wf}\right)}{B\mu\left(\ln\frac{r_e}{r_w}+S-\frac{3}{4}\right)}$$

Note: [1]

t	–	time, d
q	–	flow rate, m^3/d
Others	–	shown in Appendix 1

Appendix 5

Late Time Solution for a Well in a Two-Region Composite Reservoir with Closed Circular Boundary

We say that the flow regime is boundary dominated if the productivity index remains the same at any time. Boundary-dominated flow period will occur after the pressure transient reaches all boundaries in a closed system. Pseudo-steady state flow period is only one example of a boundary-dominated flow period, which exists for all closed systems. It is a regime encountered for the constant production rate case. At this stage, the well yields entirely from the fluid, and rock elastic expansion causes the average pressure drop of reservoir. For a well producing at a constant rate in a two-region composite reservoir with a closed circular boundary, the flow rate at any r (Ambastha, 1988) can be written as

$$q(r) = -\left[\left(r_{\mathrm{e1}}^2 - r^2\right)\left(\frac{\pi h \varphi C_{\mathrm{t}}}{B}\right)_1 + \left(r_{\mathrm{e2}}^2 - r_{\mathrm{e1}}^2\right)\left(\frac{\pi h \varphi C_{\mathrm{t}}}{B}\right)_2\right]\frac{d\bar{p}}{dt} \quad r < r_{\mathrm{e1}} \tag{A-5.1}$$

$$q(r) = -\left(r_{\mathrm{e2}}^2 - r^2\right)\left(\frac{\pi h \varphi C_{\mathrm{t}}}{B}\right)_2 \frac{d\bar{p}}{dt} \quad r > r_{\mathrm{e1}} \tag{A-5.2}$$

Flow rate at the well can be written as

$$q_{\mathrm{w}} = -\left[\left(r_{\mathrm{e1}}^2 - r_{\mathrm{w}}^2\right)\left(\frac{\pi h \varphi C_{\mathrm{t}}}{B}\right)_1 + \left(r_{\mathrm{e2}}^2 - r_{\mathrm{e1}}^2\right)\left(\frac{\pi h \varphi C_{\mathrm{t}}}{B}\right)_2\right]\frac{d\bar{p}}{dt} \quad r < r_{\mathrm{e1}} \tag{A-5.3}$$

If $h_1 = h_2$, $B_1 = B_2$, and $\varphi_1 = \varphi_2$, Eq. (A-5.3) can be written as

$$q_{\mathrm{w}} = -\frac{\pi h \varphi}{B}\left[\left(r_{\mathrm{e1}}^2 - r_{\mathrm{w}}^2\right)C_{\mathrm{t1}} + \left(r_{\mathrm{e2}}^2 - r_{\mathrm{e1}}^2\right)C_{\mathrm{t2}}\right]\frac{d\bar{p}}{dt} \tag{A-5.4}$$

Assuming $r_{\mathrm{e1}}^2 \gg r_{\mathrm{w}}^2$,

$$\delta = \left(r_{\mathrm{e1}}^2 - r_{\mathrm{w}}^2\right)C_{\mathrm{t1}} + \left(r_{\mathrm{e2}}^2 - r_{\mathrm{e1}}^2\right)C_{\mathrm{t2}} = r_{\mathrm{e1}}^2 C_{\mathrm{t1}} + \left(r_{\mathrm{e2}}^2 - r_{\mathrm{e1}}^2\right)C_{\mathrm{t2}} \tag{A-5.5}$$

Advanced Production Decline Analysis and Application

Eq. (A-5.4) can be written as

$$q_{\mathrm{w}} = -\frac{\pi h \varphi \delta}{B} \frac{d\bar{p}}{dt} \tag{A-5.6}$$

If $r < r_{\mathrm{e1}}$, we have

$$\frac{q(r)}{q_{\mathrm{w}}} = \frac{\left(r_{\mathrm{e1}}^2 - r^2\right) C_{\mathrm{t1}} + \left(r_{\mathrm{e2}}^2 - r_{\mathrm{e1}}^2\right) C_{\mathrm{t2}}}{\delta} \tag{A-5.7}$$

If $r > r_{\mathrm{e1}}$, we have

$$\frac{q(r)}{q_{\mathrm{w}}} = \frac{\left(r_{\mathrm{e2}}^2 - r^2\right) C_{\mathrm{t2}}}{\delta} \tag{A-5.8}$$

From Darcy's law, flow rate at any r, $q(r)$, can be written as

$$q(r) = \frac{2\pi K h}{\mu} \left(r \frac{dp}{dr} \right) \tag{A-5.9}$$

Integrating Eq. (A-5.4) from r_{w} to r, we have

$$\int_{r_{\mathrm{w}}}^{r} \frac{\mu q(r)}{Kr} dr = 2\pi h \int_{p_{\mathrm{wf}}}^{p_{\mathrm{r}}} dp \tag{A-5.10}$$

If $r > r_{\mathrm{e1}}$, substituting Eq. (A-5.7) into Eq. (A-5.10), we have

$$q_{\mathrm{w}} \left[\frac{\mu_1}{K_1} \int_{r_{\mathrm{w}}}^{r_{\mathrm{e1}}} \left(1 - \frac{C_{\mathrm{t1}} r^2}{\delta} \right) \frac{dr}{r} + \frac{\mu_2}{K_2} \int_{r_{\mathrm{e1}}}^{r} \frac{\left(r_{\mathrm{e2}}^2 - r^2\right) C_{\mathrm{t2}}}{\delta} \frac{dr}{r} \right] = 2\pi h \left(p_{\mathrm{r}} - p_{\mathrm{wf}} \right) \tag{A-5.11}$$

where

$$M_{\mathrm{C}} = \frac{(K/\mu)_1}{(K/\mu)_2} \tag{A-5.12}$$

Eq. (A-5.11) can be written as

$$\int_{r_{\mathrm{w}}}^{r_{\mathrm{e1}}} \left(1 - \frac{C_{\mathrm{t1}} r^2}{\delta} \right) \frac{dr}{r} + M_{\mathrm{C}} \int_{r_{\mathrm{e1}}}^{r} \frac{\left(r_{\mathrm{e2}}^2 - r^2\right) C_{\mathrm{t2}}}{\delta} \frac{dr}{r} = \frac{2\pi h K_1 \left(p_{\mathrm{r}} - p_{\mathrm{wf}} \right)}{\mu_1 q_{\mathrm{w}}} \tag{A-5.13}$$

Integrating Eq. (A-5.13), we have

$$\ln r_{\mathrm{e1}D}-\frac{C_{\mathrm{t1}}r_{\mathrm{e1}D}^{2}}{2\delta}+\frac{M_{\mathrm{C}}C_{\mathrm{t2}}}{\delta}\left[r_{\mathrm{e2}}^{2}\ln\left(\frac{r}{r_{\mathrm{e1}}}\right)-\frac{r^{2}-r_{\mathrm{e1}D}^{2}}{2}\right]=\frac{2\pi hK_{1}\left(p_{\mathrm{r}}-p_{\mathrm{wf}}\right)}{\mu_{1}q_{\mathrm{w}}} \tag{A-5.14}$$

If $r < r_{\mathrm{e1}}$, substituting Eq. (A-5.8) into Eq. (A-5.10), we have

$$\int_{r_{\mathrm{w}}}^{r}\left(1-\frac{C_{\mathrm{t1}}r^{2}}{\delta}\right)\frac{dr}{r}=\frac{2\pi hK_{1}}{\mu_{1}q_{\mathrm{w}}}\left(p_{\mathrm{r}}-p_{\mathrm{wf}}\right) \tag{A-5.15}$$

Integrating Eq. (A-5.15), we have

$$\ln\left(\frac{r}{r_{\mathrm{w}}}\right)-\frac{C_{\mathrm{t1}}\left(r^{2}-r_{\mathrm{w}}^{2}\right)}{2\delta}=\frac{2\pi hK_{1}}{\mu_{1}q_{\mathrm{w}}}\left(p_{\mathrm{r}}-p_{\mathrm{wf}}\right) \tag{A-5.16}$$

The volumetric average reservoir is, $\overline{p}$

$$\overline{p}=\frac{1}{V}\iiint_{V}p\,dV=\frac{2}{r_{\mathrm{e2}}^{2}-r_{\mathrm{w}}^{2}}\int_{r_{\mathrm{w}}}^{r_{\mathrm{e2}}}pr\,dr \tag{A-5.17}$$

Substituting Eq. (A-5.14) and (A-5.16) into Eq. (A-5.17), we have

$$q_{\mathrm{w}}=\frac{2\pi hK_{1}}{B\mu_{1}}\frac{\left(\overline{p}-p_{\mathrm{wf}}\right)}{\beta} \tag{A-5.18}$$

where

$$\beta=\ln r_{\mathrm{e1}D}-\frac{r_{\mathrm{e1}D}^{2}}{2r_{\mathrm{e2}D}^{2}}-\frac{1-\dfrac{r_{\mathrm{e1}D}^{2}}{2r_{\mathrm{e2}D}^{2}}-\omega_{\mathrm{C}}\left(2-\dfrac{r_{\mathrm{e1}D}^{2}}{2r_{\mathrm{e2}D}^{2}}-\dfrac{3r_{\mathrm{e2}D}^{2}}{2r_{\mathrm{e1}D}^{2}}+\dfrac{2r_{\mathrm{e2}D}^{2}}{r_{\mathrm{e1}D}^{2}}\ln\dfrac{r_{\mathrm{e2}D}}{r_{\mathrm{e1}D}}\right)}{2\left[1+\dfrac{\omega_{\mathrm{C}}}{M_{\mathrm{C}}}\left(\dfrac{r_{e2D}^{2}}{r_{e1D}^{2}}-1\right)\right]} \tag{A-5.19}$$

At late time for a closed reservoir, we have

$$q_{\mathrm{w}}=-\frac{Ah\varphi C_{\mathrm{t}}}{B}\frac{d\overline{p}}{dt} \tag{A-5.20}$$

Integrating Eq. (A-5.20), we have

$$\frac{2\pi hK_{1}}{\mu_{1}Bq_{\mathrm{w}}}\left(p_{\mathrm{i}}-\overline{p}\right)=\frac{2K_{1}t}{\varphi_{1}\mu_{1}\left[r_{\mathrm{e1}}^{2}C_{\mathrm{t1}}+\left(r_{\mathrm{e2}}^{2}-r_{\mathrm{e1}}^{2}\right)C_{\mathrm{t2}}\right]} \tag{A-5.21}$$

Combining Eq. (A-5.18) and (A-5.21), we have

$$p_{\mathrm{wfD}} = \frac{2\pi h K_1}{\mu_1 B q_{\mathrm{w}}}\left(p_{\mathrm{i}} - p_{\mathrm{wf}}\right) = \alpha_1 t_D + \beta \tag{A-5.22}$$

where

$$\alpha_1 = \frac{2}{r_{\mathrm{e1D}}^2 + \dfrac{\omega_{\mathrm{C}}}{M_{\mathrm{C}}}\left(r_{\mathrm{e2D}}^2 - r_{\mathrm{e1D}}^2\right)} \tag{A-5.23}$$

Van Everdingen and Hurst (1949) indicated that according to the superposition principle, the relationship between the pressure solution with a constant flow rate and the flow rate solution with a constant bottom hole flowing pressure (BHFP) could be determined

$$\overline{p}_{\mathrm{wfD}}(s)\overline{q}_{\mathrm{D}}(s) = \frac{1}{s^2} \tag{A-5.24}$$

Taking the Laplace transform of Eq. (A-5.22) and substituting it into Eq. (A-5.24), we have

$$\overline{q}_{\mathrm{D}}(s) = \frac{1}{s^2\left(\dfrac{\alpha_1}{s^2} + \dfrac{\beta}{s}\right)} = \frac{1}{\alpha_1 + s\beta} = \frac{1}{\beta\left(\dfrac{\alpha_1}{\beta} + s\right)} \tag{A-5.25}$$

Taking the Laplace inverse transform of Eq. (A-5.25), we have

$$q_{\mathrm{D}} = \frac{1}{\beta} e^{-\frac{\alpha_1}{\beta} t_{\mathrm{D}}} \tag{A-5.26}$$

Fetkovich dimensionless decline time in a two-region composite reservoir with a closed circular boundary can be written as

$$t_{\mathrm{Dd}} = \frac{\alpha_1}{\beta} t_{\mathrm{D}} = \alpha t_{\mathrm{D}} \tag{A-5.27}$$

Fetkovich dimensionless flow rate can be written as

$$q_{\mathrm{Dd}} = \frac{\beta}{L^{-1}\left[\overline{p}_{\mathrm{wfD}}\right]} \tag{A-5.28}$$

If $\frac{\omega_C}{M_C} = 1$, Eq. (A-5.19) can be simplified to

$$\beta = \ln r_{e1D} + (1 - M_C)\frac{r_{e1D}^2}{r_{e2D}^2}\left(\frac{r_{e1D}^2}{4r_{e2D}^2} - 1\right) - M_C\left[\ln\left(\frac{r_{e1D}}{r_{e2D}}\right) + \frac{3}{4}\right] \tag{A-5.29}$$

Eq. (A-5.23) can be simplified to

$$\alpha_1 = \frac{2}{r_{e2D}^2} \tag{A-5.30}$$

If $\omega_C = M_C = 1$, homogeneous reservoir, Eq. (A-5.29) can be simplified to

$$\beta = \ln r_{e2D} - \frac{3}{4} \tag{A-5.31}$$

Appendix 6

Late Time Solution for a Well in a Two-Layer Commingled Reservoir with Closed Circular Boundary

For large values of time (Tariq and Ramey, 1978), or consequently, small values of the argument $\sigma \to 0$, the following approximations can be made

$$I_0(\sigma) = 1 + \left(\frac{\sigma}{2}\right)^2 \tag{A-6.1}$$

$$I_1(\sigma) = \frac{\sigma}{2} + \left(\frac{\sigma}{2}\right)^3 \tag{A-6.2}$$

$$K_0(\sigma) = -\ln\left(\frac{\sigma}{2}\right) \tag{A-6.3}$$

$$K_1(\sigma) = \frac{1}{\sigma} \tag{A-6.4}$$

Substituting Eq. (A-6.1) through (A-6.4) into Eq. (11.59) and applying analytic inversion, we have

$$p_{\mathrm{wfD}} = \frac{2t_D}{\sum_{j=1}^{2} \omega_j r_{\mathrm{ejD}}^2} + \frac{\sum_{j=1}^{2} \frac{\omega_j^2}{\kappa_j} r_{\mathrm{ejD}}^4 \left(\ln r_{\mathrm{ejD}} - \frac{3}{4}\right)}{\left(\sum_{j=1}^{2} \omega_j r_{\mathrm{ejD}}^2\right)^2} \tag{A-6.5}$$

Taking the Laplace transform of Eq. (A-6.5), we have

$$\overline{p}_{\mathrm{wfD}} = \frac{2}{s^2 \sum_{j=1}^{2} \omega_j r_{\mathrm{ejD}}^2} + \frac{\sum_{j=1}^{2} \frac{\omega_j^2}{\kappa_j} r_{\mathrm{ejD}}^4 \left(\ln r_{\mathrm{ejD}} - \frac{3}{4}\right)}{s\left(\sum_{j=1}^{2} \omega_j r_{\mathrm{ejD}}^2\right)^2} \tag{A-6.6}$$

Van Everdingen and Hurst (1949) indicated that according to the superposition principle, the relationship between the pressure solution with constant flow rate and the flow rate solution with constant bottom hole flowing pressure (BHFP) could be determined by

$$\overline{p}_{\mathrm{wfD}}(s)\overline{q}_{\mathrm{D}}(s)=\frac{1}{s^2} \tag{A-6.7}$$

Substituting Eq. (A-6.6) into Eq. (A-6.7), we have

$$\overline{q}_{\mathrm{D}}(s)=\frac{1}{\beta\left(\frac{\alpha_1}{\beta}+s\right)} \tag{A-6.8}$$

where

$$\alpha_1=\frac{2}{\sum_{j=1}^{2}\omega_j r_{\mathrm{ejD}}^2} \tag{A-6.9}$$

$$\beta=\frac{\sum_{j=1}^{2}\frac{\omega_j^2}{\kappa_j}r_{\mathrm{ejD}}^4\left(\ln r_{\mathrm{ejD}}-\frac{3}{4}\right)}{\left(\sum_{j=1}^{2}\omega_j r_{\mathrm{ejD}}^2\right)^2} \tag{A-6.10}$$

Taking the Laplace inverse transform of Eq. (A-6.8), we have

$$q_{\mathrm{D}}=\frac{1}{\beta}e^{-\frac{\alpha_1}{\beta}t_{\mathrm{D}}} \tag{A-6.11}$$

Fetkovich dimensionless decline time in a two-layer commingled reservoir with a closed circular boundary can be written as

$$t_{\mathrm{Dd}}=\frac{\alpha_1}{\beta}t_{\mathrm{D}}=\alpha t_{\mathrm{D}} \tag{A-6.12}$$

Fetkovich dimensionless decline flow rate can be written as

$$q_{\mathrm{Dd}}=\frac{\beta}{L^{-1}\left[\overline{p}_{\mathrm{wfD}}\right]} \tag{A-6.13}$$

If $r_{e1} = r_{e2}$, Eq. (A-6.9) can be simplified to

$$\alpha_1 = \frac{2}{r_{eD}^2} \tag{A-6.14}$$

Eq. (A-6.10) can be simplified to

$$\beta = \sum_{j=1}^{2} \frac{\omega_j^2}{\kappa_j}\left(\ln r_{eD} - \frac{3}{4}\right) \tag{A-6.15}$$

If $\kappa_1 = \kappa_2$, $\omega_1 = \omega_2$, homogeneous reservoir, Eq. (A-6.10) can be simplified to

$$\beta = \ln r_{eD} - \frac{3}{4} \tag{A-6.16}$$

References

Abdassah, D., Ershaghi, I., 1986. Triple-porosity systems for representing naturally fractured reservoirs. SPE Formation Eval. 1 (2), 113–127.

Agarwal, R.G., Gardner, D.C., Kleinsteiber, S.W., 1998. Analyzing well production data using combined type curve and decline curve concepts, SPE49222.

Agarwal, R.G., 1979. Real gas pseudotime—a new function for pressure buildup analysis of MHF gas wells, SPE8279.

Agarwal, R., Al-Hussainy, R., Ramey, Jr., H.J., 1970. An investigation of wellbore storage and skin effect in unsteady liquid flow: I. analytical treatment. SPE J. 10 (3), 279–290.

Ahmed, T., 2010. Reservoir engineering handbook. Gulf Professional Publishing.

Al-Ghamdi, A., Ershaghi, I., 1996. Pressure transient analysis of dually fractured reservoirs. SPE J. 1 (1), 93–100.

Al-Hussainy, R., Ramey, Jr., H.J., 1966. Application of real gas flow theory to well testing and deliverability forecasting. JPT 18 (5), 637–642.

A1-Hussainy, R., Ramey, Jr., J.J., Crawford, P.B., 1966. The flow of real gases through porous media. JPT 18 (5), 624–636.

Ambastha, A.K., 1988. Pressure transient analysis for composite systems. PhD Thesis. Stanford University.

Amini, S., Ilk, D., Blasingame, T.A., 2007. Evaluation of the elliptical flow period for hydraulically fractured wells in tight gas sands—theoretical aspects and practical considerations, SPE106308.

Anderson, D.M., Mattar, L., 2003. Material balance time during linear and radial flow, PETSOC2003-201.

Anderson, D., Mattar, L., 2004. Practical diagnostics using production data and flowing pressures, SPE89939.

Anderson, D.M., Mattar, L., 2007. An improved pseudo-time for gas reservoirs with significant transient flow. J. Can. Petrol. Technol. 46 (7), 49–54.

Anderson, D.M., Stotts, G.W.J., Mattar, L., et al., 2010. Production data analysis challenges, pitfalls, diagnostics. SPE Reservoir Eval. Eng. 13 (3), 538–552.

Arévalo-Villagrán, J.A., Wattenbarger, R.A., 2001. Production analysis of long term linear flow in tight gas reservoirs: case histories, SPE71516.

Arévalo-Villagrán, J.A., Gutiérrez-Acosta, T., 2005. Analysis of long term behavior in tight gas reservoirs: case histories, SPE95117.

Arps, J.J., 1945. Analysis of decline curves. Trans. AIME 160, 228–247.

Bahadori, A., Vuthaluru, H.B., 2010. A simple decline curve analysis approach for evaluating gas reserves and predicting future production in gas reservoirs, SPE139057.

Bahadori, A., 2012. Analysing gas well production data using a simplified decline curve analysis method. Chem. Eng. Res. Design 90 (4), 541–547.

Blasingame, T.A., Lee, W.J., 1986. Properties of homogeneous reservoirs, naturally fractured reservoirs, and hydraulically fractured reservoirs from decline curve analysis, SPE15018.

Blasingame, T.A., Lee, W.J., 1986. Variable rate reservoir limits testing, SPE15028.

Blasingame, T.A., Johnston, J.L., Lee, W.J., 1989. Type curve analysis using the pressure integral method, SPE18799.

Blasingame, T.A., McCray, T.L., Lee, W.J., 1991. Decline curve analysis for variable pressure drop/variable flow rate systems, SPE21513.

Advanced Production Decline Analysis and Application

Bourdet, D., Whittle, T.M., Douglas, A.A., Pirard, Y.M., 1983. A new set of type curves simplifies well test analysis. World Oil 196 (6), 95–106.
Bourdet, D., 1985. Pressure behavior of layered reservoir with cross flow, SPE13628.
Bourdet, D., Ayoub, J.A., Pirard, Y.M., 1989. Use of pressure derivative in well test interpretation. SPE Formation Eval. 4 (2), 293–302.
Bourdet, D., 2003. Well Test Analysis: The use of advanced interpretation model—handbook of petroleum exploration & production. Elsevier Science B.V.
Callard, J.G., 1995. Reservoir performance history matching using rate cumulative type curves, SPE30793.
Camacho Velázquez, R., 2005. Pressure-transient and decline-curve behaviors in naturally fractured vuggy carbonate reservoirs. SPE Reservoir Eval. Eng. 8 (2), 95–112.
Camacho Velazquez, R., Mario, A., Vasquez-Cruz, R., Castrejon-Aivar, V., Arana-Ortiz, V., 2005. Pressure transient and decline curve behaviors in naturally fractured vuggy carbonate reservoirs. SPE Reservoir Eval. Eng. 8 (2), 95–112.
Carter, R.D., 1981. Characteristic behavior of finite radial and linear gas flow systems — constant terminal pressure case, SPE9887.
Carter, R.D., 1985. Type curves for finite radial and linear gas flow systems: constant terminal pressure case. SPE J. 25 (5), 719–728.
Chen, Y., Dang, L., 2004. Modern reservoir engineering. Petroleum Industry Press, Beijing.
Cinco-L.H., Samanieco, V.V., Dominguez, A., 1978. Transient pressure behavior for a well with a finite conductivity vertical fracture. SPE J. 18 (4), 253–264.
Clossman, P.J., 1975. An aquifer model for fissured reservoirs. SPE J. 15 (5), 385–398.
Compiling Group of Handbook of mathematics, 2010. Mathematics Handbook. Higher Education Press, Beijing.
Da Prat, G., Cinco-Ley, H., Ramey, Jr., H.J., 1981. Decline curve analysis using type curves for two porosity systems. SPE J. 21 (3), 354–362.
Dietz, D.N., 1965. Determination of average reservoir pressure from buildup surveys. JPT 17 (8), 955–959.
Doublet, L.E., Pandie, P.K., McCollum, T.J., et al., 1994. Decline curve analysis using type curves analysis of oil well production data using material balance time: application to field cases, SPE28688.
Duong, A.N., 1989. A new approach for decline curve analysis, SPE18859.
Ebrahimi, M., 2010. Enhanced estimation of reservoir parameters using decline curve analysis, SPE133432.
Ehlig-Economides, C.A., Ramey, Ir., H.L., 1981. Transient rate decline analysis for wells produced at constant pressure. SPE J. 21 (1), 98–104.
Ehlig-Economides, C.A., Hegeman, P., Vik, S., 1994. Guidelines simplify well test interpretation. Oil Gas J. 92 (29), 33–40.
Ehlig-Economides, C.A., 2000. Pressure transient analysis. University of Houston.
El-Banbi, A.H., Wattenbarger, R.A., 1998. Analysis of linear flow in gas well production, SPE39972.
Fetkovich, M.J., 1980. Decline curve analysis using type curves. JPT 32 (6), 1065–1077.
Fetkovich, M.J., Vienot, M.E., Bradley, M.D., et al.,1987. Decline curve analysis using type curves case histories. SPE Formation Eval. 2 (4), 637–656.
Fetkovich, M.J., Fetkovich, E.J., Fetkovich, M.D., 1996. Useful concepts for decline curve forecasting, reserve estimation, and analysis. SPE Reservoir Eng. 11 (1), 13–22.
Fraim, M.L., Wattenbarger, R.A., 1987. Gas reservoir decline curve analysis using type curves with real gas pseudopressure and normalized time. SPE Formation Eval. 2 (4), 671–676.

Frederick, J., Kelkar, M., 2005. Decline curve analysis for solution gas drive reservoirs, SPE94859.

Gentry, R.W., 1972. Decline curve analysis. JPT 24 (1), 38–41.

Gerami, S., Pooladi-Darvish, M., Hong, H., 2007. Decline curve analysis for naturally fractured gas reservoirs: a study on the applicability of pseudo-time and material balance pseudo-time, IPTC11278.

Govier, G.W., 1979. Theory and practice of the testing of gas wells, fourth ed. Alberta Energy Resources and Conservation Board, Calgary, Guide G-3.

Gringarten, A.C., Ramey, Jr., H.J., Raghavan, R., 1974. Unsteady state pressure distributions created by a well with a single infinite conductivity vertical fracture. SPE J. 14 (4), 347–360.

Gross, H.E., 1938. Decline curve analysis. Oil Gas J. 37 (18), 55–56, 58, 61.

Guppy, K.H., Cinco-Ley, H., Ramey, Jr., H.J., 1981. Transient flow behavior of a vertically fractured well producing at constant pressure, SPE9963.

Hagoort, J., 2003. Automatic decline curve analysis of wells in gas reservoirs. SPE Reservoir Eval. Eng. 6 (6), 433–440.

Han, Y., 2006. Study on gas well performance analysis in low-permeability and tight reservoirs. PhD Thesis. China University of Geosciences, Beijing.

Héctor, P., 2006. On a well-test pressure theory of analysis for naturally fractured reservoirs, considering transient interporosity matrix, microfractures, vugs, and fractures flow, SPE104076.

Her-Yuan, C., Teufel, L.W., 2000. A new rate time type curve for analysis of tight gas linear and radial flows, SPE63094.

Horne, R.N., 1995. Modern well test analysis: a computer aided approach. Petroway, Inc., Palo Alto, CA.

Hower, T.L., Collins, R.E., 1989. Detecting compartmentalization in gas reservoirs through production, SPE19790.

Hurst, W., 1934. Unsteady flow of fluids in oil reservoirs. J. Appl. Phys. 5 (1), 20–30.

Ilk, D., Rushing, J.A., Perego, A.D., et al., 2008. Exponential vs. hyperbolic decline in tight gas sands understanding the origin and implications for reserve estimate using arps decline curves, SPE116731.

Issaka, M.B., Ambastha, A.K., 1998. Decline curve analysis for composite reservoirs. J. Can. Petrol. Technol. 37 (6), 17–24.

Ji, B., 1995. The fundamentals of seepage flow theory used on the production decline equations. Acta Petrolei Sinica 16 (3), 86–91.

Jiang, L., Chen, Z., 1985. Well test analysis theory. Petroleum Industry Press, Beijing.

Johnson, R.H., Bollens, A.L., 1927. The loss ratio method of extrapolating oil well decline curves. Trans. AIME 77, 771–775.

Johnston, L., 1992. Variable rate analysis of transient well test data using semi-analytical methods. M.S. Thesis. Texas A&M University, College Station, TX.

Khanamiri, H.H., 2010. A non-iterative method of decline curve analysis. J. Petrol. Sci. Eng. 73 (1–2), 59–66.

Larkey, C.S., 1925. Mathematical determination of production decline curves. Trans. AIME 71, 1322–1328.

Lee, W.J., Wattenbarger, R.A., 1996. Gas reservoir engineering. Henry L. Doherty Memorial Fund of AIME, Society of Petroleum Engineers.

Lefkovits, H.C., Hazebroek, P., Allen, E.E., Matthews, C.S., 1961. A study of the behavior of bounded reservoirs composed of stratified layers. SPE J. 1 (1), 43–58.

Lei, Q., Wan, Y., Sun, H., et al., 2008. Production dynamic analysis theory and practice. Petroleum Industry Press, Beijing.

Li, K., Horne, R.N., 2003. A decline curve analysis model based on fluid flow mechanisms, SPE83470.

Liu, N., 2008. Practical modern well testing method, fifth ed. Petroleum Industry Press, Beijing.

Londono, Fabio E., Blasingame, T.A., 2002. Simplified correlations for hydrocarbon gas viscosity and gas density—validation and correlation of behavior using a large scale database, SPE75721.

Marhaendraja, T., Blasingame, T.A., 2001. Decline curve analysis using type curves—evaluation of well performance behavior in a multiwell reservoir system, SPE71517.

Mattar, L., Hawkes, R.V., Santo, M.S., et al., 1993. Prediction of long term deliverability in tight formation, SPE26178.

Mattar, L., McNeil, R., 1998. The flowing gas material balance. J. Can. Petrol. Technol. 37 (2), 52–55.

Mattar, L., Anderson, D.M., 2003. A systematic and comprehensive methodology for advanced analysis of production data, SPE84472.

Mattar, L., Moghadam, S., 2009. Modified power law exponential decline for tight gas, PETSOC 2009-198.

Mattar, L., Anderson, D., Stotts, G., 2006. Dynamic material balance oil or gas-in-place without shut-ins. J. Can. Petrol. Technol. 45 (11), 7–10.

Mattar, L., Fast-RT [EB/OL], 2012. http://www.fekete.com/Software/RTA/Media/Webhelp/Index.hlp.

Matthews, C.S., Russell, D.G., 1967. Pressure buildup and flow tests in wells. Storm Printing Corp., Monograph Series, SPE of AIME, Dallas, TX.

McCray, T.L., 1990. Reservoir analysis using production decline data and adjusted time. M.S. Thesis. Texas A&M University, College Station, TX.

McNeil, R., Jeje, O., Renaud, A., 2009. Application of the power law loss ratio method of decline analysis, PETSOC2009-159.

Mead, H.N., 1956. Modifications to decline curve analysis. Trans. AIME 207, 11–16.

Meunier, D.F., Kabir, C.S., Wittmann, M.J., 1987. Gas well test analysis: use of normalized pseudo variables. SPE Formation Eval. 2 (11), 629–636.

Miller, F.G., 1962. Theory of unsteady state influx of water in linear reservoirs. J. Inst. Petrol. 48 (467), 365–379.

Moghadam, S., Jeje, O., Mattar, L., 2011. Advanced gas material balance in simplified format. J. Can. Petrol. Technol. 50 (1), 90–98.

Mutalik, P.N., Joshi, S.D., 1992. Decline curve analysis predicts oil recovery from horizontal wells. Oil Gas J. 90 (36), 42–48.

Nobakht, M., Mattar, L., Moghadam, S., et al., 2012. Simplified forecasting of tight shale gas production in linear flow. J. Can. Petrol. Technol. 51 (6), 476–486.

Nobakht, M., Clarkson, C.R., 2012. A new analytical method for analyzing linear flow in tight shale gas reservoirs: constant flowing pressure boundary condition. SPE Reservoir Eval. Eng. 15 (3), 370–384.

Olarewaju, J.S., Lee, W.J., Lancaster, D.E., 1991. Type and decline curve analysis with composite models. SPE Formation Eval. 6 (1), 79–85.

Oliver, H., Viturat, D., Fjaere, O.S., et al., 2013. Dynamic data analysis. KAPPA.

Ozkan, E.,1988. Performance of horizontal wells. PhD dissertation. University of Tulsa, Tulsa, Oklahoma.

Palacio, J.C., Blasingame, T.A., 1993. Decline curve analysis using type curves analysis of gas well production data, SPE25909.

Papay, J., 2001. Generalization of decline curve analysis. Erdoel Erdgas Kohle/EKEP 117 (4), 195–200.

Petroleum and Natural gas industry standards of China, People's Republic of China, 2010. Natural gas reserves calculation. SY/T6098.

Pratikno, H., Rushing, J., Blasingame, T.A., 2003. Decline curve analysis using type curves fractured wells, SPE84287.

Pratikno, H., 2002. Decline curve analysis using type curves—fractured wells. M.S. Thesis. Texas A&M University, College Station, TX.

Raghavan, R., Cady, G., Ramey, Jr., H.J., 1972. Well test analysis for vertically fractured wells. JPT 24 (8), 1014–1020.

Raghavan, R., 1993. Well test analysis. Prentice Hall Petroleum Engineering Series, Englewood Cliffs, New Jersey.

Raghavan, R., Ohaeri, C.U., 1987. Unsteady flow to a well produced at constant pressure in a fractured reservoir. SPE Formation Eval. 2 (2), 186–200.

Rahman, N.M.A., Mattar, L., Zaoral, K., 2006. A new method for computing pseudo-time for real gas flow using the material balance equation. J. Can. Petrol. Techno. 45 (10), 36–44.

Rahman, N.M., Mattar, L., Zaoral, K., 2004. A new method for computing pseudo time for real gas flow using the material balance equation, PETSOC2004-182.

Ramsay, H.J. Jr., 1968. The ability of rate time decline curves to predict future production rates. M.S. thesis, University of Tulsa, Tulsa, OK.

Robertson, S., 1988. Generalized hyperbolic equation, SPE18731.

Rodriguez, F., 2004. Well test characterization of small- and large-scale secondary porosity in naturally fractured reservoirs, SPE90287.

Roeser, H.M., 1925. Determining the constants of oil production decline curves. Trans. AIME 71, 1315–1321.

Sageev, A., Da Prat, G., Ramey, Jr., H.J., 1985. Decline curve analysis for double porosity systems, SPE13630.

Seshadri, J., Mattar, L., 2010. Comparison of power law and modified hyperbolic decline methods, SPE137320.

Shih, M.Y., 1994. Decline curve analysis for horizontal wells. M.S. Thesis. Texas A&M University, College Station, TX.

Slider, H.C., 1968. A simplified method of hyperbolic decline curve analysis. JPT 20 (3), 235–236.

Spivey, J.P., Gatens, J.M., Semmelbeck, M.E., et al., 1992. Integral type curves for advanced decline curve analysis, SPE24301.

Standing, M.B., Katz, D.L., 1942. Density of natural gases. Trans. AIME 146, 140–149.

Stehfest, H., 1970. Algorithn368, numerical inversion of laplace transforms. Commun. ACM 13 (1), 47–49.

Sun, H., 2003. Maximum effective hole mathematical model and exact solution for commingled reservoir. Chinese J. Chem. Eng. 11 (5), 550–554.

Sun, H., 2011. A balance equation of gas materials with recharge capacity and their performance prediction. Acta Petrolei Sinica 32 (4), 683–686.

Sun, H., Xu, Y., 2011. A new method for low permeability gas well performance prediction using pressure transient analysis: theory and practices, SPE144952.

Sun, H., 2012. Modern well testing analysis and productivity evaluation of complex gas reservoir. Petroleum Industry Press, Beijing.

Sun, H., Shi, Y., 2013. Integrating pressure transient analysis and production analysis for dynamic reserve evaluation in tazhong I Ordovician Carbonate Gas Field, Tarim basin, IPTC16469.

Tariq, S.M., Ramey, Jr., H.J., 1978. Drawdown behavior of a well with storage and skin effect communicating with layers of different radii and other characteristics, SPE7453.

The People's Republic of China National Standard, 2004. Petroleum and natural gas reserves classification. GBT19492.

Thompson, T.W., Mattar, L., 2000. Gas rate forecasting during boundary dominated flow, PETSOC2000-046.

Towler, B.F., Bansal, S., 1993. Hyperbolic decline curve analysis using linear regression. J. Petrol. Sci. Eng. 8 (4), 257–268.

Uraiet, A.A., Raghavan, R., 1980. Unsteady flow to a well producing at a constant pressure. JPT 32 (10), 1803–1812.

Van Everdingen, A.F., Hurst, W., 1949. The application of the laplace transformation to flow problems in reservoirs. Trans. AIME 186, 305–324.

Warren, J.E., Root, P.J., 1963. The behavior of naturally fractured reservoirs. SPE J. 3 (3), 245–255.

Wattenbarger, R.A., El-Banbi, A.H., 1998. Production analysis of linear flow into fractured tight gas wells, SPE39931.

Wu, Y.-S., 2007. A triple-continuum pressure transient model for a naturally fractured vuggy reservoir, SPE110044.

Xu, S., 1995. FORTRAN algorithms procedures set, second ed. Tsinghua University Press, Beijing.

Zhuang, H., 2013. Dynamic well testing in petroleum exploration and development. Petroleum Industry Press, Published by Elsevier Inc.

Index

E

F

R

S

T

Kolibri 360 GmbH
Ferdinand-Jühlke-Straße 7
99095 Erfurt, Deutschland
produktsicherheit@kolibri360.de

Dedication

This book is dedicated to my three boys (Quinn, Gavin, and Tristan). These three little men have supported me all along the way while I poured my heart and soul into creating this book and the growth of my practice. I went back to college for my Master's degree while on maternity leave with my third son, Tristan. The importance of education has been at the heart of our household from the time they were little. As they grew, their excitement and pride in me, their mom, fueled my continued growth in the industry. I wanted to be an example, a role model for the boys to set their goals high and continually strive for more. Even if it is not in esthetics, I wanted them to see the value in education, the devotion it takes to create something you are proud of and, what it means to influence others positively. They believed in my mission to make the esthetic industry a better place for the providers and the patients. I hope that they have seen the difference between a career versus a job with the passion invested in my craft. May they find a career that they are wholeheartedly dedicated to. I was told many years ago; that if you follow your heart, the money will follow, and you will never work a day in your life. This is so cliche as an adult, but, it resonated with me in my youth and has become a powerful truth in my adult years. It is easy to tell the difference between an educator and a salesperson. The educator believes in the task and can back it up with proof.

Boys, my message to you is; to be the leader, the educator, and the innovator. Believe in yourself and your dreams. The only thing standing in your way of success is a lack of effort. Get out there and show the world you are a shark. The shark doesn't wake up on Monday, complaining it's a Monday. Better yet, the shark is up early and biting stuff reminding everyone he is the shark. When someone tells you no, find a way, don't settle, don't let your dreams die because someone else does not share your vision. Believe in yourself and your future, and go get your dreams and desires.

I love you three,

Love Mom

ESSENTIALS OF NEUROMODULATION

A Comprehensive Guide for Aesthetic Practitioners

TARA DELLE CHIAIE MSN, FNP-BC, APRN
Private Practice, Hampton NH

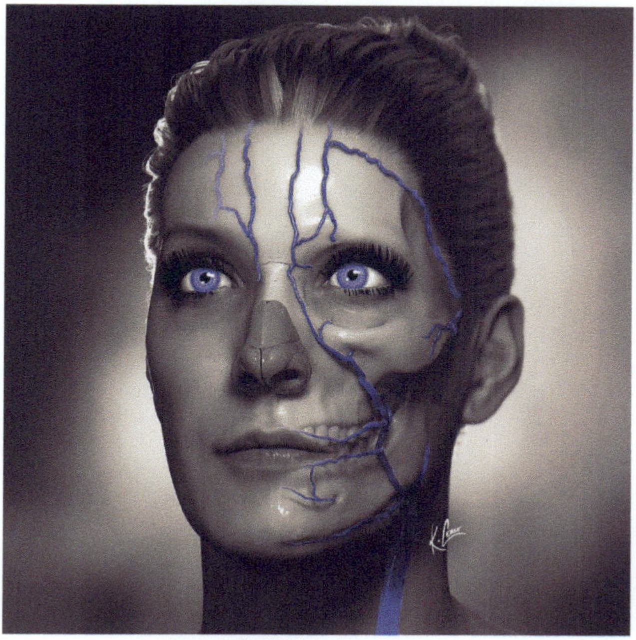

DCM

DELLE CHIAIE COSMETIC MEDICINE

ELSEVIER

ACADEMIC PRESS
An imprint of Elsevier

ESSENTIALS OF NEUROMODULATION

Contents

About the author

Tara Delle Chiaie grew up in Plaistow, New Hampshire, and Merrimac, Massachusetts. Her mom was a cosmetologist, and her dad, a carpenter as well as a fisherman. Tara loved traveling to New York as a kiddo to attend hair shows with her mom. However, when she was not at hair shows, Tara could often be found fishing in a taffeta dress with black patent leather shoes. Tara loves beauty and glam but also enjoys the rugged outdoors. She was a ballerina for much of her youth and loved dancing in the nutcracker. Tara attended nursing college while working as a ballerina. Her studies eventually took her abroad to Australia's beautiful land, focusing on psychology, philosophy, and statistics.

Tara has been in medicine since 1999. She officially became a Registered Nurse in 2002. In 2011 Tara graduated from the accelerated program at the University of New Hampshire (UNH) as an Advanced Practice Registered Nurse (APRN) and immediately became nationally recognized through the American Nurses Credentialing Center (ANCC) as a Board Certified Nurse Practitioner. She is further Board Certified in Anti-Aging Medicine by the American Academy of Anti-Aging Medicine. Growing up in the beauty industry, Tara found it was a great union to blend beauty with medicine. She has an astute sense of safety while her experience guides her practice to produce beautiful and natural results. She has been teaching nursing and medicine for 15 years. Sharing her knowledge and helping others define their talent brings her great joy. Tara's goal is to continually fine-tune the art of bringing one's inner beauty to the surface. This happens by increasing her own education and ensuring other health care providers are delivering the same quality of care.

In 2013 Tara opened her Cosmetic business, Delle Chiaie Cosmetic Medicine. Although esthetics is her passion, she found a significant flaw in the health care system and then opened a concierge medical practice—Delle Chiaie Concierge Medicine. Although Tara created her initial business with a strict business plan, she always kept one ear to the ground, listening and recognizing her environment. This keen awareness was to ensure her organic growth within the industry by being mindful of patient needs and economic trends. After seeing several hundreds of clients that needed correction work, she quickly realized it was an opportunity to begin teaching others. Creating DCCM™ Academy was her way to share her experience and knowledge with others while safeguarding patients from the ill-informed or the undereducated provider.

The purpose of this book is to set an industry standard in cosmetic injectables. Standards must be redefined in terms of patient safety and academics for entry into this booming field.

Tara is an active member of the ANA—American Nurses Association, ISPAN—International Society of Plastic and Aesthetic Nurses, ANCC—American Nurses Credentialing Center, AAMEP—American Academy of Medical Esthetic Professionals, and American Academy of Anti-Aging Medicine. She holds 53 additional certifications in advanced esthetic training and has traveled the world to improve her skill and exposure to the ever-growing esthetic industry. Tara is a National speaker for Candela Medical Corporation, speaking about her business brand and development and their technology and how she has implemented the technologies to help grow her business in terms of improved patient outcomes. Candela recognized the value in her unique proprietary formula for combination therapies with neuromodulators, filler, and laser technology. Therefore she became a global leader and speaker.

Acknowledgments

I want to thank my very first injectable representative, Christine Brickett. I am the injector that I am in part because of her. Skill, education, and personality all make a difference, but I truly believe that relationships can also change the trajectory of one's path. This amazing woman, Christine (Chris) Brickett, changed my path. After my first course on injectables, I called the drug company, and they kindly connected me to the rep in my territory. Chris was so very kind on the phone, and I told her I just finished my first course yet, I felt it was not enough to begin injecting independently, and I was nervous and didn't feel ready. Chris told me "I can help you, but you need to be willing to put in the time and money for the proper education." I paused for a minute, thinking of my three boys I was raising independently. She sensed my reservation and told me if I couldn't commit, then maybe I wasn't ready. That immediately sparked my drive. My response was crystal clear; I said, "I will find a way." Together we did find a way for me to gain the experience needed and begin developing long-lasting relationships in the industry. She quickly connected me to several providers that had been doing well in the industry to help advance my knowledge and skill in the esthetic field of injectables. That is when I felt like the real learning began. In my first year, she had introduced me to five different injectors to train with. Each one of these injectors had a training piece to their practice so I was able to private pay for personalized, hands-on training. Had she not believed in me, I may never have had the fortune to be where I am today in this amazing industry. It was her experience and guidance that helped connect me to the right people in the industry. Chris showed me the difference between being a point and shoot injector to becoming an artful provider producing safe and beautiful results. She pushed me to demand more of myself both in terms of outcomes and safety. I was very lucky to have such a savvy and intelligent rep. I owe much of my success today to her. Her wisdom, persistence, kindness, and brilliant nature are the very shoulders I stand on today. Don't ever underestimate the power of a relationship. We all need each other and can grow from one another. Thank you, Chris; I will always remember where I came from.

Disclaimer

Tara Delle Chiaie of Delle Chiaie Cosmetic Medicine, PLLC, has taken great care to ensure the accuracy of the content within this book. The information is current and based on best practices utilizing research as the foundation coupled with practical application. The author of this instructional material and teaching are not responsible for errors or omissions or for any consequences from applying the information in this book and making no warranty, expressed or implied, with respect to the accuracy, completeness, or currency of the contents within this course material. Application of the information from this manual on particular situations remains the sole professional responsibility of the practitioner performing the task. Some drugs mentioned have FDA clearance, but it is still up to the individual provider to ensure each drug's safety.

The author has done her due diligence to ensure accuracy that any section and all pages are in accordance with current best practices. Ongoing research in the field of medicine will create a constant flow of change and recommendations. It is up to the practitioner to ensure ongoing training and understanding of current trends and best practices. The readers of this material are encouraged to read each package insert and follow prescribing guidelines for each medication being utilized.

CHAPTER 1
Introduction

The human face is more than just a physical space; facial features are directly linked to our psyche, contrary to popular belief. Our facial features demonstrate our identity, and our expressions speak a language, unlike any other species. The slightest flip up of an eyebrow has significant meaning in the unspoken language. Any mother can relate to this very concept as they can appreciate the silent impact that the *hairy eyebrow* they dart at their misbehaving child can have. A mother's disapproving face can stop a child dead in their tracts, and the mother did not even have to utter a word. When I am not freshly TOXED, I speak silent volumes to my children publicly, and they know exactly what I mean. The point here is; our features matter despite the naysayers that claim that looks do not matter. Being a part of a multibillion-dollar industry based on looks and image strongly suggests that looks, in fact, do actually matter, and consumers are willing to pay for it. If looks did not matter, then why were filters created for photos? Why are there so many body shaping and slimming devices, as well as slimming clothing on the market that is all succeeding?

Our facial features can help exude confidence, or to the contrary, our aging features or disfigurement can be the grim reaper of our very soul. A person slowly deteriorates over time, not just physically but emotionally; every time they look in the mirror, they see their life slipping away both physically and metaphorically. With each glance in the mirror, they see their life, youth, and beauty wasting away. A patient told me once that she felt like the filler and neuromodulator that I injected was as if I filled her back up with her own sense of power and her zest for life. The patient stated that as she saw the wrinkles and volume loss taking over her image all the while, she felt as if she was losing her identity. She further continued, "it was as if I injected her identity back into her with every syringe. The filler and neuromodulator injected were like you were injecting me back into life. Since my treatment, I have been doing my hair every day, going to the gym, dressing my best, and have been much more efficient with my tasks." Injections are a powerful treatment to our patients, not just physically but socially and emotionally.

Essentials of Neuromodulation
https://doi.org/10.1016/B978-0-323-89920-8.00012-6

What cosmetic providers do for their patients is more than just physical. The work is essential to the emotional well-being of the patients. It is necessary to the emotional happiness and confidence of the clients to help them look and feel refreshed, youthful, and beautiful. It eludes to the concept of mind, body, and spirit, or shall we say mind, face, and spirit in the esthetic industry. When a person is happy with themselves physically, emotionally, and functionally, they give off different signals to others. Confrontation, sad, elusive, confident, humble are just a few signals portrayed from one person to the next based on looks. This concept, yet again, references the importance of facial expressions, and how we communicate with others is directly affected by our features. This concept is tied to the success and confidence of the individual. One repairs an aging deck or an aging pool, so it seems to make sense that one would make repairs to our aging faces to maintain appearances. What motivates individuals to address their aging self is unique, varied, colorful, and rich with personal history. It is up to the cosmetic providers to develop an acute sense of looking through their lens to have a glimpse of how they view themselves and what their purpose is in their esthetic journey.

The esthetic industry is booming, and medical spas are popping up at alarming rates all across the world. Nurses and doctors are leaving the traditional bedside role to become a part of this new wave of medicine. Over the past few decades, neuromodulators have taken the esthetic industry by storm and become well adopted into mainstream culture for both patients and providers. Twenty years ago, the consumer's only option was plastic surgery or a CO_2 laser to fight back against poor lifestyle decisions such as sunburns and cigarettes, coupled with the dreaded hands of time. Nowadays, consumers have a myriad of treatment options to age gracefully without going under the knife. There are so many avenues to access information, and it can be very confusing to consumers as well as our fellow peers and colleagues looking to break into the industry. What treatment is best? What product is best? Who is the best? Is a plastic surgeon better than a cosmetic nurse? Who is best to train within the esthetic industry? These are just a few of the burning questions from both new injectors and patients alike. Let this book help to ease into understanding the esthetic industry, the consumer, and the educational process necessary to become beyond proficient at cosmetic injections.

Grandview Research (2020) reported that in 2019 the non-surgical cosmetic industry was an 8.88 billion dollar industry. With the recent rise in zoom meetings, telehealth, and social media of every platform, people are becoming increasingly more conscientious of their appearance. With the onslaught of video calls, it is as if the patients are forced to look in the mirror all day long. As they peer down at the computer screen, they can see the hands of time pulling them downward. Certain features can be enhanced or exacerbated by the positioning of the face. Often with the video conference calls, patients are looking in a downward position, thus exacerbating the negative look of the nasolabial folds or the jowls on their faces.

The funny filters or the unrealistic filters one can put on themselves can only take a person so far in life. Snap chat filters and apps such as Face Tune can take the average person with severe signs of aging, and presto; the app has now made them a clone of the Kardashians. Nevertheless, let us pause for a minute and think about what the external filter is doing. Filters are actually just a filtered reality of oneself. A new sense of reality must be achieved in order to get to a point in society with the realization that a filter will not do anyone any good. In-person encounters do not come with a filter, therefore, making it essential to represent an honest presentation in person and online. Once a person goes to meet their new Tinder date in real life, the reality is now the presence of a lie if filters were utilized on the profile picture. Unless, of course, they meet in the darkest of darkest speak EZE, then the darkness will mascarade as a filter. However, utilization of filters is helping the esthetic industry forge forward. Injectables help to create, so to speak, an internal filter for the patient. Instead of filtering the photo, patients can now filter themselves. This way, the patient has rejuvenated themselves from the inside out and can have confidence that their new date from the app will appreciate the honest person who shows up and recognizes them.

Open up any social media platform, and it is evident that the trend for minimally invasive procedures is on the upswing. Between 2000 and 2014 cosmetic surgery showed a 12% decline compared to minimally invasive procedures, which demonstrated an upward shift of 154%, according to Khunger (2015). In 2015, 13.5 billion dollars was spent on non-surgical and surgical procedures (Khunger, 2015). Then in 2016, there was an increase in spending by 1.5 million dollars in the industry (Khunger, 2015). Social media and influencers are definitely aiding in driving the growth of the industry. The Aesthetic Society (2019) reported that out of 17.7 million minimally invasive procedures performed, 15.9 million were minimally invasive procedures like neuromodulation and dermal fillers.

According to The Aesthetic Society (2019), Botox treatments have been the number one leading non-surgical procedure since 1999. More than ONE BILLION dollars was spent on injectables in 2019 (The Aesthetic

Society, 2019). It is no wonder why Botox has become a common household name, and everyone has heard of the wrinkle busting medication internationally and right here in the United States.

Cosmetic medicine or esthetic medicine are terms used loosely to describe a myriad of treatment options, both surgical and non-surgical; they extend from plastic surgery to injectables, lasers, and so much more. With the click of a few buttons, consumers can now search for nearly anything they want online from the comfort of their own home. With a plethora of information at their fingertips and a booming field of surgical and non-surgical treatments available, the world is their oyster. Consumers can now have their facial features augmented without having to take on the increased risk of going under the knife for cosmetic surgery. Perhaps the patient needs a little enhancement but is not ready to have a surgical facelift just yet; the non-surgical side of injectables can help to extend their youthful looks before they go under the knife. High-risk practices, such as unapproved products (bought illegally), off-label use of products, inadequate medical facilities, and poorly supervised staff performing procedures and subpar training, are the mainstay of increased complications. This is a very competitive and fast-paced industry. It takes a lot to get ahead; it takes even more to develop mastery of skills. At the end of the day, this is still medicine, and the practice and art of injectables is a medical specialty. Cutting corners will only harm the patients and, in turn, harm our industry.

Surgery has its place but comes with significantly more risk than injectables. With that said, injectables are equally, if not more dangerous, if the injector is not adequately educated. Minimizing patient risk is the best way for providers to care for their patients; this can be achieved by education and minimally invasive procedures. Neuromodulators and dermal fillers, coupled with lasers, have opened the doors for providers to improve patient outcomes and offer them various treatment options that best fit their budget, lifestyle, and esthetic goals. Even with non-surgical options, providers can minimize the effects of aging, lift, plump, soften, and even correct genetic asymmetries. With that said, neuromodulators and other injectables do come with their own inherent risk both by the product and the injector providing the service. Although the risk of harm is minimal compared to surgery, there is still a significant risk, and that needs to be addressed with thorough education. The best way to mitigate or reduce the risk of harm is by proper education, coupled with an in-depth understanding of anatomy both at rest and in motion. Practitioners should seek out post-secondary education and guided practice with experienced and licensed practitioners to become proficient.

Even though Botox is a common name among households and coffee shops, there is still confusion surrounding the purpose and meaning for patients as well as new injectors. The following chapters' goal is to demystify the stigma, define the terms, and provide the reader with a clear understanding of how to inject neuromodulators with a comprehensive anatomical understanding. The intention is for injectors to learn more knowledge while becoming leaders and innovators, not just injectors. It is necessary to reset the standard of care and education in the injectables industry. Gone are the days where trophies are doled out because of a weekend course attendance. In order to elevate the bare minimum of initiation into the industry, a paradigm shift must occur, and it starts with the foundational concept of education. An 8-h course is not enough to demonstrate understanding, let alone proficiency or mastery. Courses should be created with follow up advancing the knowledge and skill of the student. A stepwise approach would be best paired with a mentorship program or a fellowship, just like a surgeon endures. Pause here for a minute and reflect on the mastery necessary for a surgeon. The surgeon has several years dedicated to a fellowship after the initial training has been completed before they can work independently. It is time to have esthetic courses with mentorships and fellowships to develop mastery in the art of injectables. Studies demonstrate that it takes 10 years and 10,000 h of practice to the craft or study to become a master. Look at Mozart or Steve Jobs or Bill Joy; they all dedicated 10 years and 10,000 h of blood, sweat, and tears into their craft before it finally amounted to something. Surgical options are still an active and necessary part of the industry; however, it is the non-surgical field of esthetic medicine growing exponentially. When injectors are learning how to inject, they must first have a stronghold on what products are being injected. Then they must develop an intimate understanding of where to inject and how it interacts with the tissues and the patient's overall facial expressions.

To be a proficient cosmetic injector, one must understand anatomy on a different level other than what has been previously taught in the basics of nursing school or medical school. Injection anatomy calls for an in-depth understanding of learned anatomical behaviors and characteristics. Dr. Arthur swift defines injection anatomy as its own learning curve and feels it is different from surgical anatomy. This concept is understood by understanding the tissues response to the product that is being placed, the risks one takes while injecting, and the behavior of the needle and the extrusion force of the product. The basic knowledge of the anatomical location is not enough to be successful in the esthetic industry as a cosmetic injector. The injector must understand all of the planes of the injected structures along with the depths. When it comes to the muscles, their direct effect on the face's dynamic expressions and how it may infer with a person's ability to blink, swallow, and masticate is the greatest concern for the injector. It is of equal importance to have an in-depth understanding of the medications (products) you are injecting and each product's characteristics, syringe, and needle. The face is the canvas, and the product and syringes are the paintbrushes. As the reading continues along, repeating themes will begin to emerge. The goal is not just to show injectors how to perform the injection but for the reader to gain an in-depth understanding of why and, even more importantly, how to avoid negative consequences, called severe adverse events or AE's.

Successful Tips to limit serious adverse events or displeasing results.

1. Know your injection anatomy.
2. Understand dynamic and static expression.
3. Understand the equal and opposing areas that can aid in the remedy to complication post treatment.
4. Know the depths and relative dimensions of the structures you are injecting.
5. Educate yourself with more than just 1 weekend course.
6. Comprehensively understand the pharmokokinetics and pharmacodynamics of the neuromodulator in use.
7. Proper and in-depth facial assessment is key in great outcomes.
8. Strict adherence to state board and regulatory parameters.
9. Know your Scope of Practice for the state you are practicing.
10. Have a comprehensive understanding of the aging process.

With the rise in minimally invasive procedures, we are now seeing trends of complications rising from the otherwise safe procedures. With longer follow-up periods, scientists and statisticians can wrap their heads around the true safety and danger of minimally invasive procedures. Soft gel dermal fillers have presented to be the most dangerous esthetic treatments. The development of occlusions, blindness, stroke, granulomas, infections, and disfigurement of the soft tissue is a catastrophic event post dermal filler treatment. If the word invasive is part of the sentence, it must go without saying that there is still a risk. The greatest risk with minimally invasive procedures comes from the untrained professional or the at-home injectable party. It is imperative that the provider be well educated and additionally trained in esthetic medicine.

Furthermore, the provider needs to be well versed in facial anatomy and safety protocols for the treatment they are performing. Even though providers and researchers are now seeing more and more complications given the sheer number of treatments currently performed, this still has yet to bring about the need for change in injectors' academic preparedness. At home, parties and the YouTube trained professional has left a vulnerable population of patients to suffer and hide in shame with disfigurement and judgment after receiving what they thought was a minimally invasive procedure. The health care providers' job is to be the expert and be sure that a healthy respect is maintained as injectables are still medicine. Esthetics is a softer side of medicine but none the less, still medicine. Esthetic medicine incorporates an art form of beauty and symmetry along side the rigid protocols of medicine. Educating providers within the industry and holding the craft with the highest regard is of the utmost importance to deliver natural and beautiful results. The more injectors understand the risk of complications and how to prevent the adverse outcomes, the better the education will be for our newcomers. Through this process will allow for the development of safer techniques, improved outcomes, and detailed protocols. Eventually, standardizing treatments and safety while still allowing for the artistic creativity of beauty imposed by the individual injectors.

Social media is a wonderful and potent tool in the field of esthetics. Unfortunately, it has also become a negative platform, portraying a false sense of filtered reality. Social media has given a false sense of expectations to consumers and eager new injectors, dismissing the education and understanding necessary to provide safe and exceptional results with non-surgical esthetic procedures. Social media has become a place of filtered reality influencing the perception of what an injector can achieve. There is an enormous disparity between reality and results with social media these days. Many dating apps have disappointed customers when someone other than the image portrayed on the profile picture shows up to the date. The same process is to be considered when cosmetic providers showcase their work. Providers should not be bowing down to this deceptive and deceitful style of marketing. Better yet, providers should focus on the patient instead of being the cliche one uppers of social media. Filtering work to lore customers, it is an unethical and deceitful practice. An injector's goal on social media should be to showcase their work to the public and inspire colleagues. The experienced providers need to set the tone and relay the truth to the public (the patients) on social media, not the filtered version of results. It is time injectors utilize social media to showcase the unfiltered talent and retrain the consumers to expect more realistic outcomes.

Some treatments offered by esthetic providers have minimal downtime or even only take a mere 10 min to perform. However, it certainly does not mean a lack of time was invested in education and further training to deliver the seemingly quick 10-min treatment. An extensive amount of time and money goes into the educational foundation to perform such a task in 5 or so minutes. Erickson purported that it takes 10 years and 10,000 h of practice and dedication to the craft to become a master. With a quick yet eye-catching 60-s video clip, a provider can showcase just how easy and quick a neuromodulator treatment can be. The quick yet eye-catching videos can be extremely misleading to patients and colleagues looking to enter the field. When providers make it look so easy, consumers will begin to question why it is so expensive. It is more than a quick injection in 5 or so minutes. Understanding the dose, the product, the patient goals, the relationship of nearby structures, and unique features to the individual patient makes the treatment a success. Much more goes into the art of neuromodulation and the esthetic procedures than meets the eye on social media.

As for fellow colleagues looking to break into the industry, they begin to think that it is easy and can make so much money in this field. They miss the element of time, dedication to education, and devotion to the industry that is completely lost on the whimsical, eye-catching social media post. Don't be mistaken; social media is necessary and can be very powerful to help elevate an esthetic practice and showcase the work. However, a balance needs to be struck between valuing the education and the ability to showcase the esthetic industry's beauty. Be sure to represent the industry with the highest of regards. Esthetic medicine is medicine, after all. It is well known that social media has helped elevate awareness and excite consumers for treatments, but it perhaps has given a bit of a false sense of ease. With that said, weekend training courses are in abundance, as are poor outcomes and a lack of respect for the industry. Sixteen hours of education is not enough to be an expert or develop mastery within the field of injectables or any type of medicine for that matter. My goal is to begin resetting a standard of entry-level academic programs to get a foot in the door into the esthetic industry. I am working with nursing programs to begin implementing the basics of neuromodulation to set the tone and the standard for academics in the field.

Discounting a medical provider's talent and the sanctity of the grueling educational process is an outright appalling tactic. Non-surgical and minimally invasive procedures such as neuromodulators are still medicine. At no point should a discount have ever even come into play. Think about the surgeons' office patients; they are not offered a buy one get one half of breast augmentation or a buy one get one oophorectomy. I have yet to hear of a cardiac surgeon offering a free carotid endarterectomy with a quadruple bypass. Why is the cosmetic medicine industry compelled to discount their talent, service, and medicine when others in medicine would not dream of offering a flash sale. When these flash sales occur, it can attract the wrong type of client, setting the practice and the injector up for malpractice, at which point an attorney may be needed. Do not be fooled; an attorney is not ever going to give you a special on two litigations over one. Be mindful of the talent, time money spent on the education necessary to perform tasks such as neuromodulators. Respect the profession and the degree necessary to practice medicine.

It is said that a cosmetic injector has a much more luxurious lifestyle and is stress-free. The idea that the esthetics business is a life of luxury is not the case at all. If an injector values and respects the license they hold, they will always be a little nervous and extremely hungry for new knowledge. When I decided to move my career away from emergency medicine and into the esthetic medicine industry, I thought I was giving up a punishing schedule to have a more luxurious life. The thought was that my job would be to point and shoot with a few flicks of a syringe. Little did I know, I would be on call over vacations, nights, and weekends after injecting the patient. The patient is my responsibility the minute I inject them. Our industry is such a small niche and so highly specialized that an emergency room or a primary care provider's office is not the place for the patient to be should an issue arise post treatment. The safest place for a patient experiencing an adverse event is back in the hands of the original provider or, at minimal a provider specializing in this particular field of medicine. Patients can suffer from lymphatic stalling after neuromodulation treatment. The patient may then rush to the emergency room, thinking they are having an allergic reaction or a stroke due to the puffiness from edema. The patient will end up with a huge bill, an unnecessary CAT scan or MRI, and further unnecessary tests and expenses. I thought I was giving up difficult patients in the emergency room and grueling schedules for a comfortable life. I was wrong. However, I love my career, and I would not change it for the world. I just want you, newcomers, to be prepared for the next steps. Days are longer and the education to get into the industry is steep. Patients are savvy, high maintenance, and needy but equally as impressive and vulnerable. Each patient has a special place in my heart; they become your extended family. I adore and respect each one of them. They will complain, but it is up to the provider to learn from the complaint and figure out why they feel the way they do post-treatment. As the injector becomes skilled, the focus is not on the complaint of the patient any longer. The focus is now on how and why the muscles and tissues interact to make the patient feel a particular way. At the beginning of an injector's career, a complaint is terrifying.

At some point, a switch is flipped, and the injector will no longer look at a complaint as negative feedback. The crowning moment is when the injector can hear the words and turn the complaint into an opportunity to better understand the craft and set proper expectations for the patient. The injector will then better serve their patients in terms of education, outcome, safety, and expectations. Patients sometimes teach us more than we learn from a course on injectables. Listen and observe the movements of every patient. Each one of them has a valuable lesson to teach the injector. We need to be humble enough to hear it, smart enough to hear it, and talented enough to deliver the results.

References

Grandview Research. (2020). *Aesthetic medicine market size, share & trends analysis report by procedure type (invasive procedures, non-invasive procedures), by region (North America, Europe, APAC, MEA, LATAM), and segment forecasts, 2019–2026.* doi: 978-1-68038-733-9. https://www.grandviewresearch.com/industry-analysis/medical-aesthetics-market (Accessed August 2020).

Khunger, N. (2015). Complications in cosmetic surgery: A time to reflect and review and not sweep them under the carpet. *Journal of Cutaneous and Aesthetic Surgery, 8*(4), 189–190. https://doi.org/10.4103/0974-2077.172188.

The Aesthetic Society. (2019). *Aesthetic Plastic Surgery National Databank.* https://www.surgery.org/sites/default/files/Aesthetic-Society_Stats2019Book_FINAL.pdf. (Accessed May 2020).

CHAPTER 2

Statistics in esthetics

<u>**Fun facts and statistics about the esthetic industry**</u>

Botox has been the #1 esthetic procedure performed since 1999.

More than ONE BILLION dollars was spent on injectables in 2019.

The top five nonsurgical procedures in 2018 were:

- **Botulinum toxin** (1,801,033 procedures)
- **Hyaluronic acid** (810,240 procedures)
- **Nonsurgical fat reduction** (174,244 procedures)
- **Photo rejuvenation** (135,624 procedures)
- **Chemical peel** (129,596 procedures)

From 2017 to 2018 Neuromodulators grew as an industry with an increase of 16.3%
From 2014 to 2019 Neuromodulators saw an increase of 35.8%
From 2017 to 2018 Dermal fillers grew as an industry of 12.2%
From 2014 to 2019 Dermal filler saw an increase of 58.4%

Woman ages 18–34 are responsible for 16.4% of consumption of the treatments
Woman ages 35–50 are responsible for 32.1% of consumption of the treatments
Woman ages 65+ are responsible for 12.9% of consumption of the treatments

Comparatively speaking, men accounted for 9% of Botox consumption in the United States (The Aesthetic Society, 2019). Furthermore, they are responsible for 4.9% of dermal filler utilization (The Aesthetic Society, 2019).

Statistical data helps providers understand the market industry they are practicing within. Understanding the market will guide providers to proper marketing strategies and lead the provider to the next proper training. Statistics from reputable sources is where the provider should be collecting their

Essentials of Neuromodulation
https://doi.org/10.1016/B978-0-323-89920-8.00028-X

information. Social media has a way of louring one in, but it is often a false sense of reality.

Reference

The Aesthetic Society. (2019). *Aesthetic Plastic Surgery National Databank*. https://www.surgery.org/sites/default/files/Aesthetic-Society_Stats2019Book_FINAL.pdf. (Accessed May 2020).

CHAPTER 3

Understanding esthetic terminology

Let us dig right in and define some of the esthetic industry's common terminologies before we start. Botox is the common term most patients think is the actual service. However, Botox is the branded prescription drug used for neuromodulation (inhibiting particular muscles or muscle groups). Botulinum toxin-A (BoNT-A) is a collective name for neuromodulators, and there are several neuromodulators on the market, such as Botox, Dysport, Xeomin, and Jeuveau. As readers begin to delve through the literature, some articles may use the term neuromodulator, and other articles may use neuromodulation. The term neuromodulator and neuromodulation can be used interchangeably.

However, Botox, Xeomin, Jeuveau, and Dysport may not be used interchangeably both in the literature and in the clinical application of the drug. Patients used to ask for providers to help them get rid of their wrinkles. Now, patients make Botox appointments. Yes, we are still getting rid of the wrinkles, but the service has been replaced with the drug name. It is common for patients to think Botox is the catch-all for anti-aging. Often, a patient will come in with volume loss, and they think they need Botox. Consumers equate Botox with anti-aging and feel like it is the cure-all. They have not yet had enough personal experience to understand that Botox stops the muscles from moving, and filler is for volume.

Understanding the interchangeable terms

Neuromodulation = Neuromodulator

Neuromodulator = Neuromodulation

Neuromodulation = Botulinum toxin A

Botulinum toxin A = Neuromodulation

Neuromodulator = Botulinum toxin A

Essentials of Neuromodulation
https://doi.org/10.1016/B978-0-323-89920-8.00010-2

$$\text{Botulinum toxin A} = \text{Neuromodulator}$$

$$\text{Botulinum toxin A} = \text{BoNT} - \text{A}$$

$$\text{BoNT} - \text{A} = \text{Botulinum toxin A}$$

$$\text{BT} - \text{A}- = \text{BoNT} - \text{A}$$

$$\text{BoNT} - \text{A} = \text{BT} - \text{A}$$

$$\text{Botulinum toxin} = \text{BTX}$$

$$\text{BTX} = \text{Botulinum toxin}$$

$$\text{Botulinum toxin} = \text{BoNT} = \text{BTX}$$

$$\text{BoNT} = \text{Botulinum toxin} = \text{BTX}$$

$$\text{Gram} + \text{anerobic bacterium Botulinum toxin A} = \text{BoNT} - \text{A}$$

$$\text{BoNT} - \text{A} = \text{Gram} + \text{anerobic bacterium Botulinum toxin A}$$

To streamline the terminology, Neuromodulator and BoNT-A will continuously be utilized when referring to all of the FDA forms of Botulinum toxin A for esthetic purposes. When there is a discussion of where Botulinum toxin comes from, BoNT will continuously be used (Fig. 3.1; Table 3.1).

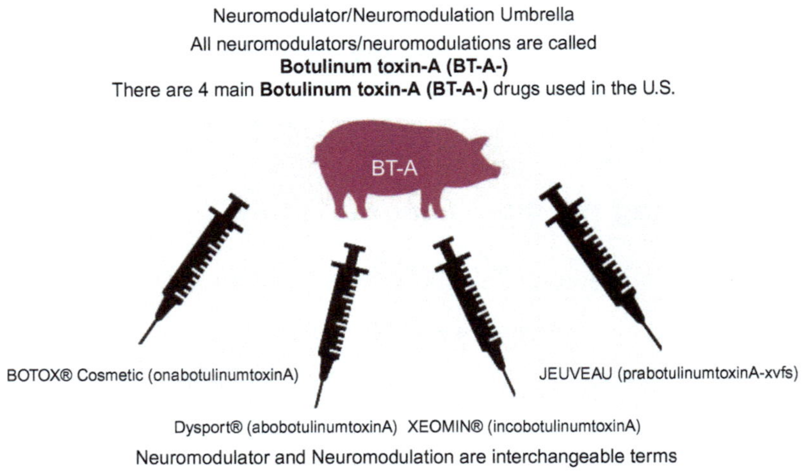

Fig. 3.1 Neuromodulation umbrella.

Table 3.1 FDA approved Botulinum toxin A.

Types of Botulinum toxin A (BoNT-A) FDA approved in the United States		
Trade name	**Generic name**	**Abbreviation**
Dysport	AbobotulinumtoxinA	ABO
Botox	OnabotulinumtoxinA	ONA
Jeuveau	PrabotulinumtoxinA-xvfs	PRAB
Xeomin	IncobotulinumtoxinA	INCO

CHAPTER 4

What is botulinum toxin

The question may have presented itself as to what a neuromodulator is and how it works? BoNT-As are acetylcholine release inhibitors. However, knowing that BoNT-As are acetylcholine release inhibitors is not enough to thoroughly understand the mechanism of action of BoNT-As. To understand the acetylcholine release inhibitor concept, we must first begin to understand the origination point of neuromodulators. That is to understand botulinum toxin (BoNT)s as a whole.

Botulinum toxin is produced by the anaerobic Gram-positive spore-forming bacteria of the genus *Clostridium*. It is a unique and complex mixture of proteins that contain botulinum neuromodulation (BoNT) along with several other non-toxic proteins. BoNTs are reportedly the most potent toxins known to mankind and can cause botulism. Botulism will leave an animal or human in a flaccid paralyzed state should it be ingested orally. Ironically enough BoNT containing medicines are used in traditional medicine and cosmetic medicine safely and with great success without completely paralyzing their host. Science has provided enough research to narrow down and isolate BoNTs to be delivered precisely to avoid harming the host entirely and just causing temporary paralysis of targeted muscles or muscle groups. One who contracts botulism does not inject it into their muscles for wrinkle reduction; better yet, they consume it due to the contamination of a food source with *Clostridium botulinum*. *Clostridium* spores can be found worldwide in food, soil, freshwater, saltwater, sediment, and various animals' intestinal tracts.

There is notably, seven serotypes of BoNT produced by different strands of *Clostridium botulinum* A–G. Man kinds nervous system is susceptible to the strands A, B, C, E, F, and G and unaffected by D. It is types A and B that is widely used in clinical application due to the long-lasting effects compared to other types. The following chapters will focus heavily on botulinum toxin type A and briefly touch upon type B to learn esthetic medicine.

Essentials of Neuromodulation
https://doi.org/10.1016/B978-0-323-89920-8.00006-0

Understanding the nervous system

To better understand how BoNT-A works is part of learning to become a proficient and eventually excellent cosmetic injector. We all learn differently, so some sections of this book will be more manageable to digest than others. The book is meant to be reread and used as a lifelong reference. For those visual learners, it may be the pictures that are most beneficial, or for the statisticians, the graphs and charts may be of most significant use in the beginning. No matter the learning style or comfort level, all of the information provided in this book is necessary to achieve mastery in esthetic medicine. It will all begin to come together at some point. Please do not fret when it does not seem to sink in the right out of the gate. Learning takes time, exposure, practice, and lots of repetition.

Let us begin to understand the nervous system bit by bit to break down the complete mechanism of action of BoNTs. The nervous system is a very compartmentalized yet complex complete system. It is responsible for coordinating the human body's internal actions, electrically, chemically, functionally, and sensory wise. Communicating messages chemically and electrical are sent through the nervous system to connect the different body systems. For example, if one has a bright light shown in their eyes, the light triggers a response in the body via the nervous system to constrict the pupil and excite the reflex to scowl. There is such a delicate balance of systems run by the nervous system's electrical and chemical signals.

There are two main nervous system divisions, the central nervous system (CNS) and the peripheral nervous system (PNS). The CNS consists of the brain and the spinal cord. The PNS includes the spinal nerves and the cranial nerves and senses, such as taste, smell, vision, hearing, balance, and the autonomic nervous system (ANS). The ANS is further subdivided into two groups called the sympathetic ANS and the parasympathetic ANS. The sympathetic nervous system is responsible for an individual's fight or flight response. The PNS is responsible for the bodies rest and digest, or feed and breed as some refer to it. One can infer the PNS functions and develop an understanding of what functions it is responsible for with the very mention of peripheral. Sensory and motor axons make up the majority of the PNS function. The sympathetic nervous system excites the body as the parasympathetic nervous system tries to restore harmony within the body. Botulinum toxin action is primarily happening within the ANS.

Moving along to developing an understanding of BoNT-A's action, it is necessary to be specific within the nervous system and develop an understanding of the terminals and the chemical and electrical action. At the junction between the muscle and the nerve is a particular space where the magic of BoNT-As occurs, and it is called the neuromuscular junction. Let us move forward to the beginning, understanding the drug's connection and how it is working within the body to impact the muscles and the nervous system.

As we were refreshed with the previous paragraphs' nervous system function, we were reminded that muscles are connected to the brain via the nervous system's route for communication. The nervous system is a highly complex network of neurons. Neurons are long cells that pass the information along using either electrical or chemical signals. Chemical signals pass between neurons and muscles through synapses. Synapses are considered special linking cells. Neurotransmitters such as acetylcholine are used to pass the message along through the synapses. Bont-A works by blocking the release of acetylcholine at the neuromuscular junction, thus preventing the transmission of messages from the nervous system to the muscle.

Botulinum toxin injections do not cross the blood-brain barrier. Its effect is only on the peripheral nerves, thus increasing its safety profile.

Acetylcholine plays a vital role in the effect of BoNT-A. Therefore, providers must understand the role of acetylcholine in the use of BoNT-A. We must continue to break down the nervous system's function so that a complete understanding of the BoNTs' mechanism of action can be concluded. That leads us to break down the question of what is acetylcholine? If BoNT is an acetylcholine release inhibitor, we now need to understand acetylcholine's role within the nerve. Acetylcholine acts to transmit nerve impulses within the parasympathetic nervous system. Remember that the parasympathetic nervous system is a part of the autonomic nervous system. The autonomic nervous system is responsible for the induction of secretion, *contracts smooth muscles*, and dilates blood vessels.

Acetylcholine is a neurotransmitter housed at the neuromuscular junction. It is stored at the end of the neuron tucked inside of a vesicle. The process of muscle movement occurs due to the snare complex that occurs once a message is received. There are three main proteins involved in the snare complex; Syntavin (connects to the nerve membrane), Synaptobrevin (connects to the vesicles), and Snap-25 (helps the other proteins link up). These proteins join together to cause the vesicle to move toward the nerve membranes in order for fusion to occur. Then the acetylcholine can be released across the synapses and passes to the muscle fibers. Now, the muscles will contract and shorten in the face of the message being received.

Mechanism of action of BoNT-A

Now that we understand that botulinum toxin A is an acetylcholine release inhibitor, we can move forward with our learning. BoNT-A is a neuromuscular blocking agent used for the temporary improvement of dynamic rhytids. Having gained the knowledge that BoNTs are neuromuscular blocking agents will have great value regarding proper medical history and patient selection for esthetic injections with BoNTs. BoNT-A and BoNT-E work by cleaving the synaptosome-associated protein (SNAP-25), a presynaptic membrane protein required for the fusion of neurotransmitter-containing vesicles.

Botulinum toxin A rapidly binds with the neuron membrane, and then it splits into the active light chain and a heavy chain. Botulinum toxin A has a high affinity to cleave snap 25. The affinity to cleave snap 25 prevents the SNARE proteins from binding so that acetylcholine can not reach or pass through the neuron membrane. Therefore, the acetylcholine does not reach the muscle, thus inhibiting the muscle from contracting, causing a neuromuscular blocking effect. BONT's direct effect is the inhibition of the cholinergic autonomic innervation of the exocrine glands and smooth muscles.

This mechanism of action has laid the foundation for the development of the toxin as a therapeutic tool in esthetic medicine and several other fields of medicine related to neuromuscular disorders.

Step by step on how BoNT-A work

It is suggested that there are three phases in the BoNT process:

Phase 1: Binding
Phase 2: Internalization
Phase 3: Blockade

1. BoNT-A is injected into the muscle.
2. The accessory proteins are disassociated from the BoNT-A via the action of proteases.
3. The active chain is released.
4. **Phase 1** is called the binding phase. The binding phase is when the active BoNT is bound to the peripheral cholinergic nerve terminals within the neuromuscular junction with a high affinity and specificity.
5. Then internalization occurs where the BoNT-A molecule is in the nerve terminal through receptor-mediated endocytosis. This is considered to be **Phase 2** of the internalization process.

6. The light chain is released into the cytoplasm. The dissociation of the disulfide bond accompanies the release of the light chain.
7. The light chain performs its metalloproteinase activity that regulates the exocytosis of acetylcholine vesicles. The targets are the Snare complex.
8. **Phase 3** is the blockade phase. The blockade phase is when Snap 25 is now cleaved, thus blocking the release of acetylcholine. Cleavage of the snare proteins (the light chain cleaves SNAP-25), which results in the blockade of neurotransmitter release and, therefore, neuroparalysis. This inhibition of the synaptic exocytosis and incapacitated neuronal transmission blocks the release of acetylcholine at the neuromuscular junction, which directly blocks the muscular contraction.
9. Now the muscle can not contract as the message delivery system has been shut down with the inhibition of acetylcholine release (Fig. 4.1).

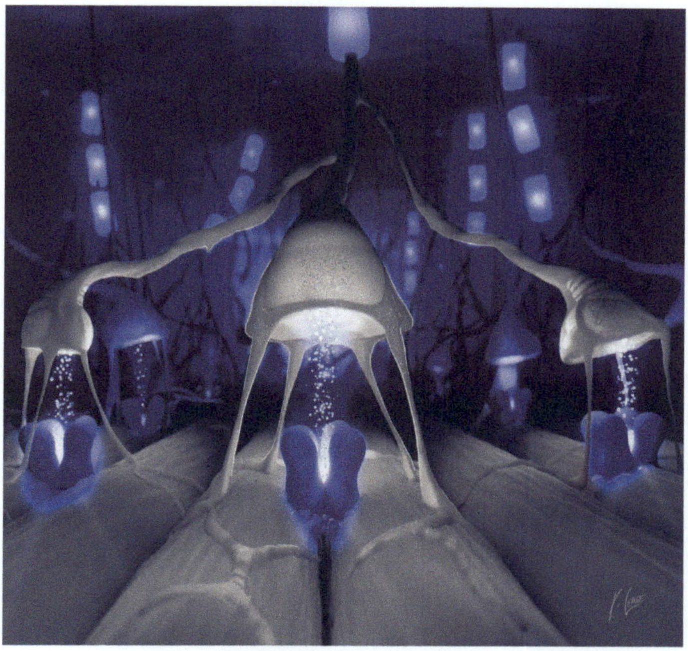

Fig. 4.1 As seen in this photo, the presynaptic membrane releases the acetylcholine down to the acetylcholine receptor in the muscles. The release of acetylcholine allows the muscle to contract. Once acetylcholine is blocked from being released, the receptor is not infused with acetylcholine, thus prohibiting the message of contractility to occur within the muscle.
What we have at the end of this process is chemical denervation.

Nerve terminal with release of acetylcholine

Some may still be asking what the previous chapter meant ... Essentially, BoNT-A works on muscles by inhibiting the muscles motions due to interference with the nervous systems' chemical reaction. The primary effect of BoNT-A is on the nerve, and the secondary effect is on the muscle; the last effect is on the skin. Alternatively, shall we say the last effect is on the patient's mood because after the BoNT-A has taken effect, they love their results and reflection, promoting a happy mood. When the muscle can not compress and squeeze the skin, it does not weaken the elastin, collagen, and hyaluronic acid, thus preventing the onset of the deep-set at rest lines (otherwise known as static rhytids) and wrinkles seen on the aging face. When BoNT-A is injected into the muscle, the muscle remains in working order while the nerves impulse is paused.

To break it down even further and put it in layman terms, BoNT-A disrupts the communication between one's emotion, and the muscle, along with the nervous system. It is said to decrease the muscle's action potential. If one is feeling disgruntled, it may be common to furrow the brows based on that emotion. When a patient has had a neuromodulator injected into the muscle between the brows where the physical manifestation of emotion may show up, the ability to furrow will be diminished. The decrease in animation is because the neuromodulator impedes the muscle's ability to respond to the emotion as the nervous system has been temporarily disabled. The results are temporary and are a reversible inhibition of neurotransmitter release.

Botulinum toxin A takes approximately one and a half to two hours to bind to the nerve. The paresis effect occurs over the next 2–7 days after injections, peaking at about 7–10 days. Completing the nerve impulse impedance occurs at 2 weeks. At this time is when a touchup or retreatment is appropriate. No matter how much the patient begs and pleads, DO NOT retreat before 2 weeks. The rule of 2 weeks is due to the incomplete mechanism of action before 2 weeks time frame. If the patient has retreated before the 2-week mark, an unintended result may ensue.

The medication peaks at about 6 weeks and will start to decline slowly until full reinnervation is restored. Reinnervation is when a slight movement is becoming noticeable again in the previously treated areas with BoNT-A. The decline of the medication's effect varies; however, it is typically noted to remain in effect for 3–4 months. It has been reported to last 6–12 months; however, this is rare. At which time, muscle movement will begin to return.

The recovery of one's ability to furrow or squint again will return, typically 3–4 months post-treatment with BoNT-A. Recovery occurs through proximal axonal sprouting and muscle reinnervation by the formation of a new temporary neuromuscular junction. Eventually, the original neuromuscular junction regenerates. The recovery phase happens in two stages. The first stage involves exocytosis, meaning the nerves resprout, which aids in the neurons' recovery. The second phase is the sprouts regress, and the original terminal is restored. BoNT-A is entirely reversible with time making it extremely safe as a therapeutic and or cosmetic therapy. Reports are reproducible in that there is no atrophy in the nerve endings upon blockade of neurotransmitter even in the face of repeat treatments (Dolly & Aoki, 2006).

The onset of action is one and a half hours to bind to the nerve.
The onset of clinical results is 2–7 days.
The final chemical denervation is 14 days.
Duration for facial esthetics is 6–12 weeks (product and dose vary).
Duration for hyperhidrosis is 6–9 months.

Review of the variety of BoNT-As commercially available in the United States

BoNT-A products have several variations in preparation, which means they are not necessarily interchangeable, and their intended use must be modified based on product choice, reconstitution, and the intended esthetic goal. The dose and volume play a role in the outcomes in the neuromodulation treatment (in other words, the volume of reconstitution of the product).

Currently, the US FDA has four approved products on the mainstream market in cosmetic practice.

1. Botox Cosmetic is the trade name for onabotulinumtoxinA and is a prescription drug made by Allergan. BOTOX Cosmetic (onabotulinumtoxinA) is indicated in adult patients for the temporary improvement in the appearance of:
   - **(i)** Moderate to severe glabellar lines associated with corrugator and procerus muscle activity.
   - **(ii)** Moderate to severe lateral canthal lines associated with orbicularis oculi activity.
   - **(iii)** Moderate to severe forehead lines associated with frontalis activity.

     Allergan (2020) (https://www.botoxcosmetic.com/what-is-botox-cosmetic/botox-cosmetic-history)

2. Dysport (abobotulinumtoxinA) is a prescription drug for injection made by Galderma. It is an acetylcholine release inhibitor and neuromuscular blocking agent indicated for the temporary improvement in the appearance of moderate to severe glabellar lines associated with procerus and corrugator muscle activity in adult patients less than 65 years of age.

   Dysport (2020) (https://www.dysportusa.com/healthcare-professionals#importantsafetyinformation)
3. Merz Aesthetics makes Xeomin (incobotulinumtoxinA) for injection. It is used for intramuscular use and is indicated for the temporary improvement in the appearance of moderate to severe glabellar lines with corrugator and procerus muscle activity in adult patients.

   Xeomin (2020) (https://www.xeominaesthetic.com/professionals/)
4. Jeuveau (pronounced Jū vō) (prabotulinumtoxinA-xvfs) for injection is made by Evolus and goes by Newtox. Jeuveau is a prescription medicine injected into muscles and used in adults for a short period (temporary) to improve moderate to severe frown lines between the eyebrows (glabellar lines). It is not known if Jeuveau is safe and effective for use in children.

   Jeuveau (2020) (https://jeuveau.evolus.com)

   Each formulation can vary by the following characteristics:

1. Amount of neuromodulator
2. Units and or concentration of the product
3. Composition
4. Accessory proteins
5. Toxin complex size
6. Molecular weight
7. Biological activity
8. pH
9. Storage
10. Risk of antigenicity
11. Indication for use
12. Geographical distribution
13. Chemical properties

Each FDA approved formulation of BoNT-A has specific indications for reconstitution, limited indications and are not universal worldwide. Each neuromodulator is manufactured differently and has unique characteristics, which makes the products not universally interchangeable. These drug formulations are not interchangeable in preparation nor delivery and or efficacy. When treating a patient with four units of Botox, it is equivalent to four units of Xeomin, four units of Jeuveau, and ten units of Dysport as long as they are reconstituted according to the branded drugs package insert. There is a variety of reconstitution options which can change the formula unit by unit.

BoNT-A effect is correlated with the concentration, dose, placement of the product, amount of spread, and is relative to the depth of the procedure and the size and health of the target muscle. The products vary mainly in diffusion and spread; however, efficacy is still being debated. Most literature reports and demonstrates that Botox is still the leader concerning efficacy and longevity. The onset of action is also not universally validated and is mainly dependent on reports by the practitioner and the patient. The patient's genetics, target muscle size, and thickness, skin condition regarding the thickness or amount of laxity all play a role in the satisfaction rate of BoNT-A. The onset and outcome of treatment are further influenced by the practitioner's choice of a particular drug, the reconstitution process, placement, depth of injection, and dosing.

Given that BoNT-A is a metalloprotease drug, it is necessary to have an ample supply of zinc at the nerve terminal in order for effect. So, a patient's zinc level also plays a role in the effect and duration of treatment. There is more to the treatment than just learning injection points. The patient's overall health and function are equally as important as understanding where to inject and how to inject. Despite increased public demand for the BoNT-A, the pharmacological understanding of the commercially available products' differences is still vastly poorly understood.

Surprisingly enough, despite the overwhelming statistics on consumer utilization, there is still some debate on the reconstitution of BoNT-A. The impact of reconstitution has on the denervation process is still debatable. Although well studied, the makers of BoNT-A may have overstated the fragility of the product initially with the new research that has since come out after the products received their FDA approval. When the product is reconstituted with normal saline versus 0.9% benzyl alcohol saline, the patient experiences more discomfort. When the BoNT-A is reconstituted with the preserved saline with benzyl alcohol, the patients experienced less discomfort, and it was reported that there was less of an incidence of post-injection infection. Many studies have refuted that the preservatives break down the BoNT-A and that the preserved saline may be a better option. Most injectors around the globe do reconstitute their BoNT-A with the preserved saline over the non-preserved saline. Greater detail will be provided later in the text on how to reconstitute each product properly. All FDA and commercially prepared BoNT-A's will have a package insert on the FDA approved areas for use and adequately reconstituting the product for injection.

Cosmetic and therapeutic uses of BoNT-A

BoNT-A is used in various medical settings for a host of diseases, symptoms, and beauty treatments.

Cosmetic uses

- Hyper functional facial lines
- Brow ptosis
- Hyperhidrosis
- Facial contouring
- Lateral canthal lines
- Beautification by way of symmetry

Therapeutic use in ophthalmology

- Strabismus
- Blepharospasm
- Nystagmus

Therapeutic use in neurology

- Hemifacial spasm
- Facial asymmetry
- Oromandibular dystonia
- Cervical dystonia

Therapeutic use in otolaryngology

- Vocal ties
- Stuttering
- Spasmodic dysphonia
- Oromandibular dystonia

Therapeutic use in pain

- Migraine
- Tension headaches
- Knee
- Shoulder
- Neuropathic pain
- Chronic low back pain

Therapeutic use in autonomic dysfunction

- Freys syndrome
- Sialorrhea
- Rhinorrhea

Therapeutic use in gastroenterology

- Gastroparesis
- Delayed gastric emptying
- Chronic anal fissure
- Achalasia

References

Allergan. (2020). *Botox Cosmetic history*. https://www.botoxcosmetic.com/what-is-botox-cosmetic/botox-cosmetic-history. (Accessed March 2020).

Dolly, J. O., & Aoki, K. R. (2006). The structure and mode of action of different botulinum toxins. *European Journal of Neurology*, *13*(Suppl. 4), 1–9. https://doi.org/10.1111/j.1468-1331.2006.01648.x. 17112344.

Dysport. (2020). *For health care professionals*. https://www.dysportusa.com/healthcare-professionals#importantsafetyinformation. Accessed March 2020.

Jeuveau. (2020). *Ready for a modern-made tox?*. https://jeuveau.evolus.com (Accessed July 2020).

Xeomin. (2020). *A uniquely purified choice for frown lines*. https://www.xeominaesthetic.com/professionals/. (Accessed March 2020).

Further reading

Stone, H. F., Zhu, Z., Thach, T. Q. D., & Ruegg, C. L. (2011). Characterization of diffusion and duration of action of a new botulinum toxin type A formulation. *Toxicon*, *58*(2), 159–167. Science Direct https://doi.org/10.1016/j.toxicon.2011.05.012. Accessed May 2020.

CHAPTER 5
History of neuromodulators

In the early 1800s, Justinus Kerner and John Muller first describe botulism. It was referred to as "sausage poison" Then *Clostridium botulinum* (BoNT) was later isolated by Emile van Ermengem in 1895. Emile was a bacteriologist professor in Belgium who discovered the bacteria after a dinner that caused an outbreak of illness, at which time botulism was identified. Eighty years before the outbreak, there were the beginning rumblings of chatter about food born illness. Botulism received its name due to the Latin word for sausage, which is botulus. The name botulus was given because the food that was suspected of causing the outbreak was pork filled sausage. Fast forward to 1946, and at the time, the purification process of onabotulinumtoxinA (Botox) occurred. Dr. Edward J Schantz succeeded in purifying BoNT-A in crystalline form cultured *Clostridium botulinum* and isolated the toxin. In 1950, Dr. Vernon Brooks discovered the toxins' true potential when he concluded that it prohibited acetylcholine release and thus impeded motor nerve innovation in a hyperactive muscle. In 1980 Dr. Alan B Scott revolutionized the world of ophthalmology with the treatment of strabismus. It was only 9 years later that onabotulinumtoxinA received its first FDA approval. The utilization of onabotulinumtoxinA in strabismus was just one of many FDA approvals; this particular brand would get FDA approval.

The 1989 FDA approval set the stage for many great things to come with this unique and extremely deadly toxin. In 1989, onabotulinumtoxinA received the FDA approval for therapeutic use in Strabismus and Blepharospasm and hemifacial spasm in patients younger than 12 years. That same year it was rebranded as Botox, which is the drug name consumers recognize. Through expanded research into esthetic uses, Allergan later commenced clinical trials to study the treatment's safety and efficacy for moderate to severe frown lines (Allergan, 2020).

Then in 2002, the first treatment was approved for cosmetic use. On 15 April 2002, the FDA approved Botox as a temporary esthetic treatment for moderate to severe frown lines in adults and requested the product be marketed as Botox Cosmetic to distinguish esthetic from therapeutic uses.

Essentials of Neuromodulation
https://doi.org/10.1016/B978-0-323-89920-8.00016-3

It is the first treatment of its kind to be approved for esthetic use in three different areas. Today it is the only product FDA-approved to temporarily make moderate to severe frown lines, crow's feet, and forehead lines look better in adults (Allergan, 2020). The FDA has approved others; however, Botox is the only one with three cosmetic use indications.

History of onabotulinumtoxinA (Botox), incobotulinumtoxinA (Xeomin), abobotulinumtoxinA (Dysport), prabotulinumtoxinA-xvfs (Jeuveau)

1895—*Clostridium botulinum* (BoNT) was isolated by Emile van Ermengem.

1946—Purification process of onabotulinumtoxinA (Botox). Dr. Edward J Schantz succeeded in purifying BoNT-A in crystalline form cultured *Clostridium botulinum* and isolated the toxin.

1949—Discovery that botulinum toxin blocks neuromuscular toxin by Dr. Burgen's ASV group.

1950—Discovery that botulinum toxin blocks the release of acetylcholine from motor nerve innovation when injected into a hyperactive muscle by Dr. Vernon Brooks.

1980—Strabismus was treated in humans for the first time by Dr. Alan B. Scott.

1989—FDA approval for use of Strabismus and Blepharospasm and hemifacial spasm in patients aged younger than 12 years. This was done under the trade name of Botox.

1990—BoNT-A received US FDA approval for treatment of cervical dystonia.

2001—United Kingdom approved onabotulinumtoxinA (Botox), synthesized by Allergan, for axillary hyperhidrosis. Canada approved onabotulinumtoxinA (Botox) for axillary hyperhidrosis, focal muscle spasticity, and cosmetic treatment of wrinkles at the brow line.

2002—US FDA approval of onabotulinumtoxinA (Botox) Cosmetic to temporarily improve the appearance of moderate to severe frown lines between the eyebrows (glabellar lines).

2011—US FDA approved incobotulinumtoxinA (Xeomin) for temporary improvement in the appearance of moderate to severe glabellar lines in adult patients.

2004—FDA approved for Axillary Hyperhidrosis under the trade name of Botox.

2009—FDA approved a new botulinum toxin A; abobotulinumtoxinA under the trade name of Dysport. It had been used in the United Kingdom since 1994. It was approved for the use of treatment of adults with cervical dystonia and Glabella region.

2010—US FDA approved onabotulinumtoxinA (Botox) to treat spasticity in the flexor muscles of the elbow, wrist, and fingers in adults with stroke, traumatic brain injury, or the progression of multiple sclerosis.

US FDA announced the approval of incobotulinumtoxinA (Xeomin) for the treatment of adults with cervical dystonia, to decrease the severity of abnormal head position and neck pain in both botulinum toxin-naïve and previously treated patients and for blepharospasm in adults previously treated with onabotulinumtoxinA (Botox).

US FDA approved onabotulinumtoxinA (Botox) injection to prevent headaches in adult patients with chronic migraine.

2011—US FDA approved onabotulinumtoxinA (Botox) injection for the treatment of urinary incontinence due to detrusor overactivity associated with a neurologic condition (e.g., spinal cord injury, multiple sclerosis) in adults who have an inadequate response to or are intolerant of an anticholinergic medication.

2013—US FDA approved onabotulinumtoxinA (BOTOX) for the temporary improvement in the appearance of moderate to severe lateral canthal lines, known as crow's feet. This is the only FDA-approved drug treatment option for lateral canthal lines.

2015—US FDA approved abobotulinumtoxinA (Dysport) for the treatment of upper limb spasticity (ULS) in adult patients to decrease the severity of increased muscle tone in elbow flexors, wrist flexors and finger flexors. US FDA approved incobotulinumtoxinA (Xeomin) for the treatment of upper limb spasticity (ULS) in adult patients.

2017—US FDA approved onabotulinumtoxinA (Botox) for the use of forehead rhytids.

2019—US FDA approved prabotulinumtoxinA-xvfs (Jeuveau) for injections into facial muscles to improve the appearance of moderate to severe frown lines between the eyebrows.

Reference

Allergan. (2020). *Botox Cosmetic history*. https://www.botoxcosmetic.com/what-is-botox-cosmetic/botox-cosmetic-history. (Accessed March 2020).

Further reading

Jabbari, B. (2016). History of botulinum toxin treatment in movement disorders. *Tremor and Other Hyperkinetic Movements (New York, NY)*, *6*, 394. https://doi.org/10.7916/D81836S1.

CHAPTER 6

Safety of neuromodulators

After seeing how long BoNT-A has been around, it is easy to say it is a well-studied drug and has a proven safety track record. Further understanding, it does not cross the blood-brain barrier and works locally at the peripheral nerve terminal continues to prove its safety in the use of esthetics. However, there is always the risk with the BoNT-A drug in the untrained provider's hands. Nevertheless, in the hands of skilled and trained injectors, BoNT-A's are very safe and can bring consumers a lot of joy, control over the aging process, and confidence. The FDA held a 13.5-year review of adverse events with the use of Neuromodulators. Throughout this review, it was identified that there were 36 severe adverse events, six of which were more of a user error (Coté, Mohan, Polder, Walton, & Braun, 2005). During this time, there were no deaths related to the use of neuromodulators. The results concluded that 25% of patients would bruise when having their crows feet treated (Coté et al., 2005). 15% of patients will have a headache post-injection (Coté et al., 2005). Very rarely will someone report the headache as severe and debilitating. 3–5.4% of patients suffer from lid ptosis from inadvertent injection into the levator palpebrae. 5% of patients will suffer a lateral tail droop (Coté et al., 2005).

Little nuisances within the injecting craft can make a world of difference in the outcomes and comfort during the injections. The use of fixed needles to the syringe to gain access to the product is contraindicated as the needle will become blunt very quickly, and then the injections will be more painful and will also cause more trauma to the patient's tissue, increasing bruising. As trained professionals versed in injection techniques, we can anticipate a bruise and see it as a usual occurrence. The esthetic patient does not see the bruise as a worthy and necessary risk to the treatment. 99% of the patients will expect their injector to penetrate the skin 60 times during treatment and miss every single one of their vessels. Here is an excellent time to remind the reader to correctly set the treatment expectations, duration, and downtime, including bruising. Coté et al. (2005); study demonstrated that only 36 patients in 13.5-year review had an adverse event. However, the adverse

Essentials of Neuromodulation
https://doi.org/10.1016/B978-0-323-89920-8.00017-5

events were minor comparatively speaking to other treatments such as dermal filler implantation.

Caution with the silicone stoppers

It is thought that when the needle punctures through the silicone stopper of the BoNT-A bottle with the same needle intended for injection that the needle becomes infiltrated with silicone molecules, which is then inadvertently injected into the patient's tissue. Silicone can create a foreign body reaction within the patient's tissue and have adverse long term sequela. Some research has identified that the presence of silicone granulomas and siliconoma can be the direct result of utilizing the same needle that has gone through the silicone stopper (Stone, Zhu, Thach, & Ruegg, 2011).

Possible side effects with ALL neuromodulators

*Swelling at the injection site
*Bruising at the injection site
*Discomfort (temporary) at the injection site
Dry mouth
Tenderness in the neck
*Headache
Double vision
Dizzy
*Faint
Itching
Rash
Wheezing
Asthma flair
Allergic reaction
Hives

***Indicates the most common yet benign side effects**

Possible *emotionally* charged side effects

Not effective
Not strong enough
Results not as expected

Contraindications with the use of ALL neuromodulators

(Absolute contraindications)

1. Hypersensitivity
   Hypersensitivity to any botulinum toxin preparation or any of the components in the formulation. Severe and immediate hypersensitivity reactions have been reported. These reactions include anaphylaxis, serum sickness, urticaria, soft-tissue edema, and dyspnea. If such a reaction occurs, further injection of botulinum toxin should be discontinued, and appropriate medical therapy immediately instituted. One fatal case of anaphylaxis has been reported in which lidocaine was used as the diluent, and consequently, the causative agent cannot be reliably determined.
   (Allergan, 2020; Dysport, 2020; Jeuveau, 2020; Xeomin, 2020)
2. Infection or rash
   Active infection or rash at the injection site is generally a contraindication in almost any esthetic procedure unless, of course, the injectionist is treating the complication.
   (Allergan, 2020; Dysport, 2020; Jeuveau, 2020; Xeomin, 2020)
3. Pregnancy or lactating
   Currently, neuromodulators are listed as a pregnancy category C. There have been reports that woman have inadvertently received neuromodulation while pregnant, and to this date, no teratogenicity has been reported. The reported cases have had uneventful deliveries. Given the procedures' elective nature and that, proper research has not been performed to concretely demonstrate pregnant and lactating women's safety. In animal studies, fetal abnormalities were reported in the growth and development phases of pregnancy. It was reported that a decrease in fetal weight along with skeletal ossification occurred. The expert opinion is that it is not worth the patient's risk, the unborn fetus, or a professional's license to treat a known pregnant or lactating mother.
   (Allergan, 2020; Dysport, 2020; Jeuveau, 2020; Xeomin, 2020)
4. Allergy to cows milk *protein* with the use of Dysport is an absolute contraindication. A true cows milk allergy is different from what most people will report as an allergy to milk, otherwise called lactose intolerant. Dysport is formulated with a trace amount of a protein that is contained in cows milk. Therefore, if the patient is allergic to cows milk, they may not receive Dysport.
   (Dysport, 2020)

5. If the patient has a known sensitivity to human albumin, they may not use any commercially available neuromodulators as it is reported that they all contain human albumin. Even though there is a warning of the possibility to spread viral diseases such as Creutzfeldt Jakob disease (a fatal brain degenerative disease), there has not been a single reported case of the transmission to date. It only comes as a warning at this point to injectors. The idea that neuromodulators contain human byproduct is also an essential piece of information or point to make for particular religious preferences. Individual religious preferences will not want any injections or consume anything that has components of another human. It is advised to specifically ask about someone's religious preference so that the provider can be sure that they are appropriately educated and treating the patient according to their beliefs.

   (Allergan, 2020; Dysport, 2020; Jeuveau, 2020; Xeomin, 2020)
6. Neurological disorder such as:
   - Amyotrophic lateral sclerosis (ALS, aka Lou Gehrig's disease)
   - Myasthenia gravis
   - Lambert Eaton

The use of any neuromodulators may very well exacerbate the clinical effects. Individuals with peripheral motor neuropathic diseases may be at an increased risk of clinically significant effects, including generalized muscle weakness, diplopia, ptosis, dysphonia, dysarthria, severe dysphagia, and respiratory compromise from any of the commercially available neuromodulations. Typically patients with these types of delimitating diseases are not seeking out treatment for esthetic reasons. With that said, medicine is changing so rapidly, and lives are being extended, and cures are happening daily.

(Allergan, 2020; Dysport, 2020; Jeuveau, 2020; Xeomin, 2020)

Contraindications for the injector to consider

1. Thick sebaceous skin—Thick sebaceous skin can be prohibitive in achieving beautiful results with neuromodulators. The thick skin limits the ability to substantially minimize glabellar lines with the use of neuromodulation. Patients with thick sebaceous skin will often need more neuromodulation than the average dosing along with concomitant treatments such as skin resurfacing, prescription-strength skincare such as ZO

Skin health. DCCM™ strictly uses the ZO Skin health line because it is aggressive yet promotes skin health. The focus should be on the whole patient, not just one wrinkle.

2. Prior CO_2 resurfacing or eye surgery is undoubtedly cause for pause. It is essential to stop all neuromodulators 3 months before the patient's eyelid surgery. The surgeon must have a complete and proper scope of the patines natural and unopposed facial function. A patient who has recently had a CO_2 treatment or aggressive skin tightening treatments also causes pause. Allowing the patient to have ample healing time is necessary. The skin can be swollen and tight, with some of these procedures masking the muscle movement's accurate anatomical representation. They are also at greater risk of infection during their healing phase. A slow restorative journey is typically the safest for the patient and the provider.
3. Upcoming special events—This is a problematic situation as many patients come in a little too late to have things done. It is a common misconception that esthetic services deliver quick results. The continued misconception is that they are relatively inexpensive and do not pose any risk of downtime. The harsh reality is that neuromodulation comes with the risk of bruising and takes 2 weeks to heal. The minimal risk of bruising is really the only sequela in the hands of a trained professional. If the injector needs to tweak a treatment, the final result is 4 weeks from the initial injection. It is a delicate balance here with special events. It is common practice to begin a brides treatment plan about a year from the date. A wedding is too essential of a day to begin practicing treatments. The photos will be her only moments from that day, and if she had a botched esthetic procedure, the injector would be thought of in a negative light every time that album is picked up. All of her friends and family will be told the horrific story of how the injector ruined her one special day in her life. A bride is very stressed out and will typically have unrealistic expectations of perfection. It is advised that the injectionist use their judgment with regards to upcoming special events.
4. Cardiovascular disease

   There have been a limited number of reports following the administration of botulinum toxin A of adverse events involving the cardiovascular system. The reports included arrhythmia and myocardial infarction; some cases did end with fatal outcomes. The patient reported having issues post neuromodulation and suffered a cardiac event with previous risk factors, including pre-existing cardiovascular disease. In expert

opinion, the literature is not supportive at this time to demonstrate that the neuromodulation caused the event. However, with the use of any drug, it is advised that caution be used when administering the medication to patients with pre-existing cardiovascular disease. One fatal case was reported in the absence of cardiac history post neuromodulation. After further review, it was found that the provider reconstituted the neuromodulator with lidocaine instead of saline or preservative saline. This particular isolated instance of injector error was not related to the neuromodulator.
(Allergan, 2020; Dysport, 2020; Jeuveau, 2020; Xeomin, 2020)

5. Pre-existing conditions at the injection site
Caution should be used when botulinum toxin treatment is used in the presence of inflammation at the proposed injection site(s), ptosis, or when excessive weakness or atrophy is present in the targeted muscle(s). Scaring, trauma, swelling, and many underlying issues at the injection site can lead to an adverse event.
(Allergan, 2020; Dysport, 2020; Jeuveau, 2020; Xeomin, 2020)

6. Breathing or swallowing disorders
Treatment with neuromodulators can result in swallowing or breathing difficulties. Patients with pre-existing swallowing or breathing difficulties may be more susceptible to these complications. In most cases, this is a consequence of the weakening of muscles in the injection area that is involved in breathing or oropharyngeal muscles that control swallowing or breathing. Deaths as a complication of severe dysphagia have been reported after treatment with botulinum toxin. These reports of death were not in the esthetic client. They were reported under the use of therapeutic medical treatment. Much higher doses are given in these patients, and they also receive the neuromodulator in higher risk areas that control pertinent muscles related to the stability of neck muscles and airway. Dysphagia may persist for several months and require the use of a feeding tube to maintain adequate nutrition and hydration. Aspiration may result from severe dysphagia and is a particular risk when treating patients in whom swallowing or respiratory function is already compromised.
Treatment with botulinum toxins may weaken neck muscles that serve as accessory muscles of ventilation. The inadvertent injection into the wrong area may result in a critical loss of breathing capacity in patients with respiratory disorders who may have become dependent upon these accessory muscles. There have been post-marketing reports of severe

breathing difficulties, including respiratory failure. When treating the esthetic client's platysmas band, this is an inherent risk.
Patients with smaller neck muscle mass, and patients who require bilateral injections into the sternocleidomastoid muscle for cervical dystonia treatment are at greater risk for dysphagia. Limiting the dose injected into the sternocleidomastoid muscle may reduce the occurrence of dysphagia. Injections into the levator scapulae may be associated with an increased risk of upper respiratory infection and dysphagia.
Patients treated with botulinum toxin may require immediate medical attention if they develop swallowing, speech, or respiratory problems. These reactions can occur within hours to weeks after injection with botulinum toxin.
(Allergan, 2020; Dysport, 2020; Jeuveau, 2020; Xeomin, 2020)

7. Human albumin and transmission of viral diseases
Allergy to human albumin (it is remote with the use of all neuromodulators)*. All of the commercially available products contain albumin. Human albumin is a derivative of human blood. Based on effective donor screening and product manufacturing processes, neuromodulation carries an extremely remote risk of transmitting viral diseases. A theoretical risk for transmission of Creutzfeldt-Jakob disease (CJD) is also considered extremely remote. No cases of transmission of viral diseases or CJD have ever been reported for albumin. Religious preference may play an essential role in the use of neuromodulators, given that there is a trace amount of human blood in the product.
(Allergan, 2020; Dysport, 2020; Jeuveau, 2020; Xeomin, 2020)

8. Immunogenicity
As with all therapeutic proteins, there is a potential for immunogenicity. Treatment with BoNT-A may result in the formation of neutralizing antibodies known to reduce the effectiveness of subsequent treatments by inactivating the toxin's biological activity. Xeomin claims that because they do not use any accessory proteins, users can not develop immunogenicity. Currently, the recommendation for treating the immune response is waiting 6 months to 1 full year without utilizing any neuromodulation treatments. After the washout period of 6 months to 1 year, the patient is advised to go back to their provider and initiate treatment. Clinical trials have shown this pause therapy to effectively clear out the system and allow neuromodulation to be effective again.
(Allergan, 2020; Dysport, 2020; Jeuveau, 2020; Xeomin, 2020)

9. Zinc deficiency
We learned earlier that BoNT-A is a metalloprotease drug; this means that BoNT-A's are metal-dependent, specifically Zinc dependent. There is a subgroup among the metalloproteases, and that is gluzincin. The word even has Zinc right in it, alluding to the importance of Zinc in the role of BoNT-A release of acetylcholine to disrupt the message center. There must be adequate Zinc within the host in order for the BoNT-A to bind and prohibit the release of acetylcholine. Zinc deficiencies are most prevalent in developing countries. However, in the United States, the range of statistics is from 12% to 40% demonstrating zinc deficiencies (Lebeda et al., 2010). Even with an adequate nutritional intake of Zinc, it is still possible for one to be deficient. Zinc deficiency can occur due to poor absorption of the nutrients ingested. There is heaps of relevant information coming to light on the gut biome. After much reading, it appears that a zinc deficiency can cause an alteration in the gut biome. Conversely and interestingly enough, a person with an altered gut biome will have trouble absorbing the nutrients. The concept of Zinc deficiency brings about a moment of pause and highlights the need for further research on nutrients and the gut biome.

Most developers of neuromodulators will advise that the office has an Epi-pen on hand as well as liquid Benadryl, ice, and sniffing salts. It is at a minimum that an esthetic practice have a basic first aid kit on hand. The state boards will not take exception to the first aid kit during an inspection.

There have been no systemic reaction reports and no reported long-term complications or hazards using neuromodulation for approved esthetic purposes. Severe adverse reactions, including excessive weakness, dysphagia, and aspiration pneumonia, with some adverse reactions associated with fatal outcomes, have been reported in patients who received botulinum toxin injections for unapproved uses. In these cases, the adverse reactions were not necessarily related to the distant spread of toxin but may have resulted from the administration of botulinum toxin to the site of injection and adjacent structures. In several of the cases, patients had pre-existing dysphagia or other significant disabilities. There is insufficient information to identify factors associated with an increased risk for adverse reactions associated with the unapproved uses of botulinum toxin. The safety and effectiveness of botulinum toxin for unapproved uses have not been established.

A lethal dose of botulinum toxin would be 25–35, one hundred unit vials. Diluted that would equate to 2500–3000 units of BoNT-A. These studies come from reports of childhood deaths that were treated with high dose therapy for cerebral palsy. Cosmetic procedures have a maximum cumulative dose recommendation generally to not exceed 360 units in a 3-month interval.

Drug to drug interactions

Many patients are at the mercy of multiple doctors and, at times, can be experiencing polypharmacy. Due diligence is required for the community's safety, and part of due diligence is taking a thorough medical history. Present the form in layman's terms so that the reader can truly understand the language. Medical terms are a different language, and health care providers can not assume that the patient has the same level of understanding. Break it down for the patient.

1. Avoid concurrent treatment with aminogylcosides or other agents interfering with neuromuscular transmission. It is thought that the aminoglycoside could potentiate the weakness from botulinum toxin A injections.

Examples of an aminoglycoside:

Tobramicin
Gentamicin
Paromycin
Amikacin
Neomycin

2. Anticholinergics—these are drugs that block acetylcholine.

Examples of anticholinergics:

Amitriptyline (Elavil)
Benztropine (Cogentin)
Chlorpheniramine (Actifed, Allergy & Congestion Relief, Chlor-Trimeton, Codeprex, Efidac-24 Chlorpheniramine, etc.)
Chlorpromazine (Thorazine)
Clozapine (Clozaril)
Cyclobenzaprine (Amrix, Fexmid, Flexeril)
Dicyclomine (Bentyl)
Diphenhydramine (Advil PM, Aleve PM, Bayer PM, Benadryl, Excedrin PM, Nytol, Simply Sleep, Sominex, Tylenol PM, Unisom, etc.)
Doxepin (Adapin, Silenor, Sinequan)
Hydroxyzine (Atarax, Vistaril)
Meclizine (Antivert, Bonine)
Nortriptyline (Pamelor)

Olanzapine (Zyprexa)
Orphenadrine (Norflex)
Oxybutynin (Ditropan, Oxytrol)
Paroxetine (Brisdelle, Paxil)
Prochlorperazine (Compazine)
Protriptyline (Vivactil)
Pseudoephedrine HCl/Triprolidine HCl (Aprodine)
Scopolamine (Transderm Scop)
Thioridazine (Mellaril)
Tolterodine (Detrol)

Anticholinergic drugs with a lesser effect:

Alprazolam (Xanax)
Amantadine (Symmetrel)
Baclofen
Carisoprodol (Soma)
Cetirizine (Zyrtec)
Cimetidine (Tagamet)
Clorazepate (Tranxene)
Codeine
Colchicine
Digoxin (Lanoxicaps, Lanoxin)
Diphenoxylate (Lomotil)
Fluphenazine (Prolixin)
Furosemide (Lasix)
Hydrochlorothiazide (Esidrix, Dyazide, HydroDIURIL, Maxzide & literally scores of other medications for high blood pressure)
Loperamide (Imodium)
Loratadine (Alavert, Claritin)
Maprotiline
Nifedipine (Adalat, Procardia)
Ranitidine (Zantac)
Thiothixene (Navane)
Tizanidine (Zanaflex)

3. Curare like non-depolarizing blockers-muscle relaxers.
4. Lincosamides—A class of antibiotics treating Gram-positive bacteria and Protozoans.

Examples of this drug:

Clindamycin

5. Polymyxins are antibiotics that work on gram-negative bacteria.

Examples of polymixins:

Polymyxin B and polymyxin E (also known as colistin)

6. Quinidine is a class 1 antiarrhythmic drug and is used as an antimalarial drug.

7. Magnesium sulfate—used to treat low levels of magnesium.

8. Anticholinesterases—these drugs are used in Parkinson's and prevent the neurotransmitter acetylcholine breakdown by the enzyme acetylcholinesterase.

Examples of this drug are:

Physostigmine

Neostigmine

9. Succinylcholine chloride.

These drugs can potentiate the effect of the toxin.

References

Allergan. (2020). *Botox Cosmetic history*. https://www.botoxcosmetic.com/what-is-botox-cosmetic/botox-cosmetic-history. (Accessed March 2020).

Coté, T. R., Mohan, A. K., Polder, J. A., Walton, M. K., & Braun, M. M. (2005). Botulinum toxin type A injections: Adverse events reported to the US Food and Drug Administration in therapeutic and cosmetic cases. *Journal of American Academy Dermatology*, *53*, 407–415. https://doi.org/10.1016/j.jaad.2005.06.011. Accessed May 2020.

Dysport. (2020). *For health care professionals*. https://www.dysportusa.com/healthcare-professionals#importantsafetyinformation. (Accessed March 2020).

Jeuveau. (2020). *Ready for a modern-made tox?*. https://jeuveau.evolus.com. (Accessed July 2020).

Lebeda, F., Cer, R., Mudunuri, U., Stephens, R., Singh, B., & Adler, M. (2010). The zinc-dependent protease activity of the botulinum neurotoxins. *Toxins*, *2*(5), 979–997. https://doi.org/10.3390/toxins2050978.

Stone, H. F., Zhu, Z., Thach, T. Q. D., & Ruegg, C. L. (2011). Characterization of diffusion and duration of action of a new botulinum toxin type A formulation. *Toxicon*, *58*(2), 159–167. Science Direct https://doi.org/10.1016/j.toxicon.2011.05.012. Accessed May 2020.

Xeomin. (2020). *A uniquely purified choice for frown lines*. https://www.xeominaesthetic.com/professionals/. (Accessed March 2020).

CHAPTER 7

Common myths about neuromodulators

Common myths

Common myths about Neuromodulators are vast and sometimes silly from the patient's point of view. It is essential to know the common myths and fears that patients have so that injectors can demystify them and earn the trust and respect along with their business.

FAQs from the patient

Will I be frozen?
Paralyzed is a better term.

I heard it is excruciating.
Proper technique and reconstitution decrease the pain associated with injections.

Will it get rid of my deep lines?
No, that is reduced by lasers and volume restoration as well as effective at-home skin care.

Will I become addicted?
Absolutely! In the hands of a skilled provider, patients will have a natural and youthful appearance. However, in all seriousness, technically, the answer is no.

Is it safe?
Yes, it is. It is temporary and does not cross the blood-brain barrier preventing it from ever causing botulism. 100% of the studies conducted showed complete re-innervation.

What is the potential side effects?
Bruising, headache, unsatisfactory results due to poor consultative process. It is of the utmost importance to set up proper expectations for the patient.

Essentials of Neuromodulation
https://doi.org/10.1016/B978-0-323-89920-8.00014-X

Can I go to the gym after?
Not for 24–48 h after as you can increase the metabolization of the product, therefore decreasing its efficacy.

Is there downtime?
Minimal downtime includes physical restrictions such as the gym, facials, massages and the wearing of hats and helmets.

Is it snake venom?
No, not at all.

I heard you are injecting botulism into me?
Patients will receive injections of an isolated protein from the toxin *Clostridium botulinum*.

Is it permanent?
No, it will last roughly 3–4 months. Re-sprouting will begin to occur at about 6 weeks.

Can it be used to correct the volume loss in my face?
Neuromodulation is only used to block the acetylcholine at the neuromuscular junction, preventing the muscles from moving. Volume loss can be slowed down, however, with the proper use and placement of Neuromodulators. For example, it is thought that the constant muscle activity creates a shearing effect and aids in the break down and redistribution of fat pads in the face.

Should I start Neuromodulators after the lines appear?
You may, however, prevention will yield a more satisfied client with their appearance and the overall treatment.

If I just have a facelift, I will not need Neuromodulators.
This statement is false. A facelift is often best when done in combination with other treatment modalities. Lifting older skin that is not at optimal health performance will not yield the best result. It is ubiquitous for surgeons to offer skin tightening and or resurfacing procedures with the facelift. These treatments work synergistically with one another. Aging is happening at the level of the skin, vessels, muscles, and fat pads. Ultimately, the aging face is becoming compromised at every layer. Therefore, a one-trick pony does not exist yet. Neuromodulation often is best when used in combination. Another combination will be a facelift and dermal filler to add volume.

Will I look worse afterward if I decide to stop Neuromodulators?
The answer is no. Technically, Neuromodulation is not causing any damage to tissues, so anatomically, they will not be any worse than they had never been treated with neuromodulators. After 10 years of use, if you stop neuromodulators, one may perceive that they look worse because they have not seen the accurate representation of their aged muscle and skin position in those 10 years. Studies have shown those that have neuromodulation have a decrease in static rhytids compared with non-user.

CHAPTER 8

Anatomy

Beginning to understand facial anatomy

There are six main anatomical concerns for anatomy when we are injecting for cosmetic purposes (Figs. 8.1–8.7)

Skin
Muscles
Vessels
Fat
Nerves
Bone

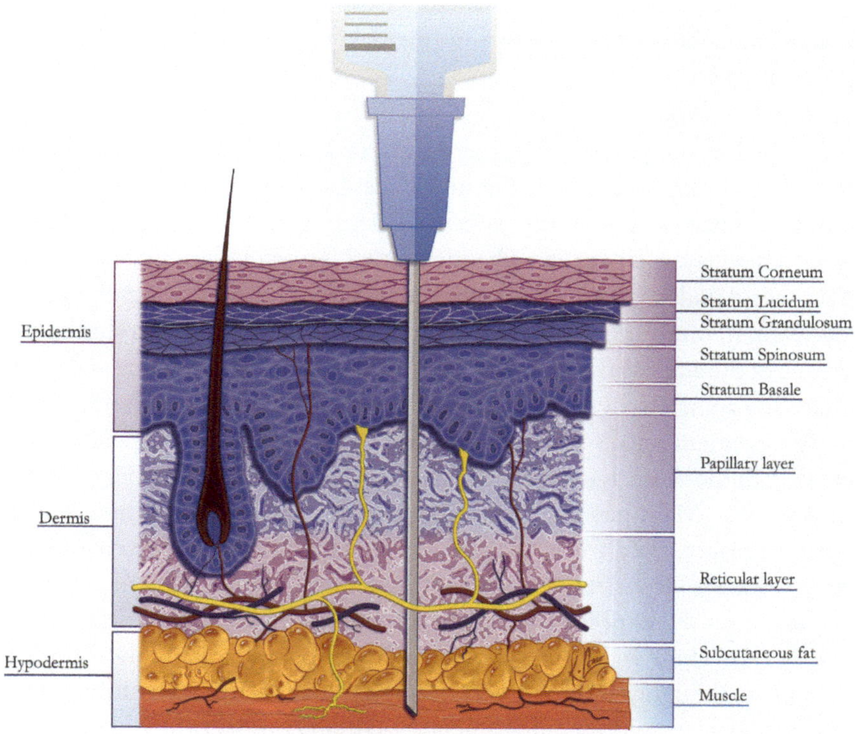

Fig. 8.1 Layers of the skin.

https://doi.org/10.1016/B978-0-323-89920-8.00005-9

Soft tissue

Four distinct tissue planes are recognized on the face:

1. Integumentary system (skin, hair, nails)
2. Hypodermis (subcutis, otherwise known as the subcutaneous layer or fibroadipose tissue)
3. Superficial musculoaponeurotic system (SMAS)
4. Parotid-masseteric fascia

Within the skin, there are two distinct layers

1. Epidermis
2. Dermis (blood, nerve endings, sebaceous glands)

Within the epidermis, there are four distinct layers

1. Stratum corneum—outer most layer and is exposed to the elements of the world
2. Stratum granulosum
3. Stratum spinous
4. Stratum basale

Within the dermis, there are two distinct layers

1. Papillary layer
2. Reticular layer

As outlined in the previous list, there are four layers to the soft tissue of the face. The four layers are; skin, the subcutaneous layer of fibro adipose tissue, superficial musculoaponeurotic system (SMAS), and the parotid mesenteric fascia. These layers all age in a chaotic pattern leaving hollows in some parts of the face decades earlier than in other areas of the face. It is common to see a hollowing in one region, and then about a decade later, there is a shift in the volume of the adipose tissue. The shifting of fat pads leads to jowling and other displeasing aging characteristics. The focus of neuromodulation should be on the aging patient's goals alongside the individual's unique assessment. When referring to the word aging, begin thinking in terms of clinical aging concepts and processes. Many patients believe that neuromodulators can tighten and lift when, in reality, they are responsible for preventing muscle movement. If a patient requests lift and tightening at a consult for neuromodulation, the injector must reset the expectations; otherwise, the patient will be dissatisfied.

It is a well-known fact that skin is the largest organ of the body. The skin accounts for roughly 15% of a person's body weight. It is part of the integumentary system, and the integumentary system consists of skin, hair, nails, and exocrine glands. Skin is broken down into three layers, the epidermis, the dermis, and the subcutaneous tissue. Each of these layers is further broken down within each layer. Surprisingly, the skin is only a few millimeters in thickness, yet 15% of the body weight. The outermost layer is the epidermis, followed by the dermis and the third layer is the subcutaneous structure. In terms of neuromodulation, the esthetic injector must understand where the muscles lay and how the skin can affect dosing, product placement, and the patient's overall satisfaction. To practice holistically, injectors must understand each layer of the skin and how they function in tandem with the muscles and fat pads within the face. Understanding how the skin behaves in response to intrinsic and extrinsic factors related to aging is vital in the face of esthetic treatments. An emerging understanding of the skin's functions can help injectors better deliver BoNT-A treatment effectively to patients.

We uncovered early on that the aging process is happening on multiple multidisciplinary systems as we age. Therefore, we must understand the integumentary system, how we understood the nervous system, and the muscular anatomy.

The skin's primary function is to protect the body from external physical force, both chemically and biologically. It also prevents water loss in conjunction with its thermoregulatory system. Later in the reading, the treatment of hyperhidrosis with neuromodulation will be reviewed. Therefore, the thermoregulatory function of the skin will be an essential concept to understand when treating hyperhidrosis. Patients that sweat profusely on there forehead do see a reduction in the sweat production of their forehead when treated with neuromodulation of the frontalis, procerus, and corrugators.

The outermost layer of the skin is the epidermis, and this layer is always rejuvenating itself. The renewal process slows as we age, and that is where chemical peels come into play in the esthetic arena. A chemical peel will do the slowed epidermal layer's job in terms of manual sloughing of the outermost layer. The basal cells within the epidermal layer give rise to the epidermal layer's renewal process. The epidermis consists of four layers; the basal layer (stratum germinativum), squamous cell layer (stratum spinosum), granular cell layer (stratum granulosum), and lastly, the cornfield layer (stratum corneum). Given the focus is on esthetics, another critical cell to comprehend in the epidermis is the melanocyte. This cell is responsible for pigment within the skin. Melanin is produced by the melanocytes in response to protect the skin from ultraviolet radiation from sunlight. Melanin is the darker spots seen in outer layers of the skin. Melanin is a response to the activation of the melanocyte in an attempt to shield itself from external forces such as ultraviolet light.

As clients begin to volumize and iron out the wrinkles on their faces with neuromodulation, the next request is often to address sun damage. Developing a greater understanding of melanocytes and melanin will help to elevate treatment outcomes. In order to deliver exceptional results in regards to pigment problems, the esthetic provider needs to understand how the melanocytes work in combination with the basal cell layer. The melanocytes primarily reside within the basal layer of the epidermis. When patients complain of dull, sun-damaged skin, typically, the health and function have slowed naturally. Chemical peels give rise to the manual exfoliation to turn over the dead cells and present the healthier layer waiting underneath. Peels can restore the patient's glow and aids in sloughing off sun damage or hyperpigmentation.

There are specific cells known as the keratinocytes within the epidermal layer along with the melanocytes. The keratinocytes make up nearly 80% of this layer. These are essential cells in the protective function of the epidermis. The epidermal layer is the reason clients are dissatisfied with their topicals products. Topical products often need to make it to the dermal layer to elicit change, but the epidermal layer acts as a barrier. The keratinocytes in the epidermis originate in the basal layer and eventually migrate to the skin's surface. These cells are eventually shed and replaced by new healthier cells.

Moving deeper into the layers of the skin is the dermal layer. It is made up of primary collagen. Collagen is a protein that acts as fibrous structural support to the skin. The dermis has two layers; papillary and reticular layer. The papillary layer resides just beneath the epidermal layer. Eccrine, apocrine sweat glands, and pilosebaceous units reside in the dermal layer, specifically the dermal-epidermal junction. When patients are treated in the axilla for hyperhidrosis, this area is the target for effect. The dermal layer makes up the bulk of the skin and is responsible for collagen, hyaluronic acid, and elastin formation.

Moving beyond the skin layers, as described previously in this chapter's beginning, there are four planes to the face. We will not spend much time on the remaining three as these layers are a more advanced topic regarding neuromodulation. Circle back to understanding the four tissue planes in greater depth as master emerges with the use of neuromodulation. Then begin to move forward in an advanced knowledge base of esthetics.

Let us now look at what structures are beneath the skin. These layers will become increasingly more critical when injecting filler and performing skin resurfacing and tightening procedures. The focus is on the superficial skin in regards to gaining an understanding of neuromodulators. However, if each one of these structures and layers is essential when talking about facial esthetics and addressing the aging face.

Hypodermis

This layer is referred to as the hypodermis, the subcutis, or the subcutaneous layer of the fibroadipose tissue. They all are referring to the layer just beneath the dermis. It is not a part of the skin but has finger-like projections that connect to the dermis.

The hypodermis layer is made up of fat, otherwise known and lipocytes. The fat is a layer that adds volume to the appearance of the skin. The thickness of the subcutaneous layer varies from patient to patient and over different parts of the face. The fatty layer above the upper lip is often very thin. However, when we look at the malar eminence (the apple of the cheek), that area has much greater subcutaneous tissue volume. This subcutaneous layer also stores energy. The fatty layer of the skin will be most valuable in the discussion of dermal filler implantation. Nonetheless, it is necessary to understand the entire system as injectors emerge from novice to proficient in their emerging knowledge acquisition of the anatomy and the industry.

Typically, the two subcutaneous tissue sections can be broken down into the medial compartment and the lateral compartment. The lateral compartment is in the preauricular region (just before the ear). According to many anatomists, this area comprises a tiny, thin layer that is dense and highly vascular. The superficial temporal artery and vein snuggle in tight to the preauricular space. The medial part is softer, more pliable as well as has significantly more volume in nature. The zygomatic ligament and perforators create a boundary by the malar eminence separating the medial portion from the malar compartment. This layer is what adds volume to the face giving a youthful appearance. As the name mentions, this layer is more fibrous than fatty. The word fibrous means that it is a bit more rigid and less buoyant than other subcutaneous tissue types. This tissue is not uniform throughout the face. Different sections of the face will present with varying thickness. As we age, this layer will change along with the degree of adiposity. The fat content of the cheek mass and the malar fat pad is primarily made up of this subcutaneous adipose tissue type. The fibrous septi separates the subcutaneous fat into compartments. The deflation of these fat pads occurs laterally first then progress to the medial portion of the face.

With a keen eye, providers can begin to pick up the deflation when observing the face from the side. The deflation in the lateral face mimics bonnet straps. If a bonnet was placed on the patient and the ties for the bonnet would represent deflation areas. This deflation is noticeable at about age 40, and then the medial portion starts to deflate about 10 years later. However, it is common to see a malar fat pad split earlier in life as the fibrous bands holding the fat in place become weak and allow the fat to shift even though it may not have deflated just yet. Patients are aging on a global scale, and it is not homogenous. Therefore, the esthetic rejuvenation process should mirror the process in order to create soft, natural, and beautiful results. The fat deflates in regions instead. The malar fat pad split is often confused by novice injectors and patients by feeling that they need to have their tear troughs inflated. When in reality it is a need for cheek augmentation with fillers instead of tear trough filler. As injectors progress into dermal filler implementation, the subcutaneous layer, the superficial muscle aponeurotic system, and the parotid masseteric fascia will play a more significant role in importance. At the beginner level of injecting neuromodulators, it is nice to be familiar with these planes as it will help injectors better assess and properly consult patients about developing effective anti-aging treatment plans. If we are chasing the patient's complaint, it will often lead to dissatisfied customers as they are indeed analyzing their face as a whole and do not understand the restorative process in depth as trained professionals. They cannot articulate their needs. At this point, the injector steps up to the plate and becomes the expert understanding the relationship of assessment, goals, and the product's ability to perform the task the patient desires.

There are three building blocks to the health and age of a patients skin; elastin, collagen, and hyaluronic acid. Much of the esthetic industry talks a lot about elastin, collagen, and hyaluronic acid as the foundation concepts regarding restorative esthetic treatment and procedures. What are these three building blocks in the skin, and how do they help injectors assess and treat the patient?

1. **Elastin** is a 70 kDa elastic protein found in the body, mainly in the skin and connective tissue, characterized by its elastic qualities. It provides the skin with the flexibility and elasticity necessary to regain form when stretched, contraction, or compression. Elastin aids in the skin's ability to get back to baseline after being placed under stress, such as pinching or animation. Elastin-rich structures calcify over time, decreasing their ability to recoil. In terms of skin, it is harder for the skin to maintain it is elastin or, in lay terms, its recoil ability. It is a protein that is very

similar to collagen. Collagen is the primary structural support system in elastic fibers. Elastin, hyaluronic acid, and collagen are all equally important to one another.

2. **Collagen** is defined as the tough fibrous protein. This protein is found in bone, cartilage, and, most notably, for injectors, the skin. It can also be found in other connective tissues within the body. Collagen provides the body's structures with the ability to withstand forces that stretch or lengthen them. It is a natural structural protein in a fibrous form, found in a mammal's body, including humans, in abundance: around 30% of all body proteins are collagen. These are responsible for keeping the body tissues connected and supported and, therefore, commonly found in our body's strong structures like bones, tendons, ligaments, and, of course, skin. Specifically relevant to the skin, collagen provides strength, flexibility, and resistance. In other words, it is the presence of collagen that gives skin its firmness. To better understand collagen in action, when the skin is cut, the skin produces collagen to repair the damaged tissues and replaces the old tissue with new collagen. It is the body's natural form of stitches. Due to its specific properties, collagen's medical uses extend to treat joint mobility problems, severe burns and even create lab skin substitutes.

   Collagen and elastin work together to keep the skin smooth, supple, and flexible. Fibroblasts produce both proteins in the dermis, and, as we grow old, the body loses the ability to produce them, so cell structures weaken. As it loses strength and elasticity, the skin grows thinner and less resistant to damage, while it tends to sag, stretch, and wrinkle; this is aging. A proper diet is vital in helping to maintain youthful collagen and elastin levels. Sources rich in vitamin A, zinc, vitamin C, and other oxidants have a preventive function in aging, helping the body to retain or produce these proteins.

3. **Hyaluronic acid** seems to be the new big hit in town with the onset of reversible hyaluronic acid filler treatments. However, technically it is not new since this carbohydrate is also produced naturally by our body. Hyaluronic acid is found mostly in the eyes, joints, and skin (50% of hyaluronic acid in the body is found in the skin, both in deep underlying areas and visible epidermal layers). This substance plays a huge role in cell growth and renewal. It is responsible for retaining water and lubricating the body's movable parts: hyaluronic acid molecules can retain up to 1000 times its weight in water. Therefore, hyaluronic acid's secret is keeping the skin moisturized and hydrated.

It is the dual action of HA and collagen that maintains the skin's layers and structure fresh and healthy. While collagen works on skin firmness, hyaluronic acid nourishes and hydrates the collagen to keep it supple and flexible, filling the skin matrix (formed by collagen and elastin) with water and giving it a fuller, firmer, and youthful appearance. Hyaluronic acid has been used mainly to treat joint disorders, but its water-retaining and hydrating properties have generated a growing interest in the cosmetic industry. Nowadays, it is used in an injectable form (Juvederm, Vollure, Volbella, Restylane; to name a few) to smooth fine lines and wrinkles and create fullness around the injected location (for example, to create volume in fine lines in the lips). As we age, hyaluronic acid levels in the body decrease since our bodies are incapable of maintaining the same concentrations of this substance.

Consequently, our body loses the ability to hold water and maintain hydration levels. Ultraviolet radiation and pollution play a dramatic role in the reduction of this ability. Our skin becomes drier and, therefore, more susceptible to wrinkling and loss of natural radiance. Hyaluronic acid has gained much attention in the last years due to its regenerative properties, especially when combined with vitamin C.

Fat pads

The fat pads create contour and fullness in the facial features. As we age, the fat pads become thin and begin to make the great descent to the lower face. Hollows begin to form under the eyes as fat pads fall from the cheeks. Drooping skin will create unwanted pressure on fragile tissues and enhance a deep line due to the new pressure and shift change from the falling fat pad. The fat from the upper face moves to the lower portion of the face-pulling, everything downward. Most clients will describe this as they look sad and tired. There are two separate subcutaneous fat pad layers; one is the superficial, and one is deep. The superficial fat pad layer is right below the dermis. The deep subcutaneous fat is located below the superficial fascia (SMAS). As we age, some superficial subcutaneous fat will be redistributed to visceral organs; other superficial subcutaneous fat compartments descend. The deep fat does not change position.

There is notably seven superficial fat compartments

1. Superficial nasolabial fold
2. Medial cheek

3. Middle cheek
4. Lateral cheek
5. Superficial superior temporal
6. Superficial inferior temporal
7. The jowl fat compartment

Deep fat compartments are located deep to the SMAS and arranged around muscles, nerves, and ligaments

Lateral soof
Medial soof
Deep medial cheek fat
Deep lateral cheek fat

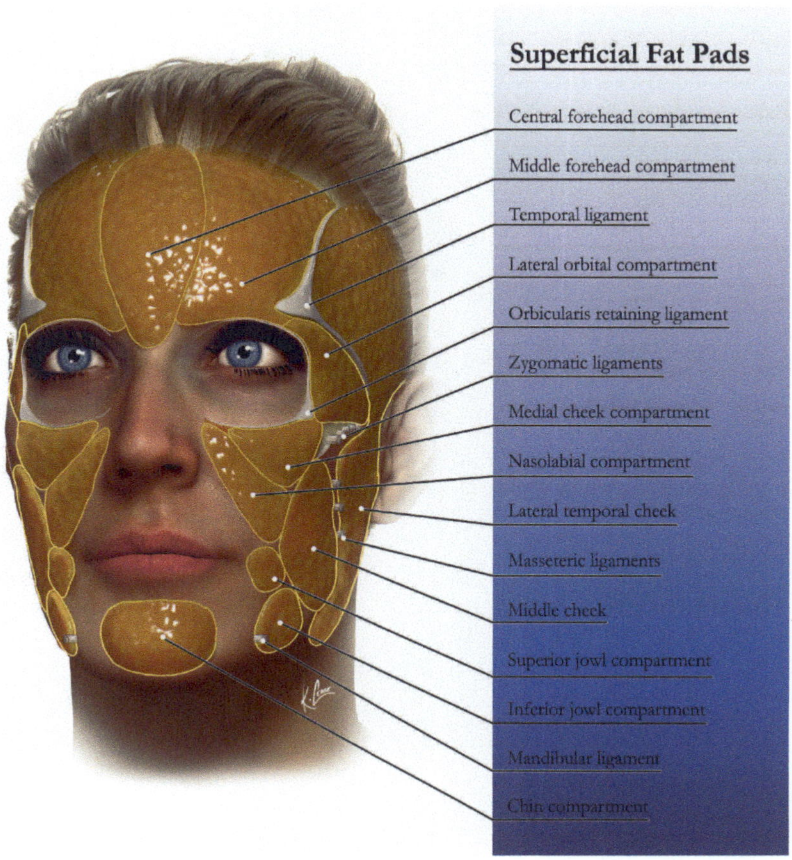

Fig. 8.2 Superficial fat pads.

Muscles of the face

The muscles of the face are our most significant concern when injecting neuromodulators. It is essential to understand where they lay within the tissue and where they connect, as this is how injectors understand how to prevent adverse outcomes instead of managing them (Table 8.1). Our last focal point is, of course, how they make a face animate and which position are structures pulled in, up, down, medial, or lateral. With each neuromodulator injection, there is an indirect effect on a surrounding or opposing

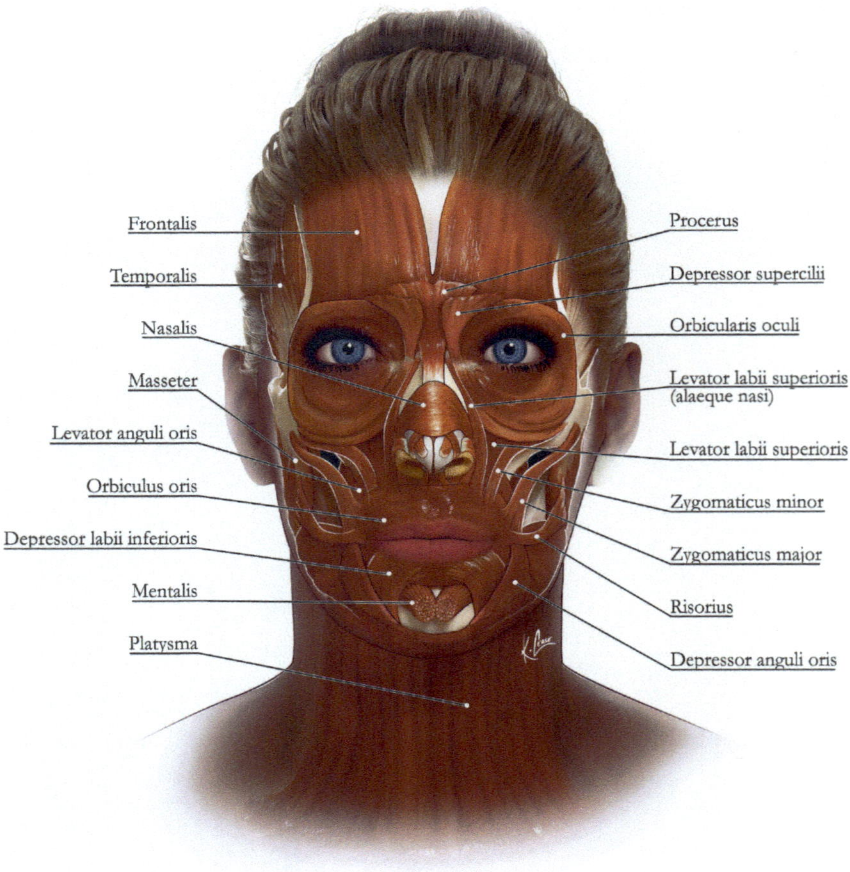

Fig. 8.3 Muscles of the face.

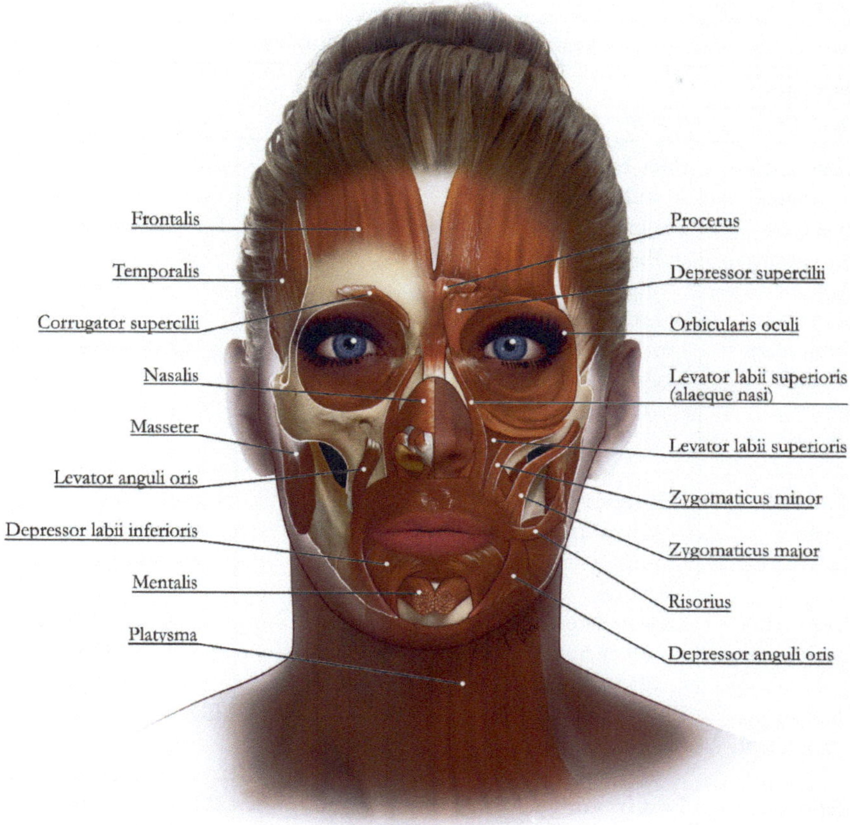

Fig. 8.4 Deep muscles of the face.

structure. In the next section, as we describe the muscles function and its connections, it will be a concept beginners will want to go back to repeatedly. By grouping the muscles, the reader should be able to begin to understand areas of neuromodulator treatment better. This way, learners can begin to see the unique and intricate relationship between the groups. With every muscle that is injected, another muscle group will be either directly or inadvertently affected. Truly comprehending what muscle groups work together is the difference between being injectors and being a great injector. Each patient will respond differently to treatment, making the anatomy extremely important in creating beautiful and natural results.

Table 8.1 Depressor and elevator muscles.

Depressor and elevator muscles			
Depressor muscles	**Action**	**Elevator muscles**	**Action**
Corrugator supercilii	Pulls brow down and medially	Frontalis	Elevates brows
Procerus	Lowers brow Pulls the medial angle of the brow down	Levator anguli oris	Elevates the angle of the mouth
Orbicularis oculi	Forcefully closes eyes	Levator palpebrae superioris (Muellers muscle)	Elevates upper eyelid
Depressor labi inferioris	Depresses the lower lip and pulls it outward lateral	Masseter	Elevates and protrudes the mandible
Platysmal	Lowers lateral lower lip and tenses the neck	Menatlis	Elevates and dimples chin and protrudes the lower lip
Depressor nasii septi	Pulls down the tip of the nose when smiling	Nasalis	Elevates nasal skin, draws the ala of the nose to the septum
Depressor anguli oris	Lowers corners of the mouth (frown)	Levator labii superioris alaeque nasi	Raises the upper lip and dilates the nostril
Orbicularis oris	Purses/puckers lips Contracts the lips	Orbicularis oris	Raises lower lip
Depressor supercilii	Lowers brow	Zygomaticus major	Pulls the angle of the mouth up and back to aid in smiling
		Zygomaticus minor	Elevates the upper lip

Circumorbital group

Orbicularis oculi—The orbicularis oculi muscle is a sphincter muscle of the eyelids. It is a broad and flat muscle spreading into three regions. A sphincter muscle closes circumferentially. Therefore, the orbicularis oculi muscle plays a vital role in facial expression and various ocular reflexes such as frowning, squinting, and eye closure. There is a palpebral portion that can be contracted actively and passively. There is an upper and lower section of the palpebral fibers. The upper portion acts as a depressor, and the lower portion acts as an elevator. Due to the sphincter nature of this muscle, it is considered a depressor muscle.

Corrugator supercilii—This muscle works in combination with the orbicularis oculi muscle to draw the eyebrows in medially and downward to shield the eyes in bright sunlight when squinting. Because the corrugator supercilii muscle is responsible for the frowning and squinting action, it is directly related to the development of static vertical wrinkles in the glabellar region. The corrugator supercilii is a depressor muscle as it draws the tissue downward.

Levator palpebrae superioris—The levator palpebrae superiors muscle is responsible for elevating the upper eyelid. The levator muscle is part of the elevator muscle group. Not only is this muscle responsible for elevating and lowering the upper lid, but it also aids in the expression of fear. The palpebral apertures widen in the prescience of fear or excitement when the eyes are widened. The widening of the apertures is a direct result of the increased sympathetic nervous system.

Nasal muscle group

Procerus—The procerus muscle pulls down the medial portions of the eyebrows and results in wrinkling over the bridge of the nose. Due to the downward pull of this muscle, it is part of the depressor muscle group. If this muscle is hyperactive, a static rhytid will develop over time directly on the bridge of the nose at the insertion point. The procerus is an active muscle in the emotion exhibited in frowning, squinting, and or concentration. The glabellar complex is a leading treatment area in esthetic treatments. The depressor supercilii sits on both sides of the procures and aids in pulling the brows outward, creating a widening in between the brows.

The depressor supercilii can be unintentionally and intentionally injected. In the right candidate, it can be a fantastic treatment. However, in the wrong patient, it can create a very disfiguring treatment. Understanding the patient's goals and how the tissue will react when the muscles have been treated with BoNTA is essential in creating superior outcomes.

Nasalis—The nasalis muscle comprises two components, the transverse section and the alar portion. The transverse section merges with the procerus at the bridge of the nose and sits on the cartilaginous part of the nose. As the transverse muscle extends down the sides of the nose, it attaches to the maxilla just above the lateral incisive fosa. The transverse part is responsible for compressing the nasal aperture. It is also responsible for the expression that is illicit in the face of a bad smell. Some refer to the wrinkling of the nose in an upward fashion as the "bunny lines." The alar part of the nasalis draws the nares and posterior part of the columella down and laterally. It results in widening the nares and elongating the nose. When the alar portion of the nasalis is treated with BoNT-A, it can slim the flair of the alar.

Depressor septi—This muscle sits deep within the tissues. It attaches to the periosteum and connects with the nasal spine. The depressor septi pull the columella, the tip of the nose, and the nasal septum downwards. The depressor septi is a depressor muscle as it may be recognized as such by its name. As the septi is activated, it can aid in the flailing of the nasal alar. As this happens, the tip of the nose dips. A common aging problem is to see the nasal tip begin to droop with age. Adding a few units of BoNTA to the base of the nasal spine can elevate the tip to a more youthful appearance.

Levator labii superioris alaeque nasi (LLSAN)—This muscle is the target for treatment in the gummy smile. The LLSAN is a superficial elevator muscle of the nose and the upper lip utilized in oral-facial expression. Its origination point is on the upper portion of the maxilla and then inserts into the alar cartilage and levator labii superiors (LLS). It starts on the sides of the noses and runs lengthwise down the sides to the upper medial to the lateral portion of the lip. The constant pull on this muscle will deepen the nasolabial folds with time. It is an excellent treatment spot for the prevention of the nasolabial folds. The LLSAN pulls the upper lip up to display teeth, and sometimes, with a hyperdynamic muscle, the gums will also show. When the lip is pulled upwards wards with activation of this muscle, the nares will flare. It is sometimes accidentally injected when treating the bunny lines.

In the wrong candidate, the patient will have their upper lip hang over their top teeth. The patient section is of most importance.

Buccolabial muscle group

Upper lip (elevators, retractors, and evertors)

Levator labii superioris alaeque nasi (LLSAN)—This muscle is the target for treatment in the gummy smile. This is a superficial elevator muscle of the nose and the upper lip utilized in oral-facial expression. Its origination point is on the upper portion of the maxilla and then inserts into the alar cartilage and levator labii superiors (LLS). It starts on the sides of the noses and runs lengthwise down the sides to the upper medial to the lateral portion of the lip. The constant pull on this muscle will deepen the nasolabial folds with time. It is an excellent treatment spot for the prevention of the nasolabial folds. The LLSAN pulls the upper lip up to display teeth, and sometimes, with a hyperdynamic muscle, the gums will also show. When the lip is pulled upwards with activation of this muscle, the nares will flare. It is sometimes accidentally injected when treating the bunny lines. In the wrong candidate, the patient will have their upper lip hang over their top teeth. The patient section is of the utmost importance when treating this area.

Levator labii superioris (LLS)—This superficial muscle works in conjunction with the upper lip's other muscles to elevate and evert the upper lip. This muscle is an elevator muscle of the upper lip. The levator labii superior is much like the LLSAN modifies the nasolabial fold. This muscle is responsible for making an expression of sadness or seriousness. The LLS starts on the maxilla and inserts into the upper lip's skin between the levator anguli oris and the LLSAN.

Zygomatic major—The zygomatic major muscle is another superficial elevator muscle that inserts on the lateral aspect of the zygomatic bone posterior to the zygomatic minor muscle. This muscle inserts into the orbicularis oris, the DAO, and the levator anguli oris. The zygomatic major muscles draw the angle of the mouth up and outward laterally when smiling or laughing. It attaches to the lateral portion of the zygomatic arch, and at times it will snuggle under the orbicularis oculi muscle. The zygomatic major muscle is another vital muscle to avoid when injecting the lateral canthal lines. We also need to be careful when treating the DAO as the smile can be interrupted in an unnatural and displeasing way due to the insertion point in the DAO.

Zygomatic minor—The zygomatic minor is yet another superficial elevator muscle. The zygomatic minor muscle elevates the upper lip outward and upward, allowing to show the maxillary teeth. This muscle sits beneath the orbicularis oculi on the medial portion of the zygomatic bone just in front of the zygomatic major muscle and connects to the subcutaneous tissue of the upper lateral lip muscle. The zygomatic muscle is essential in the function of smiling. This muscle can be inadvertently injected when the lateral canthal lines are treated in the ill-prepared injector's hands.

Levator anguli oris—The levator anguli oris raises the mouth's angle in smiling and contributes to the depth and contour of the nasolabial fold. This muscle lies deep and lateral to the levator labii superiors. It originates in the canine fossa of the maxilla and connects the modiolus of the lips. The levator anguli oris is not typically a muscle that would be injected with neuromodulation. However, due to the proximity of other muscles treated, it is a muscle to be mindful of while injecting.

Risorius—The risorius pulls the corner of the mouth laterally in numerous facial activities, including grinning and laughing. The risorius muscle originates in the fascia over the parotid gland. This muscle is closely related to the master muscle as it course along to the corner of the mouth just above the master muscle.

Lower lip (elevators, retractors, and evertors)

Depressor labii inferioris (DLI)—The depressor labii inferioris draw the lower lip downwards and a little laterally in the action of chewing or otherwise known as a masticatory activity. It is further thought that the DLI may assist in the eversion of the lower lip as well. The DLI is a muscle that injectors want to avoid entirely. The DLI is responsible for the expressions of irony, sorrow, melancholy, and doubt, to name a few. When treating the DAO incorrectly, it can lead to the neuromodulator being introduced to the DLI. Because it is a depressor muscle when it is injected with neuromodulation, the lateral portion of the lower lip from the corner to midline will be forced upward, and then action of the pink of the lip will curl inwards. The result can look like a palsy event. Only time will heal this mistake.

Depressor anguli oris (DAO)—The depressor anguli oris muscle draws the angle of the mouth downwards and laterally in opening the mouth and in expressing sadness or questioning. With age, the DAO is responsible for the volume loss and the development of creases in the marionette region. This muscle always pulls downward and creates a shearing effect deepening the crease in this area and disrupting the fat position. Treating this area with neuromodulation will help to prevent the progression of a rapid breakdown in this area.

Mentalis—The mentalis muscle raises the lower lip and is responsible for the skin's wrinkling on the chin. It raises the lower lip base and, therefore, helps in protruding and everting the lower lip in drinking and expressing doubt or disdain. In patients with a jaw that is thrusting forward as in an overbite, this muscle often stays active and will result in a cobble stoning effect on the chin.

Compound sphincter muscle group

Orbicularis oris—The orbicularis muscle consists of four quadrants the upper, lower, left, and right. This muscle is responsible for oral competence. The orbicularis muscle can create the kissing face or a puckering of the lips and pull the lips toward the teeth to complement the expression of confusion or disbelief. The orbicularis oris is a sphincter muscle, which is also confirmed to be a part of the depressor group of muscles. It aids in speech as well as oral competence. If treatment with neuromodulation is done incorrectly here, it can create difficulty with speech, drinking through a straw, and whistling. The injections should be superficial and always in the muscle's belly but not too deep to affect the deeper portion of the muscle, which can affect a patients' oral competence.

Three other muscles are a part of this group but do not play a significant role in the art of neuromodulation; therefore, we will make a note of them but will not spend much time on them. They are the incisivus labii superioris, incisivus labii inferioris, and the buccinator.

Muscles by facial groups and areas

Circumorbital and palpebral muscle group
- Orbicularis occuli
- Corrugatur supercilii

Buccolabial muscle group
- Levator labii superioris alaeque nasi

Zygomatic major
Zygomatic minor
Risorius
Levator labii superioris
Menatlis
Depressor labii inferioris
Depressor anguli oris
Buccinator
Pterygomandibular raphe
Orbicularis oris
Pars peripherals
Pars marginalis
Incisivus labii superioris
Incisivus labii inferioris
Platysma

Nasal muscle group

Procerus
Nasalis
Depressor septi

Skeletal structure

The skeletal structure is the support, or shall we say the framework for all facial layers to rest. The craniofacial skeletal structure undergoes much stress throughout the years, which produces many changes over time. Bone loss and position changes are relative to the dynamics of expansion and loss, along with musculoskeletal changes. As muscles atrophy over time, the craniofacial structures reposition as the muscle and ligaments act like anchors or support systems for the skeleton. In aging, the mandible begins to drop forward, so the mentalis becomes hyperdynamic in an effort to hold it back in position. Patients' facial height, width, and depth all change with time. The chin can widen, the gonial notch demineralizes and is pushed forward, the maxilla changes aiding in the formation and accentuation of the nasolabial folds. The orbits increase in size by widening and increasing the opening, which leads to the dreaded dark tear trough. As the maxilla decreases in size, this leads to a shift in the malar fat pad, accentuating the nasolabial fold. There is also a loss of support of the upper lip, creating a lengthening of

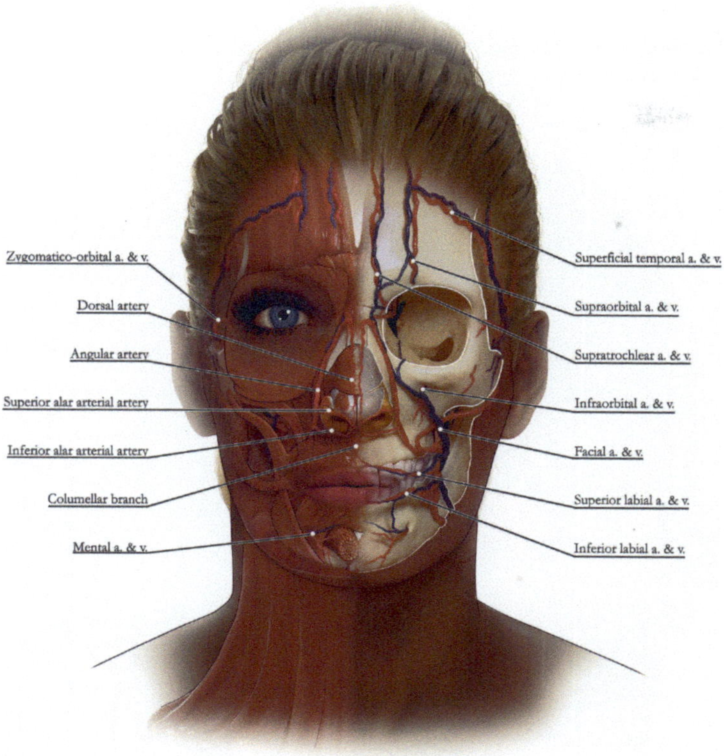

Fig. 8.5 Vessels of the face.

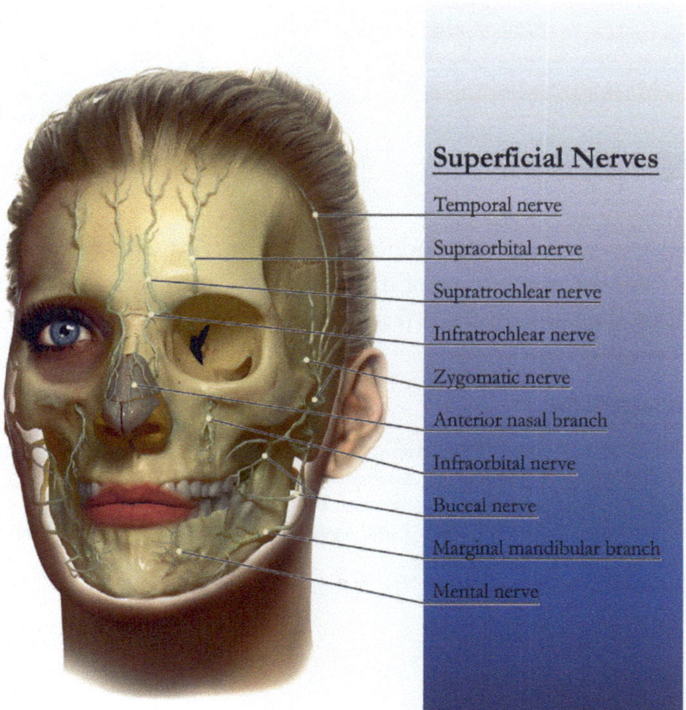

Fig. 8.6 Superficial nerves.

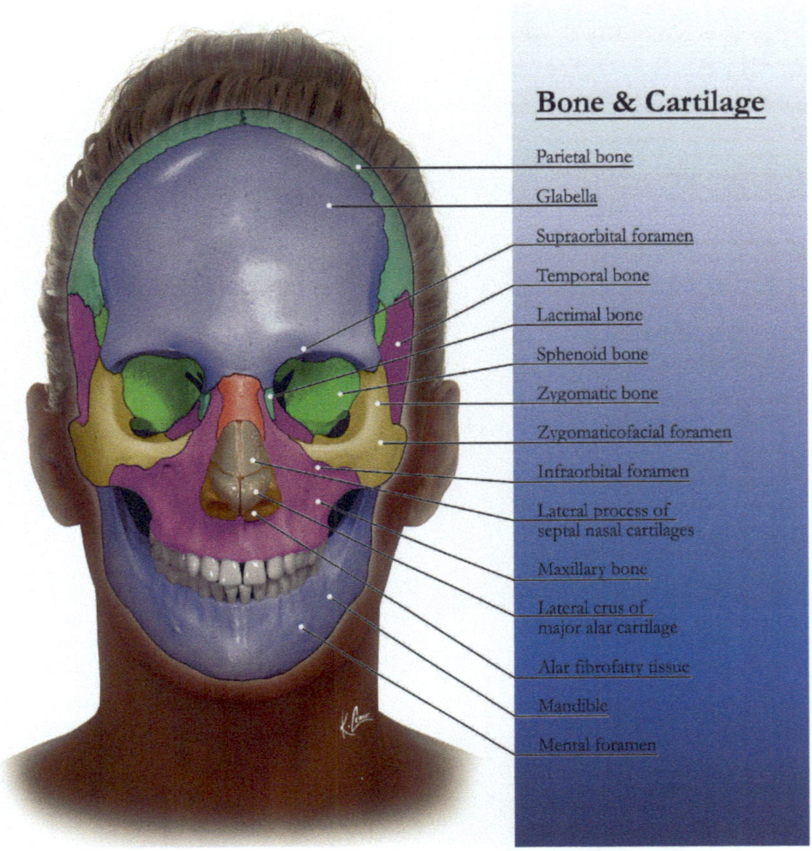

Fig. 8.7 Bone and cartilage.

the upper lip with the maxilla bone reabsorption. As teeth start to decay, this will also change the structure of the craniofacial structures, further creating volume loss and shifts in facial features' contours.

Further reading

Kolarsick, P., Kolarsick, M., & Goodwin, C. (2011). Anatomy and physiology of the skin. *Journal of the Dermatology Nurses' Association*, *3*(4), 1–11.

Netter, F. (2014). *Atlas of human anatomy* (6th ed.). Philadelphia, PA: Elsevier.

Pessa, J. E., & Rohrich, R. J. (2012). *Facial topography: Clinical anatomy of the face*. St. Louis, MO: Quality Medical Publishing.

Standring, S. (2016). *Gray's anatomy: The anatomical basis of clinical practice*. Philadelphia, PA: Elsevier Limited.

CHAPTER 9

The esthetics of aging

The aging face

Now that the mechanism of action of BoNT-A, the nervous system, the soft tissues, and the muscles of the face have been reviewed, let us begin to put the pieces together by looking into how the face ages so that proficiency can be developed in the art of assessment to develop unique yet effective treatment plans with the use of neuromodulation.

The onset and pace of aging are multifactorial and unique to different individuals, and factors such as lifestyle, gender, and ethnicity play an essential role in the development of prescriptive treatment plans. Given the dynamic relationship to each structure involved in the aging process (bone, ligaments, fat, muscles, fasciae, and skin), it can be challenging to pinpoint the culprit for most aging processes. It is essential to understand the underlying structures and how they age before offering esthetic advice to the patient. Often the patient will have a budget that does not meet the reality of a singular rejuvenating treatment. It is crucial to begin down the esthetic journey with the patient by building trust. This way, a partnership can be developed of equal trust resulting in a long term relationship. The patients can learn from their provider, fostering confidence in their decisions based on the treatment plan carved out for them. If the provider comes on too strong, they will feel as if they are being sold. Also, pointing out too many areas of concern can make the client feel hopeless. When a trusting relationship is created, providers can begin to educate them on their aging process slowly. Patients will feel more comfortable with their decisions and less focused on the dollar amount. All providers should produce a prescription of treatments to the patients based on their needs, not based on their interpretation of the patient's budget. Providers prescribe and then let the patients decide if they want or can afford the plan for rejuvenation.

The effects of aging on the face are complex and multi-dimensional (Fig. 9.1). Injectors must analyze this carefully and systematically ask the questions; is the skin thick or thin, rough or smooth, dry or oily, how much fat has shifted, and how do the muscles aid in the aging process? The facial assessment will be unique to every patient. There is no way to provide a

Essentials of Neuromodulation
https://doi.org/10.1016/B978-0-323-89920-8.00034-5

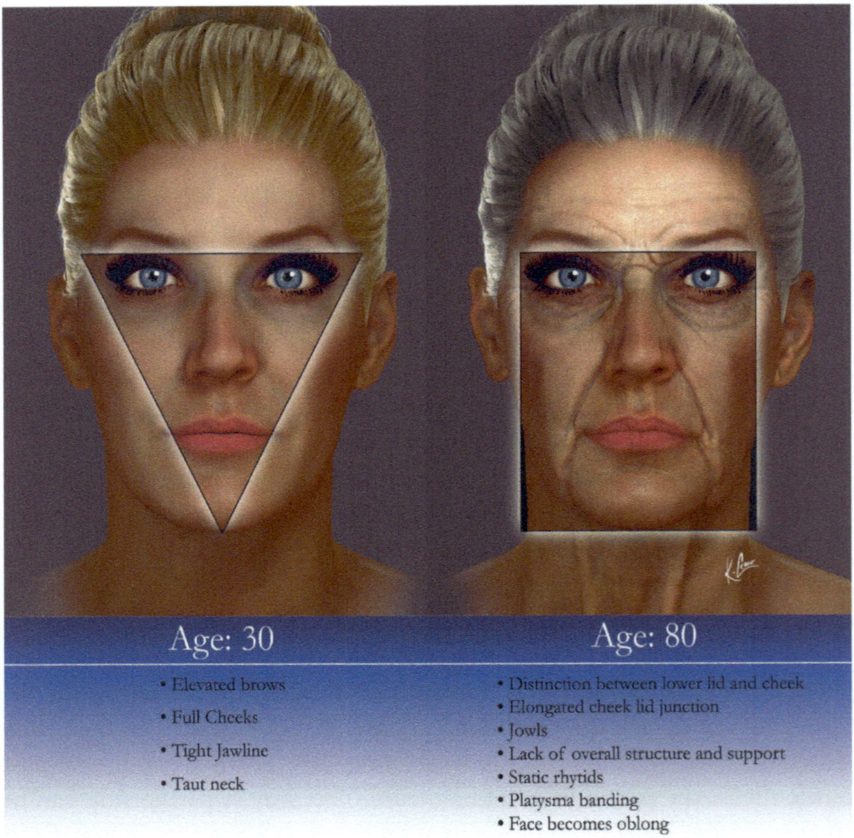

Fig. 9.1 The aging face.

cookie-cutter treatment plan based on the millions of unique presentations of facial features that will present in the clinic for treatment. Address and assess every patient individually. The way a novice injector injects will not be the way they inject in 5 years from now. This concept is based on the idea that as the novice injector injects more and more and analyzes more and more faces, a more in-depth knowledge of the anatomic relationships and the complex emotional view of the patient will develop. Having every patient come back for a 2-week follow-up is best for many reasons. One reason is so that proper after photos can be taken to showcase the work performed. Another reason for the follow-up visit is to develop a real understanding of the treatment's effect. Having the ability to review the treatments' effect will allow for further growth in the industry and shorten the learning curve time frame. Before and after photos help in growing new businesses. When a patient comes back at the 2 weeks follow up, it is an opportunity to capture photos and see the results of the actions taken during treatment in real-time.

Your assessment will evolve over time, and so will your skill and technique.

How does the face age?

(1) The dermis thins and tiny vessels become more visible with time.
(2) Bony structures can become more prominent as fat shifts or are lost.
(3) The cheeks and the mouth area hollow.
(4) The philtral columns will sink and or deflate. They, at times, will widen as well.
(5) The upper lip lengthens.
(6) Corners of the mouth become sad appearing and create deficits in the marionette region.
(7) The nasal tip drops.
(8) The lower 1/3 of the face drops and jowls form, enhancing a hollowing of the pre-jowl sulcus.
(9) The lower mandibular region narrows and pushes forward.

Aging occurs in two ways; intrinsically and extrinsically

Intrinsic aging is due to our genetic makeup and ethnicity. Extrinsic aging is due to sun damage, smoking, pollutants, and ultimately poor lifestyle choices, all of which add to exposure.

Patients can help to curb the aging process by paying attention to extrinsic factors. Providers can manipulate the internal environment to help patients reverse the hands of time with neuromodulators, fillers, lasers, chemical peels, and various other non-surgical options. The exposome influences skin changes. This refers to the internal and external factors and the interactions with an individual from birth to death. Keeping in mind that severe outside influences that rapidly progress the aging process are the use of tobacco, pollutions, lack of sleep, stress, nutrition, temperature, radiation (sun and computers).

Aging is a vicious process of wear and tear over time while being exacerbated by the exposure to daily oxidants. Muscles create friction and shearing, which causes a breakdown and a loss of volume in the face. The force of muscle activity also creates epidermal injuries such as static rhytids. Our bones begin to shift and demineralize, which creates even more shifting of tissues. It is important to understand anatomy and then how that anatomy ages over time.

Common aging features are drooping skin, flattened cheeks, fullness is lost in the upper face, and resides in the lower face at the mouth and jawline. A deepening of the nasolabial fold, enhanced jowls, harsh static rhytids are also a part of the more commonplace aging features. The face can take on a sad, tense, and tired appearance. We lose volume, and muscles weaken, connective tissue loosens and allows for more significant sag. Aging is part genetics and part lifestyle; prevention is more manageable than repair and restoration.

Youthfulness is expressed through shape, structure, and contour

Skin in the younger years is soft, plump, hydrated, and looks full. The cells of the skin turn over rapidly, improving the overall tone and texture of the skin (Figs. 9.2 and 9.3). However, once we age, we experience a loss in facial glands, which in turn, we have a decrease in oil production, immediately dehydrating the skin. We then lose collagen, hyaluronic acid, and elastin. The loss of collagen, hyaluronic acid, and elastin depletes the skin's building blocks, minimizing the health and youthfulness of our skin.

The loss of the three building blocks will lead to the dynamic lines of expression that become sad and angry emotions. Static lines will be created due to the muscles' abuse and forceful, repetitive action on the skin. With time our expressions leave a story on our face of how our anatomy works. Elastin is essential for our skin's ability to recoil, and once we lose elastin, our recoil ability is sharply stunted.

Facial esthetic treatments have evolved and now include correcting age-related volume loss in the midface. Rejuvenation has shifted from the two-dimensional approach to a three-dimensional approach. The typical youthful face follows the shape of an ogee curve. The ogee curve is the central facial triangle or the double soft-S curves seen on the youthful face from an oblique angle. However, with age, the mid-face flattens, and the skin begins to sag. The deflation of the cheek makes the nasolabial fold much more prominent. Then a concavity begins to form under the eye in the tear trough. As gravity pulls these structures down the pre-jowl sulcus area is hollowed, and the pre-jowl region becomes full and prominent.

Aging is happening on a variety of levels. The skin surface becomes discolored; the skin can become thicker or thinner. The texture changes, pores become larger and more pronounced, and old scar tissue changes how the

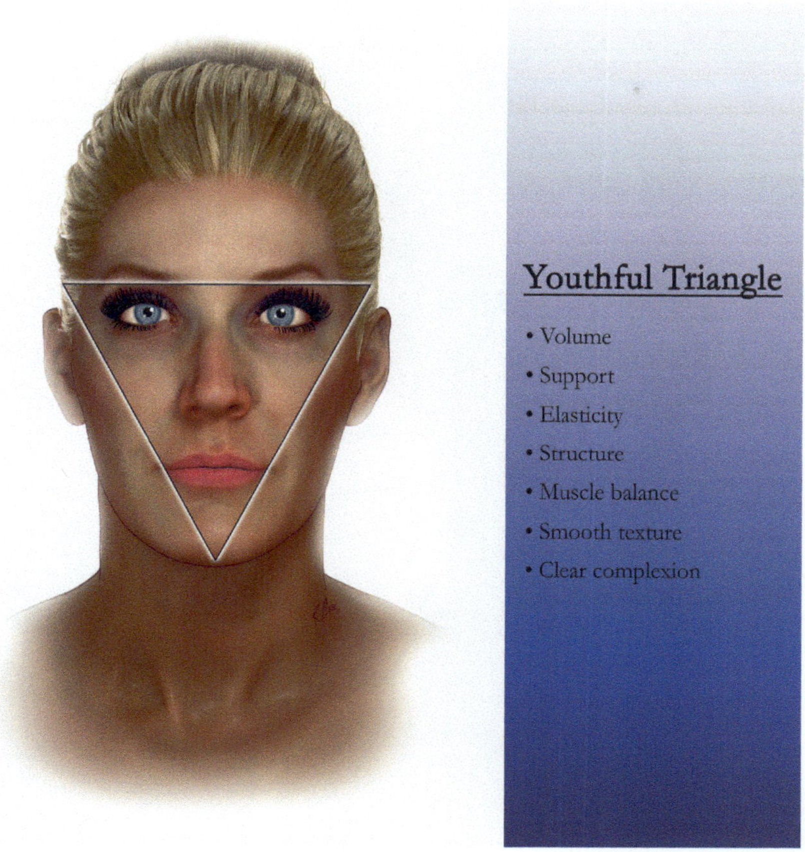

Fig. 9.2 Typical attributes of a youthful face include; a full upper face above a sloping lower midface.

skin lies or puckers. Permanent lines and wrinkles are set into hyper-dynamic areas. Gravity grips and begins the pull downward, causing fat pads to deplete and drop into the nasolabial area and the jowls.

In more recent esthetic literature regarding aging has focused on the anatomical facial fat compartment. This has revolutionized the concept and approach of adding volume to specific deflated soft-tissue compartments with dermal filler. However, it has added depth to the muscle's effect on the facial fat compartments. Dermal fillers with the understanding of the facial fat pads have allowed injectors to create a more individualized youthful restoration to the face. The overall improvement in understanding facial fat pads has compartmentalized the treatment strategies and prescriptive plans for facial rejuvenation. Rejuvenation and restoration treatment plans should

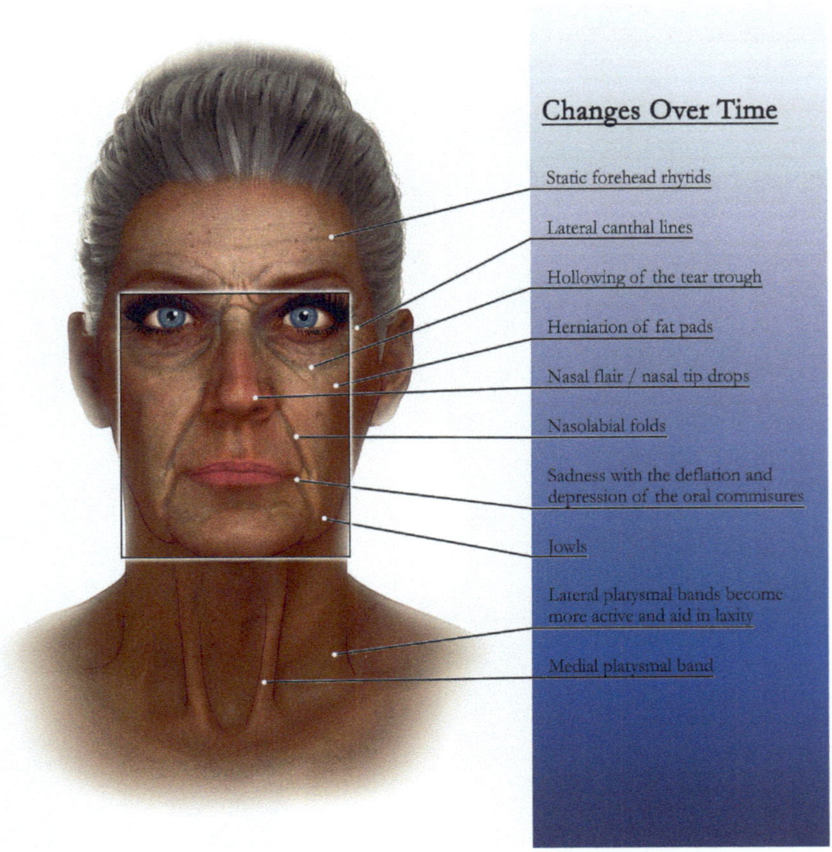

Fig. 9.3 As the face ages, fat atrophy often occurs in the mid-face area and settles at or below the jaw.

always start by looking at the patient's skeletal structure, evaluating the fat pads, and evaluating the skins' overall appearance. Looking for movement and static lines can give the injector insight into how the patient is aging. Aging is a layered concept and often starts laterally and progresses midline. As we begin to retire the youthful characteristics, we should be preparing to build progressively to optimize the natural beauty of the unique features the patient possesses. With an astute analysis and proper diagnosis, the treatment plan for correction and restoration can be created.

The skeletal aging process can be addressed with soft gel-like hyaluronic acid fillers. Reversible dermal fillers improve the patients' safety profile, and the providers, given the risk, are decreased in regards to adverse events. The use of dermal fillers will address bone loss areas, allowing the skin to redrape over the re-volumized bone structure. The injector can then address skin tone and texture with the use of chemical peels, prescriptive skincare, and or the utilization of light-based, heat-based, or radiofrequency devices. The last or it could be the first step in addressing the face as a whole is the use of neuromodulators to stop the hyperdynamic expressions using neuromodulators. Morphological changes occur on every level and layer of the face; the skin, the skeleton, soft tissue, retaining ligaments, and fat compartments. The intrinsic and extrinsic variables are what make each patient's presentation unique. The prescriptive plan for rejuvenation must include a proper diagnosis and a rejuvenation plan that is obtainable with the tools the injector has available in the office.

A patient's restorative plan should be a stepwise approach and should address the internal and external factors affecting aging. This plan should also address the face with thoughtful consideration to all the layers of the face. Create an organized and detailed plan for patients to be educated and empowered in order to make the right decisions for themselves. A treatment plan should be written out and provided for the patient. Find a way to brand the prescriptive treatments so patients can be a reference point for the patient, and if it is eye-catching and pretty, they will keep it with them and show their friends. In Fig. 9.4 you can see the treatment cards that DCCM™ uses at check out for the front desk staff's convenience. The treatment card is scanned into their record and then given to the patient as a point of reference. It is branded so that the plan sticks out in the eye and the minds of the patient. This will be discussed in greater detail when we move to the assessment portion of the reading.

Throughout the aging process, 1% of dermal collagen is lost per year and hyaluronans, especially after 50. Fillers can be curative and even preventative. Studies show that fillers have a synergistic effect in stimulating collagen production. Maintenance is essential with upkeep and unique and specialized at-home care. Fillers and neuromodulators are not a replacement for a facelift, but with impressive advancements in laser devices and other non-surgical options, a facelift can be postponed. Many surgeons have found a way to combine the facelift with other minimally invasive procedures to create an even more natural look. The current trends show that surgeries are falling in the statistics, while non-surgical options continue to rise. Extrinsic factors such as smoke or smoking, alcohol, nutritional intake, gut biome, and even stress all cause brown spots, rough texture, and premature aging on all facial levels.

Name:__________________ Date:__________________ TX:__________

Cost: __________

Follow Up: Phone ❑ In Person ❑

Service	Cost
❑ Botox / Dysport __________	________________
❑ Dermal Fillers	________________
❑ Facials	________________
❑ Fat Reduction	________________
❑ Hair Removal / Restoration	________________
❑ Laser	________________
❑ Skin Resurfacing	________________
❑ Skin Tightening	________________
❑ Vaginal Rejuvenation	________________
❑ ZO Skincare	________________

Notes: ____

6 0 3 . 7 8 3 . 1 0 8 7

Fig. 9.4 DCCM™ prescription treatment card. (Top) cover. (Bottom) back side.

Genetics, or otherwise known as intrinsic factors, also affects all the skin layers and contributes to the thinning of the dermis, dryness, and elasticity loss. Combine poor genetics with poor lifestyle choices, and a stellar recipe for rapid and early onset aging has been developed. However, this recipie is what creates an esthetic client. When patients seek injectors services and are still smoking, not utilizing sunscreen, or are in tanning beds, they may not be the best client for services. This type of client will have the most need, yet the results most likely will not be the best due to their lifestyle choices. These lifestyle choices can impact the outcome of esthetic treatments and or impede longevity. At times it is best to counsel the client on lifestyle choices before we pick up the syringe.

Aging muscles

Lying beneath the fat pads is the muscles that are repeatedly in motion through chewing, swallowing, talking, laughing, and so many other facial activities, causing the volume shift and or loss in the fat pads. As the fat is lost, the muscles' repetitive motion will directly lead to static deep lines. Creases will begin to form at the corners of our mouth, in-between patients' eyebrows and the corners of their eyes, to name a few.

The facial muscles do get weaker over time. The loss in muscle tone, combined with the skin's thinning, can make it very hard to treat a client with just neuromodulation. As patients progress in their aging process, injectors are forced to think of it as a unit of carefully woven together subsets. As the face ages, we also need to be mindful and often need to change the approach to the injection plan. It is thought by many varying esthetic injectors and esthetic surgeons that the aging of muscles is not all that well understood. It has been reported that muscles lengthen with age, increase in tone, meaning at rest. Muscles are more contracted in youth. Aging and weakening muscles are why the development of static rhytids in highly dynamic areas increases with age.

Bones are the foundation of exterior features. Everything rests on that gentle foundation. Typically a youthful appearance will have full high cheekbones. Age changes the dimensions of the bones, and the position shifts. The orbital rim will widen, creating a hallow look to the under eyes; there is also a significant change in the brow bone's angle, allowing the tail of the eyebrows to drop low and create hooding skin in the lids. Bones are continuously changing from the time we are born. The height of the midface will reduce as we age simultaneously with the angle of the anterior face.

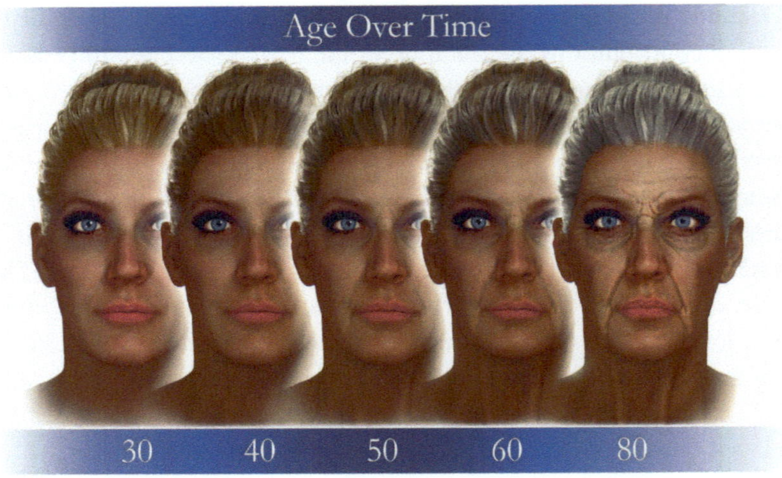

Fig. 9.5 Age over time.

Ultimately bone resorption changes the boney landscape, which redistributes all other structures of the face (Fig. 9.5).

Patient selection

Patient selection is crucial in esthetics. It is appreciable that there is a high rate of body dysmorphia and unrealistic expectations in this industry. Body dysmorphia and unrealistic expectations bring about an interesting topic for discussion. That is, the injector gets to choose the patient as well as the patient choosing the provider. Esthetic medicine is a talent based business coupled with a medical or nursing license. If an injector feels a patient may not be the right candidate for neuromodulation, the injector must not treat them. There is a slew of reasons why a patient would not be the right candidate. Injectors need to assess the patient not just from a physical standpoint but also from a psychological perspective. Being sure they are of sound mind and are reasonable with their goals and the anticipated outcome is key to the injector's success of the outcomes. It is far more effective to say no to a patient than to continually chase a look that is not achievable. In the beginning phases of an injector's career, it is common to feel so excited to have patients in the treatment chair looking to experiment and treat everyone.

Nevertheless, we need to hit the breaks and genuinely assess the patient to be sure they are appropriate candidates. Do not get so excited and blindsided by the opportunity or the dollar amount. Injectors must only treat the right candidate. A significant amount of education will be necessary to share at the patient consult. Sharing the anticipated treatment goals and realistic outcomes are all part of the process and will only aid in the patient's overall satisfaction with the treatment. Patient satisfaction was highlighted previously, focusing on patients having a false sense of understanding of the product's intended use. Their lack of understanding, coupled with the false sense of reality portrayed by social media platforms due to misrepresentations and filtered results, can increase patient dissatisfaction. If the injector is quick to perform the treatment in the wrong candidate, it may end up costing them more in the long run. It is better not to treat than to be in the business of chasing results or having to offer refunds. Esthetic medicine is not a get rich quick business. Esthetic medicine is medicine, and the patient's needs should always come first. Inject with full knowledge of anatomy, and in-depth understanding of the mechanism of action of the drugs being used, and a foundation of morality. Think on all levels before beginning to inject. Look at the situation from the patient's perspective and develop sound treatment plans based on science, research, and integrity. Injecting to get rich can ensure failure. The sense of integrity and morality is transparent to the patients. Patients are savvy and will see through this mindset. This mindset can also hinder the development of meaningful relationships within the industry that will be discussed later in the reading.

Patient assessment

Before we begin to get into the art of assessment, it is a good idea to review some terms that are more friendly to the patients' ears. A consult is a vulnerable time for them and can lead to a feeling of worthlessness or hopelessness if injectors do not choose positive words over negative words. Think for a minute about the difference between these two statements.

This deep fold here on the lower face can be filled with filler, that will be $1800.

Alternatively, another version would be:

The nasolabial fold can be softened by redraping the skin of the cheeks and then adding a bit of volume directly to the affected area. This way, a natural repositioning of the tissue will occur due to the placement of the product. The total would be $1800 to address these two areas. The cost did not change, but the presentation did. It is essential to use words that do not make the patient feel bad. It is the injector's job to create a sense of peace for the patient as many of them feel badly about their aging process; "I should not have drank," "I should not have laid in the sun," "I should have started earlier." The patients are already beating themselves up. It is the job of an injector to soften that blow. An injector is not just filling and freezing better yet; they offer patients a sense of restored youth; we are often empowering them to look and feel their best. Many patients arrive at the doorstep of rejuvenation clinics because they have just endured some of their most challenging life moments, and they are ready to make a change for the better. The first foot in the door is their path to making a positive change in their lives, both physically and emotionally. If the injector can not tap into the emotional side of the reason they are in the treatment chair; the results will always be mediocre. Even though the injections may be perfect, the patient also needs the confidence to recognize their beauty. The job of an injector sometimes his 90% psychological and 10% treatment.

Create a journey for the patients and create an environment where they are relaxed, educated, and can begin developing their confidence. Empower them to take charge and not look to the past. Better yet, show them what the future can look like in trusting hands. Continued education in esthetics, dermatology, and an advanced understanding of anatomy will help empower the injector, empowering the patient. Being a leader and an innovator, being kind, and sharing knowledge are just a few keys that can help separate one injector from another. Point and plunge injectors are not the goals that should be set to become a success in the industry. Let us move on to those kinder, gentler terms to share with the patient during the assessment and consultative process.

A consistent assessment with an individualized treatment plan will yield success.

Consider sharing the following ideas with the patients:

Refill with fillers
Resurface with lasers and chemical peels
Relax with neuromodulators
Redrape with fillers or surgery
Retighten with elastin boosting treatments such as the profound device by Candela Medical.

Consider the face like a bed. The volume loss is the mattress, and the skin is the sheets. Once the mattress is volumized, it is essential to put fresh unstained sheets on the new mattress. Reversing the hands of time does not happen over night, nor does it happen on one facial plane. Be sure to make the patient aware of this so that then a collaborative approach to creating a treatment plan to meet their goals. When the injector and the patient are in harmony with their understanding of one another, patient satisfaction is increased.

Over the years, the field of esthetic medicine has moved away from assessing patients in a one-dimensional view. We now view patients at rest, in motion, and from all sides at rest and in motion. With each new product or device technology development, there comes a new way for injectors to view their patient during the assessment. Another view on assessment yields a different prescription for treatment. The advancement of technology, combined with the evolution of the assessment process, only enhances the injector's ability to create better and more beautiful, natural results. The art of assessment should consist of animation, rotation, and tilting (Fig. 9.6). Ultimately, any treatment's success will come from the injector and the patient viewing the facial features from all angles and in all forms of animation in all angles. It is advised that consistent photos are captured of the patient at rest and in motion from all angles. It is common practice to have photos from a side profile, 3/4 angle, and straight on.

(1) Animation is necessary for all esthetic treatments. Watch the patient animate while they are talking to see them in their natural animated state. Then, have them squint, raise their eyebrows, frown, and smile. Be sure to take note of both upper and lower face animation changes. Take note of asymmetries at rest and with animation. When the treatment begins, be mindful not to overcorrect the at rest asymmetry as the correction can create a new asymmetry in animation. It is a delicate

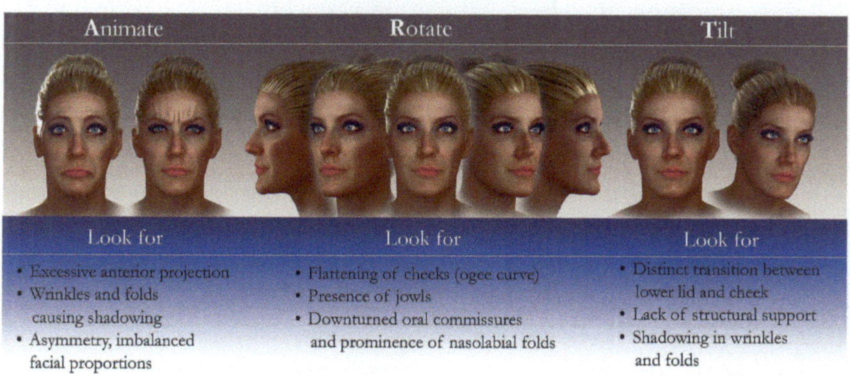

Fig. 9.6 The art of assesment.

balance with the treatment. Pointing out an asymmetry to the patient should be a delicate and considerate conversation when a patient has a stronger nasolabial fold on one side versus the other. It is best not to point out the heaviness of the fold. Instead, discuss with the patient that the muscle is a bit stronger on one side versus the other side, causing the difference. Alternatively, another may approach it from another position on the face altogether. It could be helpful to note the difference in the fullness of the patient's cheek. The injector might say that one cheek bone sits a bit higher than then the other side. So, if the opposite side is volumized, the nasolabial fold can then be softened. Please note to the patient that all patients have subtle asymmetries, and they make each patient unique.

(2) Rotation is necessary to view from all sides of the face.

Have the patient rotate from side to side and view these positions both in rest and animation. Take note of the severity of rhytids. Are they moderate or severe? When the patient turns to the side, this is an opportunity to evaluate the canthal fan; is it a full fan, upper, middle, or a lower fan? An iPad or tablet can be helpful in the assessment process as well as a mirror to engage the patient in the thoughts and observations as the treatment plan is being created. When they can begin to understand the evaluation process, the multistep approach will often be an easier transition for them. If the patient cannot understand why the injector is doing what they are doing, they will unlikely convert from one syringe to multiple syringes or multiple modalities. We never want to sell to the patient. We want to educate them so that they can make informed decisions as well as informed consent.

(3) Tilting is extremely important to highlight the malar fat pad split when the tear troughs and cheek volume are a concern. Ask the patient to put the chin down in order to assess this portion of the face. This tilted position is also essential to notate the central laxity developing in the patient's face. The patient should tilt their face down, bringing the chin to the chest, and then have them lookup. Do their eyebrows raise? The injector can assess for ptosis in this position and focus on the levator palpebrae superioris.

Physical assessment

Let us begin to identify key assessment features regarding neuromodulation and those specific muscles in the treatment zone.

Assessment of the upper face

When assessing the Glabella region, always assess at rest and in animation. It is also wise to assess the patient naturally, observing them talking during their consult. Capturing the standard animation is critical in successful treatment.

If the patient presents with a downward pull that forms a line at the bridge of the nose, then a dose of neuromodulator will need to be placed low in the procerus. Another example is if the patient presents with at rest creases or creases with animation between the eyebrows, then the corrugator, both medially and laterally, will need to be treated.

It is imperative to note the pull of the lateral tail of the brow. If the patient has a significant pull at the tail, and this injection is missed, they will still present with movement in this area at 2 weeks. If the opportunity to see them back at 2 weeks is missed, the relationship an outcome could be compromised. Leaving even a little movement could shorten the life of the treatment, thus creating a dissatisfied patient.

There is an important consideration when treating only the glabella region, which is to assess the lateral portion of the frontalis about mid way up the forehead; this has been referred to as the Mephisto sign. To assess this region, the injector will want to immobilize the glabella complex with two fingers and then have the patient raise their brows. If it is noted that the patient has an exaggerated arch in the brow during the assessment, then it will be necessary to treat the main pull of the arch on the frontalis.

Assessment of the frontalis

First and foremost, does the patient even have a functioning frontalis muscle?

Have the patient close their eyes, gently open the eyes. If the patient moves the frontalis during this gentle movement, they are not the right candidate for forehead treatment. Given that the frontalis muscle is the only elevator muscle for the brows, and if that is taken out by treating the frontalis, they will feel heavy and develop hooding of the lids. They will say they feel like they have to continually lift the brow, lift their brow to put makeup on, or have a headache. No matter how they describe it. They will not like it.

In a suitable candidate, one will want to asses within the treatment zone and beyond the temporal suture lines to gain insight into the entire area. Does the patient have a split in the midline of their forehead, or are the lines consistent across the forehead's entire span?

Assessment of the orbicularis oculi

The crow's feet are an early indicator of premature aging. The area only has a small bit of subcutaneous fat in the area; the orbicularis oculi muscle is attached to the dermis in this area. This area's repetitive motion will cause deep lines "crows feet" around the eye, otherwise known medically as the lateral canthal lines.

To assess if the patient is a good candidate, have the patient smile big and then squeeze the eyes shut. The injector will want to observe the muscle fibers where the orbicularis oculi muscle pulls down in any area.

If the patient has a herniation of the lower eyelid fat pad, typical surgical blepharoplasty is the only route. There has been some off label success with specific Radio Frequency Microneedling devices as an alternative to a surgical blepharoplasty.

If the patient has fine lines under the eyes, these may be treated with Neuromodulation; however, it can increase the aperture of the eye. Significant skin laxity in the treatment area would be a contraindication for this treatment. Perform a simple snap test to see how quickly the skin snaps back. Pinch the skin gently just below the lash line and see how quickly it snaps back. The pinch should be gently performed with just the fingertips. If the snap is quick, then it is safe to treat in the area. A general rule of thumb is that if there is a situation where there is a question on whether or not to perform a snap test, then the patient is likely not a good candidate.

Brow assessment

The brows have only one elevator, and that is the frontalis. There are three depressors for the brows. The procures, corrugator, and the orbicularis oculi. The procures pulls the medial aspect of the eyebrow down to produce the horizontal line across the bridge of the nose. The corrugators draw the brow downward and inward. The action from the corrugator muscle is responsible for producing the vertical wrinkles in the glabellar complex.

Static versus dynamic lines

We have mentioned static and dynamic lines repeatedly throughout the text so far. A static line is a line seen while the patient is at rest. Static lines have been etched in overtime with the over use or natural use of the muscle or muscle groups in that area. The dynamic lines are lines only seen when the patient animates. Over time, the dynamic lines are the cause for most static lines in areas of animation. Static lines can be seen in the absence of an area that is highly mobile. These static lines come from intrinsic and extrinsic factors such as we noted before, sun exposure, genetics, smoking, alcohol. These static lines not associated with animation are due to loss of hyaluronic acid, decrease in elastic and collagen. A good example of these particular static lines is when the crepey appearing skin is stretched and the lines are still present. Patients will often think that having their neuromodulation treatments that the static lines will go away. This type of thinking is not correct, as it is not the intended use of neuromodulation. Therefore, setting up proper patient expectations is key to their satisfaction. It is essential to educate the patient that neuromodulators are working on the muscle and directly affect the muscle and that the static lines are not the intended target. Static lines benefit most from skin resurfacing treatment in combination with their neuromodulation. Youthful skin will commonly have some restoration in the static lines abut do not over-promise that as an intended outcome, or the patients will be very dissatisfied customers. The older population will often have much more stubborn, and deep-set static lines and will almost always need synergistic modality to properly restore the skins' health and function.

Setting patient expectations

We have touched upon this throughout the book here and there. Patient expectations can sometimes be more challenging than the actual treatment. The challenge is to visualize what the patient sees as satisfactory lift or revolumization versus what can be achieved by the injector and the product. Eventually, injectors will develop a solid understanding of the treatments reproducibility and the products chosen. The patients will not always have a realistic expectation of what talent lies within the injector accompanied by the arsenault of tools they can access. There is a wonderful device that can enhance the foundation of the patient's skin; it is called the Profound Radiofrequency device. This device helps to restore collagen, elastin, and hyaluronic acid within the skin. It is a product that can create about 37 to 65% of a facelift. The Profound Radio Frequency device treatment is called a non-surgical facelift. Saying non-surgical facelift to the patient sets the injector up for failure unless the rest of the conversation is carefully masterminded. It is expected that the patient often only hears facelift. Their interest is immediately peaked, but by using the word facelift, we may have inadvertently created a false sense of reality, making the happiness of the outcome unobtainable.

Patient outcomes are not quantifiable and or measurable in a tangible fashion. Outcomes can often be described from a qualitative nature. A qualitative understanding of outcomes makes it extremely difficult to pinpoint an anticipated outcome for patients receiving injectables or other esthetic procedures. Utilizing before and after photos can help provide a sense of realistic expectations for the patient. Selecting the proper before and after photos is imperative. The best before and after photo to share with the patient to provide a visual reality for them; is one of a patient similar in age, goals, skin quality, weight, and volume loss. Show the most common results. It is nice to share the home run outcomes but pepper them into the mix of more realistic results. This is an attempt to showcase the work's reality, not an opportunity for a gimmick or a sleazy sales pitch.

Given that the esthetic industry is not an industry that fairs well to refunds, being truthful out of the gate will be most rewarding to the practice. A laser can not be undone, nor can the product be sucked back out, so be sure it is what the patient needs and wanted. Ensuring the patient is happy is the best way to deliver results that will not yield a refund request. The best tool is a series of progressional photos as the treatment journey moves along to showcase a patient's progression. Take pictures before and after every treatment and every visit. Quality and consistent pictures are a saving grace with regard to patient complaints.

Understanding how patients interpret their outcomes qualitatively can help injectors direct the conversation of anticipated results. Quantifying patients anticipated outcomes could be challenging for sure. The expectation of outcomes is more of a concept than a concrete piece of evidence, so as providers, we need to tease out a way to quantify the outcome to improve satisfaction. The healing process is an essential part of the overall outcomes. The gap in perception is vast with regards to outcomes. Think for a minute about wealth means? Wealth to different individuals may be a billion dollars to one, whereas wealth to another, maybe a small hacienda with an abundance of family members in the homestead. Directly ask the patient what they want for an intended outcome. The patients are generally much more optimistic about the results than the experienced injector. Injectors tend to be less optimistic about the results so that there is room for the patient's emotional variability in the anticipated outcome.

Make a clear note of the interview process with regard to the questions asked in the consult and the patient's answers. Ask them about their facial appearance in general, about the quality, luster, hydration, their aging process. Patients will often infuse emotional verbiage in their description of their physical appearance. Emotional terms are critical to notate as it can help understand the emotions behind the patients aging process and give the injector further insight into the possibility of some underlying body dysmorphia. The psychological well-being of confidence, distress, self-esteem, and self-consciousness are all emotional hurdles to the patient's interpretation of the esthetic outcome's success. Injectors can hit a home run, but if the patient is not willing or capable of seeing their beauty, the harmonious, balanced, and beautiful outcome will be wasted. Patients long to feel normal in their skin at all ages. At age 30, patients will pick apart their flaws and will not appreciate their beauty until a decade later. When the patient turns 40, they long to go back to the beautiful and youthful image of 10 years ago. It is not easy to appreciate what we have until we do not have it any longer. After the patient has been augmented, the flip side of this coin is that they are now uncomfortable in their new face, and it can take them a few days to get used to it.

It is always interesting to listen to patients talk about their significant others' opinions on their choice to partake in the esthetic services. When a patient notes that her husband does not know that she has work done, then a bruise will be a real big problem for this patient. As medical professionals, we understand the bruise is the absolute relative risk for the procedure. If the treatment is a secret, the bruise becomes a part of the experience. The emotional satisfaction of the results can be skewed at times because of the bruise. It is also common for patients to think that the injector is bad because they bruised. Education on downtime and the reality of bruising should be discussed and emphasized so that the patient may make an informed decision on the proper timing of the treatments. If a patient has a big event coming up, it is best not to treat the patient right before the big event.

Discussing proper expectations during the results and healing phase is directly related to their end satisfaction. The more informed they are, the less nervous they are; therefore, they will judge the process less critically. Be sure to let them know that they will need to come back for a follow up to ensure both parties are satisfied with the results. Remember, this is a partnership, not a dictatorship. Sometimes, the patient thinks they have all the power, but the power should balance between both parties. It is the patient's goal and budget, and it is the provider's executive abilities to produce the results and keep the patient safe while creating a more youthful or restored look.

Interpreting patient outcomes in the minimally invasive world is highly reliant on qualitative interpretations. It is recommended to develop a tool for the office with a few keywords on it for the patient to describe their facial features, describe the emotions attached to the aging process, and then have them use descriptors of how they anticipate their outcome. For example;

If I were to describe my features. They would be described as squaring of the lower face, enlarged pores, redness, and volume loss. My emotions and or fears surrounding my aging process are fear of not being loved or taken seriously, fear that my aged appearance will lessen others' respect, and I fear that if I do not age in a preservative fashion, my career as an injector will be shortened. When I describe my anticipated goals for the outcome after my treatment, I am looking for clear skin, less laxity, tighter pores, a more youthful, healthy, and vibrant appearance. I do not want to look run down, tired, and blotchy. By this quick interview process, you could get into my head and grab onto my feelings and emotions and create a tangible plan for my outcome. Risk-benefit analysis is an excellent way to tease out the patient's version of their anticipated outcomes.

The consultative process is very time-consuming, but it is necessary. Keep in mind that the consultation process is time out of the schedule while developing a treatment plan. Free consults are not advised. The consult is where injectors are providing advice, and the treatment plan is created. Time is money. Perhaps for the first year, offer complimentary consults but quickly move away from this practice as it can discredit the educational process. When a patient goes to the doctor and they recommend a treatment, but the patient does not have the treatment, the doctor still charges for the consult. Esthetic medicine is medicine and should be treated as such. Education in this industry is not cheap; protect it by charging for the time and knowledge given to the patient during the consult. It seems daunting and over the top to be spending so much time on the consultation, but it is worth the time and effort. A consult will not yield much profit initially, but a well done and thorough consultation will pay off in the long run. Messaging with an unhappy patient and having to fight a credit card company because the disgruntled patient reported the credit card's charge could take up way more time than 30 min of the day. Spend the 30 min in the consultation room, teasing out the variables to ensure the patient's satisfaction. We are health care providers at the end of the day; our goal should always be our clients' safety and happiness.

In the notes, notate it as such:

Reason for seeking esthetic services:

(1) Reshape the squaring of the lower face
(2) Enlarged pores

(3) Redness
(4) Volume loss
Emotions used to describe aging:
(1) Fear of love loss
(2) Fear of job security
(3) Fear of lack of respect
The common theme here is fear
Goals for anticipated outcomes:
(1) Clear skin
(2) Less laxity
(3) Smaller pores
(4) Youthful appearance
(5) Healthy skin/vibrant
(6) More awake and less tired

Now, let us discuss how one would go about addressing these goals to put a plan into action. The jaws squaring can be treated with fillers in the upper face, Neuromodulation of the masseters, along with skin tightening devices or fat reduction devices depending on the causative system creating the squaring off of the jaw. Enlarged pores can be corrected with microneedling and laser treatments as well as little microdroplets of neuromodulation throughout the face; these should be coupled with monthly facial services in the office with good prescriptive skincare sold at the office. The volume loss would be addressed with the use of hyaluronic acid gel dermal fillers. When all of these goals are met, the injector wipes out the emotions of aging. Once the patient feels restored, they feel less fearful of the aging process as they have taken back control over the aging process. When it is left untreated, it can feel very hopeless and chaotic. Putting them back in the driver's seat can help improve the clients' emotional well-being, thus increasing the overall satisfaction of their outcomes. During the patient's follow-up appointment, be sure to ask how they feel before asking about how they think they look. Ask them specifically about their goals utilizing their own keywords written down at the initial visit. Utilize the goals provided as an example. Ask how they feel about their skin? Note a positive change seen easily by the injector and the patient. Be specific with the questions about their skin utilizing the words they chose in their initial evaluation. Referring back to the example previously, redness, laxity, vibrance, tighter were of concern. In that case, utilize those words precisely as this is how to take a qualitative concept and put some tangibility to the results.

Consult process

The consulting process certainly includes setting patient expectations, but many housekeeping needs to be addressed during the consultation process. A consult should begin with an in-depth medical history. See Appendix 1 for a sample medical intake for neuromodulation.

The medical history should include a patient's past medical and surgical history, previous esthetic services, medications, allergies, social history, surgeries, implants, and previous esthetics services. Furthermore, for thorough documentation, occupation and occupational history can play a role in the treatment selection of services. Having the patient's pharmacy name and number on file and their primary care provider's name is of the utmost importance. At times the injector will need to prescribe medication such as anti-virals, antibiotics, or steroids for the patient, making this information necessary. The patient should sign all consents before the commencement of pictures and services. The consent should have the before and aftercare recommendations within the consent and be a separate form so that the patient can leave with a copy. Once all of the necessary paperwork is complete, have the patient face double cleansed and prepped for treatment. With a headband or headwrap on securing the hair away from the patient's face and ears, it is now time for photos. Photographs should be taken consistently with the same lighting and foot placement to ensure the after photos are in the same position. Many name-brand companies of neuromodulators have foot dials for the very purpose of consistent photos. A painted matte flat black wall will work, or a black matte curtain with a ring light in front. There are several apps on the commercial market for use. One program over the other is not necessarily superior. For the most part, it is which one works best for the individual and unique practice and technological abilities or limitations. Evaluate the software from the standpoint of ease and the ability to interface with the booking or online charting software system. The RX photo has gained significant popularity. This particular photo software is HIPPA compliant, user, friendly with many exciting features for marketing. The patient's face should always be midline, with eyes looking straight ahead for most of the photos. One exception is when treating the tear trough area, it will be necessary to obtain one picture looking straight ahead, one looking down, and looking up at the ceiling with the head in a neutral position. With the change in eye positioning, one can elicit the festoons and capture them on camera. We discussed earlier the importance of trying to quantify a qualitative feeling. Pictures can help to quantify patient outcomes. However, if the pictures are not consistent, this is not a useful tool. Note-taking is essential during the consultative process. Watch their body language, which areas do they touch most frequently. Listen to their word choices, especially the words that are emotionally charged, such as, " I look sad," "I look old," "I look like my mother." The word sad, old, and mother are critical emotional elements to the successful development of a treatment plan. These are the three key emotional trigger words related to what needs to be changed in the patient's mind. With the syringe, injectors are changing things both physically and emotionally. The syringe's success is directly connected to an injector's ability to get inside the head of the patient and understand them psychologically while figuring out the emotional motivators and hang-ups. If someone states that they fear they will look like their mother, ask them for a picture of their mother and have them point out their fears. Beauty is in the eyes of the beholder. Figuring out what beauty means to every patient will be the key to success and of course, in combination with award-winning results. Most injectors start with neuromodulation then branch off to filler, chemical peels, lasers, and the addition of esthetics services to provide lashes and facials. Even when in the infancy stages, a wholesome understanding of aging and treatment options should be understood. The patients signed consent long before they got into the treatment chair, but now it is time to have a verbal consent. The consult's consent portion is when the products, procedures, goals, risks, benefits, maintenance, and cost are all verbally mapped out. The written consent and verbal consent go hand in hand. In order to be genuinely informed, one must first be educated. Take the time to educate the patient to be genuinely informed, so their signature has more merit on the consent. As we become experts in the industry, we can lose detailed educational components to the treatment plan. Patients are well educated these days and have already done some homework on Google and other social media platforms. They have an agenda and a plan of their own before they even walk in the door. At times, it can take real rewinding to properly educate and inform the patient about what is best for them, not what is best for their friends or family members. It is common for patients to think because a friend had filler in their chin and that they also need filler in their chin. Figuring out what the patient's perception of beauty is difficult. The patient's friend who had the chin filler probably looks great, so now the patient also wants to look great. It is up to the injector to be the expert and figure out which treatment is best for them to look great. Patients will hear what they want and retain about 10% of what was said.

Now that the paper work photos are done let us now focus on the patient's anatomy.

Does the patient have:

Static lines?
Dynamic lines?
Do they raise their brows even at a resting position?
An elongated upper white portion of the upper lip?
Rhytids along the vermillion border?
Photodamage?
Volume loss?
Malar fat pad split?
Do they have hooding of the upper lid?
Gummy smile?
Dimpled chin?
Static mental crease?
Reabsorption of bone?
Increased bigonial distance (squaring of the lower face)?
Broad jaw?
Clicks or pops with the opening of the mouth?
Jowls?
Have depressed oral commissures?
Nasal tip droop?
Gingival show?
Deep nasolabial folds?
Asymmetries?
What is the shape of their face?
Oval
Triangular
Square
Round

Concomitant procedures and their efficacy

Neuromodulation is a fantastic esthetic treatment and is consistently the leader year after year in non-surgical interventions to prevent aging. However, it can be enhanced with concomitant procedures. Knowing that the face is a multidimensional system, and the aging is not occurring in a systematic approach will help the injector better assess and plan for the patient's needs. Aging, in terms of the patient's perspective, is seen as chaotic and uncontrollable. The patient will often ask for neuromodulation of the "11 lines" in their upper face. The well-informed injector realizes how and why the lines arrived, but most often, the patient does not understand the physiological changes. Here is where it is essential to properly consult the patient to set realistic expectations for the treatment. When treating the dreaded 11 lines in-between, the eyes ask the patient if they want the static line or the dynamic lines gone. Keep in mind; the patient most likely will not understand static versus dynamic, so it will be necessary to use words like at rest or in animation. If the patient is treated with neuromodulation in the glabellar region and wanted the static line gone, they will be unhappy with their neuromodulation results. Neuromodulation is working on the muscle, not the skin. There can, however, be a secondary effect of smoothing of the skin, but that is not the intended purpose of neuromodulation. Neuromodulation stops the muscle from compressing the skin and breaking it down, further exacerbating the static rhytids. More often, the static lines in the glabellar region are deep and challenging to get rid of, so injectors must counsel the patients on what neuromodulation can genuinely achieve. In the case of a patient with static rhytids, neuromodulation should be done to initiate a preventative plan that will prevent the further breakdown of the skin in that area by the blockade of the muscle. After neuromodulation has been initiated, then it is appropriate to explore resurfacing or collagen inducing therapies.

The treatment options are vast, and the injector must understand the options available to the patient, not just what the injector offers. Whether injectors offer other modalities for concomitant procedures or not in their office, it is 100% necessary to understand all of the options available to the clients. The quickest way to lose a patient or be taken advantage of is by not being the expert in the room. If the patient knows more than the injector from social media and google versus the degree the injector has earned, then a trusting relationship likely will not be developed. Consumers need to be confident that the injector is the expert in order to care for their needs. It is a delicate balance of power with the patient. They need to be included in the treatment plan and have a say, but they still need confidence; they are safe in the expert's hands. Find the balance as it may vary from patient to patient. It will be a juggling act while figuring out the personalities of the clients. At times the injector will need to be the piranha and, at other times, a silent observer like a chameleon. Each patient will need something different. Psychology should be a part of the entry-level education for injectors as this is not traditional medicine and requires a slightly different approach to meet the patients' needs. It is recommended that every new injector take a course on psychology to understand the patients better. Understanding the patient's thoughts and wishes is of equal importance to the outcome. The importance of the assessment is essential, especially when other modalities are not offered. The injector always needs to be rooted in morality to deliver the best possible care, and sometimes that does mean not providing any care.

The aging process is brutal in so many ways regarding volume loss, photodamage, static and hyperdynamic muscle movement, jowling, and many other complaints. If the problem is photodamage, the patient should have intensed pulse light therapy (IPL), not neuromodulation. The patient may benefit from neuromodulation, but if it is not what they seek, it will not make it the correct treatment for them just because it is what is offered at the practice. Pairing neuromodulation with fillers, kybella, CoolSculpting, Profound RF skin tightening, ZO skin health prescriptive skincare are all synergistic in the outcomes. At first, novice injectors will only see the muscles. Over time, they will begin to see the skin changes, the brown spots, the enlarged pores, the jowling, and the ligament shifts creating volume loss and displacement of tissues. The focus while learning to inject neuromodulation will be on the muscle and the lack of animation initially. As one progresses in their learning, a natural emersion of a greater understanding of the facial assessment will develop. The more faces the injector sees in a consult for neuromodulation, the more they will begin developing a greater understanding of in-depth anatomy and facial assessment. Moving beyond the proper dose and landmark combined with the needle's right depth will yield the next phase in the learning curve. The individual anatomical representations of anatomy will pose the most comprehensive challenge yet. Early on in an injector's career, they notice the value of concomitant procedures. As wonderful as neuromodulators are, they are limited in their ability to address aging as a whole.

Figuring out the next treatment to bring into the practice can be overwhelming, with so many platforms influencing the decision process. Turn to the statistics for a basis for narrowing down the options. After a thorough review of non-surgical esthetic procedures' statistical trends, then survey the existing patients within the practice. If the statistical data suggested that vaginal rejuvenation was the leading non-surgical treatment, but the existing clients all asked for photofacial, then perhaps the national statistics should be the jumping-off point, but the real data is the survey form the existing patients. Existing patients are a captive audience to the injector's marketing. However, if the patients are not in need, they will not consume the service.

Start slow and small. Become a master in one or two treatments before offering them all and only giving marginal results. The patients within the practice are the best people to poll to figure out the next best service.

References

Benedetti, J. (2019). *Structure and function of the skin.* Merck Manuals Consumer Version. https://www.merckmanuals.com/home/skin-disorders/biology-of-the-skin/structure-and-function-of-the-skin. (Accessed 11 August 2019).

Farkas, J. P., Pessa, J. E., Hubbard, B., & Rohrich, R. J. (2013). The science and theory behind facial aging. *Plastic and Reconstructive Surgery. Global Open*, *1*(1), 1–8. https://doi.org/10.1097/gox.0b013e31828ed1da.

Grey, A. (2015). Spirituality seeking theology. *The Catholic Library World*, *1*(1), 191.

CHAPTER 10

Injection preparation

Now the exciting part is moving toward learning how to prepare the products and inject the patients. Developing long-lasting relationships with the patients will set injectors apart and excel individual injectors to the next level. Loyal patients can have a detailed yet honest conversation about their real opinion on their treatment. How did it make them feel, what do they see, what does the injector see? Do not ever minimize what they feel or see; it is their face and their experience. The job of the injector is to educate the patient and set realistic expectations through knowledge sharing.

Pause for a minute and review which direction each muscle is moving before neuromodulation is placed. Think of it as Neuromodulators are going to flip the script; if the muscle goes up, then in the face of neuromodulation, the muscle will go down.

Reconstitution

All products come in a vacuum-sealed glass bottle and are in powder form, meaning they are lyophilized. It is advised that the instructions on each product for reconstitution be carefully read. There is a Federal Drug Administration guideline (FDA), and then there is, frequently, a varying practical application.

Authors Recommendations for reconstitution

(1) Attach the 27 1 1/2 gauge needle to a 3 mL syringe.

(2) Clean off the top of the medication bottle and the sterile saline or bacteriostatic saline bottle with alcohol.

(3) Draw up 2.5 mL of bacteriostatic saline into a 3 mL syringe with the 27 gauge needle to use Botox, Jeuveau, Xeomin. Draw up 3 mL sterile saline or bacteriostatic saline for the use of Dysport.

(4) Insert the needle at a 45-degree angle into the medication bottle. Hold firmly to be sure the needle will not plunge straight down to the bottom and strike the glass. Remember, the bottles are vacuum-sealed. Therefore, the needle will be pulled in rapidly and drive straight down

Essentials of Neuromodulation
https://doi.org/10.1016/B978-0-323-89920-8.00030-8

if not firmly holding onto the syringe. Once all of the saline has been placed in the medication bottle, remove the syringe.

(5) Record the date and time of reconstitution on the side of the bottle. Take note that the vials are indicated for single use according to the maker's FDA approval guidelines and further recommendations.

(6) Gently swirl the medication to mix. The exception to this rule is with the use of Xeomin. Xeomin package instructions require the provider to flip the bottle upside down and place it on the counter upside down for 15 min before use.

Botox 100 unit mix with 2.5 mL of bacteriostatic saline
Xeomin 100 unit mix with 2.5 mL of bacteriostatic saline
Jeuveau 100 unit mix with 2.5 mL of bacteriostatic saline
Dysport 300 unit mix with 3 mL of bacteriostatic saline

All medication prescribing guidelines recommend reconstituting with 0.9% preservative free saline. The Aesthetic Society (2019) found in a study that bacteriostatic saline use caused less pain during injections and demonstrated fewer infections.

Each product depending on how it is reconstituted will yield different units of measure per 0.1 cc. The following Tables 10.1–10.4 will help progress the novice injector to advanced. Many injectors are given just the tables and the thought process has been taken out of then equation. It is imperative to be able to think independently and be able to understand the concepts of drug reconstitution and the yielding dose (Tables 10.5 and 10.6). Below each drug table is the formulas on how to arrive at the resulting dose depending on the reconstitution.

Table 10.1 Dilution instructions for Jeuveau 100 unit vial.

Diluent[a] added to 100 unit vial	Resulting dose units per 0.1 mL
10 mL	1 unit
8 mL	1.25 units
5 mL	2.0 units
4 mL	2.5 units
[a]2.5 mL	4 units
2.0 mL	5 units
1.0 mL	10 units

[a]Preservative free 0.9% sodium Chloride Injection, USP only.

Table 10.2 Dilution instructions for Botox Cosmetic vials (100 units and 50 units).

Diluent[a] added to 100 unit vial	Resulting dose units per 0.1 mL	Diluent[a] added to 50 unit vial	Resulting dose units per 0.1 mL
10 mL	1 unit	5 mL	0.5 units
8 mL	1.25 units	3 mL	1.5 units
5 mL	2.0 units	2.5 mL	2 units
4 mL	2.5 units	2 mL	2.5 units
[a]2.5 mL	4 units	[a]1.25 mL	4 units
2.0 mL	5 units	1.0 mL	5 units
1.0 mL	10 units	0.5 mL	10 units

[a]Preservative free 0.9% sodium Chloride Injection, USP only.

Table 10.3 Dilution instructions for Xeomin vials (50 units, 100 units, 200 units).

Diluent[a] added to 50 unit vial	Resulting dose units per 0.1 mL	Diluent added to 100 unit vial	Resulting dose units per 0.1 mL	Diluent added to 200 unit vial	Resulting dose units per 0.1 mL
0.25 mL	20 units	0.25 mL	–	0.25 mL	–
0.5 mL	10 units	0.5 mL	20 units	0.5 mL	40 units
1 mL	5 units	1 mL	10 units	1 mL	20 units
1.25 mL	4 units	1.25 mL	8 units	1.25 mL	16 units
2 mL	2.5 units	2 mL	5 units	2 mL	10 units
2.5 mL	2 units	2.5 mL	4 units	2.5 mL	8 units
4 mL	1.25 units	4 mL	2.5 units	4 mL	5 units
5 mL	1 units	5 mL	2 units	5 mL	4 units

[a]Preservative free 0.9% sodium Chloride Injection, USP only.

Table 10.4 Dilution instructions for Dysport (300 units and 500 units).

Diluent added to 500 unit vial	Resulting dose units per 0.1 mL	Diluent added 300 unit vial	Resulting dose units per 0.1 mL
1 mL	50 units	0.6 mL	50 units
2 mL	25 units		
2.5 mL	20 units	1.5 mL	20 units
3 mL	15 units	2.5 mL	12 units
5 mL	10 units	3.0 mL	10 units

Preservative free 0.9% sodium Chloride Injection, USP only.

Table 10.5 Authors suggested dosing guidelines.

Authors suggested dosing guidelines			
Region	**Muscle**	**Botox, Xeomin, Jeuveau dose**	**Dysport dose**
Glabellar	Corrugator supercilii		
	Medial	6–24	20–60
	Lateral	4–12	10–30
	Depressor supercilii	4–12	10–30
	Procerus	4–12	10–30
		Total dose 20–60	Total dose 50–100
Crows	Orbicularis oculi	12–30	30–70
Lower lids	Orbicularis oris	1–6	2.5–10
Horizontal forehead rhytids	Frontalis	6–30	5–50
Lateral brow lift	Orbicularis oculi vertical & lateral fibers	2–14	5–25
Bunny lines	Nasalis	2–8	5–20
Vertical lip lines	Orbicularis oris (upper and lower)	3–12	5–25
Chin	Mentalis	2–10	5–20
Down turned mouth	Depressor anguli oris (DAO)	4–8	5–20
Horizontal neck lines	Platysma	10–20 per line	20–50 per line
Platysma neck bands	Platysma SMAS	10–20 per band	20–50 per band
Hyperhidrosis	Axillae, palms, soles	50–200	N/A

Table 10.6 Authors advanced esthetic suggested dosing guidelines.

Advanced injection dosing guideline				
Condition	**Muscle**	**Injections**	**Dose per site Botox, Xeomin, Jeuveau**	**Dose per site Dysport**
Tension headache	Frontalis	4–6 injections	2–2.5 units	5 units
	Procerus	1–2 injections	4–5 units	10–12.5 units
	Corrugator	1–2 injections	4–5 units	10–12.5 units

Temporo-mandibular Disorder (TMD)	Anterior temporalis	1 injection anterior portion	2–8 units in anterior portion	5–20
		1 injection medial portion	4–6 units mid portion	10–15 units
	Masseter	Stay close to the angle Inject bilaterally 3 points of injections	4–10 units Start low and increase PRN	10–25 units
Trismus	Temporalis	3–4 injections	4–6 units	10–15 units
	Masseter	2–3 injections	4–6 units	10–15 units
	Lateral pterygoid	Massage muscle with finger		
Masseteric hypertrophy	Masseter	2–3 injections	10–16 units	25–40 units
Down turned mouth	Depressor anguli oris	1 injection on the hyperactive side or if both down turned treat both sides accordingly	2 units	5 units
Gummy smile	Levator labii superioris	2 injections	2–4 units	5–10 units
Bar code lip lines	Orbicularus oris Vermillion border of the upper and lower lip	2–8 injections	1 unit	2.5 units
Down turn nasal tip	Depressor septi nasi	1 injection	2 units	5 units
Neck bands Nefrittiti neck lift	Platysmal	Depends on the number of bands	30 units	75 units

Example of calculation for reconstitution of Botox

This formula can be substituted for Jeuveau and Botox but not for Dysport.

100 units of Botox divided by 2.5 (saline) and then multiply by 0.1
= 4 units of Botox per 0.1 mL.

$$(100\text{ units}/2.5\text{ mL})*0.1\text{ mL} = 4\text{ units}$$

or

$$(100/2.5)*0.1 = 4$$

or a further example:

100 units of Botox divided by 5 (saline) and then multiple by 0.1
= 2 units of Botox per 0.1 mL.

$$(100\text{ units}/5\text{ ML})*0.1\text{ ML} = 2\text{ units}$$

or

$$(100/5)*0.1 = 2$$

Examples of calculation for reconstitution of Xeomin

This formula can be substituted for Jeuveau and Botox but not for Dysport

100 units of Xeomin divided by 2.5 (saline) and then multiply by 0.1
= 4 units of Xeomin per 0.1 mL

$$(100\text{ units}/2.5\text{ mL})*0.1\text{ mL} = 4\text{ units}$$

or

$$(100/2.5)*0.1 = 4$$

or a further example:

100 units of Xeomin divided by 5 (saline) and then multiple by 0.1
= 2 units of Xeomin per 0.1 mL

$$(100\text{ units}/5\text{ ML})*0.1\text{ ML} = 2\text{ units}$$

or

$$(100/5)*0.1 = 2$$

Examples of calculation for reconstitution of Jeuveau

This formula can be substituted for Botox and Xeomin but not for Dysport

100 units of Jeuveau divided by 2.5 (saline) and then multiply by 0.1
= 4 units of Jeuveau per 0.1 mL

$$(100\text{ units}/2.5\text{ mL})*0.1\text{ mL} = 4\text{ units}$$

or

$$(100/2.5)*0.1 = 4$$

or a further example:

100 units of Jeuveau divided by 5 (saline) and then multiple by 0.1
= 2 units of Jeuveau per 0.1 mL

$$(100\ \text{units}/5\ \text{ML})*0.1\ \text{ML} = 2\ \text{units}$$

or

$$(100/5)*0.1 = 2$$

Examples of calculation for reconstitution of Dysport

This formula can NOT be substituted for Botox, Jeuveau and Xeomin this is ***ONLY*** for Dysport

300 units of Dysport divided by 3 (saline) and then multiply by 0.1
= 10 units of Dysport per 0.1 mL

$$(300\ \text{units}/3\ \text{mL})*0.1\ \text{mL} = 10\ \text{units}$$

or

$$(300/3)*0.1 = 10$$

or a further example:

300 units of Dysport divided by 1.5 (saline) and then multiple by 0.1
= 20 units of Dysport per 0.1 mL

$$(300\ \text{units}/1.5\ \text{ML})*0.1\ \text{ML} = 20\ \text{units}$$

or

$$(300/1.5)*0.1 = 20$$

This creates less potential spread effect as you are minimizing the volume of saline or bacteriostatic.

Storage and handling

Unopened and not reconstituted Botox and Dysport and Jeuveau should be stored at 36–46 °F. Xeomin does not need to be stored in a refrigerator until after it is reconstituted. Botox Cosmetic, Jeuveau, Xeomin, and Dysport can be mixed with bacteriostatic saline and is suitable for 6 weeks. On label, reconstitution is with 0.9% saline.

All neuromodulators will come with a package insert and will recommend to reconstitute the product with normal saline. Once reconstituted, it is recommended that the product be utilized within 6–24 h. It is necessary to read the package insert frequently on each product to keep current on best practice. However, it is common practice to reconstitute the neuromodulator products with bacteriostatic saline. When the product is reconstituted with bacteriostatic saline, the shelf life will increase from 24 h (with normal saline) to 6 weeks (with bacteriostatic saline). However, once the injector has strayed from the on-label guidelines, it is no longer on the FDA approved use of the product.

Novice areas and advanced areas for neuromodulation

Commonly treated areas

- Lateral Canthal lines
- Frontalis Muscle
- Glabellar region
- Procerus
- Orbicularis Oculi

Advanced areas

- Vertical Rhytids above the lips
- Bunny Lines
- DOA
- Mentalis (Peau d' orange)
- Lower orbicularis oculi
- Platysmal Bands
- Masseter
- Depressor Supercili

Novice (N) areas and advanced areas (A)

- (N) Forhead wrinkles—Frontalis muscle to treat the transverse rhytids on the forehead
- (N) Frown Lines—Glabeller Complex Injections
- (N) Eyebrow Shaping—Lateral and Medial Arch
- (N) Crows Feet—Lateral Canthal Rhytids
- (A) Bunny lines
- (A) Gummy Smile—Gingival display
- (A) Nasal tip ptosis

(A) Marionette lines
(A) Depressor Anguli Oris Injection
(A) Neck bands—Platsymal bands
(A) Dimpled Chin: Hypertrophic Mentalis Injections
(A) Hyperhidrosis: Axillae
(A) Dental Application

References

The Aesthetic Society. (2019). *Aesthetic plastic surgery national databank*. https://www.surgery.org/sites/default/files/Aesthetic-Society_Stats2019Book_FINAL.pdf. (Accessed May 2020).

Further reading

Allergan. (2020). *Botox cosmetic history*. https://www.botoxcosmetic.com/what-is-botox-cosmetic/botox-cosmetic-history. (Accessed March 2020).

Dysport. (2020). *For health care professionals*. https://www.dysportusa.com/healthcare-professionals#importantsafetyinformation. (Accessed March 2020).

Jeuveau. (2020). *Ready for a modern-made tox?*. https://jeuveau.evolus.com. (Accessed July 2020).

Xeomin. (2020). *A uniquely purified choice for frown lines*. https://www.xeominaesthetic.com/professionals/. (Accessed March 2020).

CHAPTER 11

Neuromodulation injection technique procedural steps

1. Treatment area
2. Indications with area defined
3. Emotions expressed with animation
4. Muscles in the treatment area
5. Muscles to be treated in the targeted zone
6. Special consideration
7. Complimentary locations to increase the efficacy
8. Treatment goals
9. Documentation
10. Equipment
11. Assessment
12. Pre-procedure steps
13. Technique
14. Results
15. Duration
16. Follow up
17. Complications

Areas for injection

1. Glabellar complex
2. Frontalis
3. Eyebrow shaping—lateral and medial arch
4. Crows feet-lateral canthal rhytids
5. Bunny lines
6. Gummy smile-gingival display
7. Smokers lines
8. Nasal tip ptosis
9. Marionette lines
10. Depressor Anguli Oris Injection
11. Neckbands—Platsymal bands

Essentials of Neuromodulation
https://doi.org/10.1016/B978-0-323-89920-8.00011-4

12. Dimpled chin: hypertrophic mentalis injections
13. Hyperhidrosis: Axillae
14. Dental application

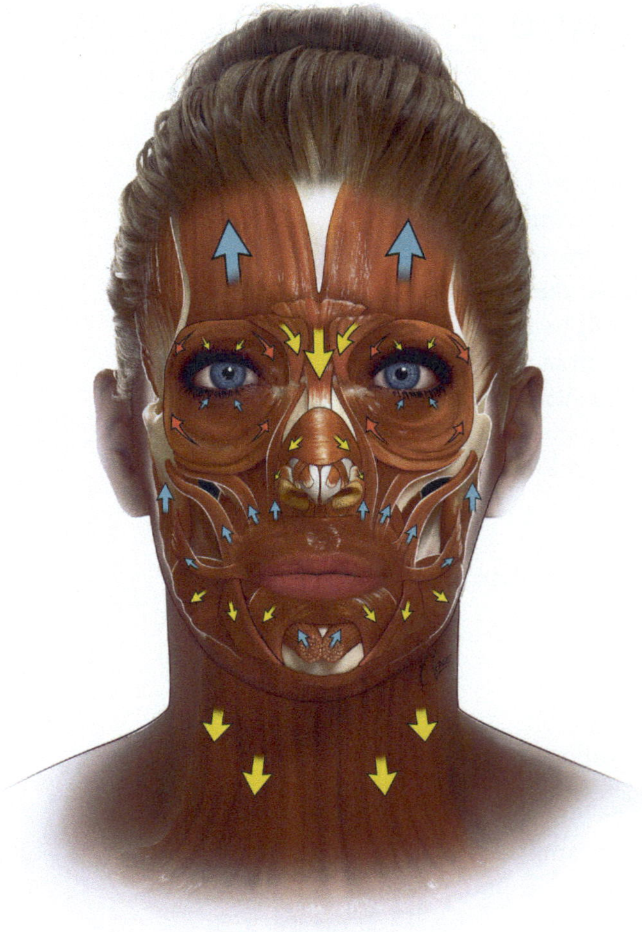

Fig. 11.1 Elevator and depressor muscle diagram. Neuromodulation will cause the opposite effect once placed in the muscle. Depressors become elevated and elevators become depressed.

Neuromodulation for frown lines—Glabellar complex injections

Static and dynamic frown lines occur due to the active use of the complex depressor muscles, which consist of the procerus and the medial and lateral corrugators. This was the most commonly treated area for neuromodulation, but with the recent additional FDA approval Botox, has acquired for additional areas such as the forehead lines and the crows feet, these statistics will be rapidly changing (Fig. 11.1).

Indications for treatment

Neuromodulation for injection is indicated for the temporary improvement in the appearance of moderate to severe glabellar lines associated with corrugator and procerus muscle activity in adult patients.

1. Frown lines
2. Medial eyebrow elevation

Emotion

Disapproving, angry, squint due to bright lights, irritation, disdain, disgruntled, frustrated, concentration.

Muscles in treatment area

Proceurus
Corrugator
Frontalis
Depressor Supercilli
Orbicularis oculli
Levator palbebrae

Muscles to be treated in the targeted zone

Procerus
Corrugator medial and lateral portions
Depressor Supercilli
Lateral tail of the orbicularis oculi

Special considerations

If the patient has a deep-set crease across the bridge of the nose, the injector will want to inject a bit lower at the procerus insertion point on the nasal bridge. Alternatively, provide two injection points into the procures to help remedy the horizontal crease. It is common to find patients that use their frontalis muscle to frown. If this is the case, it is advised that the injector educate the patient on the need for concomitant treatment of the frontalis.

Complimentary locations to increase efficacy

Frontalis
Lateral canthal lines

Treatment goals

Inhibit the procerus and corrugators' activation to minimize the patient's ability to frown or furrow. Preventing muscle activity in this area will help to prevent deep-set at rest rhytids from developing. This treatment can be both preventative and restorative.

Dosing

Women 12–30 units Botox, Jeuveau, and Xeomin
Women 30–70 units Dsyport
Men 24–40 units Botox, Jeuveau and Xeomin
Men 50–70 units Dsyport

Documentation

Medical history including allergies
Medication
Dose
Lot number
Expiration date
Provider
Equipment used
Treatment area and specific dose per injection point
How the patient tolerated
Treatment plan

Equipment

Medication
Botox
Jeuveau
Xeomin
Dysport
Bacteriostatic saline or preservative-free saline
3 mL luer lock syringe
27 Gauge luer lock 1 1/2 in. needle
2 × 2 non-woven gauze

Alcohol
Ice
Bottle opener
Marking pencil
BD insulin Syringe 31 gauge ultra thin
Non-sterile gloves

Hemostat, if needed, to assist in the removal of tight needles from luer lock or removing the metal portion of the bottle of neuromodulator.

Assessment

The patient should be assessed in animation and at rest. They should further be assessed from all angles. Even if only the glabellar complex is being treated, the forehead should still be assessed at rest and with animation. It is essential to understand the relationship between the brows and the treatment area. It is also necessary to have the patient smile as well to assess how the smile can interfere with the lateral tail of the eyebrow, which is most often a part of the treatment zone for the glabellar complex.

Pre-procedure steps

1. Consultation
2. Have patient sign consent and pre and post-procedure guidelines
3. Photographs at three angles at rest and animated
   - 50 degrees both left and right
   - 45 degrees both left and right
   - Straight on
4. Have the patient comfortably seated in an upright position
5. Ensure adequate overhead lighting
6. Mark out danger zones
7. Additionally, mark out treatment areas with a marking pencil. Be sure not to inject through the marking pencil as this can create a permanent tattoo.
8. Calculate dosing according to the desired look, anatomy, and area being treated.
9. Draw up the medication, being sure not to dull the needle by touching the sides of the glass bottle, or going through the rubber stopper with the intended injection needle.
10. Cleanse the face thoroughly with a double cleanse and then chlorhexidine or alcohol (allow to dry) (Fig. 11.2).

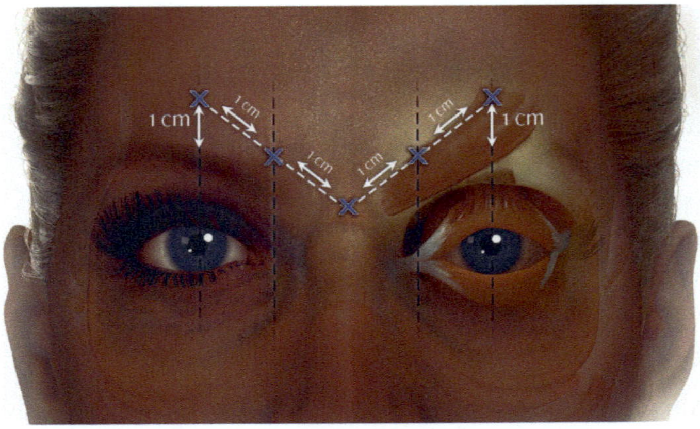

Fig. 11.2 SAFE ZONE for injection into the glabellar complex.

Technique

1. Identify the treatment danger zones.
2. Mark out the treatment zone paying close attention to the lateral edges of the corrugators. This area can be elicited by having the patient frown. Often, a bit of a divot is seen where the corrugator ends, and the levator palpebrae inserts onto the forehead above the mid pupillary line.
3. Be sure that the patient is still cleansed and has not touched the treatment area since proper cleansing performed in the pre-procedural steps.
4. Have the patient activate the muscles to be treated and begin at the procerus muscle. Inject in the middle of the procerus belly. The needle will be inserted about halfway into the skin if using a 31 gauge 5/16″ (8 mm) needle. Inject 4 units of Botox, Jeaveau, Xeominor 10 units of Dysport to the belly of the procerus. Do not hit the periosteum. If the periosteum is struck, back the needle out before pushing the product into the muscle. When the periosteum is struck, change the needle or the syringe as it will be lead to an increase in the sheering of the skin and vessels, leading to an increased risk of bruising. Should a two-point injection be necessary to the procerus, place one injection toward the bottom of the procerus and one at the top of the procerus muscle. If the patient has a deep crease that runs straight to the insertion at the nasal bridge, it will be necessary to split the injection and do half on either side of the split. So this time inject 2 units of Botox, Jeuveau, Xeomin or 5 units of Dysport bilaterally into both bellies of the procerus (Fig. 11.2).

5. The next injection point will be placed into the corrugator muscle about 1 cm inferiorly to the previous injection of the procerus muscle. Angle the needle toward the procerus muscle. The needle depth will be about 1/2 way into the skin. Slowly inject 4 units of Botox, Jeuveau, Xeomin or 10 units of Dysport at this insertion point.
6. Now move to the lateral corrugator. Have the patient contract the targeted muscles for the best visualization of the area. Insert the needle about 1–2 cm above the supraorbital ridge at the lateral margin of the corrugator muscle. It is essential that the injector angle the needle toward the procerus muscle and insert the needle 1/4–1/2 in. (depth depends on individual patient anatomy) way in and deliver 2–4 units of Botox, Jeuveau, Xeomin or 5–10 units of Dysport depending on the desired dose. Be sure to push slowly with the bevel down. The reason for injecting toward the procerus muscle is to avoid inadvertent spread or diffusion to the frontalis or levator palpebrae muscle.
7. Repeat steps five and six on the opposite side.
8. Do not press or rub on the injection sites. If the patient happens to bleed, apply gentle pressure to stop the bleeding. GENTLE!!! It is not advised that manual force and manipulation occur in the area post-injection. Manual manipulation can force the product from the intended site, and the patient may suffer from an adverse event.

Results

Within 3–10 days, patients will start to see gradual improvement of the treated area's dynamic wrinkles. At no time during the 2 weeks should more neuromodulator be injected in the same area. It takes a full 2 weeks for the product to take effect. All good things come to those who wait. At 2 weeks' time, it is strongly advised that the patient come back in for a follow-up evaluation. Discuss how the patient feels, how they like their results. Take after photos in the exact positions as the before photos.

Duration

Neuromodulators last anywhere from 2 to 4 months. The average length is about 3 months. It is essential to the clients to let them know they will not be completely frozen the entire treatment duration. It will take about 2 weeks to become fully effective, and then at about 6–8 weeks, the neuromodulation will start to "break." Break, meaning a little bit of expression or movement will start to come back, and the patient will notice this. Be sure always to keep photos for the records to compare pre and post-treatment. At times patients will call at 6 weeks and state that ALL of their neuromodulators wore off, and they need to be re-seen or re-treated. Pictures will come in handy here as it is likely they see the beginning stages of the reinnervation with baby sprouts of the nerve terminals. It is not necessary to wait till all of the patient's movement is back. Retreat at 3 months before the full expression is restored; this way, they can train the muscles.

Follow up and special consideration

It is common for patients to still demonstrate movement in the area treated after treatment. Therefore, it is necessary to assess the patient in animation and add additional units according to the previously treated muscles with movement. Additional dosing ranges from 2–4 units of Botox, Jeuveau, Xeomin per site, or 5–10 units of Dysport per site.

If the patient only had the glabella complex treated when they frown, they may activate their frontalis muscle. At which time, the injector could advise that the patient take treatment in the forehead. A patient's potential to need additional dosing at a follow up is an important discussion point to have before the initial treatment. Informed consent starts with ensuring the patient has proper expectations of the treatment goals and effect. Patients can quickly feel like they had a bad or botched injection treatment if the expectations were not identified.

If the glabellar complex is treated without the forehead; the patient's brows may seem extremely high, peaked, or spocked. If this is the case, then treat the frontalis accordingly. The patient must be informed that this can happen before treatment. The patient needs as much information upfront to make informed consent, plan their budget, and anticipate the progression of the treatment as it takes effect till they get to their end result at 2 weeks out.

If the patient is not appropriately counseled on the use of neuromodulation, they will often come back and say that the treatment did not work because they have lines still. The patient must be educated that the treatment is only working on the muscles and the motion and that it is not intended to reduce the static lines. Of course, there is a caveat; in the younger patients, it can reduce the appearance of static lines, but this is not something the injector should promise. Always teach the client that the treatment is for the hyperdynamic muscle movement, not the static line (Table 11.1).

Table 11.1 Complications and management table.

Complications	Cause	Management
Bruising	Penetration of a vessel	Arnica ointment
Swelling	Penetration of skin, inflammatory reaction	Ice
Pain	Trauma	Ice and acetaminophen
Headache	Anxiety	Acetominophen and real dark cocoa
Eye brow ptosis (eye brow droop or depression)	Injection into frontalis muscle	Do nothing You may consider adding more units to the medial corrugators, procerus and lateral tail of the orbicularis oculi. Add more only if the patient can still move the muscles in that area
Medial eyebrow splaying	To much neuromodulator Depressor Supercilli injected unintentionally or in the wrong candidate	Best course of action is time
Blepharoptsois (eye lid droop)	Inadvertent injection into the levator palpebrae	Iopidine gtts and time
No effect		Consider Zinc deficiency Report to the medical science liaison of the maker of the neuromodulation for further guidance Review medical history, are they on chelators, are they zinc deficient, do they have an underlying disease not disclosed concerning possible increased metabolization Review the reconstitution, injections, dosing and placement
Dysport allergy	Cows milk protein allergy	Antihistamine Follow up with primary care or Allergist Notate in chart new allergy

Details on Dysport allergy

If using Dysport and the patient is allergic, the patient will complain of flu like symptoms or feeling run down like they are experiencing seasonal allergies or experiencing a sinus infection. Not all patients know that they have a milk protein allergy, and treatment with Dysport uncovers the allergy. Typically no intervention is needed as it will self-resolve. However, it is appropriate to suggest over the counter antihistamines or see their primary care or allergist for further testing and lifestyle modifications.

Before and after photos

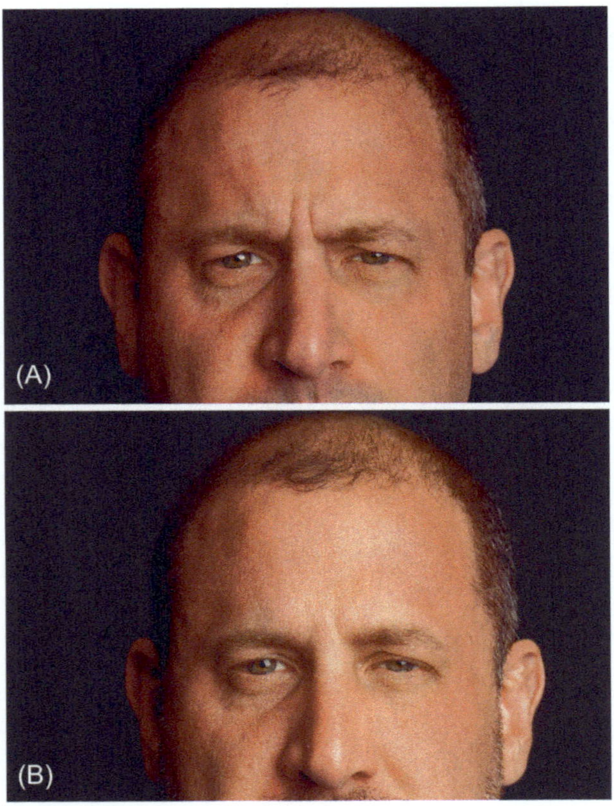

Fig. 11.3 (A) Before photo. (B) After photo.

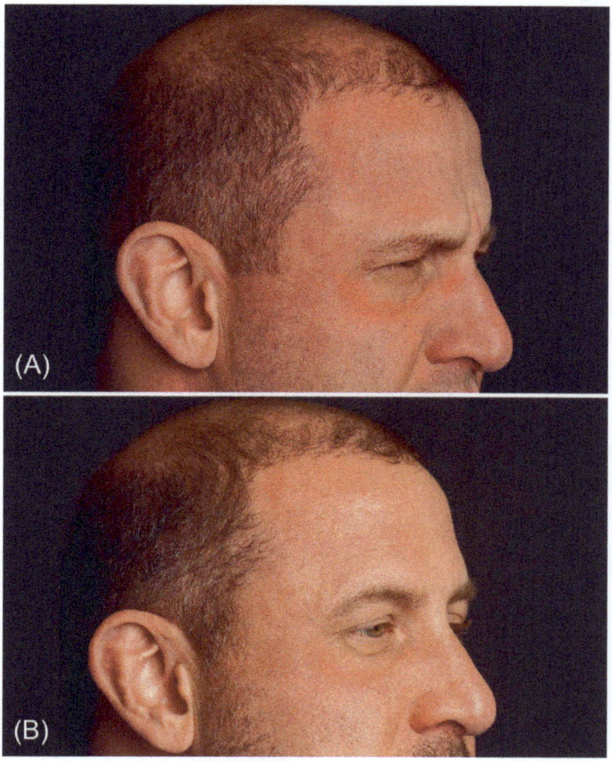

Fig. 11.4 (A) Before photo. (B) After photo.

This before and after picture in Fig. 11.3A and B demonstrates, inadvertently, an improvement within the static lines after treatment with neuromodulators. However, keep in mind, it is not the intended goal. The goal is to prevent the muscle from animating and causing the further breakdown of the skin. This male patient in his 40s has a healthy turgor to his skin, some laxity, and photodamage. The skin on his forehead is a bit thicker than females, but not quite as thick as one may commonly cross with male patents. It is appreciable to note the eyelid laxity of this male patient in Figs. 11.3A and 11.4B. Figs. 11.3A and 11.4A demonstrates the enhanced lid laxity with animation. After he received neuromodulation and attempts to scowl, note the lack of eyelid depression in both Figs. 11.3B and 11.4B. When a patient has hooded eyes, the injector needs to be cautious when treating the corrugator's lateral tail and the frontalis to not make the brow or lid any heavier. In Fig. 11.4B, during the scowl, his lash line on the left side shows his lash line, but in Fig. 11.4A, the scowl obliterates the lash line. These differences may appear subtle in the picture, but they can be night and day in terms of a feeling to the patient. When depressor muscles lose their tension, they can become very heavy, and patients feel like they are carrying them around. With appropriately placed neuromodulation, they can have a sense of a weight lifted. The patient's side profile view in animation shows a remarkable difference in the lid heaviness in this patient's after photo. In the after photo his lash line is visible, whereas in the animated before picture the lash line is obliterated. Look at the amount of scleral show in the after photo compared to the before photo. This is representative of a home run for both the patient and the injector. This patient photo is an excellent reminder to take photos and assess the patient from all angles and in animation at all angles. This patient was treated with 24 units of Botox in the glabellar region. He had 4 units in the procerus, 6 units to each medial corrugator, and 4 units to each lateral corrugator.

Treatment

Glabellar complex 24 units Botox
Photos 2 weeks apart

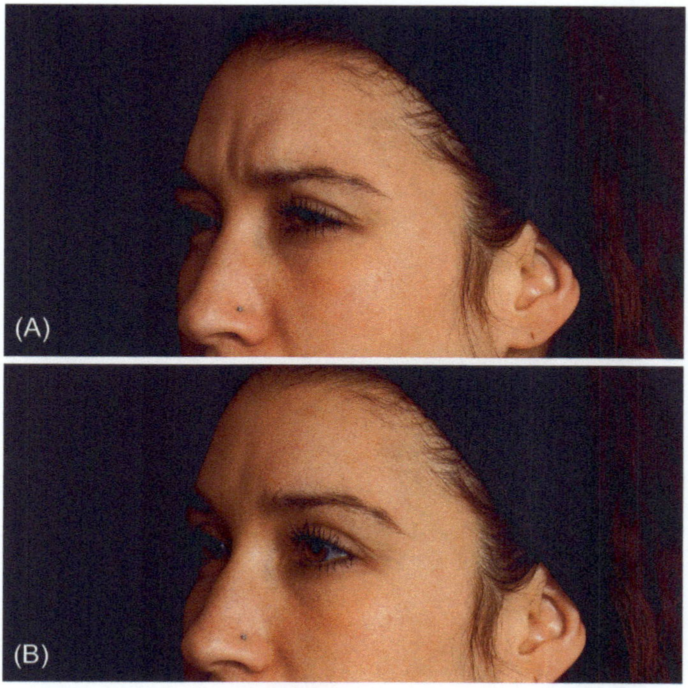

Fig. 11.5 (A) Before photo. (B) After photo.

Thirty-year-old female in good health with some hyperpigmentation from intermittent skin eruptions. The patient's goal for treatment with neuromodulation was to have a more youthful presentation with a lift of the upper face and prevent the rapid onset of aging. Fig. 11.5B demonstrates a nice opening of her eye when she is frowning. This leaves the focal point on her beautiful brown eyes while she is expressing disdain. She is able to maintain more of a wide-open look to her eyes after treatment with neuromodulation as opposed to her before photo in Fig. 11.5A. Her brow position stays more upright in animation after successful treatment with neuromodulation as noted in Fig. 11.5B. Her treatment was successful for the reduction of the frown lines, as seen in Fig. 11.5B. She received a total of 20 units of Botox. Four units in her procerus, four units in the medial corrugator. Four units on the right and four units on the left medial corrugator and then two units in each of lateral tails of the right and left corrugator. Then she received two units on each side of the later tail of the orbicularis oculi.

Treatment

Glabellar Complex 20 units Botox
Brow lift 4 units Botox
Photos 2 weeks apart

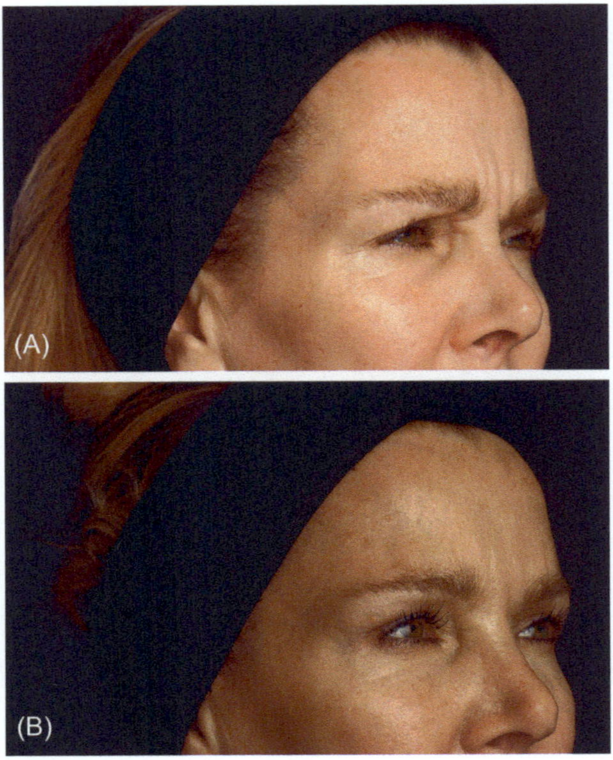

Fig. 11.6 (A) Before photo. (B) After photo.

This female patient was treated for the improvement of moderate to severe frown lines. The brightening of her eye and the lift she has received from the neuromodulation treatment is beautiful. In Fig. 11.6A, her gorgeous spider leg, long lashes are hidden under the skin fold during animation. Once the neuromodulator kicked in, we could capture the beauty of her lashes as they cascade toward her brow was uninterrupted by the laxity of the upper eyelid as seen in Fig. 11.6B. Take note that there is a noticeable reduction of the frown lines between the eyebrows and the ancillary services that the neuromodulation treatment provided in Fig. 11.6B. Look at the medial brow where the depressor supercilii rests and look at the wrinkling in that tissue space in Fig. 11.6A. Fig. 11.6A shows three strong deep creases. If this area were left untreated, it would become static lines at rest. In Fig. 11.6B, those three medial lines at the beginning of her brow are gone. This reduction of creasing changes the light's ability to reflect in different areas highlighting a youthful appearance. She received a total of 20 units of Botox for the treatment of her glabellar region. She received 4 units in her procerus, 4 units in each side of her medial corrugator, and 4 units in each side of the lateral corrugator muscles. Also, she received a small amount for a brow lift. Each lateral tail of the orbicularis oculi received 2 units per side.

Treatment

Glabellar Complex 20 units Botox
Brow lift 4 units Botox
Photos 2 weeks apart

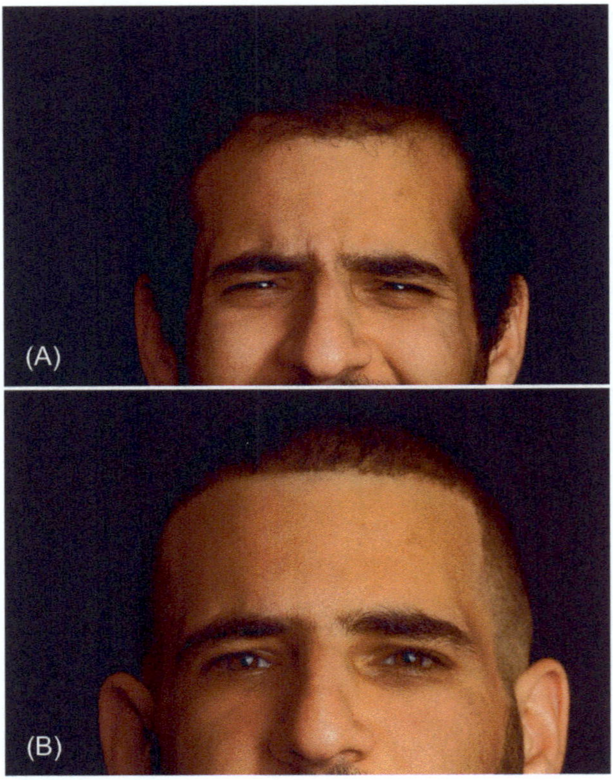

Fig. 11.7 (A) Before photo. (B) After photo.

This is a young man has thick olive toned sebaceous skin. He is a youthful male seeking treatment for prevention as well as restoration. He has developed some static lines, given the thickness and strength of his animation. In Figs. 11.7B and 11.8B you can see significant improvement in the hooding of his eyes; beyond the obvious reduction of glabella activity. Preneuromodulation in Fig. 11.7A his brows force the lids down and create a heaviness over the eye in this young patient. Yet, in Figs. 11.7B and 11.8B, while he is in animation, one can see much more of his sclera and the iris of his eye, and even his pupil. Creating openness gives the impression of youth, awake, alert, and vibrance. This is where we can begin to attach emotions to the findings just as the patient would. So far, each patient shown has presented differently, even though their goals have all essentially been the same, maintenance, prevention, and youth. A tremendous amount of hyperdynamic movement is not necessarily evident in the glabellar region in Fig. 11.7A. However, what is most prominent is the drawn down pulling

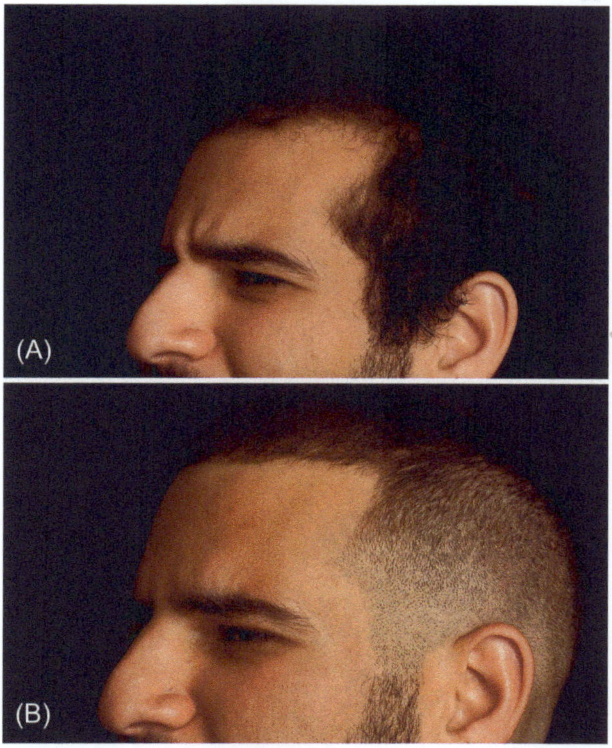

Fig. 11.8 (A) Before photo. (B) After photo.

of the brows medially. Fig. 11.8A demonstrates the dynamic expression lines better than forward-facing. His treatment was to lift the skin folds off his lash line to create openness in animation instead of the forceful contraction noted. This shows the importance of animated photos in every position with all animation variations such as a scowl, smile, resting, surprised, and sad face. This young man received 24 units total in his glabellar complex. There were 4 units in his procerus, 6 units in each medial corrugator both right and left, and 4 units in each lateral tail of his corrugator. Given the natural flattening of the brow and how it compressed so much to obliterate the patient eyes when he scowled, he was treated with 2 units of Botox in each lateral tail of the orbicularis oculi.

Treatment

Glabellar Complex 24 units Botox
Brow lift 4 units Botox
Photos 2 weeks apart

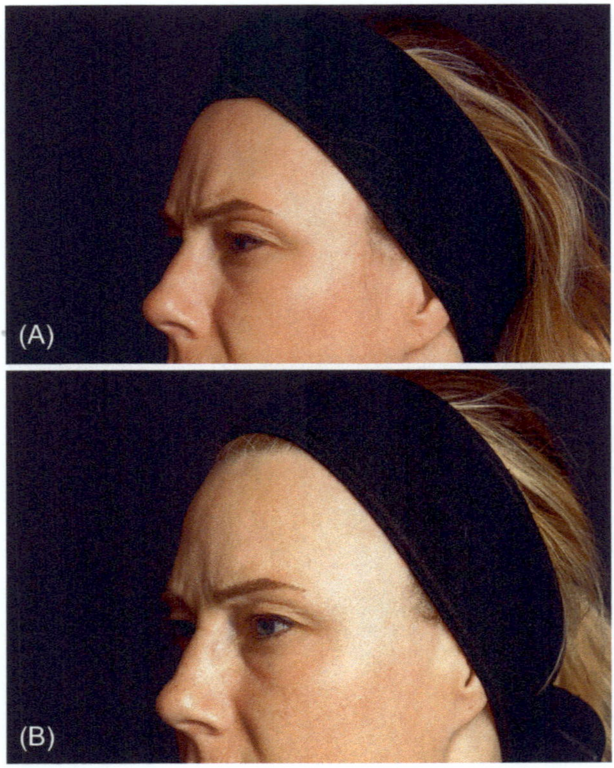

Fig. 11.9 (A) Before photo. (B) After photo.

This early 40s patient was treated for frown lines to elevate her brows and take care of the angry look she gets during concentration. Patients will often describe the glabellar region's expression as an angry face, a quizzical face, or a concentration face. We are beginning to see the repeating theme here with emotions tied to the expression on their face. As you can note between Fig. 11.9A and B represents a slight under correction. As one can see in Fig. 11.9B, there is a touch of movement in the medial and lateral corrugator. She was initially treated with 20 units of Botox. Four units were placed in her procerus, four units in each side of the medial corrugator, and four units in each lateral portion of her corrugator. Additionally, 2 units were placed on each side of the lateral tail of the orbicularis muscle. In Fig. 11.9B, movement is still noticeable, especially in the corrugator's medial portion, so she received an additional 12 units in her glabellar complex to complete the treatment to satisfaction. She constantly chews gum, which can cause an increase in the patient's dosing due to the constant mobility of all the facial muscles involved in chewing. It is important to tell clients that this side of medicine is both art and science, and at times, it will be necessary to add additional units to complete the look or compensate for the natural compensatory mechanism of a neighboring muscle.

Initial treatment

Glabellar Complex 20 units Botox
Brow lift 4 units Botox
Photos 2 weeks apart
Follow up treatment
Glabellar complex 12 units Botox

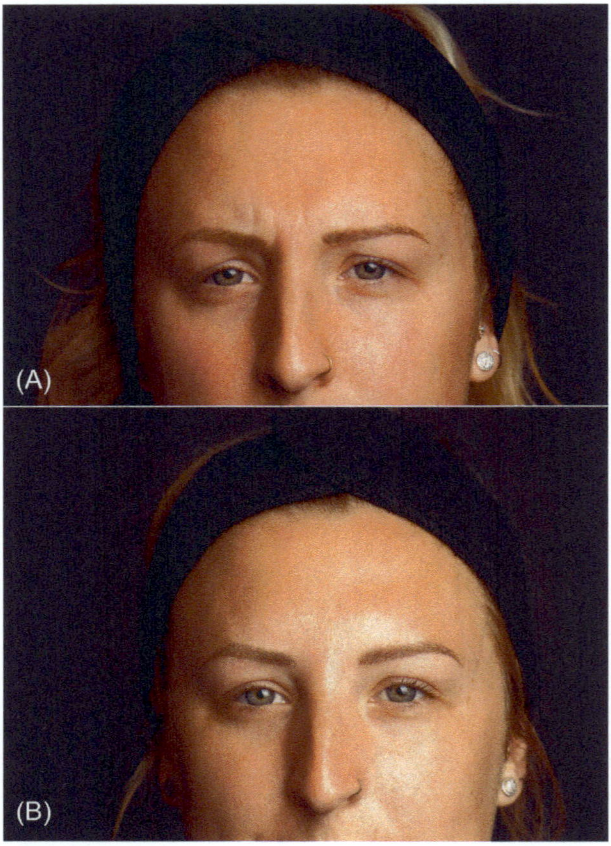

Fig. 11.10 (A) Before photo. (B) After photo.

This young female had beautiful, healthy skin that had been well kept from the sun. She has sought out neuromodulation to preserve her youth and hopes for a bit of a brow lift. She is an excellent example of a younger client looking to get into the anti-aging process early on to avoid having to repair later in life. Her mindset was to maintain and prevent aging. We did not utilize much neuromodulation for her. Eighteen units of Botox total. Four units in the procurus, 4 in each side of the medial corrugator, and two units in each lateral corrugator. She had a slight asymmetry on the left brow seen in Fig. 11.10A, so 2 units of Botox was placed in her left lateral tail of the orbicularis oculi to elevate that brow at rest and help prevent it from dropping down when she smiles and or scowls. When she scowls, the left lid becomes a touch heavier than her left as noted in Fig. 11.10A, and the lid

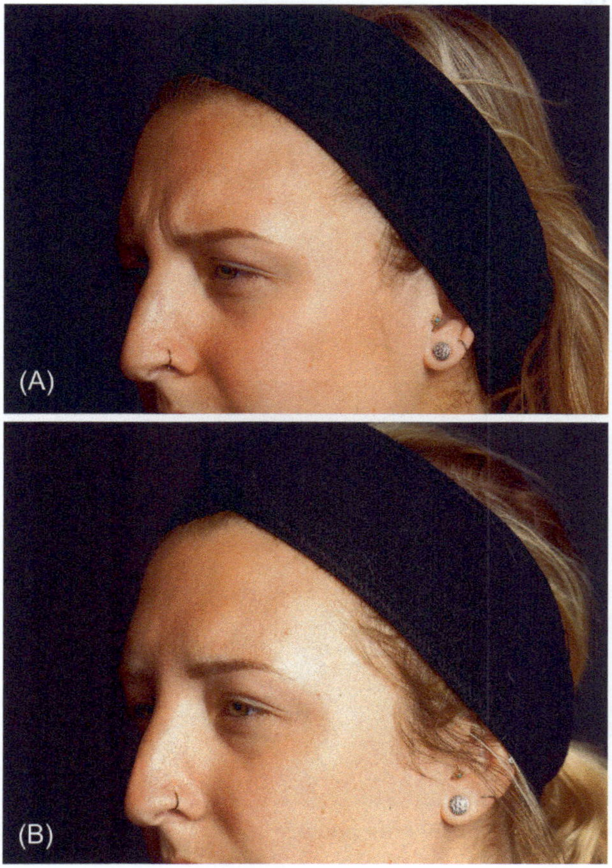

Fig. 11.11 (A) Before photo. (B) After photo.

lays over the lash line. Fig. 11.11A and B demonstrates the neoromodulation effect from an alternate angle. Assessments and photos are essential to be taken in all ages and in all forms of animation. The lateral tail of the orbicularis oculi treatment was to help lift the tissue off of the lash line and create a greater sense of alertness or a more awake look; some may refer to it as brighter.

Treatment

Glabellar complex 16 units Botox
Brow lift 2 units Botox
Photos 2 weeks apart

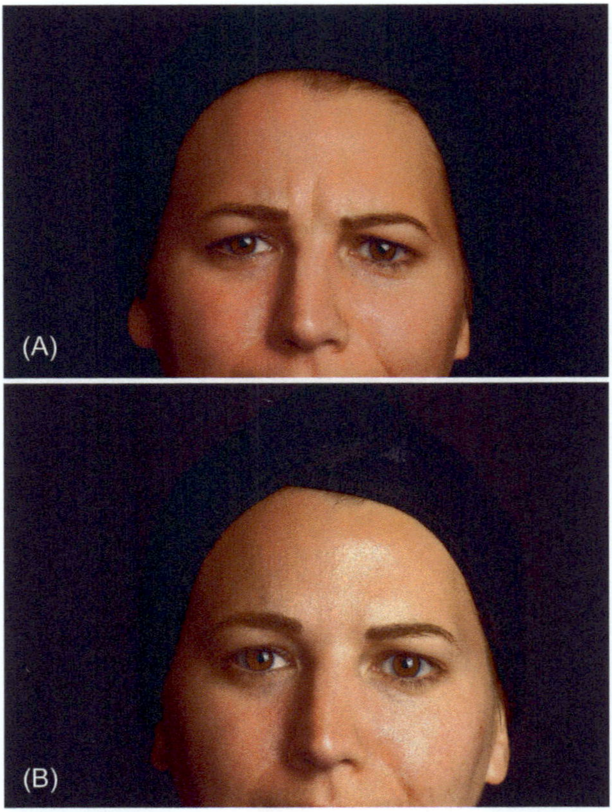

Fig. 11.12 (A) Before photo. (B) After photo.

Here is another very young lady looking to prevent the formation of static lines in her glabellar complex. Her brows are asymmetrical, and this is exacerbated upon animation seen in Fig. 11.12A. Bilaterally her lids become heavy when she scowls; noticed in Fig. 11.12A, there is much less heaviness, and the attention has been shifted from the asymmetrical inner brows seen in Fig. 11.12B. We are able to see more of her eyes in Fig. 11.12B. Overall she had a gentle softening in the area and a general improvement of the asymmetry and reduced heaviness of the lids with animation as seen in Fig. 11.12B. She received 16 units of Botox in the glabellar complex and 4 to treat the lateral orbicularis to help elevate the brow's tail. Her treatment was completed with 4 units in the procerus, 4 units in each side of the medal corrugator and 2 units in each later portion of the corrugator. She was thrilled with her appearance after treatment and is excited to repeat her injections in 3 months to help prevent the signs of aging and maintain the lifted brows.

Treatment

Glabellar complex 16 units Botox
Brow lift 4 units Botox
Photos 2 weeks apart

Neuromodulation for the frontalis-forehead wrinkles

The activation of the frontalis muscle creates the transverse rhytids on the forehead. The use of neuromodulation in this area reduces the transverse rhytids, aka forehead lines, by inhibiting the muscle contraction, which smooths the skin overlying the muscles treated. The frontal muscle is directly responsible for the patient's eyebrow shape and position. Careful attention needs to be given when treating this area. Proper injection technique in this area will yield a brow lift, and improper technique could result in an eyebrow and or lid ptosis.

Indication for treatment

Neuromodulation for injection is indicated for the temporary improvement in the appearance of moderate to severe forehead lines associated with the hyperdynamic activity of the frontal muscle in adults.

Emotions

Surprise
Shock
Disbelief
Excitement

Muscles in the treatment area

Frontalis
Orbicularis oculi
Levator palbebrae

Muscles to be treated in the target zone

Frontalis

Special consideration

1. Eyebrow position and shape. Typically women prefer high arched eyebrows, where men prefer a flat, broad eyebrow shape. Do not

over-promise the correction of asymmetry or a significant eyebrow lift as the muscle can only be elevated or depressed as much as the size allows.
2. Dermatacholsis-excessive skin laxity or baggy skin of the upper eyelid and or of the lower lid. This condition is a contraindication for the treatment of the frontalis muscle with neuromodulation. As the injector becomes more experienced, this can be reconsidered for treatment.

Complimentary locations to increase the efficacy

Glabellar Complex Region
Procerus
Corrugator
Depressor Supercilli (depending on the space between the medial brows)

Treatment goals

The goal of this area is to reduce or obliterate dynamic forehead lines. Over time the static lines can improve with continued use of neuromodulation. If an eyebrow lift is desired, then a partial dose reduction in the lateral portion of the forehead will be necessary for treatment. It is the author's suggestion that the area only be treated if the glabellar complex is also treated. If only the frontalis muscle is treated, than the frontalis will become depressed and will rest on the glabellar region, which is a naturally depressed area. By treating the glabellar with the frontalis, the injector can enhance the elevation and prevent a sense of heaviness. It is common for a 2-week follow-up to maintain and or create a good eyebrow position.

Dosing

Women 4–24 units Botox, Jeuveau, and Xeomin
Women 7.5–40 units Dysport
Men 12–24 units Botox, Jeuveau, and Xeomin
Men 7.5–40 units Dysport

Documentation

Medical history including allergies
Medication
Dose
Lot number
Expiration date
Provider

Equipment used
The treatment area and specific dose per injection point
How the patient tolerated the procedure

Equipment

Medication
Botox
Jeuveau
Xeomin
Dysport
Bacteriostatic saline or preservative-free saline
3 mL luer lock syringe
27 Gauge luer lock 1 1/2 in. needle
2 × 2 in. non-woven gauze
Alcohol
Ice
Bottle opener
Marking pencil
BD insulin Syringe 31 gauge ultra-thin
Non-sterile gloves

Assessment

The patient must be assessed at rest and in motion. Ask them to frown, smile, and raise their eyebrows. Watch the patient as they speak and assess them while they are naturally animating in conversation. Assess for dermatochalasis (skin laxity of the upper and lower eyelid). It is vital to be especially mindful of the dermatochalasis of the upper eyelid area. Further, look for eyebrow (low set droopy brows) and eyelid ptosis. Patients with these conditions often have significant transverse forehead lines as they utilize the forehead to compensate for the laxity in the brow and lid. These are the type of patients that perhaps one would treat the glabellar complex first and then consider treating the forehead at their 2-week follow-up. Neuromodulation of the frontalis muscle will soften the dynamic expression lines, but it will likely create a worsening of the lid and brow ptosis, further exacerbating the dermatochalasis. Once the ptosis is enhanced or exacerbated by the neuromodulation treatment, it can make the patients feel like that area is heavy, and it can even change their peripheral vision. As experience and wisdom are gained, providers may choose to treat this more challenging patient (Fig. 11.13).

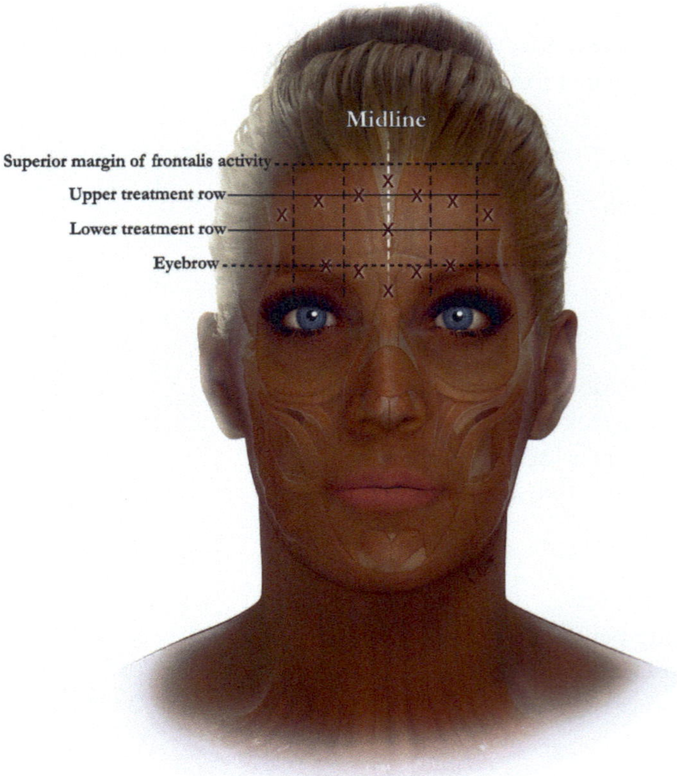

Fig. 11.13 Treatment rows for neuromodulation.

Pre-Procedure steps

1. Consultation
2. Have patient sign consent and pre- and post-procedure guidelines
3. Photographs at 3 angles at rest and animated 50 degrees both left and right 45 degrees both left and right straight on
4. Have patient comfortable seated in an upright position
5. Ensure adequate overhead lighting
6. Mark out danger zones
7. Additionally, mark out treatment areas with a marking pencil. Be sure you do not inject through the marking pencil as you can tattoo the patient
8. Calculate dosing according to desired look, anatomy, and area being treated

9. Draw ups medication being sure not to dull your needle by touching the sides of the glass bottle or going through the rubber stopper with your intended injection needle
10. Cleanse the face thoroughly with a double cleanse and then chlorhexidine or alcohol (allow to dry)

Technique

1. Identify the treatment danger zones.
2. Mark out the treatment zone paying close attention to the lateral edges of the frontalis muscle and the peak of the brow.
3. Be sure the patient is still cleansed and has not touched the treatment area since proper cleansing was performed in the pre-procedural steps. Have the patient activate the muscles to be treated and begin in the center of the frontalis muscle. Insert the needle at a 30 degree angle into the medial portion of the frontalis. You will inject into the muscle ridge not the crease. The belly of the muscle lives in the ridge not the crease. With gentle pressure you will instill 2 units of Botox, Jeuveau, Xeomin or 5 units ofDysport. It is common that a patient would need another injection point above near the hair line. A common novoice mistake would be to skip this injection and the patient is left smooth everywhere but that one spot. This is a dead giveaway that the patient had treatment. Most patients want to go unnoticed.
4. The injector will then continue along the frontalis muscle moving the injection points up or with consideration of the brow position. Each injection point will typically be 2 units of Botox, Jeuveau, Xeomin or 2.5 to 5 units of Dysport. There is not a clean cut dose or placement in this treatment area. Many considerations play a role for this special area. As you move along to the lateral sides of the frontalis muscle the spacing will be 2–4 cm apart and the smaller the forehead the closer you will be to the hairline and less injections will be performed.
5. When you are above the maximal point of eyebrow peak, place your injection 2 cm below the hairline. It may be necessary to cut the dose in half for this injection point.
6. Repeat on the contralateral side.
7. Do NOT press, rub, or manipulate the wheels of product. If the patient happens to bleed you may apply gentle pressure to stop the bleeding. GENTLE!!! It is not advised that you manipulate the product from the site as adverse events and product migration can occur.

Results

Within 3–10 days, patients will start to see gradual improvement of the dynamic wrinkles in the treated area. At no time during the 2 weeks should more neuromodulator be injected in the same area. It takes a full 2 weeks for the product to take effect. All good things come to those who wait. It is strongly advised that the patient come back in for a follow-up evaluation at 2 weeks. Discuss how the patient feels, how they like their results. Take after photos in the exact positions as the before photos.

Duration

Neuromodulators last anywhere from 2 to 4 months, the average being about 3 months. It is essential to the patients to let them know they will not be completely frozen the entire treatment duration. It will take about 2 weeks to become fully effective, and then at about 6–8 weeks' time, the neuromodulator will start to "break." Break, meaning a little bit of expression or movement will start to come back, and the patient will notice this. Be sure always to keep photos for the record so that it is a point of reference to compare. At times patients will call at 6 weeks and state that all of their neuromodulators wore off, and they need to be re-seen or re-treated. Pictures will come in handy here as it is likely they see the beginning stages of the reinnervation with baby sprouts of the nerve terminals. It is not necessary to wait till all of the patient's movement is back. Retreat at 3 months before the full expression is restored; this way, they can train the muscles.

Follow up and special considerations

Follow up appointment should always occur after 2 weeks but within 3 weeks post-treatment. At this visit, the assessment of treatment is crucial to the long term relationship with the client. Assess for effectiveness, overall satisfaction of the treatment, patient feelings on the treatment, and assessment of the brows.

Should the patient develop a spock brow or a peaked brow, aka a quizzical look, additional dosing is necessary. This look is most noticeable in animation. Let us discuss why this happens,

Under dosed

Strong frontalis

Lateral injections were omitted or not lateral enough.

The correction would be to add 1–2 units of Botox, Jeuveau or Xeomin or 2.5–5 units of Dysport at the apex of the frontalis muscle's peak above the brow.

Education at the initial consultation can help the patient expect certain expressions based on the assessment and consultative process. It is advised that they understand that a bit of a peak in the eyebrow can happen at any point during the 2-week phase. Be sure to let them know that many years have been dedicated to the craft and education as the injector. The patient should feel confident that their injector is competent and understands the anatomy to manage this should it last beyond the 2 weeks settling out phase. It is common for patients to experience this peak, spock, or surprised look temporarily during the 2-week phase, and then it settles out on its own. They may call and demand touch up. Stand firm and counsel them that this is normal and that a touch can not happen till the 2-week mark, or they could risk ptosis (Table 11.2).

Table 11.2 Complications and management table.

Complications	Cause	Management
Bruising	Penetration of a vessel	Arnica ointment
Swelling	Penetration of skin, inflammatory reaction	Ice
Pain	Trauma	Ice and acetaminophen
Headache	Anxiety	Acetaminophen and real dark cocoa
Eye brow ptosis (eye brow droop or depression)	Frontalis muscle injected to low	Do nothing You may consider adding more units to the medial corrugators, procerus and lateral tail of the orbicularis oculi. Add more only if the patient can still move the muscles in that area
Blepharoptsois (eye lid droop)	Uncommon with frontalis injections. Caused by lateral corrugator treated with penetration or migration into the levator palpebrae	Iopidine gtts and time
Spock brow	Injections in the frontalis to high or not lateral enough	Inject at the apex of the peak 1–2 units of Botox, Jeuveau, Xeomin or 2.5 units of Dysport

Continued

Table 11.2 Complications and management table—cont'd

Complications	Cause	Management
Eyebrow asymmetry	Natural May be a result of a drop and or a spock on one side or the other	If the relatively symmetrical patient presents at follow up with asymmetry. Treat the lateral frontalis "spock" 1–2 units posterior to the peak of the brow and 2 units at the lateral tail of the orbicularis oculi
No effect		Consider Zinc deficiency Report to the medical science liaison of the maker of the neuromodulation for further guidance Review medical history again, are they on chelators, are they zinc deficient, do they have an underlying disease not disclosed concerning possible increased metabolization Review the reconstitution, injections, dosing and placement
Dysport allergy		Antihistamine Follow up with primary care or Allergist Notate in chart new allergy

Before and after photos

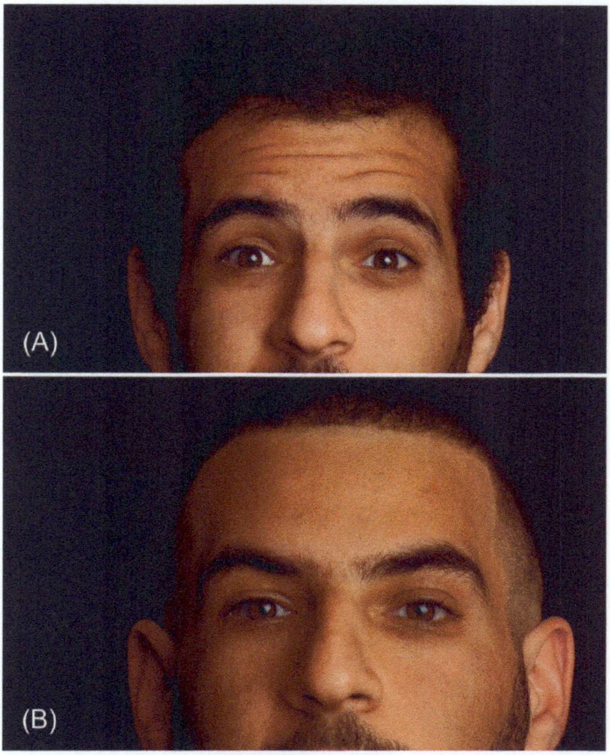

Fig. 11.14 (A) Before photo. (B) After photo.

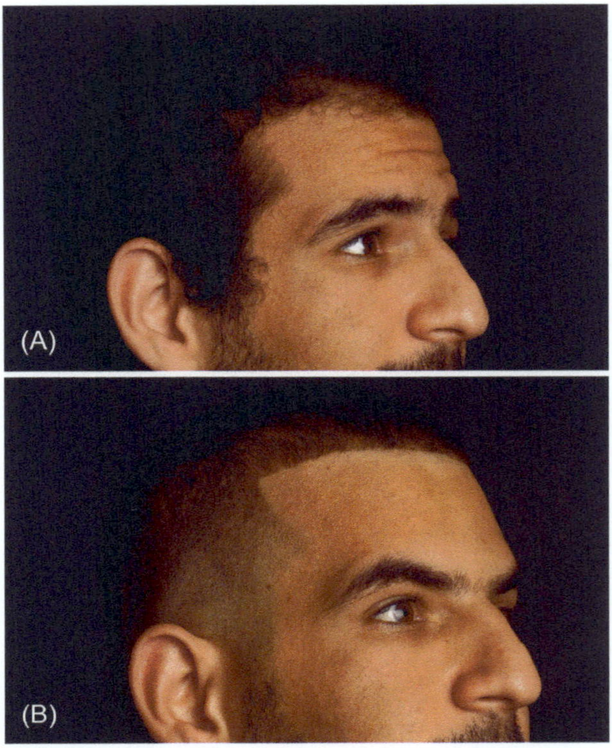

Fig. 11.15 (A) Before photo. (B) After photo.

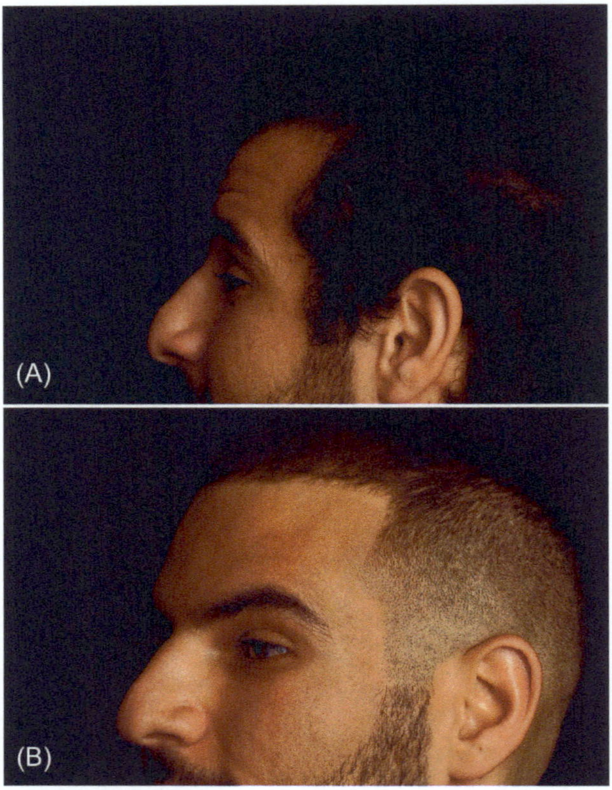

Fig. 11.16 (A) Before photo. (B) After photo.

Young male patient with thick sebaceous skin. He was treated for static and dynamic rhytids of the forehead. As we can see in Figs. 11.14A,B, 11.15A,B, and 11.16A,B, he has had a significant improvement in all angles while animating. Because he was treated in the frontalis, we also treated his glabellar complex, including his procerus, medial and lateral corrugators. We discussed his glabellar treatment in the previous section, where he received 24 units in the glabellar complex. The critical piece is noting that he was treated in his glabellar complex as if he was not, he very well could have been left with a heaviness of his brows and forehead. The procerus and corrugator muscles are depressor muscles, and the frontalis is an elevator muscle. In regards to this patient, had he only been treated in the forehead with neuromodulation, he would have felt very heavy. As his only elevator would have become a depressor muscle sitting on top of a depressor group of muscles (procerus and corrugators). So it would have been a heavy feeling and could have made him look and feel tired and run down. By treating the glabellar complex, we could counteract some of the heaviness by giving him lift in-between the brows before we paralyzed his only elevator muscle in the upper face, the frontalis. He received a total of 20 units of Botox in the frontalis muscle. Note in Fig. 11.14A, the tail of his left brow is a bit droopy. In Fig. 11.14B, it has been perked up. He was treated with 2 units at the left lateral tail to help give that section a bit of lift to even out his brows and open up that eye. Taking note of these subtle asymmetries is very important in measuring the patient's overall satisfaction. Most patients are not aware of their subtle asymmetries until after they receive their treatment. If injectors have not gently addressed the patient's asymmetry prior to treatment, it is common for the patient to feel that the treatment has created the asymmetry. Find a gentle yet subtle way to point out the asymmetry prior to the initiation of the treatment. Do not over-promise the correction of the asymmetry. Treat accordingly and attempt to correct it but do not over-promise. Asymmetries can be difficult to treat as sometimes the cause is structural or trauma-related.

Treatment

Glabellar Complex 24 units Botox
Frontalis 20 units Botox
Photos 2 weeks apart

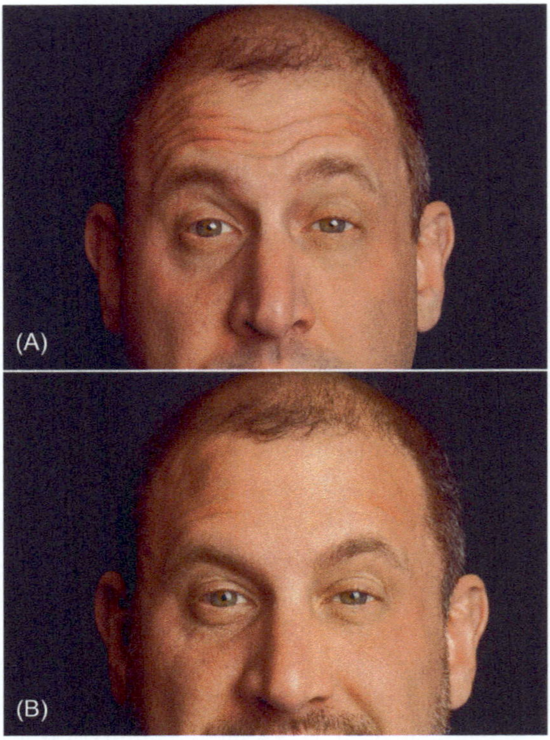

Fig. 11.17 (A) Before photo. (B) After photo.

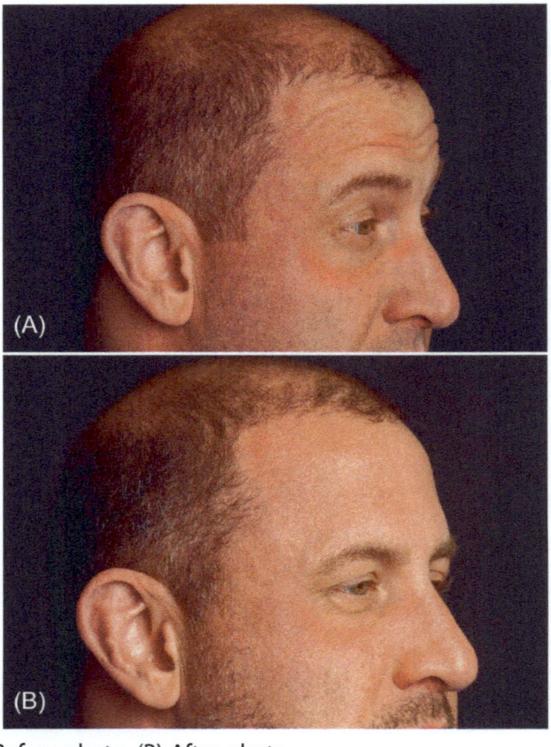

Fig. 11.18 (A) Before photo. (B) After photo.

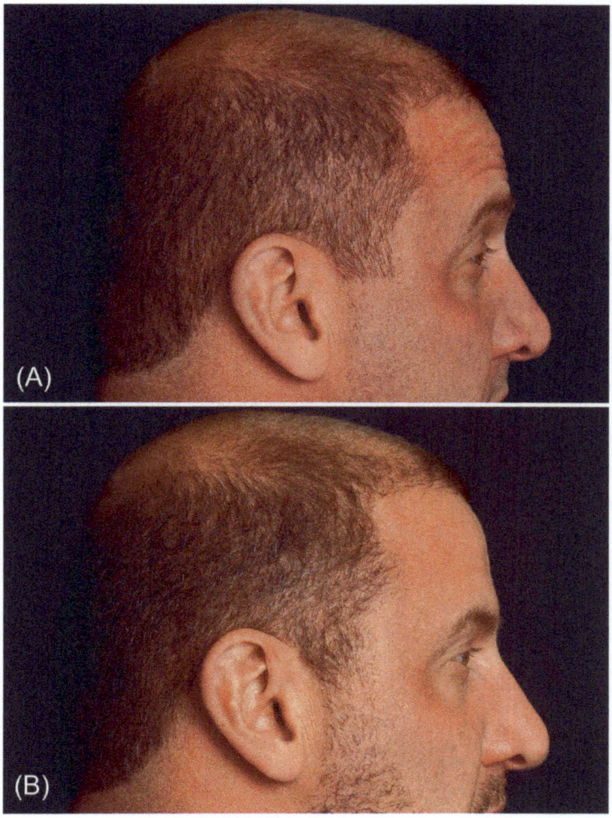

Fig. 11.19 (A) Before photo. (B) After photo.

This male patient in his 40s was seeking treatment for the glabellar complex in combination with his frontalis. He wanted his forehead treated for his severe static forehead rhytids. He was treated in the glabellar complex to combat the heavy feeling he otherwise could have experienced post-treatment. A total of 24 units of Botox was utilized in this patient's glabellar complex along with 24 units of Botox in his forehead. A single row of injections was utilized on the lateral frontalis, and a double row was used medially, where the majority of his activity was focused. Caution was used on the lateral portions of the frontalis, given his hooded lids at rest. The laxity of his upper eyelid is noticed in Fig. 11.17A even in animation. This is the type of client that needs careful attention with the frontalis treatment. If too much neuromodulator is used or the treatment is too low, they can have a severe heavy sensation of their lids can become even heavier. Take note of the patient's frontal view (Fig. 11.17B). Since he had the heavy lids to start with, a conservative approach was taken in regard to the lateral frontalis portion. The lateral portion of his frontalis muscle was still moving at 2 weeks seen in Fig. 11.17B, so additionally, 2 units were placed 1 cm up from the brow and 1 cm over from the mid pupillary line. Please take note of Fig. 11.17A, Figs. 11.18 and 11.19A as he does present with a significant amount of creasing on the right compared to the left side. This can be an indication that the right side may need more neuromodulation than the other side. Unique variations are the very reason why aesthetic medicine is referred to as both art and science. It takes an aesthetic eye to create harmony among the anatomy (Figs. 11.20–11.23).

Treatment

Glabellar Complex 24 units Botox
Forehead 24 units Botox
Photos taken 2 weeks apart

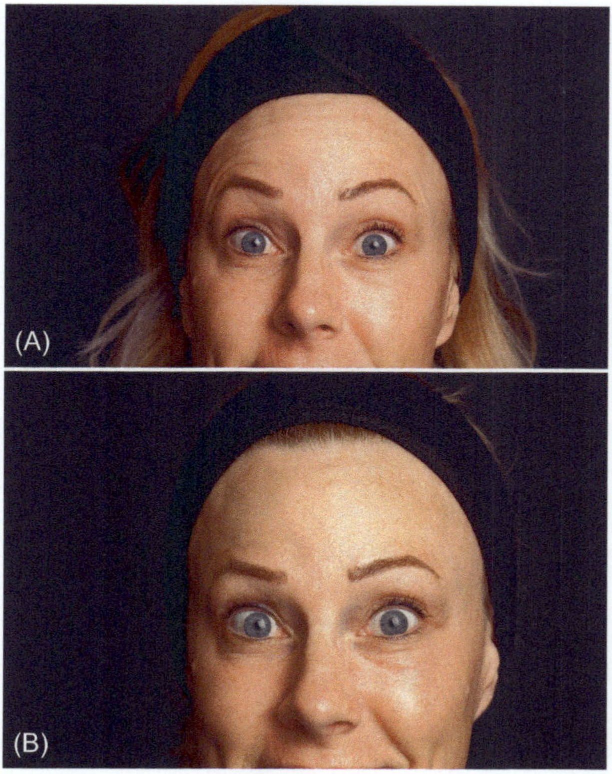

Fig. 11.20 (A) Before photo. (B) After photo.

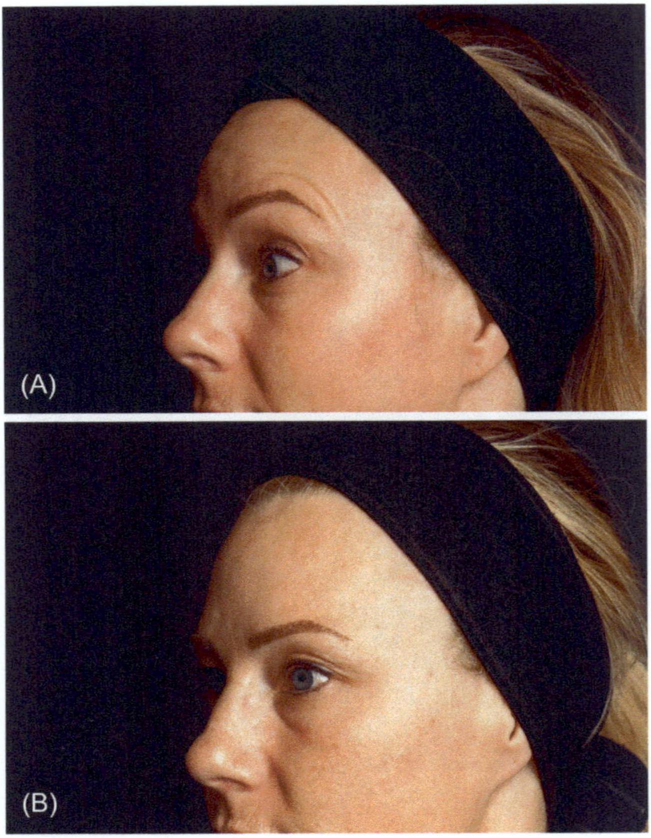

Fig. 11.21 (A) Before photo. (B) After photo.

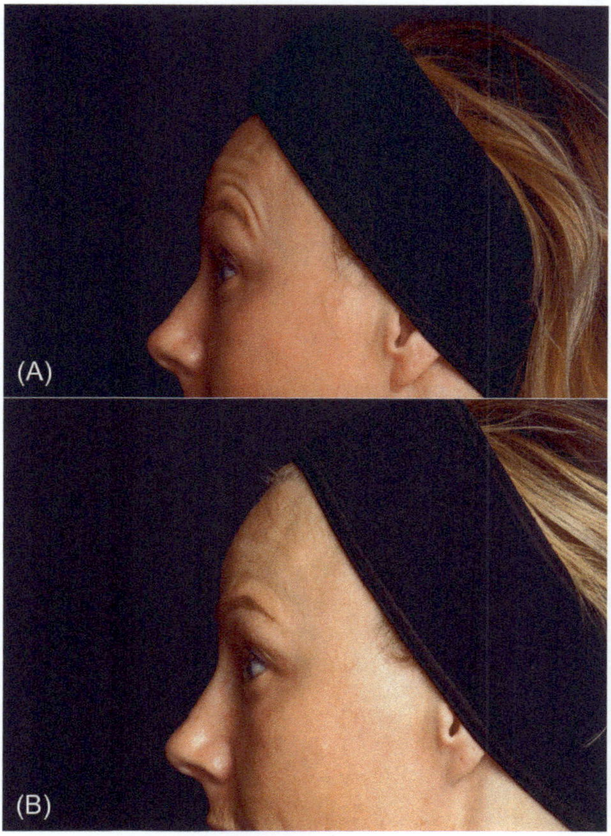

Fig. 11.22 (A) Before photo. (B) After photo.

Treatment

Glabellar Complex 20 units Botox
Brow lift 4 units Botox total
Forehead 16 units Botox
Photos 2 weeks apart

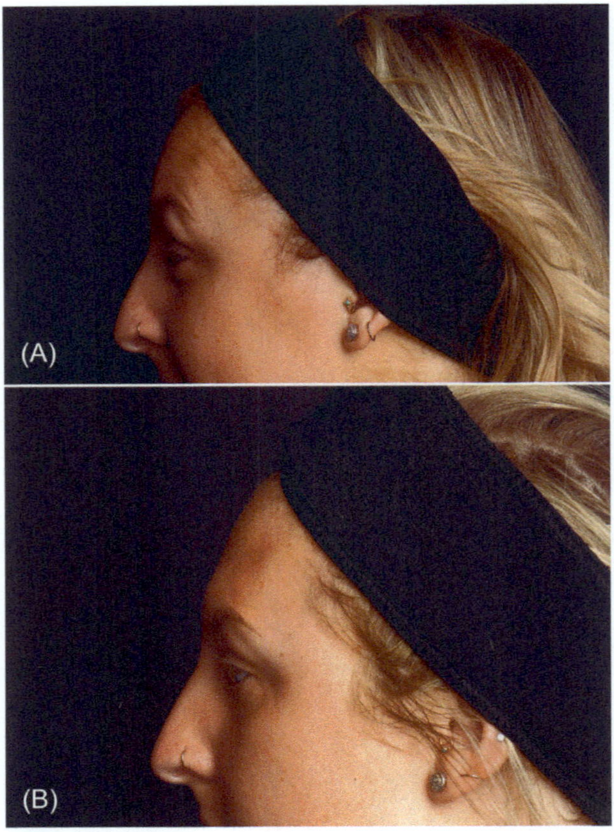

Fig. 11.23 (A) Before photo. (B) After photo.

Treatment

Glabellar complex 16 units Botox
Forehead 12 units Botox
Photos taken 2 week apart

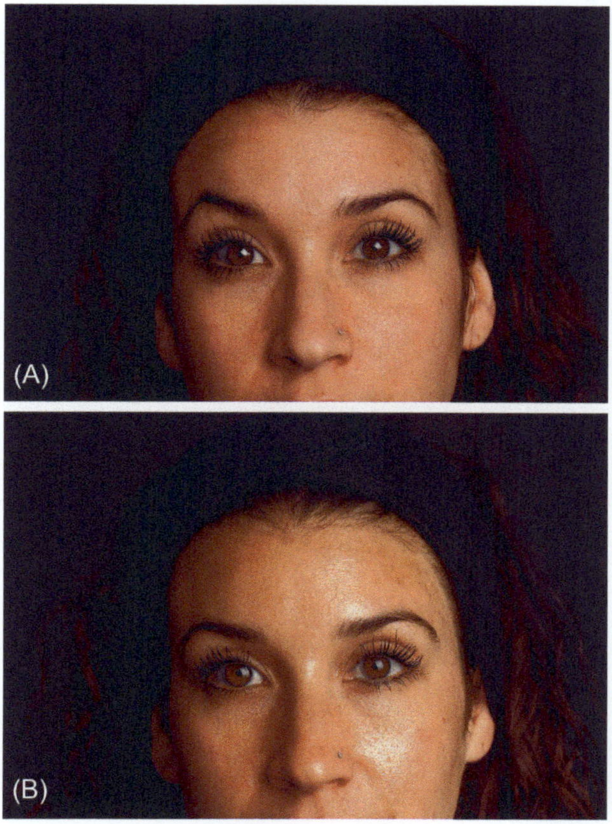

Fig. 11.24 (A) Before photo. (B) After photo.

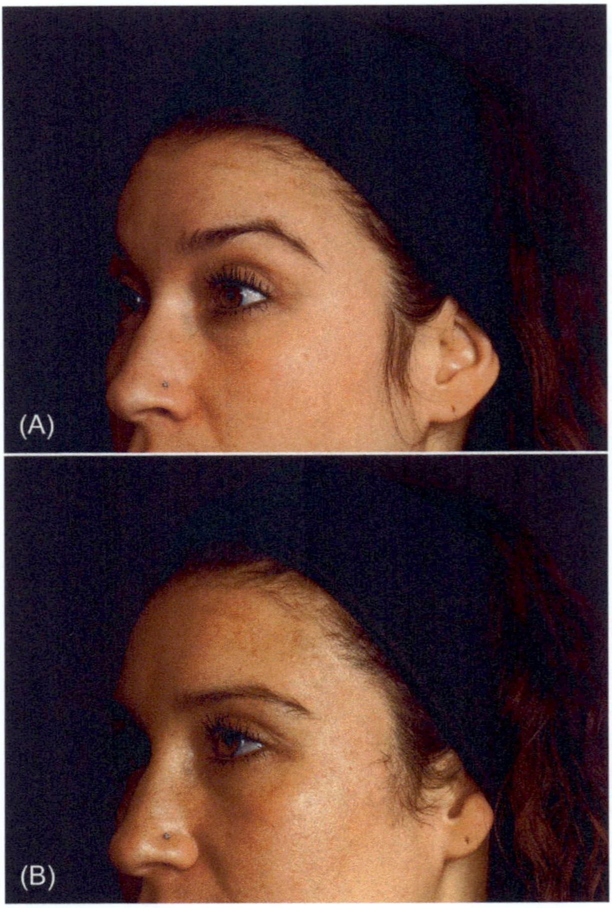

Fig. 11.25 (A) Before photo. (B) After photo.

This beautiful young female patient in her 30s has a unique peaked expression of the brows in animation when her frontalis muscle is activated. When she animates, she does not have much dynamic movement as noted in Fig. 11.24A; however, the peaks really concerned her. A lighter dose was used on her frontalis as she has a shorter forehead than average and without a significant amount of activity. A total of 12 units of Botox was adequate to help soften the sharp point in her brow with the activation of the frontalis. Note her right eye in Fig. 11.24A, one can see that the frontalis muscle actually pulls the lid up and creates a different appearance to her eye shape. Fig. 11.25 is to demonstrate the same patient in a different angle before and after treatment.

Neuromodulation technique in lateral canthal lines aka crows feet

Neuromodulation for lateral canthal lines—Crows feet injections

Static and dynamic crows feet occur due to the active contraction of the lateral portion of the Orbicularis oculi muscle.

Indications for treatment

Neuromodulation for injection is indicated for the temporary improvement in the appearance of moderate to severe lateral canthal lines associated with orbicularis oculi muscle activity in adult patients. Treatment in this area will soften the lines and can elevate the lateral brow.

1. Crows feet (lateral canthal lines)
2. Lateral brow lift

Emotion

Happy when smiling
Squinting to see or avoid a bright light

Muscles in treatment area

Orbicularis oculi
Zygomaticus major
Zygomaticus minor

Muscles to be treated in the targeted zone

Orbicularis oculi

Special considerations

The lateral portion of the orbicularis oculi muscle acts as a depressor to the lateral tail of the eyebrow. This directly contributes to the eyes' closure and more directly creates the crows feet, aka lateral canthal lines. This muscle acts both in the involuntary reflex of blinking and the voluntary closure of the lids. The orbicularis oculi muscle aids in the lymphatic drainage as well. It is important to note that in some instances, the inhibition of this muscle can create swelling under the eyes as fluid can be trapped in the trough due to the reduction of the orbicularis oculi's natural force. It typically works as a pump mechanism to push the fluid out of the area down the lymphatic system's track. When a patient has static lines beyond the targeted treatment area's scope, it is essential to discuss the treatment area and the limitations.

The patient often will not understand that the lines of expression that extend beyond the eye area and down into and over the cheeks are not part of the treatment. When the injector can talk about a treatment plan, perhaps consider laser resurfacing or the addition of dermal fillers.

Another special consideration is the lip elevator muscles. The lip elevator muscles in the treatment area are the Zygomaticus major and Zygomaticus minor. The levator labii superioris sits anterior to the Zygomaticus minor. This muscle is typically far enough out of the treatment area but should be made a note of due to its importance in the smile.

Complimentary locations to increase the efficacy

Bunny lines
Infraocular lines

Treatment goals

Complete paralysis of the orbicularis oculi muscle. Preventing muscle activity in this area will help to prevent deep-set at rest rhytids from developing. This treatment can be both preventative and restorative. If movement is left in this treatment zone, the results can wear off quicker. Do not let the patient dictate the dose based on the price. Remember the injector is the expert, the expert sets the dose, and the patient decides if the recommended dose works within their budget.

Dosing

Women total dose 24–30 units; 12–15 units per side Botox, Jeuveau, and Xeomin.
Women total dose 50–70 units; 25–35 units per side Dysport.
Men total dose 24–40 units; 12–20 units per side Botox, Jeuveau, and Xeomin.
Men total dose 50–70 units; 25–35 units per side Dysport.

Documentation

Medical history including allergies
Medication
Dose
Lot number
Expiration date
Provider
Equipment used

The treatment area and specific dose per injection point
How the patient tolerated
Treatment plan

Equipment

Medication
Botox
Jeuveau
Xeomin
Dysport
Bacteriostatic saline or 0.9% preservative free saline
3 mL luer lock syringe
27 Gauge luer lock 1 1/2 needle
2 × 2 non-woven gauze
Alcohol
Ice
Bottle opener
Marking pencil
BD insulin Syringe 31 gauge ultra thin
Non-sterile gloves

If needed, Hemostats can remove tight needles from luer lock or remove the metal portion of the bottle of neuromodulator.

Assessment

The patient should be assessed in animation and at rest. They should further be assessed from all angles. Even if you are only treating the lateral tail of the orbicularis oculi for eyebrow lift. It is essential to understand the relationship of the brows and your treatment area. The smile and apple of the cheek can change with treatment in this area as well. Therefore, it is also necessary to have the patient smile as to assess how the smile can interfere with eye opening and the apple of the cheek. If the patient is having eye surgery in the future be sure that injections stop 3 months prior to their schedule surgery date. The surgeon needs to have full expression of the muscles in order to surgically alter them permanently. Be cautious not to inject inside the orbital rim near the lateral canthus to avoid diplopia. This treatment area is rich with vessels so be sure to mark out the vessels in order to minimize bruising, ice can help to constrict vessels just prior to injecting.

Additional notation:

Lagopthalmos—incomplete closure of the eye
Excessive scleral show
Ectropian-Eversion of the eyelid margin
Xeropthalmia-decreased lacrimal flow

Pre-procedure steps

1. Consultation
2. Have patient sign consent and pre and post-procedure guidelines
3. Photographs at three angles at rest and animated
   - **(i)** 50 degrees both left and right
   - **(ii)** 45 degrees both left and right
   - **(iii)** Straight on
4. Have the patient comfortably seated in an upright position
5. Ensure adequate overhead lighting
6. Mark out danger zones
7. Additionally, mark out treatment areas with a marking pencil. Be sure not to inject through the marking pencil as it can tattoo the patient.
8. Calculate the dosing according to the desired look, anatomy, and area being treated.
9. Draw up the medication being sure not to dull the needle by touching the sides of the glass bottle or going through the rubber stopper with the intended injection needle.
10. Cleanse the patients face thoroughly with a double cleanse and then chlorhexidine or alcohol (allow to dry).

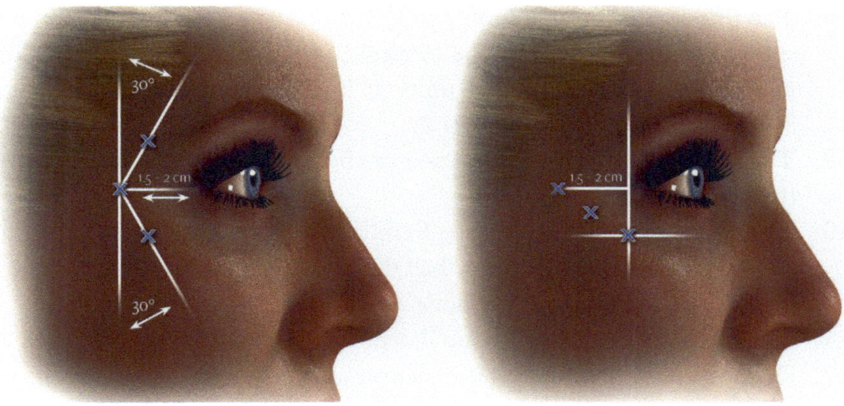

Fig. 11.26 Suggested injection pattern depending on the positioning of the canthal fan.

Technique

1. Identify the treatment danger zones.
2. Palpate the orbital rim. All injections will be placed superficially (sub-dermally) and 1 cm outside of the orbital rim. It is essential to know the lateral orbicularis oculi muscle lies on top of the Zygomaticus major and minor. When treating the top portion of the orbicularis oculi at the tail of the eyebrow, it is essential to note that the orbicularis oculi sits on top of the frontalis muscle. Lateral Limbus Line
3. Locate the orbicularis oculi muscle by having the patient force a big smile.
4. Mark out the injection points. There is a wide variety of anatomical presentations in this area. Each presentation will need an individual assessment and an individual treatment plan (Fig. 11.26).
5. Be sure that the patient is still cleansed and has not touched the treatment area since proper cleansing performed in the pre-procedural steps.
6. It is best to stand on the opposite side being treated. Have the patient turn their face to assist in the proper position. The needle should always be inserted into the skin away from the globe. Proper positioning will also help to prevent inadvertent involvement of the deep extraocular muscles that would cause diplopia (Fig. 11.27).
7. Typically there is three injection points for the crow's feet, two is common in younger patients or patients that do not like the reposition of the apple in their cheek by the lowest injection point of the crows feet.

8. The first superficial injection point will be 1 cm outside of the orbital rim and in line with the lateral canthus. Based on the assessment needs, each injection could be 2–4 units of Botox, Jeuveau, or Xeomin or 5–10 units of Dysport.
9. The second injection will be 0.5–1 cm below the first injection point at the same dosing as above.
10. The third injection will be 0.5–1 cm below the second injection point and the same dose as above or half the dose depending on the muscle's strength or the patients' aesthetic goals.
11. Repeat injections on the contralateral side.
12. Do not press or rub on the injection sites. If the patient happens to bleed, apply gentle pressure to stop the bleeding. GENTLE!!! It is not advised to manipulate the product from the site as adverse events, and product migration can occur.

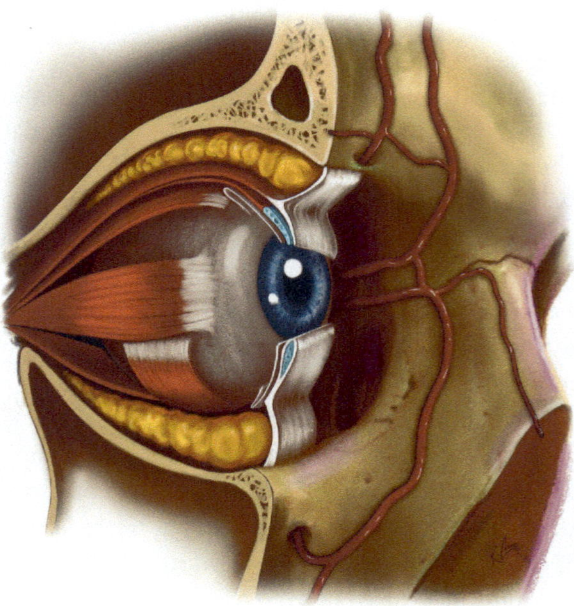

Fig. 11.27 Deep extraocular muscles of the eye.

Results

Within 3–10 days, patients will start to see gradual improvement of the dynamic wrinkles in the treated area. At no time during the 2 weeks should more neuromodulator be injected in the same area. It takes a full 2 weeks for the product to take effect. All good things come to those who wait. At 2 weeks, it is strongly advised that the patient come back in for a follow-up evaluation. Discuss how the patient feels, how they like their results. Take after photos in the exact positions as the before photos.

Duration

Neuromodulators last anywhere from 2 to 4 months. The average is about three. It is essential to inform the clients that they will not be completely frozen the entire treatment duration. It will take about 2 weeks to become fully effective, and then at about 6–8 weeks, the neurotoxin will start to "break." Break, meaning a little bit of expression or movement will start to come back, and the patient will notice this. Be sure always to keep photos for the records so that you have them as a reference point to compare. At times patients will call at 6 weeks and state that ALL of their neuromodulators wore off, and they need to be re-seen or re-treated. Pictures will come in handy here as it is likely they see the beginning stages of the reservation with baby sprouts of the nerve terminals. It is not necessary to wait until all of the patient's movement is back. Retreat at 3 months before the full expression is restored; this way, they can train the muscles.

Follow up and special consideration

At the 2 weeks follow up.

It is common for patients to still demonstrate movement in the area treated after treatment. Therefore, it is necessary to assess the patient in animation and add additional units according to the previously treated muscles with movement. Additional dosing ranges from 4–12 units of Botox, Jeuveau, Xeomin per site, or 10–20 units of Dysport per site.

Persistent crows feet.

If the patient is not appropriately counseled on the use of neuromodulation, they will often come back and say that the treatment did not work because they have lines still. Educate the patient that the treatment is only working on the muscles and the motion and that the treatment is not intended to reduce the static lines. Of course, there is a caveat; in the younger patients, it can reduce the appearance of static lines, but this is not something the injector should promise. Always teach the client that the treatment is for the hyperdynamic muscle movement, not the static line (Table 11.3).

Table 11.3 Complications and management table.

Complications	Cause	Management
Bruising	Puncture of a vessel	Arnica ointment Avoid sun exposure to reduce the risk of hemosideren staining
Swelling	Trauma	Ice
Pain	Trauma	Ice and acetaminophen
Headache	Adverse event from neuromodulation	Acetaminophen and real dark cocoa
Persistant movement	Initial dose not adequate or the muscle is stronger than anticipated	Consider Zinc deficiency Add additional dose in areas of movement If movement is persistent posterior to the crows feet treatment zone (this is a normal finding) meaning that the Zygomticus major and minor were preserved
Static lines	Late onset skin care treatments Sun exposure	Sun Screen At home skin care Skin resurfacing treatments
Diplopia	Injection in the extrocular muscles	Inject away from the globe
Impaired blink reflex	Rare	Time
Bunny lines	Can become more prominent when the orbicularis oculi becomes incompetent	Treat the nasalis muscle
Preseptal lines	Compensatory mechanism of the medial preseptal orbicularis oculi muscle	Reducing the dose of the third injection point or by adding a few droplets in the preseptal space so superficial you can see the fluid under the skin. Just like a PPD plant

Table 11.3 Complications and management table—cont'd

Complications	Cause	Management
Infrabrow lines	The superior lateral orbicularis oculi muscle becomes hyper dynamic as a compensatory mechanism	Treat the superior lateral orbicularis oculi muscle at the 2 week follow up
Photophobia	Reduced squinting from over injections	Avoid repeat treatment in this area
Lip and cheek ptosis	Inadvertent injection to the Zygomatic major and minor	Time
Lagopthalmos	High dose and too deep	Moistening drops
Ectropian	High dose and two deep	Moistening drops
Xeropthalmia	Previous Lasik surgery	Moistening drops
Globe truama	Poor injection technique	Ocular specialist stat
No effect		Consider Zinc deficiency Report to the medical science liaison of the maker of the neuromodulation for further guidance Review medical history again, are they on chelators, are they zinc deficient, do they have an underlying disease not disclosed concerning possible increased metabolization Review the reconstitution, injections, dosing and placement
Dysport allergy		Antihistamine Follow up with primary care and Allergist Notate in chart new allergy

Details on Dysport allergy

If using Dysport and the patient is allergic, the patient will complain of flu like symptoms or feeling run down like they are experiencing seasonal allergies or experiencing a sinus infection. Not all patients know that they have a milk protein allergy, and treatment with Dysport uncovers the allergy. Typically no intervention is needed as it will self-resolve. However, it is appropriate to suggest over the counter antihistamines or see their primary care or allergist for further testing and lifestyle modifications.

Before and after photos

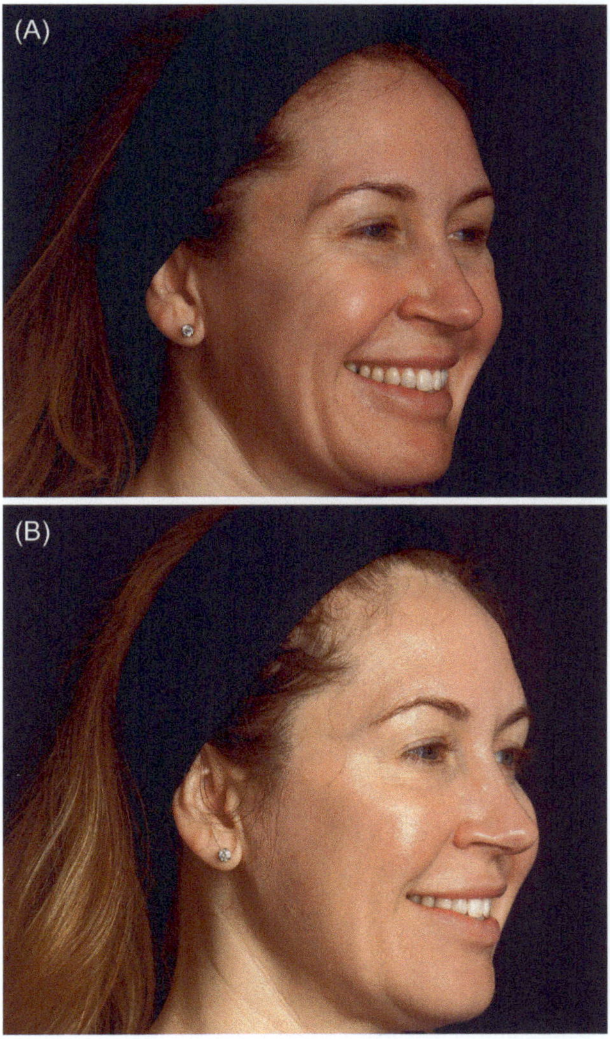

Fig. 11.28 (A) Before photo. (B) After photo.

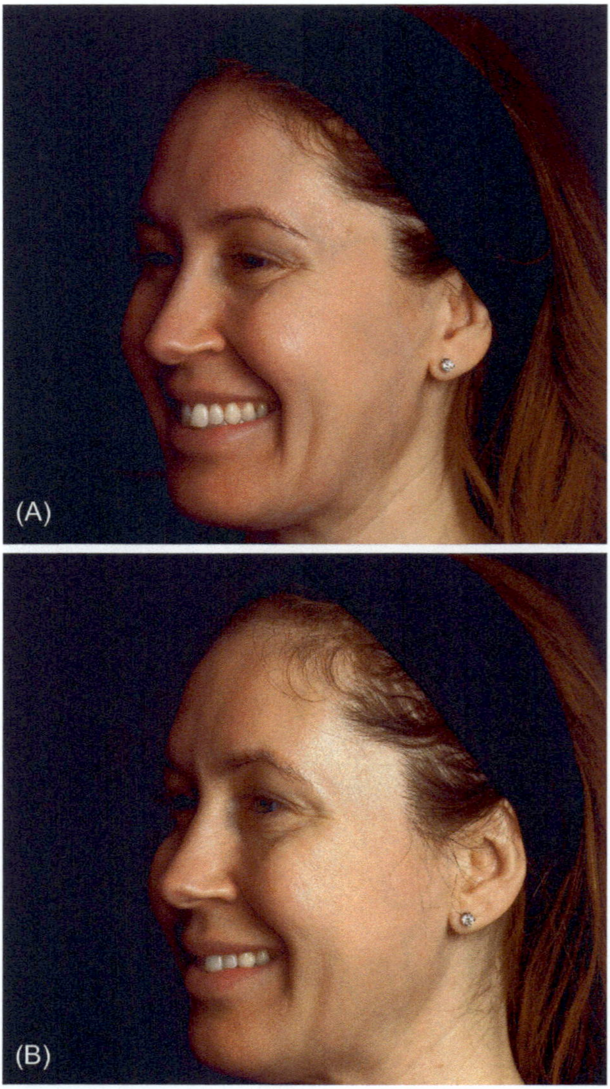
(A)
(B)

Fig. 11.29 (A) Before photo. (B) After photo.

This female patient in her 40s sought treatment of her crow's feet. Her concern was the lines she sees at rest, along with the loss of a scleral show when she smiled as seen in Fig. 11.28A. Her treatment goal was to prevent further breakdown of the skin around the eyes and prevent the extreme closure of her lids with smiling and squinting. With this treatment, she will have a brighter, more youthful look in her eyes when she smiled or squinted.

In Fig. 11.29B it is noted that she has a touch of movement left. An additional 6 units were placed on this side at her follow up. She is a sales representative, so she spends lots of time in the car with the sun beating on that side of her face.

Treatment

Total dose was 24 units of Botox
Right canthal lines 12 units Botox
Left canthal lines 12 units Botox
Additional treatment at her 2 week follow up was 6 units Botox
Photos 2 weeks apart

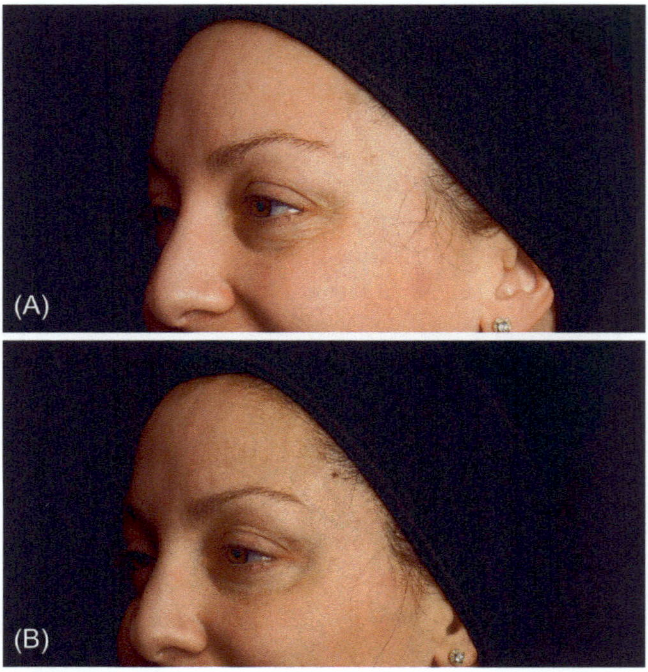

Fig. 11.30 (A) Before photo. (B) After photo.

This young female has porcelain skin with minimal sun damage. She did not have significant crows feet as evidenced in Fig. 11.30A. Her concern was with the dynamic rhytids under her eye. As she smiled, the thin skin bunched up and was bothersome to her. We treated her canthal lines to help prevent the skin's bunching directly under her lower lash line. Only 8 units of Botox were used per side, and it is evident that there was an excellent softening of the animation in the area. Treatment in this area will help prevent the static rhytids that can become quite troublesome later in her life.

Treatment

Right canthal lines 8
Left canthal lines 8
Photos taken at 2 weeks

Neuromodulation technique in eyebrow lift

Neuromodulation for lateral eyebrow lift

Brows can be asymmetrical and need help to be repositioned on the face. Low sitting brows (eyebrow ptosis) can convey sadness on the face despite the emotion. Dermatochalasis (eyelid skin laxity or Hooding) can make the patient's eyes feel heavy and tired. They often have difficulty applying mascara and eyeliner due to the sagging of the tissue into the eyelid area. These conditions can be helped with proper placement of neuromodulation to the lateral tail of the superior orbicularis oculi muscle.

Indications for treatment

Neuromodulation for injection is indicated for the temporary improvement in the appearance of eyebrow asymmetry and or eyebrow ptosis.

1. Lateral brow lift
2. Upper eyelid dermatochalasis
3. Lateral eyebrow ptosis

Emotion

Angry
Sad
Tired
Aged
Sinister

Muscles in treatment area

Orbicularis oculi
Frontalis
Levator palpebrae

Muscles to be treated in the targeted zone

Superior lateral orbicularis oculi

Special considerations

The lateral portion of the orbicularis oculi muscle acts as a depressor to the lateral tail of the eyebrow. The orbicularis oculi directly contribute to the eyes' closure and more directly creates the crows feet, aka lateral canthal lines. This muscle acts both in the involuntary reflex of blinking and the voluntary closure of the lids. It is advised that concomitant areas be treated when the request is to lift the brow. Treating the glabellar complex will elevate the medial brow and aid in the synergistic effect of the eyebrow ptosis. When injecting this area, be mindful to keep the injections superficial to not inadvertently inject the frontalis or the levator palpebrae muscle with deep injections.

Complimentary locations to increase efficacy

Crows feet
Glabellar complex

Treatment goals

We are looking for partial inhibition of the superior lateral orbicularis oculi. Do not let the patient dictate the dose based on the price. Remember the injector is the expert, the expert sets the dose and the patient decides if the recommended dose works within their budget.

Dosing

Women total dose 4–8 units; 2–4 units per side Botox, Jeuveau and Xeomin.
Women total dose 5–10 units; 2.5–5 units per side Dysport.
Men total dose 4–8 units; 2–4 units per side Botox, Jeuveau and Xeomin.
Men total dose 5–10 units; 2.5–5 units per side Dysport.

Documentation

Medical history including allergies
Medication
Dose
Lot number
Expiration date
Provider
Equipment used
Treatment area and specific dose per injection point
How patient tolerated
Treatment plan

Equipment

Medication
Botox
Jeuveau
Xeomin
Dysport
Bacteriostatic saline or 0.9% preservative free saline
3 mL luer lock syringe
27 Gauge luer lock 1 1/2 needle

2 × 2 non-woven gauze
Alcohol
Ice
Bottle opener
Marking pencil
BD insulin Syringe 31 gauge ultra thin
Non-sterile gloves

Hemostats assist in removing tight needles from luer lock or removing the metal portion of the bottle of neuromodulator if needed.

Assessment

The patient should be assessed in animation and at rest. They should further be assessed from all angles. Even if only treating the lateral tail of the superior aspect of the lateral orbicularis oculi for an eyebrow lift, it is essential to understand the relationship of the brows and the treatment area. If the patient is having eye surgery in the future, be sure that injections stop 3 months before their scheduled surgery date. The surgeon needs to have full expression of the muscles in order to surgical alter them permanently. This treatment area is rich with vessels, so be sure to mark out the vessels to minimize bruising; ice can help constrict vessels just before injecting.

Additional notation:

Lagophthalmos—incomplete closure of the eye
Excessive scleral show
Ectropion-Eversion of the eyelid margin
Xeropthalmia-decreased lacrimal flow

Pre-procedure steps

1. Consultation
2. Have patient sign consent and pre and post-procedure guidelines
3. Photographs at three angles at rest and animated
    - **(i)** 50 degrees both left and right
    - **(ii)** 45 degrees both left and right
    - **(iii)** Straight on
4. Have patient comfortably seated in an upright position
5. Ensure adequate overhead lighting
6. Mark out danger zones

7. Additionally, mark out treatment areas with a marking pencil. Be sure not to inject through the marking pencil as it can tattoo the patient.
8. Calculate dosing according to the desired look, anatomy and area being treated.
9. Draw up medication being sure not to dull your needle by touching the sides of the glass bottle or going through the rubber stopper with your intended injection needle.
10. Cleanse the patients face thoroughly with a double cleanse and then chlorhexidine or alcohol (allow to dry).

Technique

1. Identify the treatment danger zones.
2. Palpate the orbital rim. Identify the superior lateral portion of the orbicularis oculi muscle by having the patient squint or forcefully close the eyelids. An evident depression will be seen in the skin. Your first injection point will be outside of the lateral limbus line and 1 cm outside of the orbital rim.
3. Mark out the injection points.
4. Be sure that the patient is still cleansed and has not touched the treatment area since proper cleansing performed in the pre-procedural steps.
5. It is best to stand on the same side you are treating. Hold the skin taught (be sure not to pull or manipulate the skin) so that the skin does not slide away when you insert the needle. The first injection point will be performed while the patient is relaxed. The needle will be about 1.5 cm lateral to the lateral limbus line. The needle should be angled toward the frontalis muscle away from the globe. Be careful to manage a slow and steady extrusion force not to force migration into the levator palpebrae. Your depth will be subdermal and your dose will be 2 units ofBotox, Xeomin, Jeuveau, or 2.5 units of Dysport.
6. The second injection will be 1 cm medical to the first injection point closer to the lateral limbus line. Again this injection will be subdermal and only 1 unit of Botox, Xeomin, Jeuveau, or 2.5 units of Dysport.
7. Repeat the treatment on the contralateral side.
8. Do NOT press or rub on the injection sites. If the patient happens to bleed you may apply gentle pressure to stop the bleeding. GENTLE!!! It is not advised that you manipulate the product from the site as adverse events and product migration can occur.

Results

Lateral eyebrow elevation. Patient will notice an opening of the eye as well.

Within 3–10 days patients will start to see gradual improvement of the dynamic wrinkles in treated area. At no time during the 2 weeks should more neuromodulator be injected in the same area. It takes a full 2 weeks for the product to take effect. All good things come to those who wait. At 2 weeks time it is strongly advised that the patient come back in for follow up evaluation. Discuss how the patient feels, how they like their results. Take after photos in the exact positions as the before photos.

Duration

Neuromodulators last anywhere from 2 to 4 months. The average is about three. It is essential to the clients to let them know they will not be completely frozen the entire duration of the treatment. It will take about 2 weeks to become fully effective and then at about 6–8 weeks time, the neurotoxin will start to "break." Break, meaning a little bit of expression or movement will start to come back and the patient will notice this. Be sure always to keep your photos for your records so that you can compare. At times patients will call at 6 weeks and state that ALL of their neuromodulators wore off and they need to be re-seen or re-treated. Pictures will come in handy here as it is likely they are seeing the beginning stages of the reinnervation with baby sprouts of the nerve terminals. You do not need to wait till all of the patients movement is back. *Re*-treat at 3 months before full expression is restored; this way, they can train the muscles.

Follow up and special consideration

At the 2 week follow up.

If at your follow-up visit, it would be aesthetically pleasing to have a greater lift than assess the superior lateral orbicularis oculi muscle's function. Should you notice significant contractility administer more neuromodulator as you did at the previous visit. Total dose for touch up can be up to the same amount as initial visit dose.

If your patient is not appropriately counseled on neuromodulation's use, they will often come back and say that the treatment did not work because they have lines still. It is crucial that you educate the patient that the treatment is only working on the muscles and the motion and that it is not intended to reduce the static lines. Of course there is a caveat; in your younger patients it can reduce the appearance of static lines but this is not something the injector should promise. Always teach the client that the treatment is for the hyperdynamic muscle movement not the static line (Table 11.4).

Table 11.4 Complications management table.

Complications	Cause	Management
Bruising	Puncture of a vessel	Arnica ointment Avoid Sun exposure to reduce the risk of hemosideren staining
Swelling	Trauma	Ice
Pain	Trauma	Ice and acetaminophen
Headache	Adverse event from neuromodulation	Acetominophen and real dark cocoa
Persistant movement	Initial dose not adequate Muscle stronger than anticipated	Consider Zinc deficiency Add additional dose in areas of movement
Static lines	Late onset skin care treatments Sun exposure	Sun Screen At home skin care Skin resurfacing treatments
Diplopia	Injection in the extrocular muscles	Inject away from the globe Time
Impaired blink reflex	Rare	Time
Bunny lines	Can become more prominent when orbicularis oculi becomes incompetent	Treat the nasalis muscle
Worsening of Festoons	Weakening of the inferior medial orbicularis oculi muscle	Time
Epiphoria-tearing	Weekend lacrimal function if placed to medial to the mid pupillary line	Time
Lagopthalmos	High dose and too deep	Moistening drops
Ectropian	High dose and two deep	Moistening drops
Xeropthalmia	Previous Lasik surgery	Moistening drops
Globe truama	Poor injection technique	Ocular specialist stat

Table 11.4 Complications management table—cont'd

Complications	Cause	Management
No effect		Consider Zinc deficiency Report to the medical science liaison of the maker of the neuromodulation for further guidance Review medical history again, are they on chelators, are they zinc deficient, do they have an underlying disease not disclosed concerning possible increased metabolization Review the reconstitution, injections, dosing and placement
Dysport allergy		Antihistamine Follow up with primary care and Allergist Notate in chart new allergy

Details on Dysport allergy

If using Dysport and the patient is allergic, the patient will complain of flu like symptoms or feeling run down like they are experiencing seasonal allergies or experiencing a sinus infection. Not all patients know that they have a milk protein allergy, and treatment with Dysport uncovers the allergy. Typically no intervention is needed as it will self-resolve. However, it is appropriate to suggest over the counter antihistamines or see their primary care or allergist for further testing and lifestyle modifications.

Before and after photos

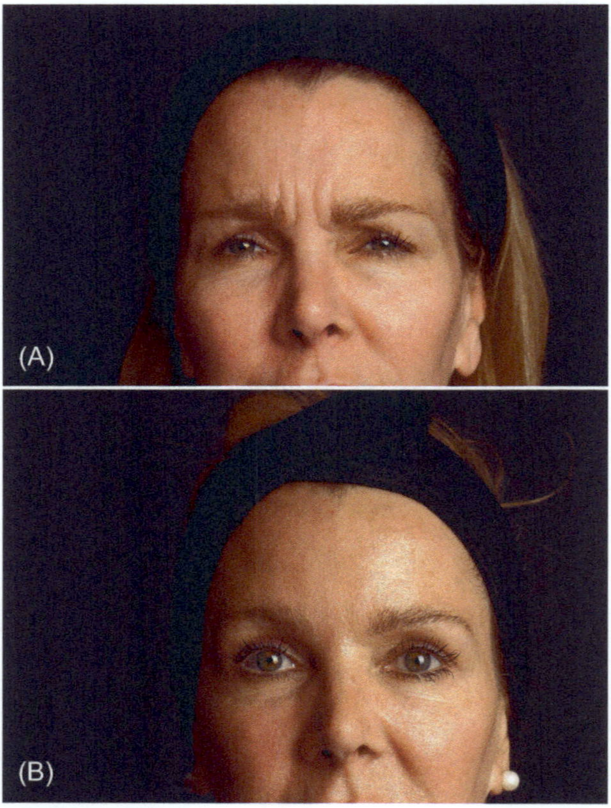

Fig. 11.31 (A) Before photo. (B) After photo.

When this patient squints or smiles the lateral portion of her orbicularis oculi compresses the corners of her eyes as seen in Fig. 11.31. If it is noticed during animation, the lateral portion of her upper lid compresses down even over her lashes seen in Fig. 11.31A, and she loses much of her iris. After treatment, when she animates the skin of her upper eyelids remain off her lashes and she has a bright eye in animation, noticed in Fig. 11.31B.

Treatment

Upper right lateral orbicularis oculi 2 units Botox
Upper left lateral orbicularis oculi 2 units Botox
Photos 2 weeks apart

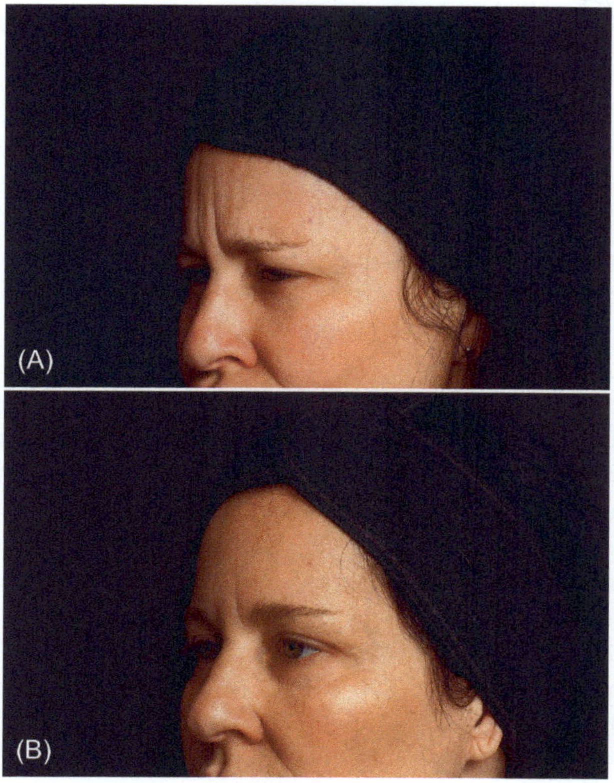

Fig. 11.32 (A) Animated. (B) At rest.

The next patient is a unique case as she has chronic migraines and is treated by a neurologist with Botox. Part of the migraine treatment includes the neurologist treating the frontalis muscle. In general the neurologists are focused on the migraine not always the aesthetic. It is common to see patients with a flattened brow or a lid ptosis post-treatment for migraines. Notice the patient's medial brow has been depressed and has forced the tail of her brow in an upwards fashion Fig. 11.32B. This is not an ideal patient that would benefit from treatment of the tail of the brow. She is flipped up a bit in Fig. 11.32B and in even more so with animation in Fig. 11.32A. This situation is an excellent reminder of the importance of assessment at rest and in animation.

Neuromodulation technique in lower eyelid wrinkles

Neuromodulation for infraocular space

Static and dynamic lines can occur in this area with or without treatment in concommitment areas. These lines may be there initially, or they may be caused by treating the orbicularis oculi muscle. This muscle can be discerned by examining for the presence of static lines. Should the patient have static lines in this area, one can be sure the neuromodulation did not cause it. It may, however, exacerbated it. Either way, it is an area that needs attention and can be bothersome to the patients.

Indications for treatment

Neuromodulation for injection is indicated for the temporary improvement in the appearance of moderate to severe infraocular rhytids. Treatment in this area prevents complete contraction of the medial lower orbicularis oculi which can also open the aperture. The infraorbital rhytids course horizontally and radiate laterally from the medial canthus to the lateral canthus.

1. Lower eyelid wrinkles
2. Plapebral aperture lengthening (distance between the upper and lower eyelid margin and creates a rounding of the eye shape)
3. Eye bulge "jelly role." This is different from a festoon

Emotion

No real emotion attached here, just aged appearance

Muscles in treatment area

Orbicularis oculi
Zygomaticus major
Zygomaticus minor
Levator Labii Superioris
Nasalis

Muscles to be treated in the targeted zone

Preseptal medial orbicularis oculi

Special considerations

Hyperdynamic function of the nasalis muscles
Severe lower eyelid dermatochalasis (skin laxity)
Abnormal snap test

Festoons
Lagophthalmos
Excessive sceral show
Ectropion
Previous eyelid surgery (blepharoplasty)

Another special consideration is the lip elevator muscles. The lip elevator muscles in the treatment area are the Zygomaticus major and Zygomaticus minor. The levator labii superioris sits anterior to the Zygomaticus minor. This muscle is typically far enough out of the treatment area but should be made note of due to its importance in the smile.

Complimentary locations to increase the efficacy

Bunny lines
Orbicularis oculi

Treatment goals

Partial inhibition of the inferior medial orbicularis oculi muscle

Dosing

Women total dose 3–6 units; 1–3 units per side Botox, Jeuveau, and Xeomin.
Women total dose 5 units; 2.5 units per side Dysport.
Men total dose 3–6 units; 1–3 units per side Botox, Jeuveau, and Xeomin.
Men total dose 5 units; 2.5 units per side Dysport.

Documentation

Medical history including allergies
Medication
Dose
Lot number
Expiration date
Provider
Equipment used
The treatment area and specific dose per injection point
How the patient tolerated
Treatment plan

Equipment

Medication
Botox
Jeuveau
Xeomin
Dysport
Bacteriostatic saline or 0.9% preservative-free saline
3 mL luer lock syringe
27 Gauge luer lock 1 1/2 needle
2 × 2 non-woven gauze
Alcohol
Ice
Bottle opener
Marking pencil
BD insulin Syringe 31 gauge ultra thin
Non-sterile gloves

If needed, hemostats assist in the removal of tight needles from luer lock or removing the metal portion of the bottle of neuromodulator.

Assessment

The patient should be assessed in animation and at rest. It is crucial to have the patient squint, smile, close their eyes tight, and make a bunny face. The hyperdyanism of these muscles can come from a variety of expressions. They should further be assessed from all angles. Pinch the skin to assess for laxity. Pinching of the skin is called a snap test. If the skin recoils quickly, the patient can have treatment. If the skin has a delayed snap test for the skin to return to normal, it is advised not to treat that particular area in the patient.

Other special instructions:

Review the medical history for surgery and whether the patient is about to have eyelid surgery and or Lasik surgery.

Lagophthalmos—Incomplete closure of the eye. Lagophthalmos is an absolute contraindication for the treatment of neuromodulation in the area.

Excessive scleral show—If the patient has a crescent shape of the sclera beneath the pupil; it is not advised to treat that the area as the neuromodulator will exacerbate the scleral show.

Ectropion-Eversion of the eyelid margin.

Xeropthalmia-decreased lacrimal flow.
Festoon—lower eye bags that require surgery. As the orbital septum weakens over time, the orbital fat pad herniates and creates an eye bag or what is called a festoon. A weakening of the orbicularis oculi over a festoon can also exacerbate the festoon.

Pre-procedure steps

1. Consultation
2. Have patient sign consent and pre and post-procedure guidelines
3. Photographs at three angles at rest and animated
   - **(i)** 50 degrees both left and right
   - **(ii)** 45 degrees both left and right
   - **(iii)** Straight on
4. Have the patient comfortably seated in an upright position
5. Ensure adequate overhead lighting
6. Mark out danger zones
7. Additionally, mark out treatment areas with a marking pencil. Be sure do not to inject through the marking pencil as it can tattoo the patient.
8. Calculate dosing according to the desired look, anatomy, and area being treated.
9. Draw up the medication, being sure not to dull the needle by touching the sides of the glass bottle or going through the rubber stopper with the intended injection needle.
10. Cleanse the patients face thoroughly with a double cleanse and then chlorhexidine or alcohol (allow to dry).

Technique

1. Identify the treatment danger zones.
2. Palpate the orbital rim.
3. Identify the lateral canthal lines and mid pupillary lines.
4. The injector should be positioned on the treatment side.
5. Treat the patient with their head resting back against the headrest of the chair and eyes closed.
6. Be sure that the patient is still cleansed and has not touched the treatment area since proper cleansing performed in the pre-procedural steps.
7. Medial lower eyelid injection. The medial lower eyelid injection will be placed mid pupillary line and 0.5 cm below the eyelid margin. Angle the needle bevel up and medially and parallel to the skin inject

extremely superficial. Proper placement and depth will result in the appearance of a fluid-filled wheal. Deposit 1 unit of Botox, Xeomin, Jeuveau, or 2.5 units of Dysport here. Be mindful of tiny vessels as a bruise in this area can become quite large.

8. The next injection will be lateral to the first injection point. Angle and dose will be the same. Yet, the injection will be a bit lower at about 1 cm.
9. Repeat the same injections on the contralateral side.
10. Do NOT press or rub on the injection sites. If the patient happens to bleed, apply gentle pressure to stop the bleeding. GENTLE!!! It is not advised to manipulate the product from the site as adverse events, and product migration can occur.

Results

Within 3–10 days, patients will start to see gradual improvement of the dynamic and static wrinkles in the treated area. At no time during the 2 weeks should more neuromodulator be injected in the same area. It takes a full 2 weeks for the product to take effect. All good things come to those who wait. At 2 weeks, it is strongly advised that the patient come back in for a follow-up evaluation of their treatment. Discuss how the patient feels, how they like their results. Take after photos in the exact positions as the before photos.

This patient was mainly concerned with the dynamic wrinkling under her eyes when she made a bunny nose or made a face like she smelled something bad, as well as when she smiled big. 2 units of Botox were placed extremely superficial and close to the lash line in the infraocular space. One unit was placed superficially like a PPD plant mid pupillary line, and the other was just a few millimeters to the lateral side and performed extremely superficial. This treatment was done on the contralateral side.

Right infraocular space 2 units Botox
Left infraocular space 2 units Botox

Duration

Neuromodulators last anywhere from 2 to 4 months. The average is about 3. It is essential to the clients to let them know they will not be completely frozen the entire duration of the treatment. It will take about 2 weeks to become fully effective, and then at about 6–8 weeks, the neurotoxin will start to "break." Break, meaning a little bit of expression or movement will start to come back, and the patient will notice this. Be sure always to keep photos for the records so that they can be refereed to for comparison. At times patients will call at 6 weeks and state that all of their neuromodulator wore off, and they need to be re-seen or re-treated. Pictures will come in handy here as it is likely they see the beginning stages of the reinnervation with baby sprouts of the nerve terminals. It is not necessary to wait till all of the patient's movement is back. Retreat at 3 months before the full expression is restored; this way, they can train the muscles.

Follow up and special consideration

At the 2 week follow up.

It is common for patients to still demonstrate movement in the area treated after treatment. Persistent wrinkling can be caused by underdosing. More neuromodulation may be added at the follow-up visit if the patient has not exhibited any of the following:

Lagophthalmos
Excessive scleral show
Ectropion

If deep static wrinkling is persistent, educate the patient on the need for repeat treatments and perhaps cheek filler and or laser resurfacing. If adjacent hyperdynamic movement is the cause, it is advised to treat additional areas (Table 11.5).

Table 11.5 Complications and management table.

Complications	Cause	Management
Bruising	Puncture of a vessel	Arnica ointment Avoid sun exposure to reduce the risk of hemosideren staining
Swelling	Trauma	Ice
Pain	Trauma	Ice and acetaminophen
Headache	Adverse event from neuromodulation	Acetominophen and real dark cocoa
Persistant movement	Initial dose not adequate Muscle stronger than anticipated	Consider Zinc deficiency Add additional dose in areas of movement
Static lines	Late onset skin care treatments Sun exposure	Sun screen At home skin care Skin resurfacing treatments

Continued

Table 11.5 Complications and management table—cont'd

Complications	Cause	Management
Diplopia	Injection in the extrocular muscles	Inject away from the globe Time
Impaired blink reflex	Rare	Time
Bunny lines	Can become more prominent when orbicularis oculi becomes incompetent	Treat the nasalis muscle
Worsening of Festoons	Weaking of the inferior medial orbicularis oculi muscle	Time
Epiphoria-tearing	Weekend lacrimal function if placed to medial to the mid pupillary line	Time
Lagopthalmos	High dose and too deep	Moistening drops
Ectropian	High dose and two deep	Moistening drops
Xeropthalmia	Previous Lasik surgery	Moistening drops
Globe truama	Poor injection technique	Ocular specialist stat
No effect		Consider Zinc deficiency Report to the medical science liaison of the maker of the neuromodulation for further guidance Review medical history again, are they on chelators, are they zinc deficient, do they have an underlying disease not disclosed concerning possible increased metabolization Review the reconstitution, injections, dosing and placement
Dysport allergy		Antihistamine Follow up with primary care and Allergist Notate in chart new allergy

Details on Dysport allergy

If using Dysport and the patient is allergic, the patient will complain of flu like symptoms or feeling run down like they are experiencing seasonal allergies or experiencing a sinus infection. Not all patients know that they have a milk protein allergy, and treatment with Dysport uncovers the allergy. Typically no intervention is needed as it will self-resolve. However, it is appropriate to suggest over the counter antihistamines or see their primary care or allergist for further testing and lifestyle modifications.

Before and after photos

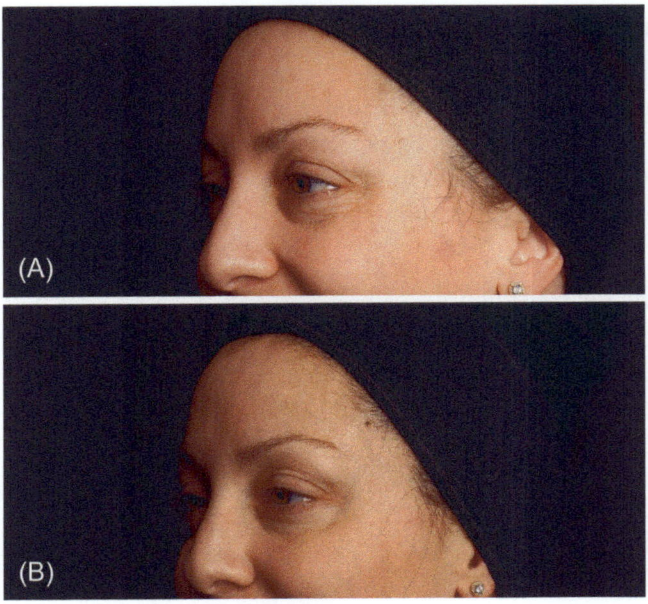

Fig. 11.33 (A) Before photo. (B) After photo.

This patient was mainly concerned with the dynamic wrinkling under her eyes when she made a bunny nose or made a face like she smelled something bad, as well as when she smiled big. Two units of Botox were placed extremely superficial and close to the lash line in the infraocular space. One unit was placed superficially like a PPD plant mid pupillary line, and the other was just a few millimeters to the lateral side and performed extremely superficial. This treatment was done on the contralateral side. As noted in Fig. 11.33B her infraorbital creasing is reduced. Typically this treatment is best served as a combination therapy. For instance treating the crow feet and the nasils muscle can help aid in the recovery of smooth skin in this area.

Treatment

Right infraocular space 2 units Botox
Left infraocular space 2 units Botox
Photos 2 weeks apart

Neuromodulation technique of the bunny lines aka nasalis lines

Neuromodulation for bunny lines

Static and dynamic lines can occur in this area with the contraction of the nasalis muscle. It can be exacerbated when the patient smiles, frowns, or squints. The bunny lines can be exacerbated by the use of neuromodulation of the crows feet. The nasalis muscle contracts the sides of the nose upward and creates the bunny line appearance.

Indications for treatment

Neuromodulation for injection is indicated for the temporary improvement in the appearance of moderate to severe nasalis rhytids, aka bunny lines.

Emotion

Foul smell
Happy
Approval

Muscles in treatment area

Orbicularis oculi
Levator Labii Superioris
Levator Labii Superioris alaeque nasi muscle
Nasalis
Procerus

Muscles to be treated in the targeted zone

Nasalis

Special considerations

Severe lower eyelid dermatochalasis (skin laxity).

Another special consideration is the lip elevator muscles. The levator labii superioris sits anterior to the Zygomaticus minor. This muscle resides the length of the side of the nose. This muscle is typically far enough out of the treatment area but should be made note of due to its importance in the smile. A transverse bunny line across the bridge of the nose can be seen at the base of the procerus. Special considerations to anatomy make concomitant treatments essential. Treating the glabellar complex along with the nasalis muscle would be ideal in a patient that presents with a static rhytid across the bridge of the nose.

Complimentary locations to increase the efficacy

Medial orbicularis oculi muscle
Lateral orbicularis oculi
Procerus
Medial and lateral corrugator

Treatment goals

Complete inhibition of the nasalis muscle

Dosing

Women total dose 4–8 units; 2–4 units per side Botox, Jeuveau, and Xeomin
Women total dose 10–20 units; 5–10 units per side Dysport
Men total dose 4–12 units; 2–6 units per side Botox, Jeuveau, and Xeomin
Men total dose 10–30 units; 5–15 units per side Dysport

Documentation

Medical history including allergies
Medication
Dose
Lot number
Expiration date
Provider
Equipment used
The treatment area and specific dose per injection point
How the patient tolerated
Treatment plan

Equipment

Medication
Botox
Jeuveau
Xeomin
Dysport
Bacteriostatic saline or 0.9% preservative-free saline
3 mL luer lock syringe
27 Gauge luer lock 1 1/2 needle
2 × 2 non-woven gauze
Alcohol
Ice
Bottle opener
Marking pencil
BD insulin Syringe 31 gauge ultra thin
Non sterile gloves

Hemostats assist in removing tight needles from luer lock or removing the metal portion of the bottle of neurotoxin if needed.

Assessment

The patient should be assessed in animation and at rest. It is essential to have the patient squint, smile, close their eyes tight, and make a bunny face. The hyperdynasim of these muscles can come from a variety of expressions. They should further be assessed from all angles. If the patient's bunny lines result from other active muscles, then it is advised to treat those areas as well. In order to assess the muscle in the treatment area, ask the patient to scrunch their nose or pretend they are smelling something foul.

Pre-procedure steps

1. Consultation
2. Have patient sign consent and pre and post-procedure guidelines
3. Photographs at three angles at rest and animated
   - **(i)** 50 degrees both left and right
   - **(ii)** 45 degrees both left and right
   - **(iii)** Straight on
4. Have the patient comfortably seated in an upright position
5. Ensure adequate overhead lighting
6. Mark out danger zones
7. Additionally, mark out treatment areas with a marking pencil. Be sure not to inject through the marking pencil as it can tattoo the patient.

8. Calculate dosing according to the desired look, anatomy, and area being treated.
9. Draw up the medication, being sure not to dull the needle by touching the sides of the glass bottle, or going through the rubber stopper with the intended injection needle.
10. Cleanse the patients face thoroughly with a double cleanse and then chlorhexidine or alcohol (allow to dry).

Technique

1. Identify the treatment danger zones.
2. Locate the nasalis muscle by asking the patient to scrunch up their nose.
3. Identify the lateral canthal lines and mid-pupillary lines.
4. The injector should be positioned on the treatment side.
5. Treat the patient with their head resting back against the headrest of the chair and eyes closed.
6. Be sure that the patient is still cleansed and has not touched the treatment area since proper cleansing performed in the pre-procedural steps.
7. This injection should be performed while the muscle is active to be precise with the placement. Inject 2 units of Botox, Jeuveau or Xeomin or 5 units of Dysport into the subdermal lateral wall of the nose directly into the nasalis muscle.
8. Repeat the same injections on the contralateral side.
9. At times a patient will need an injection point just on the dorsal portion of the nose at the point of maximum wrinkling. Inject 2 units of Botox, Jeuveau or Xeomin or 5 units of Dysport here at this site.
10. Do not press or rub on the injection sites. If the patient happens to bleed, apply gentle pressure to stop the bleeding. GENTLE. It is not advised to manipulate the product from the site as adverse events, and product migration can occur.

Results

Reduction of hyperdynamic nasalis muscle AKA bunny lines

Within 3–10 days, patients will start to see gradual improvement of the dynamic and static wrinkles in the treated area. At no time during the 2 weeks should more neuromodulator be injected in the same area. It takes a full 2 weeks for the product to take effect. All good things come to those who wait. It is strongly advised that the patient come back in for a follow-up evaluation at 2 weeks'. Discuss how the patient feels, how they like their results. Take after photos in the exact positions as the before photos.

Duration

Neuromodulators last anywhere from 2 to 4 months. The average duration is about 3 months. It is essential to the clients to let them know they will not be completely frozen the entire duration of the treatment. It will take about 2 weeks to become fully effective, and then at about 6–8 weeks, the neuromodulator will start to "break." Break, meaning a little bit of expression or movement will start to come back, and the patient will notice this. Be sure to keep photos for the records so that they can be referred to for comparison. At times, patients will call at 6 weeks and state that all of their neuromodulator wore off, and they need to be re-seen or re-treated. Pictures will come in handy here as it is likely they see the beginning stages of the reinnervation with baby sprouts of the nerve terminals. It is not necessary to wait till all of the patient's movement is back. Retreat at 3 months before the full expression is restored; this way, they can train the muscles.

Follow up and special consideration

At the 2 week follow up:

It is common for some patients to still demonstrate movement in the area treated after treatment. Persistent wrinkling can be caused by inadequate dosing. Adding more neuromodulation at the follow up will help to complete the look in some instances.

If deep static wrinkling is persistent, educate the patient on the need for repeat treatments and or laser resurfacing. If adjacent hyperdynamic movement is the cause, it is advised to treat additional areas (Table 11.6).

Table 11.6 Complications and management table.

Complications	Cause	Management
Bruising	Puncture of a vessel	Arnica ointment Avoid sun exposure to reduce the risk of hemosideren staining
Swelling	Trauma	Ice
Pain	Trauma	Ice and acetaminophen
Headache	Adverse event from neuromodulation	Acetaminophen and real dark cocoa
Persistant movement	Initial dose not adequate Muscle stronger than anticipated	Consider Zinc deficiency Add additional dose in areas of movement

Table 11.6 Complications and management table—cont'd

Complications	Cause	Management
Static lines	Late onset skin care treatments Sun exposure	Sun screen At home skin care Skin resurfacing treatments
Diplopia	Injection in the extrocular muscles	Inject away from the globe
Lip ptosis	Inadvertant injection into the Levator Labii Superioris Alaeque Nasi (LLSAN)	Time
Oral incompetence and drooling	Inadvertant injection into the Levator Labii Superioris Alaeque Nasi (LLSAN)	Time
Impaired blink reflex	Rare	Time
Epiphoria-tearing	Weakend lacrimal function if placed to medial to the mid pupillary line	Time
No effect		Consider Zinc deficiency Report to the medical science liaison of the maker of the neuromodulation for further guidance Review medical history again, are they on chelators, are they zinc deficient, do they have an underlying disease not disclosed concerning possible increased metabolization Review the reconstitution, injections, dosing and placement
Dysport allergy		Antihistamine Follow up with primary care and Allergist. Notate in chart new allergy

Details on Dysport allergy

If using Dysport and the patient is allergic, the patient will complain of flu like symptoms or feeling run down like they are experiencing seasonal allergies or experiencing a sinus infection. Not all patients know that they have a milk protein allergy, and treatment with Dysport uncovers the allergy. Typically no intervention is needed as it will self-resolve. However, it is appropriate to suggest over the counter antihistamines or see their primary care or allergist for further testing and lifestyle modifications.

Before and after photos

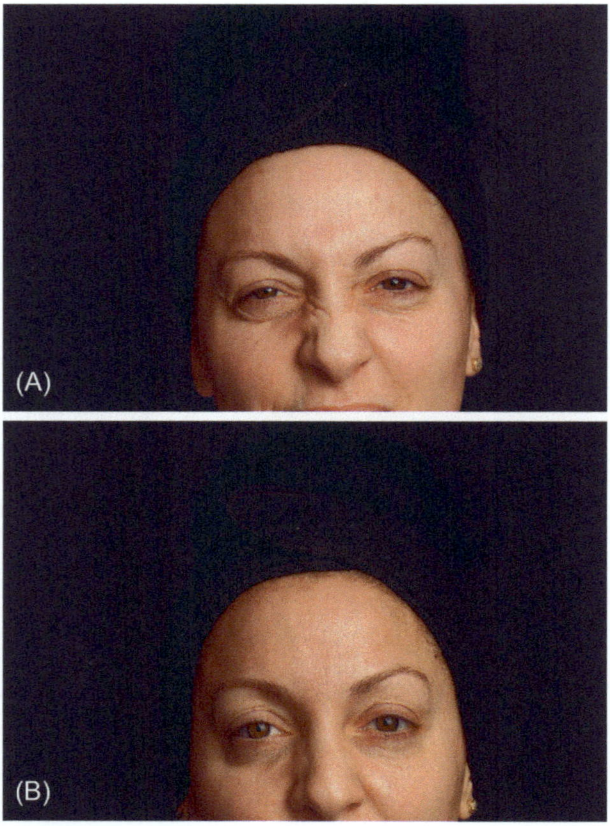

Fig. 11.34 (A) Before photo. (B) After photo.

The primary concern for this patient was static and dynamic rhytids in the infraocular space. We took a global approach to this patient. The emergence of a greater understanding of the anatomy and the muscles' relationships in more depth will move injectors away from treating the complaint, which will increase patient satisfaction. Instead, they will begin to develop a pan-facial approach to the treatment. A multifaceted approach to treatment will lead to higher patient satisfaction and duration of the neuromodulator. Two units of Botox were treated on each side of the nose. The nasalis muscle works in combination with several other muscles to create movement. The alar portion of the nasalis dilates the nares, also referred to as nasal flair. The transverse section of the nasalis compresses the nasal aperture and wrinkles the nasal skin. Many believe that this muscle can become hyperdynamic after years of treating the lateral canthal lines with neuromodulation. The human body is an amazing entity. Each system or opposing muscle will attempt to compensate for the weekend or affected side. It only makes sense that after years of taking out the natural activity of one muscle that its counterpart would begin overcompensating to help. Fig. 11.34A and B demonstrates a reduction in the infraocular rhytids as well as a significant softening of the bunny lines.

Treatment

Right upper nasalis 2 units Botox
Left upper nasalis 2 units Botox
Photos taken 2 weeks apart

Neuromodulation technique of peri-oral rhytids aka smokers lines

Neuromodulation for smokers lines

Static and dynamic lines can occur in this area with the contraction of the orbicularis oris muscle. This hyperdynamic muscle, whether a person smokes or not, will create the static appearance of rhytids. The act of chewing, drinking through a straw, and many other typical functions of the mouth can all aid in the break down of the tissue by repeated use. Treating with neuromodulation in this area will decrease the force of the muscle contraction and create a bit of a lip eversion. The eversion will help to give the patient a bit more pink show of the lip.

Indications for treatment

Neuromodulation for injection is indicated for the temporary improvement in the appearance of moderate to severe lip lines, aka smokers lines.

Emotion/movement

Angry
Tension
Aged
Whistle
Blow kisses

Muscles in treatment area

Orbicularis oris
Levator Labii Superioris
Levator Labii Superioris alaeque nasi muscle (LLSAN)
Zygomaticus Major
Zygomaticus Minor
Depressor Anguli Oris (DAO)
Depressor Labi Inferioris (DLI)
Levator Anguli Oris

Muscles to be treated in the targeted zone

Orbicularis oris

Special considerations

The orbicularis oris muscle pulls the lip tissue in centrally and will invert the lip. This muscles assist in eating, drinking, not drooling, whistling and kissing.

Complimentary locations to increase the efficacy

Depressor Anguli Oris (DAO)
Mentalis

Treatment goals

Partial inhibition of the orbicularis oris.
Complete inhibition would yield great difficulty for the patient to chew, drink through a straw, and or whistle.

Dosing

Women total dose 4–12 units; 2–6 units per side; 2–6 units upper and lower Botox, Jeuveau and Xeomin.

Women total dose 5–15 units; 2.5–7.5 units per side; 2.5–7.5 units upper and lower Dsyport.

Men total dose 4–12 units; 2–6 units per side; 2–6 units upper and lower Botox, Jeuveau and Xeomin.

Men total dose 5–15 units; 2.5–7.5 units per side; 2.5–7.5 units upper and lower Dysport.

Documentation

Medical history including allergies
Medication
Dose
Lot number
Expiration date
Provider
Equipment used
The treatment area and specific dose per injection point
How the patient tolerated
Treatment plan

Equipment

Medication
Botox
Jeuveau
Xeomin
Dysport
Bacteriostatic saline or 0.9% preservative-free saline
3 mL luer lock syringe
27 Gauge luer lock 1 1/2 needle
2 × 2 non-woven gauze
Alcohol
Ice
Bottle opener
Marking pencil
BD insulin Syringe 31 gauge ultra thin
Non-sterile gloves

Hemostat, if needed, to assist in removing tight needles from luer lock or removing the metal portion of the bottle of neurotoxin.

Assessment

The patient should be assessed in animation and at rest. Social history is essential. If the patient continues to smoke, this treatment may not be advised. Have the patient purse their lips, lick their lips, whistle, and blow a kiss.

Pre-procedure steps

1. Consultation
2. Have patient sign consent and pre and post-procedure guidelines
3. Photographs at three angles at rest and animated
   - (i) 50 degrees both left and right
   - (ii) 45 degrees both left and right
   - (iii) Straight on
4. Have the patient comfortably seated in an upright position
5. Ensure adequate overhead lighting
6. Mark out danger zones
7. Additionally, mark out treatment areas with a marking pencil. Be sure not to inject through the marking pencil as it can tattoo the patient.
8. Calculate dosing according to the desired look, anatomy, and area being treated.
9. Draw up the medication, being sure not to dull the needle by touching the sides of the glass bottle, or going through the rubber stopper with the intended injection needle.
10. Cleanse the patients face thoroughly with a double cleanse and then chlorhexidine or alcohol (allow to dry).

Technique

1. Identify the treatment danger zones.
2. Locate the orbicularis oris muscle by asking the patient to purse their lips.
3. Identify the philtral columns.
4. The injector should be positioned on the treatment side.
5. Treat the patient with their head resting back against the headrest of the chair and eyes closed.
6. Be sure that the patient is still cleansed and has not touched the treatment area since proper cleansing performed in the pre-procedural steps.
7. This injection should be performed while the muscle is at rest. Ask the patient to purse the lips and mark the peaks for injection. We do not treat the rhytids with neuromodulation; we treat the peaks as that is where the most active muscle resides.

8. Stay at least 1 cm from the corners of the upper lip and at least 2 cm away from the lower corners of the lip.
9. Depending on the lip lines' severity, two to four injection points may be necessary on the upper lip. The lower lip can be two to three injection points.
10. The safety margin for this injection is just above the pink vermillion border, up 0.5 cm.
11. Have the patient hold ice (that has been properly cleaned) on the treated area for about 30 s prior to your injections. This injection can make the patient tear. Inject subdermally into the space and deposit 1 unit of Botox, Jeuveau or Xeomin or 2.5 units of Dysport per site. The first injection will be at the peak of the cupids bow.
12. Suppose the patient needs the second point of injection. The next injection will follow the same dose and depth and will be 1 cm from the peak of the cupids bow.
13. Repeat the same injections on the contralateral side. The lower lip injections will be superficial, and be sure to avoid injection into the DLI and the DAO. Depending on the anatomy, one central injection may be required. Inject 1 unit of Botox, Jeuveau or Xeomin or 2.5 units of Dysport just out of the pink show of the vermillion border but not more than 0.5 cm up the white lip.
14. Move 1 cm right and repeat injection of 1 unit of Botox, Jeuveau or Xeomin or 2.5 units of Dysport per site. Then from the midpoint move, 1 cm left and repeat the same dose and the same boundary.
15. Do not press or rub on the injection sites. If the patient happens to bleed, apply gentle pressure to stop the bleeding. GENTLE. It is not advised to manipulate the product from the site as adverse events, and product migration can occur.

Results

Reduction of lip rhytids, lip eversion of the pink show.

Within 3–10 days, patients will start to see gradual improvement of the dynamic and static wrinkles in the treated area. At no time during the 2 weeks should more neuromodulator be injected in the same area. It takes a full 2 weeks for the product to take effect. All good things come to those who wait. It is strongly advised that the patient come back in for a follow-up evaluation at 2 weeks' time. Discuss how the patient feels, how they like their results. Take after photos in the exact positions as the before photos.

Duration

Neuromodulators last anywhere from 2 to 4 months. The average duration is about 3 months. It is essential to the clients to let them know they will not be completely frozen the entire duration of the treatment. It will take about 2 weeks to become fully effective, and then at about 6–8 weeks, the neuromodulator will start to "break." Break, meaning a little bit of expression or movement will start to come back, and the patient will notice this. Be sure to keep photos for the records so that they can be referred to for comparison. At times, patients will call at 6 weeks and state that all of their neuromodulator wore off, and they need to be re-seen or re-treated. Pictures will come in handy here as it is likely they see the beginning stages of the reinnervation with baby sprouts of the nerve terminals. It is not necessary to wait till all of the patient's movement is back. Retreat at 3 months before the full expression is restored; this way, they can train the muscles.

Follow up and special consideration

At the 2 week follow up:

It is common for some patients to still demonstrate movement in the area treated after treatment. Persistent wrinkling can be caused by inadequate dosing. Adding more neuromodulation at the follow up will help to complete the look in some instances.

If deep static wrinkling is persistent, educate the patient on the need for repeat treatments and or laser resurfacing. If adjacent hyperdynamic movement is the cause, it is advised to treat additional areas (Table 11.7).

Table 11.7 Complications and management table.

Complications	Cause	Management
Bruising	Puncture of a vessel	Arnica ointment Avoid sun exposure to reduce the risk of hemosideren staining
Swelling	Trauma	Ice
Pain	Trauma	Ice and acetaminophen
Headache	Adverse event from neuromodulation	Acetaminophen and real dark cocoa
Persistant movement	Initial dose not adequate Muscle stronger than anticipated	Consider Zinc deficiency Add additional dose in areas of movement

Table 11.7 Complications and management table—cont'd

Complications	Cause	Management
Static lines	Late onset skin care treatments Sun exposure	Sun screen At home skin care Skin resurfacing treatments Lip filler
Lip ptosis	Inadvertant injection into the Levator Labii Superioris Alaeque Nasi (LLSAN)	Time Careful placement of a corrective injection
Oral incompetence and drooling	Inadvertant injection into the Levator Labii Superioris Alaeque Nasi (LLSAN)	Time
No effect		Consider Zinc deficiency Report to the medical science liaison of the maker of the neuromodulation for further guidance Review medical history again, are they on chelators, are they zinc deficient, do they have an underlying disease not disclosed concerning possible increased metabolization Review the reconstitution, injections, dosing and placement
Dysport allergy		Antihistamine Follow up with primary care and Allergist Notate in chart new allergy

Details on Dysport allergy

If using Dysport and the patient is allergic, the patient will complain of flu like symptoms or feeling run down like they are experiencing seasonal allergies or experiencing a sinus infection. Not all patients know that they have a milk protein allergy, and treatment with Dysport uncovers the allergy. Typically no intervention is needed as it will self-resolve. However, it is appropriate to suggest over the counter antihistamines or see their primary care or allergist for further testing and lifestyle modifications.

Before and after photos

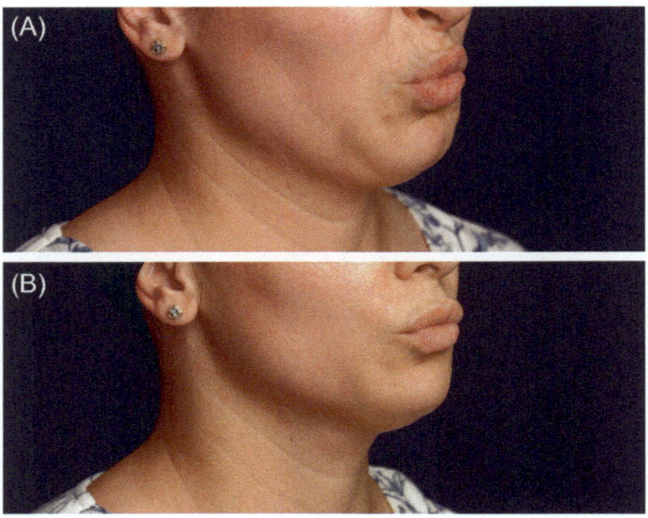

Fig. 11.35 (A) Before photo. (B) After photo.

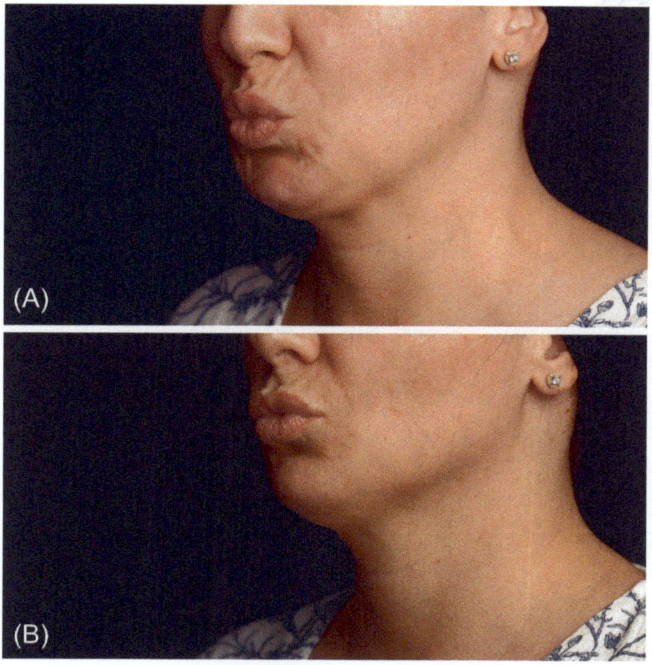

Fig. 11.36 (A) Before photo. (B) After photo.

Whether the patient smokes or not, the pesky little barcode lines above the upper lip can become a nuisance for the patients. Drinking through a straw, kissing, sun exposure, a history of cold sores can all affect the breakdown of this tissue. The orbicularis oris muscle is a powerful muscle with only a thin layer of fat over it to protect it, so it is natural that this would break down in the face of repetitive force. This area is best when treated in the preventative phase. If the patient has severe static rhytids, they will most likely need a proper prescription for rejuvenation; this would include CO_2 laser resurfacing, a series of microneedling, filler, and an at-home quality skincare line. This young patient had only 6 units of Botox placed just outside of the vermillion border to help lessen the hyper dynamism of the area. She had 4 units of Botox placed above the upper lip and two below the lower lip. This patient demonstrates a less forceful pursed-lip look in her after photo demonstrated in both Figs. 11.35A,B and 11.36A,B. Treatment in this area will help to prevent the rapid onset of aging in this area.

Treatment

Upper lip border 4 units Botox
Lower lip border 2 units Botox
Photos taken 2 weeks apart

Neuromodulation technique of gingival show AKA gummy smile

Neuromodulation for gummy smile

A gummy smile is caused by a hyperdynamic levator labii superiors alaeque nasi. Typically the gingival show is about 2 mm above the gum line. As the lip pulls up, it also inverts or rolls inward, decreasing the amount of pink show left. Overexertion of this muscle can create deeper nasolabial folds as well. Other modalities can support much of what is done with neuromodulation. With the emerging proficiency in neuromodulation, the next step should be dermal fillers and skin resurfacing with a medical-grade skincare line to support and protect their in-office treatments.

Indications for treatment

Neuromodulation for injection is indicated for the temporary improvement in the appearance of moderate to severe lip lines, aka smokers lines.

Emotion/movement

Angry
Tension
Aged
Whistle
Blow kisses

Muscles in treatment area

Levator Labii Superioris
Levator Labii Superioris alaeque nasi muscle (LLSAN)
Zygomaticus major
Zygomaticus minor
Levator anguli oris

Muscles to be treated in the targeted zone

Levator Labii Superior Alaeque nasi muscle (LLSAN)

Special considerations

Reducing the gingival show will elongate the patients' upper lip, and in the aged patient this may not be an ideal treatment; as a sign of aging is a lengthening of the upper lip.

Complimentary locations to increase the efficacy

Depressor Anguli Oris (DAO)
Mentalis
Smokers lines

Treatment goals

Reduction of the gingival show when smiling. With the treatment of this area, keep in mind, it is a partial inhibition in this area. A heavy-handed dose would make the patient feel like they suffered paralysis and would present awkward oral incompetence for eating, drinking, speaking, and kissing.

Dosing

Women total dose 2–4 units; 1–2 units per side of Botox, Jeuveau, and Xeomin.
Women total dose 5–10 units; 2.5–5 units per side units upper and lower Dysport.
Men total dose 2–8 units; 1–4 units per side of Botox, Jeuveau and Xeomin.
Men total dose 5–15 units; 2.5–7.5 units per side of Dysport.

Documentation

Medical history including allergies
Medication
Dose
Lot number
Expiration date
Provider
Equipment used.
The treatment area and specific dose per injection point
How the patient tolerated
Treatment plan

Equipment

Medication
Botox
Jeuveau

Xeomin
Dysport
Bacteriostatic saline or 0.9% preservative-free saline
3 mL luer lock syringe
27 Gauge luer lock 1 1/2 needle
2 × 2 non-woven gauze
Alcohol
Ice
Bottle opener
Marking pencil
BD insulin Syringe 31 gauge ultra thin
Non-sterile gloves

If needed, hemostats helps remove tight needles from luer lock or remove the metal portion of the bottle of the neuromodulator.

Assessment

The patient should be assessed in animation and at rest. Have the patient force a smile as hard as possible; this way, the true degree of gingival show can be displayed.

Pre-procedure steps

1. Consultation
2. Have patient sign consent and pre and post-procedure guidelines
3. Photographs at three angles at rest and animated
   - **(i)** 50 degrees both left and right
   - **(ii)** 45 degrees both left and right
   - **(iii)** Straight on
4. Have the patient comfortably seated in an upright position.
5. Ensure adequate overhead lighting.
6. Mark out danger zones.
7. Additionally, mark out treatment areas with a marking pencil. Be sure not to inject through the marking pencil as it can tattoo the patient.
8. Calculate dosing according to the desired look, anatomy, and area being treated.
9. Draw up the medication, being sure not to dull the needle by touching the sides of the glass bottle, or going through the rubber stopper with the intended injection needle.

10. Cleanse the patients face thoroughly with a double cleanse and then chlorhexidine or alcohol (allow to dry).

Technique

1. Identify the treatment danger zones.
2. Locate the Levator Labii Superioris Alaeque Nasi Muscle (LLSAN) muscle by asking the patient to smile as hard as they can.
3. Identify the point of injection and place the markings. This injection will be done without the patient animating.
4. The injector should be positioned on the treatment side.
5. Treat the patient with their head resting back against the headrest of the chair and eyes closed.
6. Be sure that the patient is still cleansed and has not touched the treatment area since proper cleansing performed in the pre-procedural steps.
7. This injection should be performed while the muscle is at rest. Ask the patient to sneer or snarl. Specific populations will relate to the Elvis Presley lip sneer he did with his left upper lip.
8. The first injection point will be at nearly the uppermost part of the nasolabial fold. About 1–1.5 cm from the base of the nasal ala is where to inject the needle. Inject the needle about halfway into the tissue. Slowly inject 1–2 units of Botox, Xeomin, Jeuveau, or 2.5–5 units of Dysport. Dosing will depend on the strength of the muscle and the desired goal for relaxation of the muscle.
9. Repeat the same injections on the contralateral side.
10. Do not press or rub on the injection sites. If the patient happens to bleed, apply gentle pressure to stop the bleeding. GENTLE. It is not advised to manipulate the product from the site as adverse events, and product migration can occur.

Results

Lengthening of the upper lip, reducing the amount of gingival show, and a softening of the nasolabial fold.

Within 3–10 days, patients will start to see gradual improvement of the dynamic and static wrinkles in the treated area. At no time during the 2 weeks should more neuromodulator be injected in the same area.

It takes a full 2 weeks for the product to take effect. All good things come to those who wait. It is strongly advised that the patient come back in for a follow-up evaluation at 2 weeks. Discuss how the patient feels, how they like their results. Take after photos in the exact positions as the before photos.

Duration

Neuromodulators last anywhere from 2 to 4 months. The average duration is about 3 months. It is essential to the clients to let them know they will not be completely frozen the entire duration of the treatment. It will take about 2 weeks to become fully effective, and then at about 6–8 weeks, the neuromodulator will start to "break." Break, meaning a little bit of expression or movement will start to come back, and the patient will notice this. Be sure to keep photos for the records so that they can be referred to for comparison. At times, patients will call at 6 weeks and state that all of their neuromodulator wore off, and they need to be re-seen or re-treated. Pictures will come in handy here as it is likely they see the beginning stages of the reinnervation with baby sprouts of the nerve terminals. It is not necessary to wait till all of the patient's movement is back. Retreat at 3 months before the full expression is restored; this way, they can train the muscles.

Follow up and special consideration

At the 2 weeks follow up:

It is common for some patients to still demonstrate movement in the area treated after treatment. Persistent wrinkling can be caused by inadequate dosing. Adding more neuromodulation at the follow up will help to complete the look in some instances.

If deep static wrinkling is persistent, educate the patient on the need for repeat treatments and or laser resurfacing. If adjacent hyperdynamic movement is the cause, it is advised to treat additional areas (Table 11.8).

Table 11.8 Complications and management table.

Complications	Cause	Management
Bruising	Puncture of a vessel	Arnica ointment Avoid sun exposure to reduce the risk of hemosideren staining
Swelling	Trauma	Ice
Pain	Trauma	Ice and acetaminophen
Headache	Adverse event from neuromodulation	Acetaminophen and real dark cocoa
Persistent movement	Initial dose not adequate Muscle stronger than anticipated	Consider Zinc deficiency Add additional dose in areas of movement
Lip ptosis	Inadvertant injection into the Levator Labii Superioris Alaeque Nasi	Time Careful placement of a corrective injection
Oral incompetence and drooling	Inadvertant injection into the Levator Labii Superioris Alaeque Nasi	Time
No effect		Consider Zinc deficiency Report to the medical science liaison of the maker of the neuromodulation for further guidance Review medical history again, are they on chelators, are they zinc deficient, do they have an underlying disease not disclosed concerning possible increased metabolization Review the reconstitution, injections, dosing and placement
Dysport allergy		Antihistamine Follow up with primary care/ Allergist Notate in chart new allergy

Details on Dysport allergy

If using Dysport and the patient is allergic, the patient will complain of flu like symptoms or feeling run down like they are experiencing seasonal allergies or experiencing a sinus infection. Not all patients know that they have a milk protein allergy, and treatment with Dysport uncovers the allergy. Typically no intervention is needed as it will self-resolve. However, it is appropriate to suggest over the counter antihistamines or see their primary care or allergist for further testing and lifestyle modifications.

Before and after photos

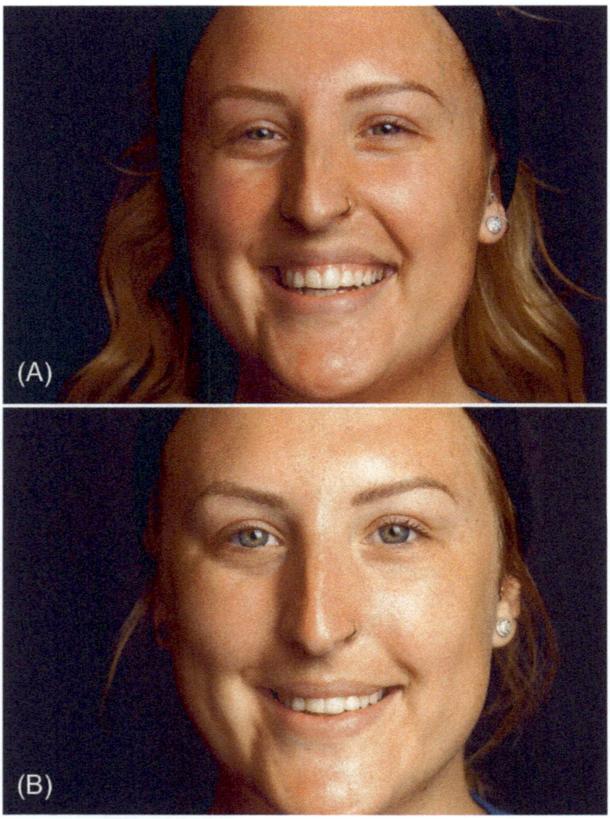

Fig. 11.37 (A) Before photo. (B) After photo.

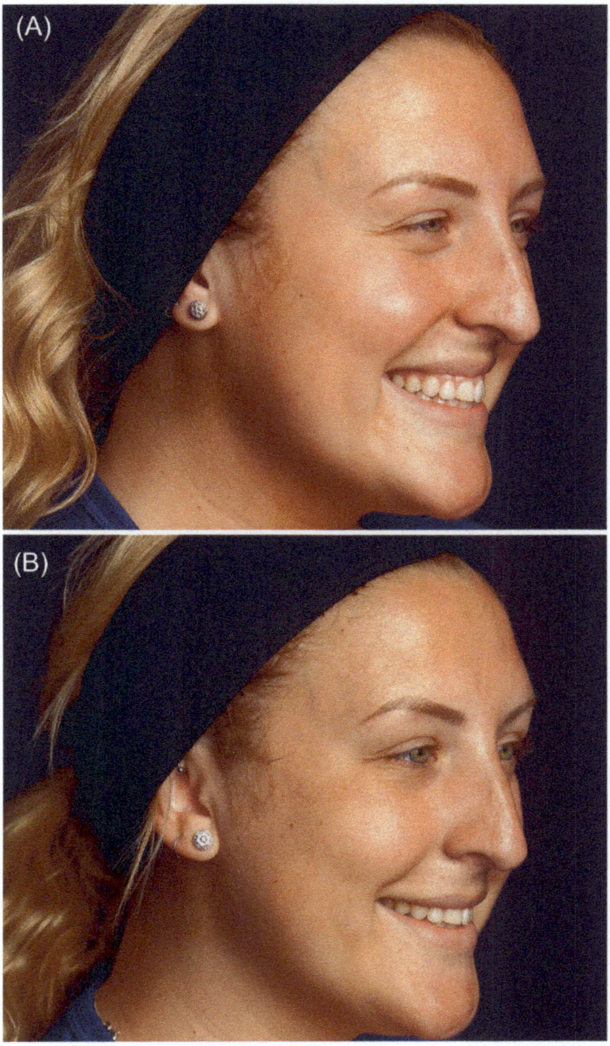

Fig. 11.38 (A) Before photo. (B) After photo.

This patient has a captivating yet forceful smile. When she smiles, she presents with a gingival show seen in Figs. 11.37A and 11.38A. She has a mild case of gingival show, but none the less it was essential to the client; therefore, we addressed it for her. Two units of Botox were placed in the right and left levator labii superioris alaque nasi (LLSAN). About 3/4 cm from the alar groove laterally is where this injection point will be placed. It is not a very deep injection, either. As noticed in Figs. 11.37B and 11.38B she has a significant reduction of the gingival show without altering her natural smile.

Treatment

2 units Botox right Levator labii Alaeque nasi
2 units Botox left Levator labii Alaeque nasi
Total dose 4 units Botox
Photos taken 2 weeks apart (Figs. 11.39 and 11.40)

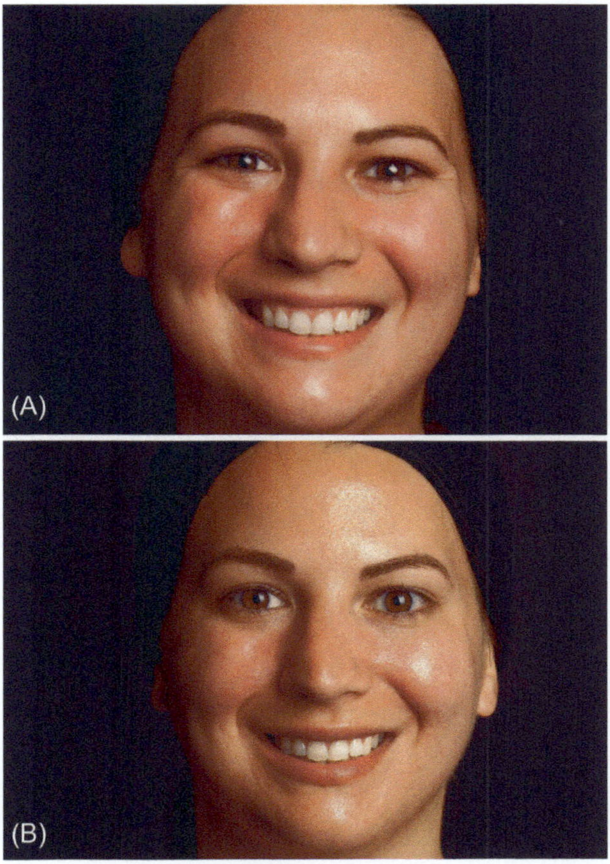

Fig. 11.39 (A) Before photo. (B) After photo.

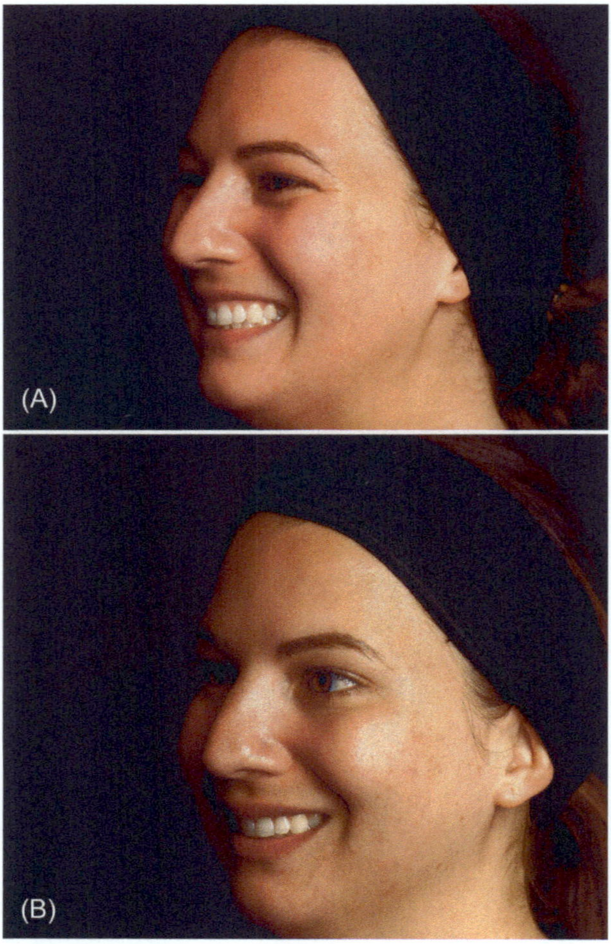

Fig. 11.40 (A) Before photo. (B) After photo.

Treatment

2 units Botox right Levator labii Alaeque nasi
2 units Botox left Levator labii Alaeque nasi
Total dose 4 units Botox
Photos taken 2 weeks apart (Fig. 11.37 and 11.38)

Neuromodulation technique of marionette lines

Neuromodulation of the depressor anguli oris

As the aging process continues to pull everything downward it also tends to pull in centrally as well. Patients will feel tired, drawn down, tense and begin to have central fullness, aka jowls and marionette hollowing with the formation of bulky folds. Some patients frown with the lower face in their regular conversations. While other patients preserve this function to show true sorrow. Watching your patients speak in the assessment naturally without force is your best indicator to begin to see the treatment plan and how each treatment and area plays off of one another. Treating this area properly can elevate the corners of the mouth.

Indications for treatment

Neuromodulation for injection is indicated for the temporary improvement in the appearance of marionette lines and down turned smile.

Emotion/movement

Angry
Tension
Aged
Sad
Joker

Muscles in treatment area

Depressor Anguli Oris (DAO)
Orbicularis Oris
Depressor Labii Inferioris (DLI)
Superior aspect of the platysma
Buccinator

Muscles to be treated in targeted zone

Depressor Anguli Oris (DAO)

Special considerations

If the patient has deflated oral commissures along with deep marionette lines, filler may be consider to help support the neuromodulation treatment. It is essential that the DLI is avoided in this area to prevent an indavertant lower lip drop. The DLI keeps the lip in an upright position and if your injection point is to medial to the commissure you will hit the DLI and drop the corner of the lip. The complexity and importance of the oral muscles needs careful attention to detail and anatomical understanding. A mm off and you can create a very unpleasant appearance, one that mimics a palsy. You can be sure you will losethat patient. Time is the only remedy in this area. In other areas the injector can attempt to inject another area or the opposite side for balance. But once there is oral incompetence, creating more oral incompetence is not ever advised.

Complimentary locations to increase efficacy

Mentalis
Obicularis oris for smokers lines
Platysmal

Treatment goals

Partial inhibition of the Depressor Anguli Oris Muscle. Due to the complex and essential natural of this muscle in the region it is advised that it be partially inhibited.

Dosing

Women total dose 2–4 units; 1–2 units per side Botox, Jeuveau and Xeomin.
Women total dose 5–10 units; 2.5–5 units per side units upper and lower Dsyport.
Men total dose 2–8 units; 1–4 units per side Botox, Jeuveau and Xeomin.
Men total dose 5–15 units; 2.5–7.5 units per side Dsyport.

Documentation

Medical history including allergies
Medication
Dose
Lot number
Expiration date
Provider
Equipment used
Treatment area and specific dose per injection point
How the patient tolerated
Treatment plan

Equipment

- Medication
  - Botox
  - Jeuveau
  - Xeomin
  - Dysport
- Bacteriostic saline or 0.9% preservative free saline
- 3 mL luer lock syringe
- 27 Gauge luer lock 1 1/2 needle
- 2 × 2 non woven gauze
- Alcohol
- Ice
- Bottle opener
- Marking pencil
- BD insulin Syringe 31 gauge ultra thin
- Non sterile gloves

Hemostat if needed to assist in removal of tight needles from luer lock or removing the metal portion of the bottle of neuromodulator.

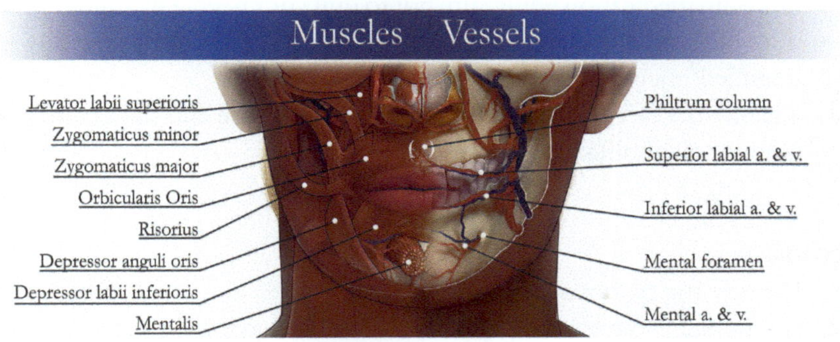

Fig. 11.41 Lower face muscles and vessels.

Assessment

The patient should be assessed in animation and at rest. Have the patient force a smile as hard as they can and have them force a down turned smile. Your zone of treatment will fall within 2 cm of the mandibular margin. When the patient forces the down turn smile a hollow will be created and then you will notice posterior to the hollow a significant bulge. This is your landmark. If the patient can not make this face on demand it is not advised to treat the area. As the act of the movement is how the injector pin points the location for anatomical boundaries.

Pre-procedure steps

1. Consultation
2. Have patient sign consent and pre and post procedure guidelines
3. Photographs at three angles at rest and animated
   - **(i)** 50 degrees both left and right
   - **(ii)** 45 degrees both left and right
   - **(iii)** Straight on
4. Have patient comfortably seated in an upright position
5. Ensure adequate overhead lighting
6. Mark out danger zones
7. Additionally, mark out treatment areas with a marking pencil. Be sure you do not inject through the marking pencil as you can tattoo the patient.
8. Calculate dosing according to desired look, anatomy and area being treated.
9. Draw up medication being sure not to dull your needle by touching the sides of the glass bottle or going through the rubber stopper with your intended injection needle.
10. Cleanse the face thoroughly with a double cleanse and then chlorhexidine or alcohol (allow to dry).

Technique

1. Identify the treatment danger zones.
2. Locate the Depressor Anguli Oris (DAO) by having the patient force a down turned smile (Fig. 11.41).
3. Identify the point of injection and place your markings. This injection will be done with out the patient animating. You will be treating the inferior portion of the DAO. A point of reference could be the medial limbus line and draw a line straight down to about 2 cm above the mandibular margin. This should be about where your injection point will end up.
4. The injector should be positioned on the treatment side.
5. Treat the patient with their head resting back against the head rest of the chair and eyes closed.

6. Be sure that the patient is still cleansed and has not touched the treatment area since proper cleansing performed in the pre-procedural steps.
7. This injection should be performed while the muscle is at rest however if the patient can hold the position you may treat in animation. The injection will be about 1 mm lateral to the oral commissure (corner of the lip) as you want to be sure to avoid the Depressor Labii Inferioris.
8. Inject subdermally and deposit 2 units of Botox, Jeuveau or, Xeomin, and 5 units of Dysopot into the Depressor Anguli Oris muscle (DAO).
9. Repeat the same injections on the contralateral side.
10. Do NOT press or rub on the injection sites. If the patient happens to bleed you may apply gentle pressure to stop the bleeding. GENTLE!!! It is not advised that you manipulate the product from the site as adverse events and product migration can occur.

Results

Reduction of downturned corners of the mouth. The patient should expect to see an elevation of the corners of their mouth. Be thoughtful not to over promise correction in this area as it often needs to be supported with filler in the upper face and in the actual marionette area.

Within 3–10 days patients will start to see gradual improvement of the dynamic and static wrinkles in treated area. At no time during the 2 weeks should more neuromodulator be injected in the same area. It takes a full 2 weeks for the product to take effect. All good things come to those who wait. At 2 weeks time it is strongly advised that the patient come back in for follow up evaluation. Discuss how the patient feels, how they like their results. Take after photos in the exact positions as the before photos.

Duration

Neuromodulators last anywhere from 2 to 4 months. The average being about 3. It is an important note to your clients to let them know they will not be completely frozen the entire duration of the treatment. It will take about 2 weeks to become fully effective and then at about 6–8 weeks time, the neuromodulator will start to "break." Break, meaning a little bit of expression or movement will start to come back and the patient will notice this. Be sure always to keep your photos for your records so that you can compare. At times patients will call at 6 weeks and state that ALL of their neuromodulation wore off and they need to be re-seen or re-treated. Pictures will come in handy here as it is likely they are seeing the beginning stages of the reinnervation with baby sprouts of the nerve terminals. You do not need to wait till all of the patients movement is back. *Re*-treat at 3 months before full expression is restored this way they can train the muscles.

Follow up and special consideration

At the 2 week follow up.

It is common and necessary for patients to still demonstrate movement in the area treated after treatment. However the goal is for reduction not complete paralysis.

If adjacent hyper dynamic movement is the cause it is advised to treat additional areas (Table 11.9).

Table 11.9 Complications and management table.

Complications	Cause	Management
Bruising	Puncture of a vessel	Arnica ointment Avoid sun exposure to reduce the risk of hemosideren staining
Swelling	Trauma	Ice
Pain	Trauma	Ice and acetaminophen
Headache	Adverse event from neuromodulation	Acetaminophen and real dark cocoa
Persistent movement	Initial dose not adequate Muscle stronger than anticipated	Consider Zinc deficiency Add additional dose in areas of movement
Lip ptosis	Inadvertant injection into the Levator Labii Superioris Alaeque Nasi	Time Careful placement of a corrective injection
Oral incompetence and drooling	Inadvertant injection into the Depressor Labii Inferiorosis	Time
Cheek abnormality	Due to injection or migration of product into the buccinator muscle	Time

Table 11.9 Complications and management table—cont'd

Complications	Cause	Management
No effect		Consider Zinc deficiency Report to the medical science liaison of the maker of the neuromodulation for further guidance Review medical history again, are they on chelators, are they zinc deficient, do they have an underlying disease not disclosed concerning possible increased metabolization Review the reconstitution, injections, dosing and placement
Dysport allergy		Antihistamine Follow up with primary care and allergist Notate in chart new allergy.

Details on Dysport allergy

If using Dysport and the patient is allergic, the patient will complain of flu like symptoms or feeling run down like they are experiencing seasonal allergies or experiencing a sinus infection. Not all patients know that they have a milk protein allergy, and treatment with Dysport uncovers the allergy. Typically no intervention is needed as it will self-resolve. However, it is appropriate to suggest over the counter antihistamines or see their primary care or allergist for further testing and lifestyle modifications.

Before and after photos

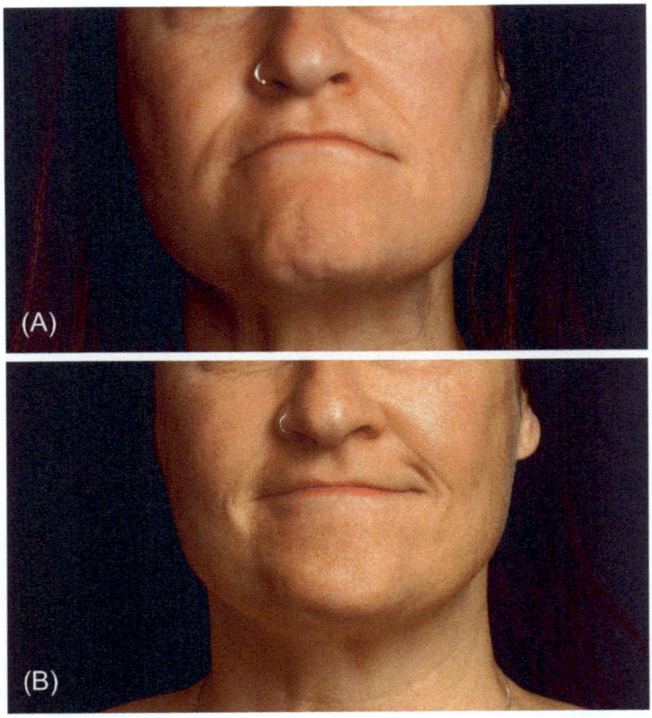

Fig. 11.42 (A) Before photo. (B) After photo.

This patient was treated for the concern of hollowing in the oral commissures as well as feeling as if her smile was becoming down turned especially in the face of animation. In the patients animated before picture it is noted that when she makes a sad face she has a significant down turning of the corners of her mouth. Over time if this is left untreated patients will have a breakdown and loss of fat in this area which creates a hollowing first and then eventually a fold will develop leaving a static crease in the skin. In Fig. 11.42B the patient is still able to express sadness but the force is decreased. Many patients will benefit form neuromodulation as well as filler for this area. Combining treatment can be the best restorative medicine. It is essential that only the Depressor anguli oris be injected otherwise if the depressor labii inferiors is accident injected the patients bottom lip on the affected side will ride up over the bottom teeth.

Treatment

2 units Botox in the right depressor anguli oris
2 units Botox in the left depressor anguli oris
Total dose 4 units Botox
Photos taken 2 weeks apart

Neuromodulation technique of the chin

Neuromodulation of the mentalis for a dimpled chin and a robust mental crease

The lower face's aging process is unique as the lower jaw begins to push forward, and the mentalis muscle will begin to engage to hold it back. The mentalis engagement will cause a prominent mental crease, and the development of a peach pit chin can occur. Dimpling of the skin is corrected by treating the mentalis muscle to help relax the muscle. It can also aid in the illusion of an elongated jaw when that muscle is relaxed.

Indications for treatment

Neuromodulation for injection is indicated for the temporary improvement in the appearance of then puckered, cobble toning, or peach pit skin of the chin. It can further elongate the jaw by dropping the chin if it is cooled in an upward sigh fashion. The mental crease will become more pronounced and will present a static line if the mentalis is not relaxed with neuromodulation.

Emotion/movement

Angry
Tension
Aged
Pout

Muscles in treatment area

Depressor Anguli Oris (DAO)
Orbicularis Oris
Depressor Labii Inferioris (DLI)
Mentalis

Muscles to be treated in the targeted zone

Mentalis

Special considerations

This treatment area can cause some unique presentations while the neuromodulator begins to work. One side can work faster than the other, along with overcompensation. Treatment of the mentalis should be considered an advanced area, as managing the patient expectations can be challenging. Much like the masseter muscle, the mentalis muscle is deep muscles, and so the injection depth will be deeper but not on the bone.

Complimentary locations to increase the efficacy

Orbicularis oris for the smokers lines
Platysmal
Depressor Anguli Oris (DAO)

Treatment goals

Partial inhibition of the mentalis muscle to improve skin texture, elongate the jawline profile, soften the mental crease.

Dosing

Women total dose 2–6 units; 1–3 units per side Botox, Jeuveau and Xeomin.
Women total dose 5–15 units; 2.5–7.5 units per side units upper and lower Dysport.
Men total dose 2–8 units; 1–4 units per side Botox, Jeuveau and Xeomin.
Men total dose 5–15 units; 2.5–7.5 units per side Dysport.

Documentation

Medical history including allergies
Medication
Dose
Lot number
Expiration date
Provider
Equipment used
Treatment area and the specific dose per injection point
How the patient tolerated
Treatment plan

Equipment

Medication

Botox
Jeuveau
Xeomin
Dysport
Bacteriostatic saline or 0.9% preservative-free saline
3 mL luer lock syringe
27 Gauge luer lock 1 1/2 needle
2 × 2 non-woven gauze
Alcohol
Ice
Bottle opener
Marking pencil
BD insulin Syringe 31 gauge ultra thin 0.5-in. needle
Non-sterile gloves

If needed, hemostats helps remove tight needles from luer lock or remove the metal portion of the bottle of the neuromodulation.

Assessment

The patient should be assessed in animation and at rest. Have the patient force a smile as hard as possible. Then have them force a downturned smile. The treatment zone will fall within 2 cm of the mandibular margin. When the patient forces the downturn smile, a hollow will be created, and then it will be noticed posterior to the hollow a significant bulge. This is the landmark for the injection to be placed. If the patient can not make this face on demand, it is not advised to treat the area. As the act of the movement is how the injector pinpoints the location for anatomical boundaries.

Pre-procedure steps

1. Consultation
2. Have patient sign consent and pre and post-procedure guidelines
3. Photographs at three angles at rest and animated
   - **(i)** 50 degrees both left and right
   - **(ii)** 45 degrees both left and right
   - **(iii)** Straight on
4. Have the patient comfortably seated in an upright position
5. Ensure adequate overhead lighting
6. Mark out danger zones
7. Additionally, mark out treatment areas with a marking pencil. Be sure not to inject through the marking pencil as it can tattoo the patient.

8. Calculate dosing according to the desired look, anatomy, and area being treated.
9. Draw up the medication, being sure not to dull the needle by touching the sides of the glass bottle, or going through the rubber stopper with the intended injection needle.
10. Cleanse the patients face thoroughly with a double cleanse and then chlorhexidine or alcohol (allow to dry).

Technique

1. Identify the treatment danger zones.
2. Locate the mentalis muscle by having the patient pout or pull up their chin.
3. Identify the point of injection and place the treatment markings. This injection will be done without the patient animating. Depending on the strength of the muscle, it may be necessary to make three points of injection. The safe zone will be in the medial portion of the chin. Assess the patient's natural fullness of their central lip. Typically injections will stay in that landmark. Another landmark that can be used is drawing two straight lines down from the outer portion of the nostril, then assessing the mentalis safety zone. Staying in this target zone with the injections will keep the treatment delivered in the safe zone minimizing adverse events. Keep in mind that the DLI is on either side of the mentalis and supports the bottom lip in an upright position.
4. The injector should be positioned on the treatment side or in front of the patient if seating allows.
5. Treat the patient with their head resting back against the headrest of the chair and eyes closed.
6. Be sure that the patient is still cleansed and has not touched the treatment area since proper cleansing performed in the pre-procedural steps.
7. This injection should be performed while the muscle is at rest.
8. Inject about halfway to the needle and deposit 2 units of Botox, Xeomin, or Jeuveau or 5 units of Dysport into the mentalis on the right side.
9. Repeat the same injections on the contralateral side.
10. Should the patient require a third dose, you will place it midline to the lip and just above the right and left side injection points at the peak of the bulge when the patient animates.
11. There are times when only a single injection point is necessary, and it is typically the medial injection point. If a single injection point is necessary, 5 units of Botox, Xeomin or Jeuveau or 10 units of Dysport are

standard. The injection will be deep and straight in with a slight angle. Always insert the needle to the desired position and then slowly press on the plunger. Once the dose has been delivered, remove the needle.

12. Do not press or rub on the injection sites. If the patient happens to bleed, apply gentle pressure to stop the bleeding. GENTLE. It is not advised to manipulate the product from the site as adverse events, and product migration can occur.

Results

Softened mental crease
Smoother skin over the chin
Elongated jawline profile

Within 3–10 days, patients will start to see gradual improvement of the dynamic and static wrinkles in the treated area. At no time during the 2 weeks should more neuromodulator be injected in the same area. It takes a full 2 weeks for the product to take effect. All good things come to those who wait. At 2 weeks' it is strongly advised that the patient come back in for follow-up evaluation. Discuss how the patient feels, how they like their results. Take after photos in the exact positions as the before photos.

Duration

Neuromodulators last anywhere from 2 to 4 months. The average duration is about 3 months. It is essential to the clients to let them know they will not be completely frozen the entire duration of the treatment. It will take about 2 weeks to become fully effective, and then at about 6–8 weeks, the neuromodulator will start to "break." Break, meaning a little bit of expression or movement will start to come back, and the patient will notice this. Be sure to keep photos for the records so that they can be referred to for comparison. At times, patients will call at 6 weeks and state that all of their neuromodulator wore off, and they need to be re-seen or re-treated. Pictures will come in handy here as it is likely they see the beginning stages of the reinnervation with baby sprouts of the nerve terminals. It is not necessary to wait till all of the patient's movement is back. Retreat at 3 months before the full expression is restored; this way, they can train the muscles.

Follow up and special consideration

At the 2 week follow up.

It is common for some patients to still demonstrate movement in the area treated after treatment. Persistent wrinkling can be caused by inadequate dosing. Adding more neuromodulation at the follow up will help to complete the look in some instances.

If deep static wrinkling is persistent, educate the patient on the need for repeat treatments and or laser resurfacing. If adjacent hyperdynamic movement is the cause, it is advised to treat additional areas (Table 11.10).

Table 11.10 Complications and management table.

Complications	Cause	Management
Bruising	Puncture of a vessel	Arnica ointment Avoid sun exposure to reduce the risk of hemosideren staining
Swelling	Trauma	Ice
Pain	Trauma	Ice and acetaminophen
Headache	Adverse event from neuromodulation	Acetaminophen and real dark cocoa
Persistant movement	Initial dose not adequate Muscle stronger than anticipated	Consider Zinc deficiency Add additional dose in areas of movement
Persistant static lines	Sun damage Poor skin routine	Laser resurfacing Repeat treatment Dermal filler Sun screen At home skin care
Lip ptosis	Inadvertant injection into the depressor labii inferioris	Time Careful placement of a corrective injection
Oral incompetence and drooling	Inadvertant injection into the Depressor Labii Inferiorosis	Time
No effect		Consider Zinc deficiency Report to the medical science liaison of the maker of the neuromodulation for further guidance Review medical history, are they on chelators, are they zinc deficient, do they have an underlying disease not disclosed concerning possible increased metabolization

Table 11.10 Complications and management table—cont'd

Complications	Cause	Management
Dysport allergy		Review the reconstitution, injections, dosing and placement Antihistamine Follow up with primary care and allergist Notate in chart new allergy

Details on Dysport allergy

If using Dysport and the patient is allergic, the patient will complain of flu like symptoms or feeling run down like they are experiencing seasonal allergies or experiencing a sinus infection. Not all patients know that they have a milk protein allergy, and treatment with Dysport uncovers the allergy. Typically no intervention is needed as it will self-resolve. However, it is appropriate to suggest over the counter antihistamines or see their primary care or allergist for further testing and lifestyle modifications.

Before and after photos

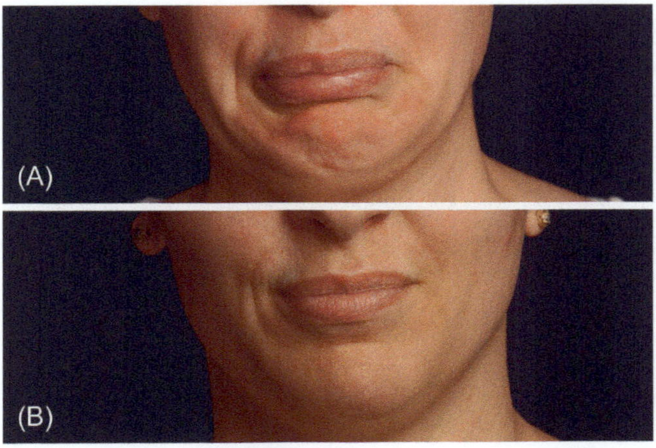

Fig. 11.43 (A) Before photo. (B) After photo.

This patient was bothered by the dimpling of her chin along with the development of a static mental crease. When she would make a downturned face or perhaps a sour face, so to speak, her chin dimpled extensively see Fig. 11.43A. Neuromodulation was used in the mentalis muscle to elongate the profile, soften the skin's dimpling, and prevent the static mental crease (Figs. 11.44 and 11.45). In her Fig. 11.43B, she cannot pull up the chin to cause the dimpling, and it can be noted that her chin is elongated, enhancing her overall profile. The safety zone here would be to draw an imaginary line from the outside corner of the nares (alar groove) straight down. The mentalis muscle resides in this space. If the treatment is to lateral to the safe zone, the depressor labii inferiors could be accidentally injected, causing the bottom lip (from the corner of the mouth to the midline) to be forced unnaturally up over the bottom teeth. The other danger zone here is the depressor anguli oris muscle. If this muscle is unintentionally injected, a corner of the mouth could be forced upward in an unnatural position. Refer back to the anatomical diagram of the lower face and note the mentalis is two columellar like muscles. They work in unison but need their own individual treatment.

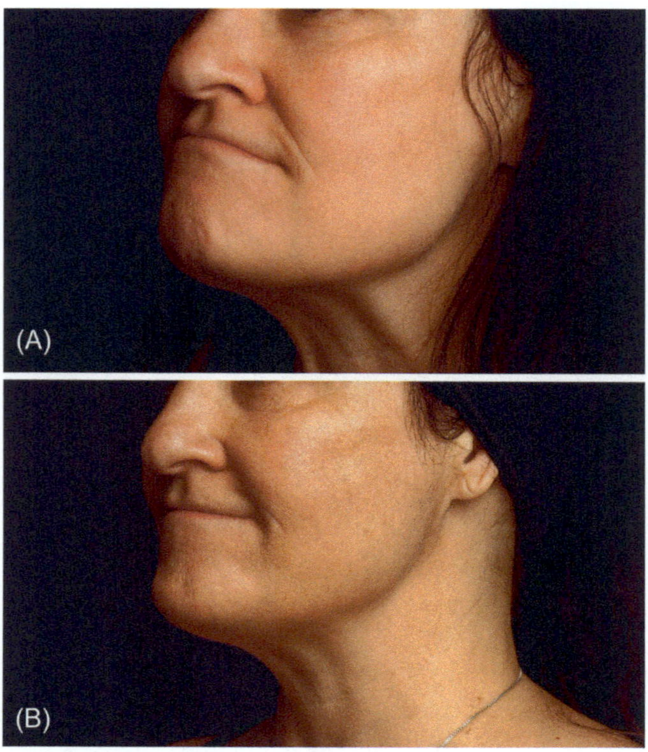

Fig. 11.44 (A) Before photo. (B) After photo.

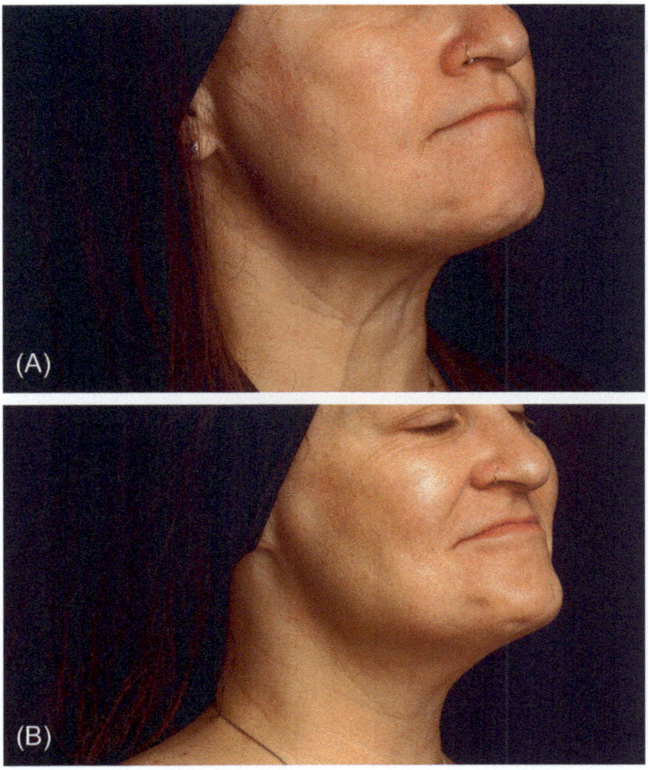

Fig. 11.45 (A) Before photo. (B) After photo.

Treatment

2 units Botox in the right mentalis
2 units Botox in the left mentalis
Total dose 4 units Botox
Photos taken 2 weeks apart

Neuromodulation technique of platysmal bands

AKA the Nefertiti neck lift

Neuromodulation of the platysmal bands is done to soften the tension in the neck and relax the bands, so they do not add stretch to the skin, increasing the rapid onset of the aged appearance of skin laxity.

Indications for treatment

Neuromodulation for injection is indicated for the temporary improvement in the appearance of neckbands.

Emotion/movement

Tension

Muscles in treatment area

Lateral Platysmal
Anterior platysma
Sternocleindomastoid
Laryngeal muscles

Muscles to be treated in the targeted zone

Lateral Pltaysma

Special considerations

It is essential to properly place the injection points. If the injection points are not accurate, it can make it extremely difficult for the patient to lift their head and swallow. Poorly placed injection points can inadvertently affect the deeper muscles or the laryngeal muscles. It tends to be easier to treat this area in animation. The band will be resistant to the needle and will feel like a rope or chord. It is thought that the constant strain and stretch from the platysmal bands are partly responsible for the skin laxity in the neck as the patient ages. By inhibiting the band's activity, it is thought that the elasticity of the skin can be preserved (Fig. 11.46).

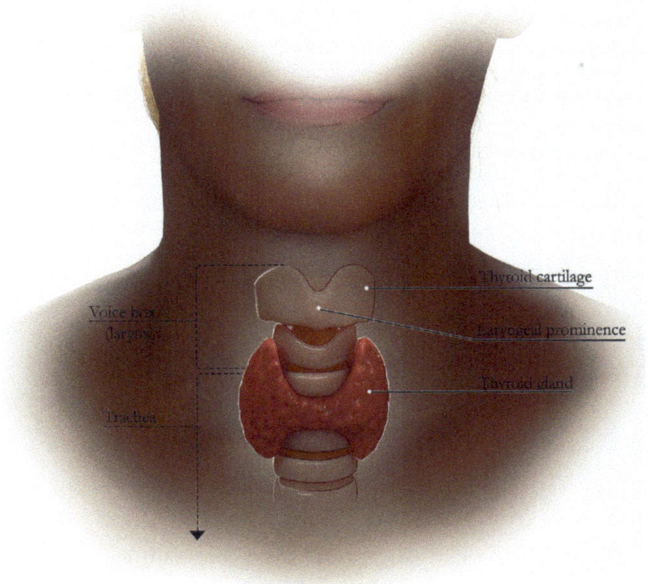

Fig. 11.46 Underlying structure when treating the platysmal bands.

Complimentary locations to increase efficacy

Depressor Anguli Oris (DAO)
Mentalis

Treatment goals

Softening of the anterior platysmal bands to create a softer looking neck without tension.

Dosing

Women total dose 12–36 units; 6 units per band Botox, Jeuveau, and Xeomin.
Women total dose 15–45 units; 7.5–22 units per band Dysport.
Men total dose 12–36 units; 6 units per band Botox, Jeuveau, and Xeomin.
Men total dose 15–45 units; 7.5–22 units per band Dysport.

Documentation

Medical history including allergies

Medication
Dose
Lot number
Expiration date
Provider
Equipment used
Treatment area and specific dose per injection point
How the patient tolerated
Treatment plan

Equipment

Medication
Botox
Jeuveau
Xeomin
Dysport
Bacteriostatic saline or 0.9% preservative-free saline
3 mL luer lock syringe
27 Gauge luer lock 1 1/2 needle
2 × 2 non-woven gauze
Alcohol
Ice
Bottle opener
Marking pencil
BD insulin Syringe 31 gauge ultra-thin 0.5-in. needle
Non-sterile gloves

If needed, hemostat helps remove tight needles from Luer lock or remove the metal portion of the bottle of neuromodulator.

Assessment

Have the patient strain the neck to elicit the bands. Have them make a "lizard neck," or say EEK. Mark the area. The safety area will be 1 cm lateral to the trachea and extends up to the commissure lines and back down to the clavicle. The placement should be 2 cm inferior to the mandible and at least 4 cm to the clavicle.

Pre-procedure steps

1. Consultation
2. Have patient sign consent and pre and post-procedure guidelines

3. Photographs at three angles at rest and animated
   (i) 50 degrees both left and right
   (ii) 45 degrees both left and right
   (iii) Straight on
4. Have the patient comfortably seated in an upright position
5. Ensure adequate overhead lighting
6. Mark out danger zones
7. Additionally, mark out treatment areas with a marking pencil. Be sure not to inject through the marking pencil as it can tattoo the patient.
8. Calculate dosing according to the desired look, anatomy, and area being treated.
9. Draw up the medication, being sure not to dull the needle by touching the sides of the glass bottle, or going through the rubber stopper with the intended injection needle.
10. Cleanse the patients face thoroughly with a double cleanse and then chlorhexidine or alcohol (allow to dry).

Technique

1. Identify the treatment danger zones.
2. Locate the Platysmal bands and mark out the safety zones.
3. The injector should be positioned on the treatment side or in front of the patient if seating allows.
4. Treat the patient with their head resting back against the headrest of the chair and eyes closed.
5. Be sure that the patient is still cleansed and has not touched the treatment area since proper cleansing performed in the pre-procedural steps.
6. This injection should be performed while the muscle is animated.
7. Have the patient animate and firmly grasp the band and directly inject the needle 2 mm into the band and inject 2 units of Botox, Xeomin, or Jeuveau or 5 units of Dysport per site. Depending on the band's length, it will be necessary to inject three to six sites about 1–2 cm from the previous injection.
8. Repeat this for each band.
9. Do not press or rub on the injection sites. If the patient happens to bleed, apply gentle pressure to stop the bleeding. GENTLE. It is not advised to manipulate the product from the site as adverse events, and product migration can occur.

Results

Softened platysmal bands reducing the tension seen within the neck.

Improved contour of the neck.

Within 3–10 days patients will start to see gradual improvement of the dynamic and static wrinkles in treated area. At no time during the 2 weeks should more neuromodulator be injected in the same area. It takes a full 2 weeks for the product to take effect. All good things come to those who wait. At 2 weeks time it is strongly advised that the patient come back in for follow up evaluation. Discuss how the patient feels, how they like their results. Take after photos in the exact positions as the before photos (Table 11.11).

Table 11.11 Complications and management table.

Complications	Cause	Management
Bruising	Puncture of a vessel	Arnica ointment Avoid sun exposure to reduce the risk of hemosideren staining
Swelling	Trauma	Ice
Pain	Trauma	Ice and acetaminophen
Headache	Adverse event from neuromodulation	Acetaminophen and real dark cocoa
Persistent movement	Initial dose not adequate Muscle stronger than anticipated	Consider Zinc deficiency Add additional dose in areas of movement
Persistant static lines	Sun damage Poor skin routine	Laser resurfacing RF micro needling Repeat treatment Sun screen At home skin care
Dysarthria	Inadvertant injection into the platysmal muscle versus the band. The muscle does have an impact on the corners of the mouth and you can see this with the articulation problems	Time

Table 11.11 Complications and management table—cont'd

Complications	Cause	Management
Dysphagia	Spread of the neuromodulator to the deeper muscles such was the laryngeal muscles or the sternocleidomastoid muscle	Time Lower the dose on the next treatment at 3–5 months and review safety land marks to identify critical structures
No effect		Consider Zinc deficiency Report to the medical science liaison of the maker of the neuromodulation for further guidance Review medical history again, are they on chelators, are they zinc deficient, do they have an underlying disease not disclosed concerning possible increased metabolization Review the reconstitution, injections, dosing and placement

Details on Dysport allergy

If using Dysport and the patient is allergic, the patient will complain of flu like symptoms or feeling run down like they are experiencing seasonal allergies or experiencing a sinus infection. Not all patients know that they have a milk protein allergy, and treatment with Dysport uncovers the allergy. Typically no intervention is needed as it will self-resolve. However, it is appropriate to suggest over the counter antihistamines or see their primary care or allergist for further testing and lifestyle modifications.

Before and after photos

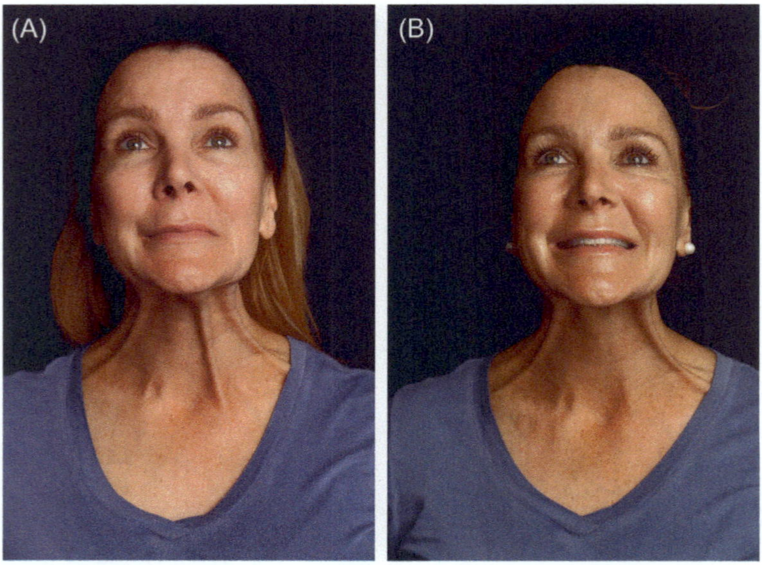

Fig. 11.47 (A) Before photo. (B) After photo.

This patient has developed some skin laxity of her neck and submittal area. She was partially treated to show the compensatory mechanism of action when only part of the area is treated. In Fig. 11.47A, five bands are noted. Her initial treatment was of the band just off-center to the left. Then in the after photo, it is noted that the remaining four bands have become even more hyperactive, see Fig. 11.47B. She received 12 units of Botox in her band just to the left of midline. A complete treatment would have included all five bands and a total of three injections with 2 units of Botox a piece in each band. Moving toward the side of the neck, it is an area to use caution in as it is necessary to avoid injecting into the Sternocleindomastoid as they can have significant heaviness holding their neck upright. Another area to take note of is the laryngeal muscles surrounding the larynx. If these muscles are injected it can lead to hoarseness and difficulty swallowing. See the picture above for underlying structures.

Treatment

Treatment 12 units of Botox in platysmal bands Photos two weeks apart

Hyperhidrosis injection technique

Hyperhidrosis is an idiopathic medical condition in which excessive sweating is produced by the sweat glands of the palms, axillae, and or the soles of one's feet. When antiperspirants are not enough to control sweating and odor, neuromodulation can help. One's thermoregulatory system is not altered in other body areas when neuromodulation is appropriately injected into the palms, soles of the feet, or the axillae. Onabotulinum toxin A is currently the only neuromodulation that is FDA approved for hyperhidrosis. Patients can expect 4–12 months' worth of relief from their excessive sweating when treating for hyperhidrosis. Results kick in within 4 to 5 days. With this treatment, the sweat glands are infused with the toxin, thus inhibiting the acetylcholine release, which in turn decreases the patient's sweet production.

Indication

Primary Axillary Hyperhydrosis Treatment is anticipated to last roughly 6 months.

Safety

Be aware that the safe dose of Botox, Xeomin, and Jeaveau in 3 months is 360 units. Be sure that the dose is calculated in combination with the patient's other aesthetic procedures to be sure they are within the safe doe threshold.

Pricing

It is not advised to price this by unit but, better yet, the treatment. The range is \$999–\$1200.

Target anatomy

Sweat glands

The sweat glands are located within the dermis. The gland is affected by the inhibition of acetylcholine, which decreases sweat production.

Reconstitution will differ here compared to the treatment of other areas. The Botox, Xeomin, or Jeuveau will be reconstituted with 5 mL of bacteriostatic saline. Resulting in 2 units per 0.1 cc. Each injection will be 0.1 mL.

Treating the axillae

Perform a Minors' Iodine or a starch test.

The patient should shave the area and be prepped and cleaned as per protocol. They should refrain from the use of deodorant for 24 h prior to the treatment.

Lay the patient flat and have them raise their arm above their head to expose the area.

Clean the area with alcohol and let dry.

Paint betadine solution on the patient's skin and allow to completely dry.

Sprinkle corn starch on the area and brush off the excess.

Wait roughly 10–20 min. The sweat will begin to slowly turn the corn starch and betadine mixture purple in the areas of increased sweating.

Once the area has turned purple or blue, outline the outer edges. The outer perimeter will be the treatment boundary, and all injections will be within this zone.

Gently wash off the betadine and starch while protecting the outline.

Always properly clean the skin before an injection.

Reconstitute a 100 unit vial of OnabotulinumtoxinA with 5 cc of Bacteriostatic saline. The larger the volume of diluent, the great the spread.

Treatment of axillae

50 units of Botox, Xeomin or Jeuveau will be used per side for a total of 100 units.

Inject 0.1 to per injection point about 20 injection points per side.

- 2 mm deep.
- 1–2 cm apart.
- Bevel up.

Lay the needle flat to the skin surface and only 2 mm deep so that the product's wheel can be visualized.

Change the needle every four to six injections to keep the patient feeling comfortable and to minimize bruising.

Repeat the treatment on the contralateral side.

*Note if betadine and starch are not available at the time of treatment, follow the hair follicle pattern as that is typically the target zone in the axillae.

Each palm, sole, or axillae will take 50 units per side.

Touch-ups may be needed, and the patient should be warned about the potential need for additional units so they can adequately plan their budget.

Treatment of the palms

The iodine starch test is not necessary when treating the entire palm.

Always correctly cleanse the skin before injection.

This treatment is much more painful than the axillae, and most patients. will require a digital nerve block or topical anesthetic before treatment.

Reconstitute a 100 unit vial of Botox, Xeomin, or Jeuveau with 5 cc of Bacteriostatic saline. The larger the volume of diluent, the great the spread.

It is advised that new injectors treat one hand at a time and start with the nondominate hand. Due to the risk of muscle weakness, it is advised to tread lightly with this treatment for beginners in neuromodulation.

50 units of Botox, Xeomin, or Jeuveau will be used per side for a total of 100 units.

Inject 0.1 mL per injection point except for the pulp of the fingertip, where most patients will be more resistant to the injection. At the pulp, inject 0.2 mL per injection point.

2 mm deep.

1–2 cm apart.

Bevel up.

100 point grid for the palms, including injections in-between the fingers and the proximal wrist fold.

Change the needle every four to six injections to keep the patient feeling comfortable and to minimize bruising.

Each palm, sole, or axillae will take 50 units per side.

Touch-ups may be needed, and patients should be warned about the potential need for additional product to adequately plan their budget. Effects will kick in in about 2 weeks and last up to 6 months and on occasion longer (Table 11.12).

Table 11.12 Complications and management table.

Complications	Cause	Management
Bruising	Puncture of a vessel	Arnica ointment Avoid sun exposure to reduce the risk of hemosideren staining
Swelling	Trauma	Ice
Pain	Trauma	Ice and acetaminophen
Headache	Adverse event from neuromodulation	Acetaminophen and real dark cocoa

Continued

Table 11.12 Complications and management table—cont'd

Complications	Cause	Management
Persistant perspiration	Initial dose not adequate, injection to deep or missed a spot	Consider Zinc deficiency Add additional dose in areas of persistent sweating
No effect		Consider Zinc deficiency Report to the medical science liaison of the maker of the neuromodulation for further guidance Review medical history again, are they on chelators, are they zinc deficient, do they have an underlying disease not disclosed concerning possible increased metabolization Review the reconstitution, injections, dosing and placement

Duration

Neuromodulators last anywhere from 2 to 4 months. The average duration is about 3 months. However, in this particular treatment it is common for it to last 6–9 months. It is essential to the clients to let them know they will not be completely sweat free the entire duration of the treatment. It will take about 2 weeks to become fully effective.

Follow up and special consideration

At the 2 weeks follow up:

It is common for some patients to still have some sweating in the area treated after treatment. Persistent sweating can be caused by inadequate dosing. Adding more neuromodulation at the follow up will help to complete the treatment along with the patients satisfaction.

CHAPTER 12

Complications

Potential complications

Complications section

Tips to Limit negative sequela or displeasing results

(1) Know your injection anatomy.
(2) Understand dynamic expression.
(3) Understand the equal and opposing areas that can aid in the remedy to complications post-treatment.
(4) Know the depths and relative dimensions of the structures you are injecting.
(5) Educate yourself with more than just 1-weekend course.
(6) Comprehensively understand the pharmacokinetics and pharmacodynamics of the neuromodulator in use.
(7) Proper and in-depth facial assessment is critical in excellent outcomes.
(8) Strict adherence to the state board and regulatory parameters.
(9) Know your Scope of Practice for the state you are practicing in
(10) Have a comprehensive understanding of the aging process

Poor patient satisfaction can be due to a number of causes; insufficient dosing, placement, improper handling, reconstitution, storage and anatomical variations of the client. Paresis is directly correlated to the dose and placement of the BoNT-A administered. This is where the concept of "the art of medicine" comes to light. It is essential to understand the drug you are utilizing but is also necessary to understand patients' individual features and specific esthetic goals.

Reconstitution is similar to Botox, Xeomin, and Jeuveau yet is vastly different regarding Dysport. All products went to FDA for approval, with the use of the preservative-free normal saline 0.9%. However, it has been reported that the use of bacteriostatic 0.9% saline with benzyl alcohol improves the longevity of the drug post reconstitution, decreases pain at the injection site, and also provides for a decrease in post-injection infections (American Society of Plastic Surgery ASPS, 2019).

Essentials of Neuromodulation
https://doi.org/10.1016/B978-0-323-89920-8.00025-4

Toxin spread

BoNT-A is generally safe in properly trained individuals partly due to its ability to stay localized to the injection. Toxin spread to unwanted areas can lead to ptosis or many other undesirable outcomes. The structures, muscle, compartments, are so compact and sit so close to unwanted and untargeted muscles it poses a greater risk for adverse effects. For example, when treating the corrugator muscle's lateral tail, it is common for the untrained injector to inadvertently inject the levator palpebrae superioris muscle and cause a lid ptosis. The inadvertent injection into the levator palpebrae superiors is because the LPSM sits directly under the tail of the corrugator at the mid pupillary line 1 cm above the orbital rim. The minimal thickness of tissue from the epidermis to bone in the glabellar region is 2.9 mm, and a 31 gauge needle, commonly used for botox is roughly 0.261 mm. With the small space demonstrated, you could see how easy it is to inadvertently injected and underserved muscle if not carefully and knowledgeable in anatomy (Chestnut, 2018).

When we discuss spread, it is the mechanical force of the product traveling from the original site. If you increase your extrusion force while plunging the BoNT-A you will increase your spread and thus potentially have negative sequela. On the other hand, diffusion would refer to the toxin's passive movement due to the kinetic dispersion outside of the original site. For example, the higher the volume used to reconstitute the BoNT-A the greater the potential for diffusion. The initial injection site would have higher concentrations of the drug than the secondary endpoint. When injected, the product will go from a very high concentration to a lower concentration no matter what the reconstitution preparation was, when we speak in terms of migration, that would be referring to the BoNT-As ability to travel along the nerve channel or by accidental hematogenous transport by accidental injection into a vessel.

Four main risk categories in botulinum toxin A injections

(1) Poor injector technique, which includes poor anatomical understanding

(2) Overdosing a patient

(3) Patient fails to adhere to aftercare instructions

(4) In rare cases, product reactions, such as sensitivity or even more rare instances of allergic reaction to the drug or an additive in the drug

If the injector improperly places the neuromodulator into the wrong muscle, an undesired effect can occur. The drugs intended effect is functional weakness; therefore, it is imperative to understand the anatomical structures and their role in dynamic expression in which you are working to only interact with the appropriate muscles intended for functional weakness. A patient may want paralysis or functional weakness for the movement between their eyebrows, but they do not want functional weakness at the arch in their brow. Dosing is also essential as you want to be sure each area is treated uniquely and individually. Too much neuromodulator can also create a sense of depression.

There are post-care instructions that the patient should adhere to, and if they do not, they could be left with an undesired esthetic look. The good news is that it is temporary and will resolve the injector's bad news is that the patient will not likely come back.

Allergic reactions to neuromodulators are infrequent. Hypersensitivity reactions have been reported to include anaphylaxis, urticaria soft-tissue edema, and dyspnea (Weiss, Silver, Lennard, & Weiss, 2007).

Neuromodulation, when performed correctly, can provide beautiful esthetic results as well as pain relief.

(1) A plethora of potential side effects due to BoNT-A injections
(2) Excessive muscle weakness
(3) Adverse effects due to the spread of the BoNT-A
(4) Tenderness
(5) Bruising
(6) Swelling
(7) Ectropion (lower lid eversion)
(8) Blepharoptosis (drooping eyelid)
(9) Burning or sting at the injection site
(10) Herpes outbreak
(11) Lagophthalmos (incomplete lid closure)
(12) Xerophthalmia (dry eyes)
(13) Epiphora (excessive tears)
(14) Diplopia (Double vision)
(15) Photophobia (light sensitivity)
(16) Lip Ptosis
(17) Hoarseness
(18) Dysphagia (difficulty swallowing)
(19) Dysarthria (difficulty articulating)
(20) Muscle weakness (treatment the paracervical region aka treatment of the platysmas bands can result in the inability to support the head as well as dysphagia).

(21) Facial asymmetry
(22) Systemic allergic reaction including anaphylaxis
(23) Urticaria
(24) Edema
(25) Dyspnea
(26) Antibody development in 1–2% of the population resulting in ineffective treatments

Immunogenicity is the ability of a protein to elicit antibody formation within the host (your patient). Immunogenicity and secondary non-responders are two different concepts and will be teased apart through the remaining chapters.

Immunogenicity—development of antibody formation

Primary Non-Responder—no response to initial or subsequent treatment

Secondary Non-responder—response to initial treatment but failure to achieve the desired outcome on subsequent treatments. The botulinum toxin A can induce an immune response similar to other proteins; thus, the host heads it as a foreign body and blocks it is transmission to the nerve terminals.

Dynamic assessment with a solid understanding of agonist and antagonist muscles is critical in good outcomes. Every time one muscle is being manipulated, it is essential to understand how will the treatment affect the opposing muscle or muscles?

As impressive as neuromodulators **are; also recognize** their **limitations**

Complications

Transient ptosis

Brow ptosis is the most common complication as injectors treat with a standardized approach versus an individual approach. Utilizing too much neuromodulation and or injecting too low in the frontalis muscle can create ptosis. Initially, when the glabellar region and the frontalis muscle are treated together, the patient will occasionally report that their brow feels heavy for the first few days. The frontalis muscle is a thinner muscle; therefore, the neuromodulation kicks in quicker than in the glabellar region. It is essential to understand the impact neuromodulators have on muscles as separate entities and a functional group. Teaching the patients this concept will help educate the patient and minimize the number of panicked phone calls the office will receive.

When injecting the frontalis muscle, do not inject too low. Stay 2–3 cm above the brow bone.

Use the bone as the landmark, not the brow. Many brows are penciled in or tattooed and this gives the injector a false landmark lending to adverse events.

If the patient does end up complaining of a heavy brow or brow ptosis, inject the corrugator muscle to help lift the area. The frontalis muscle is an elevator, and the muscles within the glabellar region are a depressor muscle. Once these areas are treated with neuromodulators; they have the opposite effect. Meaning the glabellar region becomes an elevator muscle, and the frontalis muscle becomes a depressor. Understanding the agonist and antagonist effect and the direct effect neuromodulation has on an area can keep the outcomes on point.

If it is left untreated, the brow ptosis usually corrects itself within 2–3 weeks.

Lid ptosis

Lid ptosis occurs from injecting the glabellar region to close to the orbital rim. The levator palpebrae superioris has been injected and cause the inner and outer canthus lid to droop. A droop of the eyelid is a sure way to lose a patient. Iopidine, 0.5% drops, can be prescribed with good effect. However, the patient will have to instill the drops every few hours while awake. This negative sequella can last weeks to months; however, it will feel like years to the patient.

Cosmetic application dosing

Average dosing 50–90 with Botox, Xeomin, and Jeuveau.

Dysport can be up to 150 units.

Hyperhidrosis or Torticollis dosing is 100–300 units. These two conditions are typically treated with Botox.

Glabellar region

Glabellar complex facial lines arise from the activity of the lateral corrugator and vertical procerus muscles. These muscles move the brow medially, and the procerus and depressor supercilii pull the brow inferiorly. Activation of these muscles creates a frown or "furrowed brow." The location, size, and use of the muscles vary markedly among individuals. Lines induced by facial expression occur perpendicular to the direction of action of contracting facial muscles. These can be readily identified by palpating the tensed muscle mass while having the patient frown. Medical professionals administering neuromodulators must understand the relevant neuromuscular and or orbital anatomy of the area involved and any alterations to the anatomy due to prior surgical procedures. An effective dose for facial lines is determined by gross observation of the patient's ability to activate the superficial muscles injected.

Frontalis muscle

The contraction of the frontalis muscles causes horizontal forehead wrinkles.

The frontalis muscles are two large fan-like muscles that extend from the eyebrow region to the top of the forehead. The frontalis muscle originates from the galea aponeurotica and inserts at the skin of the eyebrows and nose. Its functionality raises the eyebrows, resulting in wrinkles in forehead skin.

Understanding ptosis

The risk of ptosis can be mitigated by careful examination of the upper lid for separation or weakness of the levator palpebrae muscle (true ptosis), identification of lash ptosis, and evaluation of the range of lid excursion while manually depressing the frontalis to assess compensation.

In order to reduce the complication of ptosis, the following steps should be taken:

- Avoid injection near the levator palpebrae superioris, particularly in patients with larger brow depressor complexes.
- Medial corrugator injections should be placed at least 1 cm above the bony supraorbital ridge.
- Ensure the injected volume/dose is accurate and, where feasible, kept to a minimum.
- Do not inject toxin closer than 1 cm above the central eyebrow.

Lateral canthal lines

Lateral canthal lines arise largely from the orbicularis oculi muscles' activity around the eye responsible for blinking and eyelid closure. Forceful contraction of the orbicularis oculi results in lateral and radially oriented folds (crow's feet lines), which originate from the lateral canthus. The distribution of these radial lines differs among patients.

Managing complications

Managing complications is very tricky concerning neuromodulation. The reason being; is most mistakes can be fixed with only time. The best way to manage a complication is to avoid it at all cost. The second best tool for managing complications is to avoid them by properly educating the patient on the treatment's true potential and anticipated goal.

This patient has static and dynamic rhytids in the complex glabellar region. It is essential in this situation to educate the client on what neuromodulation can do. The injector must tease out if the patient is looking to get rid of the muscle activity that bothers them or if it is the lines that are there at rest. Figs. 12.1 and 12.2 represent the patient in animation and at rest without treatment. It is demonstrated in that she has static at rest rhytids. Given that the patient has visible static rhytids means that she would need to be counseled on what Neuromodulation can truly do for her. If this patient

Fig. 12.1 Animation photo without treatment.

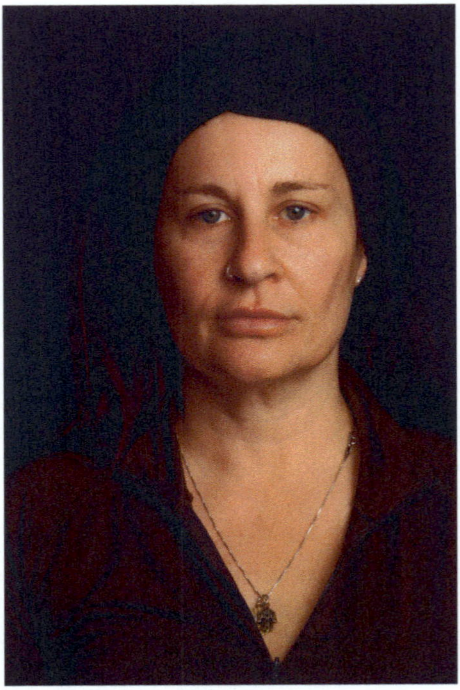

Fig. 12.2 At rest photo without treatment.

wanted the static creases gone with treatment and the injector administered neuromodulators to the area, and the static lines are still present at 2 weeks, then the patient would likely want a refund. Setting proper patient expectations is essential in all cases. However managing exceptions in unique situations is of the utmost importance. (See Table 12.1.)

Table 12.1 Understanding risk.

Muscles to inject	Intended outcome	Complication risk	Depth of injection
Frontalis	Softens horizontal rhytids	Brow ptosis	Superficial
Corrugator supercilii	Softens vertical lines in-between the brows	Medial brow ptosis or lid ptosis if it migrates to levator palpebrae muscle	Medially to superficial

Table 12.1 Understanding risk—cont'd

Muscles to inject	Intended outcome	Complication risk	Depth of injection
Depressor supercilii	Raise medial eyebrow	Lowers medial eyebrows should the frontal be injected	Superficial
Procerus	Raise the bridge of the nose, softens the horizontal rhytids on the bridge of the nose	Medial brows drop if the frontal is injected	Medially deep into the belly of the muscle
Nasalis	Softens diagonal rhytids on the side of the nose otherwise known as "bunny lines"	Drops the upper lip if the levator labii superioris alaeque nasi (LLSAN) is injected	Superficial
Orbicularis oculi	Softens the horizontal and diagonal rhytids, raise the brows and softens the "crows feet"	May cause asymmetrical smile should the zygomatic minor or major be injected	Superficial or you will increase risk of bruising and dropping the smile
Depressor septi nasi	Raise the tip of the nose	May limit the function of the upper lip should the orbicularis oculi be injected	Deep
Orbicularis oris	Softens the vertical rhytids, everts the lip	Reduces phonation	Extremely superficial
Masseter	Slims jaw line, reduces symptoms of TMD	May affect smile if the risorius muscle is injected	Very deep, needs 1/2 in. needle
Depressor anguli oris (DAO)	Lifts the corners of the mouth	If the depressor labii inferioris (DLI) is injected you may cause asymmetrical smile	Deep but lateral to avoid the DLI

Continued

Table 12.1 Understanding risk—cont'd

Muscles to inject	Intended outcome	Complication risk	Depth of injection
Mentalis	Softens the mental crease and the cobblestone appearance of the chin	May inadvertently inject DLI and cause a crooked smile	Deep inferiorly and superficial when you near the mental crease
Platysmal	Soften or eliminate bands of the neck "lizard neck"	Difficulty lifting head off the pillow, dysphagia	Pinch bands and directly inject deep into the platysmas band

References

American Society of Plastic Surgery (ASPS). (2019). 2018 Plastic Surgery Statistics Report. Retrieved from https://www.plasticsurgery.org/documents/News/Statistics/2018/plastic-surgery-statistics-full-report-2018.pdf.

Chestnut, C. (2018). Restoration of visual loss with retrobulbar hyaluronidase injection after hyaluronic acid filler. *Dermatologic Surgery*, *44*(3), 435–437.

Weiss, L., Silver, J. K., Lennard, T. A., & Weiss, J. M. (2007). Botulinum toxin. In *Easy Injections* (pp. 160–165). Butterworth-Heinemann. https://doi.org/10.1016/b978-075067527-7.10008-9.

Further reading

Allergan. (2020). *Botox cosmetic history*. https://www.botoxcosmetic.com/what-is-botox-cosmetic/botox-cosmetic-history. (Accessed March 2020).

Dysport. (2020). *For health care professionals*. https://www.dysportusa.com/healthcare-professionals#importantsafetyinformation. (Accessed March 2020).

Jeuveau. (2020). *Ready for a modern-made tox?* https://jeuveau.evolus.com. (Accessed July 2020).

Xeomin. (2020). *A uniquely purified choice for frown lines*. https://www.xeominaesthetic.com/professionals/. (Accessed March 2020).

CHAPTER 13

Return on investment (ROI)

The return on investment (ROI) seems large initially when we pull the numbers for neuromodulation. An ROI is never as clean cut as it appears on paper. There are many other factors to consider. It is actually quite a laundry list of additional factors to be considered.

Additional costs
Initial training
Text books
Supplies
  alcohol
  needles
  gloves
Medical license
Continuing education units
Liability insurance
Business owners insurance
Workers comp insurance
Lease or mortgage
Online software for booking
Record keeping
Business license
Electricity
Internet
Staff costs
Consents
Photo storage
Marketing
Legal fees
Locked fridge for product
Website creation

The start up fees can easily be 10,000–20,000 dollars in the first year and many fees on the list can be re-occurring such as education, insurance, product, supplies, utilities, staff and licensing just to name a few (Table 13.1).

Essentials of Neuromodulation
https://doi.org/10.1016/B978-0-323-89920-8.00023-0

Table 13.1 Return on investment (ROI).

Neuromodulation revenue				
Clients per week	**Monthly gross @ $14 per unit 24 units per client**	**Monthly net @ $7.90 24 units per client**	**Yearly gross @ $14 per unit 24 units per client**	**Yearly net @ $7.90 24 units per client**
1 Client	$1344	$758.4	$16,128	$9100.8
2 Clients	$2688	$1516.8	$32,256	$18,201.6
3 Clients	$4032	$2275.2	$48,384	$27,302.4
5 Clients	$6720	$3792	$80,640	$45,504
10 Clients	$13,440	$7584	$161,280	$91,008
15 Clients	$20,160	11,376	$241,920	$136,512
20 Clients	$26,880	$15,168	$322,560	$182,016

CHAPTER 14

Marketing in esthetics

Marketing

Marketing is a whole separate entity these days with the onslaught of digital marketing and google ads. Start small and slow, and do not get in over your head. Interview three or four marketing companies before deciding on which is best. It is best to scan other esthetic businesses or other businesses; in general, it does not have to be an esthetic business. Get a sense of the mood or vibe their online presence has. Deciding early on what colors and font will be used is a good place to start to begin developing the brand. It is ideal to figure out what the voice will be in order to be able to relay it to the marketing team so that they can best create a plan. Finding the voice of the brand is a struggle for most injectors or business owners. A perfect balance of confidence and humility needs to be struck. Finding the image or the online presence can be a challenge as well.

Ultimately, an online presence is the personality of the business. Is the goal of the brand to be sleek, sexy, or educational? Keeping the personality of the online presence is important for consumers viewing. Nowadays, the website and social media platforms are the new door front to a business. Think of how inviting storefronts can be, especially when the storefront is all dressed up for the holidays. Often there will be an enticing sign on the front with a special or a current trend. The physical storefront and the virtual storefront should have the same feel, message, and level of warmth. The virtual presence needs to mirror the look and feel when the client shows up on the doorstep; otherwise, they will feel duped. Marketing does extend beyond the website. How many times can the injector engage with their audience without overwhelming them? Scripted emails are helpful to the patients for delivering before and aftercare instructions. This could be a way you slip in a little incentive as well as a thank you for their loyalty. Attached are sample emails utilized at DCCM™. Once the patient books an appointment, the system should automatically send out an email to the patient based on the appointment type that has been created, along with an appointment confirmation with the cancelation policy attached. The email is meant to be informative and help the patient prepare both physically

Essentials of Neuromodulation
https://doi.org/10.1016/B978-0-323-89920-8.00007-2

and emotionally for the visit. After the patient checks out once the service is complete, they should receive another email thanking them and reminding them of the aftercare instructions that were in the consent and verbally reviewed at the visit. These emails; demonstrate to the patient the injector cares. They also serve as another step to honestly informing the patient so that consent can truly be accepted. Consent of the patient means nothing in the absence of information and education.

Examples of pre and post neuromodulation emails

Pre care

Botox pre care guidelines

Please arrive on the day of your appointment with a clean face, free of any lotions and or makeup, even eye makeup. We recommend showing up about 10 min prior to your appointment time so that you can fill out your necessary treatment consents and for us to review your medical history.

After consents are signed, a team member will take your pictures to document the skin's health and appearance pre-treatment. Once you are prepped and ready; Tara will begin your treatment, which will take about 10 min.

No dental work within 2 weeks of your treatments.

Bring a hair tie to pull your hair back from your face.

Take Arnica tabs to minimize bruising.

The healthier you eat, the better all-around.

If you feel ill, you must reschedule your appointment.

We look forward to working with you on your esthetic journey.

Want us to let you in on a little secret?

One helpful little trick to improve your Botox (neuromodulator) treatment is being sure you have adequate Zinc in your system. All neuromodulators require Zinc at the nerve terminal so that the neuromodulator can bind to it most effectively. If you feel that your neuromodulation isn't as stiff or strong as it has been in the past and we are using the same dose…consider adding zinc supplements 5 days prior to treatment and 5 days post-treatment.

Love Tara

Post care

Botox after care instructions

We want to thank you for choosing the team at DCCM™ to help meet your esthetic goals. Now that you have taken the plunge with your Botox (neuromodulator) treatment, we want to remind you about your aftercare instructions. This way, you can maximize the potential of your treatment.

Please refrain from the use of hats, helmets, and visors, as well as facials and massages. The Botox (or neuromodulator) that we put in the muscle needs to set up in the muscle in order to take effect. Restricting blood flow and putting pressure on the muscles in the treated area can cause an unwanted shift in the product and sometimes result in an unwanted outcome.

Botox (neuromodulator) takes 2 full weeks to kick in. As the treatment begins to settle, you may want to give us a call to have us add a touch more in that time frame. We will not add more in the treated area within that 2 weeks time frame. Trust in the process that it will all settle out within 2 weeks. Should you need a slight touch up at 2 weeks, we welcome you to come back in for evaluation. If you would like to add more, we request that this be done on day 14–19 so that you can get the most out of your treatment. We honor all of our work, but in order to do so, we require a follow-up visit on day 14–19 after your initial Botox(neuromodulator) appointment. Our goal is to do our best to meet your goals. We really encourage and love to see all of our clients back at 2 weeks to be sure your treatment is perfect.

If you found yourself with a bit of a bruise, we recommend using Arnica gel and tablets. Each product formulation is different; therefore, we would advise you to follow the specific directions on the back of the branded bottle you purchase. These can be found at the whole foods store right in Hampton or at most pharmacies and grocery stores.

Want us to let you in on a little secret?

One helpful little trick to improve your Botox (neuromodulator) treatment is ensuring you have adequate Zinc in your system. All neuromodulators require Zinc at the nerve terminal so that the neuromodulator can bind to it most effectively. If you feel that your neuromodulator is not as stiff or strong as it has been in the past and we are using the same dose...consider adding zinc supplements 5 days before treatment and 5 days post-treatment.

We want to thank you for choosing the team here at DCCM™. As always, please call us with any questions or concerns that may arise. We are here for you and want you to feel confident in the process just as much as in your skin!

Our number is (603)-783-1087.

Between the hours of 8 a.m. and 4 p.m. you may call the office number. However, if you feel you have a pressing question that can not wait until normal business hours; please text (603)-783-1087 so we can assist you. We hope you enjoy your results and look forward to seeing you again at your next visit.

Love Tara

In the beginning phases, social media platforms are hard as the injector does not have many before and after to utilize. The injector is often left purchasing stock photos or creating their own memes. The goal should be to have as much original content as possible. The trends are currently utilizing short video clips along with before and after photos of the injectors work. This particular form of advertising seems to be the most lucrative. Also, do not be afraid to highlight the investment in the field of esthetics with regard to the time spent in further education. Show the public and the clients that growth and training are of the utmost importance to the practice. Blogs should be done at least monthly, and posts on social platforms should be daily. It is necessary to understand each platform's audience can be targeted with proper marketing messages being delivered. Not every post has to be a sale and especially not a flash sale. A flash sale is a dead giveaway as an act of desperation and the need for patients. That should be code for the patients to run as far away as possible. Better yet, showcase the beautiful work and illustrate the talent within the hands of the injector. The customers will call when they are ready to come in.

Forcing them with a deadline and a flash sale can often backfire as those patients are the impulse buyer instead of the savvy consumer. Esthetic injectables is an expensive industry, and it is, after all their face, so let us not be pushy better yet, inform the patients while letting them make their own decision. Educating the patients will yield a happier client in the long run, most likely a client that will be a repeat customer. The patient relationship needs to be cared for and evolving constantly. A patient should be looked at as a long-term investment, not a quick grab. There are mixed opinions on buying followers on social media. The initial focus should be on organic followers, not purchased followers. Bought followers do not necessarily turn into patients. The goal of growing a practice should not be on buying followers; rather it should be on developing and maintaining long standing relationships with patients. Spend money on the existing patients than buying followers that will likely never be a client. The goal is patients, not arbitrary followers. With that said, followers are vital as they are the audience for the brand. The focus should be on the quality of followers, not the quantity. One way to engage the patients and gain more patients is through a referral program.

At the end of every checkout process, clients should receive a message of some sort that lets them know about the referral program. It reminds them that for every client they refer, they earn $50 off of their next neuromodulation treatment and their friend also receives that discount at their first visit. Specific criteria for the number of units or even cross-promotions can help grow the practice organically. The referral program is an opportunity for the patients to be rewarded for spreading the word about their injector. They feel appreciated and respected, and more often than not, they are excited to share with their friends whom they get their esthetic services. The referral program goes back to the idea of relationships, as well. If the practice is set up to be a point and click practice, it will likely not have the sustainable relationships necessary for a referral program to work. A referral program works best with loyal and happy patients. Loyal patients love the service, outcomes along with their experience with every part of the process of their injections. Every touchpoint within the practice matters to the esthetic client along their journey.

Let us get social

Delle Chiaie Cosmetic Medicine DCCM™

Twitter @surfacemed

LinkedIn Delle Chiaie Cosmetic Medicine

IG @DCCM_Tara

FB Delle Chiaie Cosmetic Medicine

Showing the clients that the art and medicine of esthetics are understood is appreciable to the audience. Find subliminal ways to educate the audience, so they feel like they learned something. Another way to engage the audience is through questions. Post a fun fact about neuromodulation. Producing fun facts and tidbits demonstrates to the audience that talent and safety are a priority. Before and after photos are key to the success of a social media account. Post the work that has consistent lighting and background for the best engagement.

Relationships in business

We all set out on this voyage of independence thinking we will have freedom form the office politics blazing new paths and making our own rules while we are the boss. But the reality of it is; even though you may be your own boss you will still need others. You will need a fellow colleague to help you through sticky situations or unique patient presentations. The product representative is a person that you should most certainly develop a relationship with. The product reps understand not only their products but they know their territory and the market demand. They further understand brand development. The concept of brand development is not just about the brand they sell, its transferable to your brand as an injector. Relationships in this industry are essential in terms of success, opportunity and growth. No matter how proficient you become, there will be a situation where you will want an extra set of eyes or an opinion on a clinical situation. It may be puzzling as to why a section on relationships is in with marketing. The concept is simple; cross promotion. Most big companies need the injectors to showcase their product; so what better way than to network. An injectors before and after photos can be just the very thing a rep needs to help sell their product or device. When you reshare injectors share their photos it is branding and marketing for the practice as well as helping a fellow colleague in the industry. The more closely injectors work with their sales reps the better the success for both individuals. Every relationship is a two way street and it is so important to help one another out. Both the injector and the sales rep are in the business of selling, the market is ultimately the same. Develop strong relationships with the product and device reps as more often than not there will be a situation where the relationship is mutually fruitful to both parties. Gone are the days when the drug reps came in and catered lunches. It's time now to earn the lunch, with integrity, education, professionals and success.

CHAPTER 15

Top 10 list for success in neuromodulation

Top 10 list for success

Setting up a proper consultation to deliver amazing results and ensuring the client has proper expectations for their neuromodulation treatment.

(1) Be the Expert!

The nurse spent years in school, learning the art of beauty, science, and medicine. The patient spent hours on google. So when the injector is fortunate enough to have a client in their exam chair, be sure to educate the patient on anatomy. Also, be sure to teach them about the limitations of the treatment. As unethical as it is, there are still providers posting before and after photos with filters. The use of filters significantly distorts the reality of outcomes. The injector's job is to be honest and set up realistic expectations of the treatment.

A prime example is when the patient has crows feet that extend well into the mid check beyond the zygoma. The injector will want to educate the patient on where the Botox will be effective and where it will not be effective. Take the opportunity to talk about resurfacing modalities and or filler if that is the better adjunctive therapy. Most patients will not understand where Botox can go and can not go. It is also a common mistake for patients to think that Botox is for wrinkles and volume loss. They do not understand the anatomy or the products the way we do. It is our job to educate them. First-time clients should anticipate being at the visit for 45 min. Take their photos, have them sign the consent, prep their face, and then plan on 20 min of consult time and then leave 10 min to treat. Find a graceful and elegant way to educate the patient to demonstrate expertise in esthetic medicine. Instilling confidence in the patient about the educational process and experience the provider has will provide a sense of consumer confidence. The feeling of consumer confidence will be the foundation to the development of a long term relationship with the patient.

Essentials of Neuromodulation
https://doi.org/10.1016/B978-0-323-89920-8.00031-X

(2) Asymmetries exist; point them out.

When evaluating the patient, be sure to take note of subtle asymmetries in their features. Use caution with word choice when pointing out the patient's asymmetries as we do not want to hurt their feelings or make them feel bad. Use positive words instead of negative word phrases. Point out the asymmetry on the more proportionate side. For example, if the patient has softer static rhytids on the right side compared to the left. You might want to say; "I notice the fine lines you are concerned about are very soft and subtle on the right compared to the left. This is common among all of our patients to have variations from side to side. It has a lot to do with lifestyles, such as sleeping patterns, driving, sunscreen, and genetics. Some things are in our control, and things like genetics and sleeping really are not in our control. My goal in chatting with you about this very subtle difference is to evaluate your progress and overall satisfaction together. These subtle differences from side to side can impact the dose of the treatment as well as the outcome." The patient will over-analyze their face after treatment and may feel that the injector created the asymmetry as the patient may not have otherwise noted it prior to treatment unless it is mentioned at the initial consult.

(3) Photos are crucial in the injector's success.

Photos are the most accurate and non-biased form of documentation a provider has. The photos are free from emotion and judgment by both the provider and the patient. RX photo is a user-friendly app to take and store patient photos. The floor should be marked so that every angle is in the exact position as before. A neat feature on the RX app is that there is a ghosting option so that the photos can be lined up exactly for before and after. Having a consistent backdrop that does not provide glare is of the utmost importance. A matte black background typically produces the least amount of glare and light reflection distortion. A simple curtain rod with a black curtain is also useful for a backdrop.

Photos should be taken at rest in all angles and animation at all angles.

Rest,
Smiling,
Pouting,
Frowning,
Raising the brows,
Pursing the lips

It is important to have a baseline of the patient at rest and in animation. For every action we take injecting the neuromodulation, we often impose an opposing reaction in another area as we paralyze one area. Take pictures of the patients in animation, even if they are not treating that particular area.

The face should be properly cleaned before photos are taken with the hair pulled back or with a head wrap. Many practices also take photos immediately after the injections. This quick additional step can help to document any bruising or skin discoloration immediately after treatment. When treating for neuromodulator, it is advised that the patient come back in 2 weeks when the effect is complete. Set a follow-up appointment for the patient after treatment on day 15 through day 20.

The before photos taken can be used as a part of the consult instead of the mirror. Use the photos and zoom in on the troubled area, this is not something that can be done in a mirror. Often, the patient will position their face in the mirror in the most flattering light in the treatment chair, but then after treatment, they will view their face in an unflattering light in the privacy of their own home.

Follow up treatments are scheduled on day 14–19 after the initial treatment date. At this visit, the effectiveness of the treatment will be evaluated. Additional units can be added at the follow up if the patient would like to tweak the treatment. Neuromodulation treatment is not an exact science, and the treatment is unique to the patient desires and goals. It is expected that patients may need more neuromodulation at the follow-up visit. Additional units will be charged accordingly to the patient at the follow-up visit. Follow-ups visits are not recommended after 19 days. Honor the work within the 14–19 day follow up time frame from the original treatment but be aware that it is common that a patient may need more units to complete the treatment. If a patient does not attend their recommended follow-up appointment within the 14–19-day time frame, advise the patient that the treatment effects can not be evaluated as they are outside of the recommended follow-up time frame. Having the patient call at 6 weeks post-treatment saying that their treatment did not work would be a clear indicator that they perhaps the patient did not listen at the consult (only 10% of information is retained at a visit) or that the injector did not properly explain how the neuromodulation works.

When the 2-week follow-up is insisted upon by the practice, it can enhance the injectors' understanding and knowledge of the treatment effect. Further, the injector will be able to collect the patients after photos. When the patients can see the before and after photos and the success of the treatment in color on screen, it creates excitement within the patient. This excitement is an opportunity for the injector to begin discussing how great they look and poses as an opportunity to ask if their beautiful photos can be used on social media. Most clients are excited to be asked to be featured on social media. It makes them feel beautiful and special.

(4) Educate

Explain to the patient about static and dynamic rhytids. Layman terminology is best used in the consultation with the patient, so they can truly understand the concepts. Patients will understand terms such as a crease, line, or wrinkle. The consultative process should be further expanded to review with the patient that the neuromodulation is only working on the muscle, and it is not working on repairing the break down in the skin caused by the muscle movement over the years. Take time to have them animate and relax in the mirror and get them to explain what they do not like. Is it the expression they want gone, or is it the static lines they want gone, or perhaps it maybe both? Here is an opportunity to teach the patient about the reality of what a neuromodulator treatment can accomplish. If the patient is not properly counseled, they will complain that the neuromodulation treatment did not work after the muscle has been properly frozen. The reason is that the patient may have had goals of the at-rest creasing being taken care of and the muscles being frozen. The education provided in the consultation is the difference between a happy client and a dissatisfied client. Sending out what to expect email is excellent before the patient's appointment and then a follow-up appointment email reiterating what was said in the consult. It is best to standardize the before appointment email as well as the follow up email. Creating a standard, informative, yet straightforward email can be drafted in most software systems and cut down on phone calls and patient ambiguity. Esthetic treatments are expensive, and often the patients that can afford them have high expectations for the treatment. It is of the utmost importance that the injector properly educates the patient in the consult and with a follow-up email. Within the informative portion of the consultation, it can be a time to discuss further areas to be treated to help in the form of prevention. Every moment is an opportunity to educate. Educating is often a softer way to sell than the cold hard sale. Patients are savior than ever with the onslaught of social media and google.

(5) Confidence Tools

Marking out the patient can be very beneficial to both the injector and the patient. Marking out the patient does not have to be done just by the novice injectors. When the injector marks out the treatment areas, the patient sees it as a plan of action, and this action can instill confidence in the patient that they made the right choice. Using a white makeup pencil to mark the face where the injections will go is used by many as a confidence tool even if they are advanced enough to perform the treatment without the markings. If the injector chooses to mark the patient, be sure that the injection points do not go through the color or permanently tattoo the patient. Advanced injectors will gather an additional confidence tool such as a pair of calipers and measure the face and point out areas of symmetry, harmony, and beauty. The calipers can also be used again to deliver confidence to the patient that the injector has a real stronghold on their craft.

(6) Ice

Ice is used twofold with esthetics clients. It is used to prevent syncopal episodes as well as a buffer to help minimize the potential for bruising. If ice is going to be used on the patients, face where they will be treated, ensure that the ice is properly cleansed before they place it on their cleaned face. An additional measure of safety will be to alcohol the treatment area after the patient applies the ice before the injections. Infection can occur with any injection despite the depths and or relatively seemingly harmless nature of a neuromodulation injection's pinpoint entry. When the patient applies ice prior to the injections, it can create vasoconstriction that can minimize the risk of bruising. The ice-aesthesia will also help to decrease the patient's amount of discomfort with each injection. Every patient will experience varying levels of discomfort during the treatment. In addition to minimizing bruising and discomfort, another reason for using ice is to aid in the comfort of the nervous patients. The nervous patients or the patients who have had syncopal episodes in the past with injections are less likely to feel anxious or syncopal with the use of ice. A patient's heart rate variability can cause a syncopal event, and the ice will help regulate the variability preventing a syncopal event by keeping the heart rate lower. As a patient becomes stressed, the heart rate will go up, and then the blood pressure will drop, thus clamping down blood flow to the extremities and forcing it to the core circulation. Vasoconstriction is when the patient will become syncopal in the chair during treatment. An effective and interesting trick that is used in emergency medicine when a patient is experiencing Supraventricular tachycardia (SVT) is a bucket of ice water. The patient presenting to the emergency room with SVT will have their face submerged into a bucket of ice and water. Shockingly enough, it is enough to lower the patient's heart rate, decrease the patient's peripheral vascular response, and restore normal oxygenation and ventilation. The concept of using ice to bring the heart down is referred to in the literature as the divers reflex. It has been reported that when divers are submerged in ice-cold water, they experience apnea, bradycardia, along with an increased peripheral vascular resistance (PVR). These three unique but very individual physiologic changes work together to preserve the patient's oxygen level Godek and Freeman (2020). So to a lesser extent, the use of ice in the esthetic setting can help prevent syncopal episodes. If the patient is anxious, have them hold the ice or place it on the nape of the neck or their chest between the breasts.

(7) Reconstitution

Reconstitution is of the utmost importance. Keeping the reconstitution consistent and documented can reduce medical errors. Medical errors in the esthetic industry can cost a pretty penny when having to refund the patient. Understanding reconstitution, spread, and extrusion force all play a role in the outcome of the treatment. Newer injectors should reconstitute according to the manufacturers recommended, and FDA approved guidelines in the package labeled as prescribers guidelines to prevent confusion in the beginning. The only caveat recommended is to substitute bacteriostatic saline as the diluent for reconstitution instead of the 0.9% preservative-free saline. When speaking with the client during the reconstitution process of the neuromodulator, do not make the mistake of saying that you are diluting the product. Dilute to the patient means something very different to them as it does to us in the medical world. Dilute to them means the product is becoming less potent. The product lot number, expiration date, and the agent used to reconstitute the neuromodulator should be documented in the chart each and every time a patient is injected.

(8) Aftercare instructions

No matter how many times the patient has been seen for treatment, it is necessary to review the aftercare instructions before leaving.

A good place to start this conversation is in the pre-scripted email to the patient before they even have treatment. Send them a what to expect email, including what to expect at the visit (this helps to cut down on anxiety and impatient patients), pre-treatment guidelines, and aftercare. Once they are in the office be sure that the pre and post-care guidelines are part of the consent. No matter how many times the patient has had neuromodulation treatment in the office, the patient must sign a new consent to treat on that particular day of treatment. Each treatment is a new treatment. A new consent should be signed even 2 weeks later at the patient's follow-ups visit if they are to get more. A proper consent is not one that the patient signs real quick on an iPad. Talk with the patient about what to expect face to face and go over the pertinent details. Then send them an aftercare visit email reiterating what had been spoken about at the visit and copy and paste the aftercare guidelines from the consent right into the email. Now with that amount of information shared, the patient can truly sign an informed consent.

(9) Follow up Call

Patients lead busy lives, and it seems that gone are the days when a card is sent in the mail or a quick little gesture of kindness is presented. A follow-up phone call the following day is a nice touch. Even if the call is simply to state that it was nice to meet them and welcome them to the practice. The injector may want to check in on the new patients to acknowledge that it was their first time seeing how they are feeling and if they had any further questions. Take notes in the chart of their child's names or an event that was discussed during the consult. This little touch is the difference between a one time client and a life long client. We must create and foster long term clients. Shopping around for the cheapest and quickest injector is not in the patient's best interest. Develop strong relationships with the clients from the get go to ensure the foundation of the relationship. This way, they become long term patients with exceptional outcomes. It is difficult to get a proper treatment plan in place when the patients bounce from practice to practice.

(10) Documentation

We began talking about documentation when it came to the photos section; however, documentation is more than just photos. When the novice injectors journey begins, it is common to remember every patient's name and each injection point. Injectors even remember and even how the tissue responded. Fast forward a few years; and on some days, it is hard to remember your own name let alone each dose placed in every injection point. Scribing a total dose in one area does not equate to proper documentation. The injector will want to document all of the injection points accurately in each area and note the exact dose used along with the reconstitution process. Document if the skin is tanned, dry, red, and any other characteristic that was assessed. Documentation helps the injector at subsequent visits; it is a legal document, so be thorough in case the treatment chart is called into court, and lastly, if another provider in the office treats the patient, the outcome is consistent. Consistency is important in developing the patient's confidence in the practice and the injectors. They have options all around them. Be the reason they never look to the competition for treatment. It is helpful when injectors document the presentation of line formation or static creasing. Also, be sure to document bruising and if the pressure was applied or ice or whatever other alleviating treatment that was initiated. The same rules apply here as they do with bedside nursing, the right patient, the right dose, the right drug, the right route, and the right time. It can also add here since it is elective, the right reason. Body dysmorphia is a real diagnosis, and it is a common diagnosis in the esthetic world. Be sure that the patient has filled out a medical history form, a consent for treatment, a photo consent if willing for the use on social media, and supply them with a copy of the pre and post care guidelines. Remember neuromodulators are a drug and require the development of a prescriptive treatment plan to care for the patient. Do not get distracted by the fact this is an elective treatment. Esthetic injectors are injecting the world's most deadly neuromodulation known to mankind into their patient. Be confident, precise, and accurate. As much as the esthetic industry pertains to beauty, it is still medicine; always remember that.

Reference

Godek, D., & Freeman, A. M. (2020). Physiology, diving reflex. In *StatPearls. Treasure Island (FL)* StatPearls Publishing. [Updated 2020 Sep 29]. Available from https://www.ncbi.nlm.nih.gov/books/NBK538245/. (Accessed October 2020).

Appendices table of contents

APPENDIX 1

Aesthetic new patient history

DCCM

DELLE CHIAIE COSMETIC MEDICINE

Aesthetic New Patient History

First Name : ____________________ Last Name: ________________________ Date: _______

Address:__________________________ City: _________________ State: _______ Zip: _____

Cell Phone: ____________________ Other: ______________________Birth Date: __________

Email:_________________________________ Occupation:___________________________

How did you hear about us?_______________________ Ethnic Background:______________

List of medications and/or vitamins you are currently taking:________________________________

____

Allergies:___________________ Are you taking Antibiotics at this time?________________

Collagen Tested:______________ Date: _________Were there complications? _____________

Primary Physician's Name & Number: ____

Pharmacy Name & Number: ____

Are you in the process of having any major dental work completed? Yes or No? If so please explain:

____

MEDICAL HISTORY

Conditions (Check all that apply)	Yes	No	Conditions (Check all that apply)	Yes	No
Myasthenia Gravis			Lupus		
Muscle Weakness			Keloid Formation		
History of Cold Sores			Autoimmune Disease		
Hepatitis			Parkinson's Disease		
Multiple Sclerosis			Amyotrophic Lateral Sclerosis		
Sensitivity/Allergy to Lidocaine			Vision Problems		
Eye Disease			Neurological Disorders		
Allergy to Beef or Dairy			Lambert-Eaton Syndrome		
Numbness			Hypersensitivity to Medications		

List and/or explain other medical conditions not listed above:______________________________

____

 Patient Initials______

Previous Hospitalizations/Operations:____

____

Have you had plastic surgery or other surgery to your face/neck area and if so when:_______________

____

WOMEN: Are you pregnant, trying to get pregnant, or lactating? ________________________

BOTOX

Have you had Botox injections before? ________________ Last treatment: _______________

What areas?_______________ Were you happy? If not, explain: ______________________

Have you ever had eyelid/eyebrow droop after Botox? ____________________________

Do your eyelids droop without sleep? ____

Do your eyelids feel extra heavy when you don't get enough sleep?________________________

Do you show a lot of upper eye lid when eyes are open?_______________________________

DERMAL FILLERS

Have you had any Dermal Fillers procedures before? ___________ Last Treatment? _____________

If yes, what filler was used? ____

What areas was it placed? ____

Were you satisfied with the results? If not, explain: _______________________________

CURRENT SKIN CARE REGIMEN

What products are you currently using on your face? ________________________________

Are you happy with your current products? Yes or No? ______________________________

Do you use a Clarisonic? Yes or No? ________________ What would you like to improve with your skin care regimen? ____

I understand the information on this form is essential to determine my medical and cosmetic needs and the provisions of treatment. I understand that if any changes occur in my medical history/health I will report it to the office as soon as possible. I have read and understand the above medical history questionnaire. I acknowledge that all answers have been recorded and truthfully and will not hold any staff member responsible for any errors or omissions that I have made in the completion of this form.

Patient Signature:____Date:______________

Provider Signature:____Date:_______________

 Patient Initials_______

IN HOUSE USE ONLY:

	Yes	No
Date:		
Details:		
Date:		
Details:		
Date:		
Details:		
Date:		
Details:		
Date:		
Details:		
Date:		
Details:		
Date:		
Details:		
Date:		
Details:		
Date:		
Details:		
Date:		
Details:		
Date:		
Details:		
Date:		
Details:		

APPENDIX 2

Pre care guidelines email

DCCM

DELLE CHIAIE COSMETIC MEDICINE

Botox pre care guidelines

Please arrive on the day of your appointment with a clean face, free of any lotions and or makeup, even eye makeup. We recommend showing up about 10 min prior to your appointment time so that you can fill out your necessary treatment consents and for us to review your medical history.

After consents are signed, a team member will take your pictures to document the skin's health and appearance pre-treatment. Once you are prepped and ready; Tara will begin your treatment, which will take about 10 min.

No dental work within 2 weeks of your treatments.

Bring a hair tie to pull your hair back from your face.

Take Arnica tabs to minimize bruising.

The healthier you eat, the better all-around.

If you feel ill, you must reschedule your appointment.

We look forward to working with you on your esthetic journey.

Want us to let you in on a little secret?

One helpful little trick to improve your Botox (neuromodulator) treatment is being sure you have adequate Zinc in your system. All neuromodulators require Zinc at the nerve terminal so that the neuromodulator can bind to it most effectively. If you feel that your neuromodulation isn't as stiff or strong as it has been in the past and we are using the same dose … consider adding zinc supplements 5 days prior to treatment and 5 days post-treatment.

Love Tara

APPENDIX 3

Post care guidelines email

DCCM

DELLE CHIAIE COSMETIC MEDICINE

Botox after care instructions

We want to thank you for choosing the team at DCCM™ to help meet your esthetic goals. Now that you have taken the plunge with your Botox (neuromodulator) treatment, we want to remind you about your aftercare instructions. This way, you can maximize the potential of your treatment.

Please refrain from the use of hats, helmets, and visors, as well as facials and massages. The Botox (or neuromodulator) that we put in the muscle needs to set up in the muscle in order to take effect. Restricting blood flow and putting pressure on the muscles in the treated area can cause an unwanted shift in the product and sometimes result in an unwanted outcome.

Botox (neuromodulator) takes 2 full weeks to kick in. As the treatment begins to settle, you may want to give us a call to have us add a touch more in that time frame. We will not add more in the treated area within that 2 weeks time frame. Trust in the process that it will all settle out within 2 weeks. Should you need a slight touch up at 2 weeks, we welcome you to come back in for evaluation. If you would like to add more, we request that this be done on day 14–19 so that you can get the most out of your treatment. We honor all of our work, but in order to do so, we require a follow-up visit on day 14–19 after your initial Botox (neuromodulator) appointment. Our goal is to do our best to meet your goals. We really encourage and love to see all of our clients back at 2 weeks to be sure your treatment is perfect.

If you found yourself with a bit of a bruise, we recommend using Arnica gel and tablets. Each product formulation is different; therefore, we would advise you to follow the specific directions on the back of the branded bottle you purchase. These can be found at the whole foods store right in Hampton or at most pharmacies and grocery stores.

Want us to let you in on a little secret?

One helpful little trick to improve your Botox (neuromodulator) treatment is ensuring you have adequate Zinc in your system. All neuromodulators require Zinc at the nerve terminal so that the neuromodulator can bind to it most effectively. If you feel that your neuromodulator is not as stiff or strong as it has been in the past and we are using the same dose ... consider adding zinc supplements 5 days before treatment and 5 days post-treatment.

We want to thank you for choosing the team here at DCCM™. As always, please call us with any questions or concerns that may arise. We are here for you and want you to feel confident in the process just as much as in your skin!

Our number is (603)-783-1087.

Between the hours of 8 AM and 4 PM you may call the office number. However, if you feel you have a pressing question that can not wait until normal business hours; please text (603)-783-1087 so we can assist you. We hope you enjoy your results and look forward to seeing you again at your next visit.

Love Tara

APPENDIX 4

Neuromodulation consent

I understand certain Neuromodulation are FDA-approved for treatment of rhytids (wrinkles) located between the eyebrow, also known as the Glabella region. Other areas are commonly treated with neuromodulation, which have not been FDA-approved, and this is considered "off-label use." Neuromodulation can relax the muscles on areas of the face and neck which cause wrinkles associated with facial expressions. I understand that neuromodulation cannot improve sagging skin or wrinkles caused by aging or sun damage and understand they are unrelated to muscle contraction. Treatment with neuromodulation can cause facial expression lines or wrinkles to essentially disappear. The primary effect of the neuromodulation injection is on the nerve terminal and the secondary effect is on the muscle. Areas most commonly treated are: (a) glabella area of frown lines, located between the eyebrows; (b) crow's feet (outside areas of the eyes); and (c) forehead wrinkles; however, neuromodulation may also be used in other facial areas. The Neuromodulation is reconstituted to a very controlled solution and, when injected into the muscles with a very thin needle, it is almost painless. I understand patients may feel a slight burning sensation while the solution is being injected. The procedure takes about 15–20 min and results generally last 3–4 months. In some individuals, the results of one treatment may last shorter or longer. With repeated treatments, the results may also tend to last longer.

Risks and complications

It has been explained to me that there are certain inherent and potential risks and side effects with any invasive procedure. In this specific instance, such risks include, but are not limited to:

1. Post treatment discomfort, swelling, redness, and bruising;
2. Double vision;
3. Weakened tear duct, rarely;
4. Post treatment bacterial and/or fungal infection requiring further treatment;
5. Allergic reaction;

6. Minor temporary droop of eyelid(s), eyebrow (s), or corner of the mouth in approximately 2% of injections. This usually lasts 2–3 weeks;
7. Occasional numbness of the forehead lasting up to 2–3 weeks;
8. Transient headache; and
9. Flu-like symptoms may occur.

Neuromodulation pre treatment instructions

It is prudent to follow some simple guidelines before treatment that can make all the **difference between a fair result or great result,** by reducing some possible side effects associated with the injections. We realize this is not always possible; however, **minimizing these risks is always desirable.**

- **Avoid alcoholic beverages** at least **24 h prior** to treatment (alcohol may thin the blood increasing the risk of bruising).
- **Avoid anti-inflammatory/blood thinning medications** ideally, for a period of **two (2) weeks before treatment**. Medications and supplements such as Aspirin, Vitamin E, Gingo Biloba, St. John's Wort, Ibuprofen, Motrin, Advil, Aleve, Vioxx, and other NSAIDS (non-steroidal anti-inflammatory drugs) are all blood thinning and can increase the risk of bruising/swelling after injections.
- **Schedule your Neuromodulation appointment at least 2 weeks prior** to a **special event which may be occurring, i.e., wedding, vacation**, etc. It is not desirable to have a very special event occurring and be bruised from an injection.
- You will need to reschedule your appointment if you are on antibiotics for an infection. It is also necessary to rescheduled your appointment if you have a fever or had a fever in the last 48 h.

Neuromodulation post treatment instructions

The guidelines to follow post treatment have been followed for years, and are still employed today to prevent the possible side effect of ptosis. These measures should minimize the possibility of ptosis almost 98%.

- **No straining, heavy lifting, vigorous exercise for 3–4 h** following treatment. It is now known that **it takes the toxin approximately 2 h to bind itself to the nerve to start its work**, and because **we do not want to increase circulation** to that area to wash away the neuromodulation from where it was injected. *This waiting period continues to be recommended by most practitioners.*

- **Avoid manipulation of area for 3–4 h** following treatment. **(For the same reasons listed above.)** This includes not doing a facial, peel, or micro-dermabrasion after treatment with Neuromodulation for 3 days. **A facial, peel, or micro-dermabrasion can be done in same appointment only if they are done before the Neuromodulation.**
- **Do not lie down or bend over for 3–4 h** following treatment. (This instruction has been employed for years by some practitioners.)
- Avoid having a facial or massage for 3 days.
- Do not wear a hat or helmet for 3 days.
- **It can take 2–10 days to take effect and 2 weeks for full effect.** It is recommended that the patient contact office **no later than 3 weeks after treatment** if desired effect was not achieved and no sooner than 2 weeks to give toxin time to work.

Makeup may be applied before leaving the office. Some practitioners recommend avoiding Retin-A, Glycolic acid, Vitamin C, and Kinerase *for 24 h* to the treated areas.

I have been counseled in these pre and post-treatment instructions and have been given a written copy of these instructions. All my questions have been answered to my satisfaction.

Photographs

I authorize the taking of clinical photographs and their use for scientific and educational purposes both in publications and presentations, as well as any social media and marketing. I understand my identity will be protected.

Pregnancy, allergies and neurologic disease

I am not aware that I am pregnant and I am not trying to get pregnant. I am not lactating (nursing). I do not have any significant Neurological disease (s) including, but not limited to, Myasthenia Gravis, Multiple Sclerosis, Lambert-Eaton Syndrome, Amyotrophic Lateral Sclerosis (ALS), Parkinson's. I do not have any allergies to the toxin ingredients or to human albumin and have never had a reaction to a neuromodulation in the past.

Payment

I understand this is an elective cosmetic procedure. I understand and agree that all services rendered to me will be charged directly to me, and I am personally responsible for payment. Payment is due at the time treatment is rendered. I understand there are no refunds on any services rendered. I further agree that, in the event of non-payment, I will bear the cost of collections and/or Court costs and reasonable legal fees, should this be required.

Results

I am aware that when small amounts of a purified Neuromodulation is injected into a muscle it causes weakness/relaxation of that muscle. This appears in 2–10 days and can last 3–4 months, but can be shorter or longer. I understand that the length of response may vary from patient to patient and from one treatment to the next.

Follow up treatments are scheduled on day 14 through 19 after your initial treatment date. At this visit we will evaluate the effect of your treatment. You can add an additional area or more to the previously treated area if the provider feels it is safe and necessary. Neuromodulation treatment is not an exact science and the treatment is unique to the patient desires and goals. It is common that patients may need more neuromodulation at the follow up visit. In a very small number of individuals, the injection does not work as satisfactorily or for as long as usual. Rarely, there are some individuals who do not respond at all. It is at the discretion of my practitioner as to whether or not a "touch-up" injection may be needed after the first 14 days of treatment. I understand, if that is the case, an additional charge may incur. I understand that I will not be able to "frown" while the injection is effective, but that this will reverse after a period of months at which time re-treatment is appropriate. Follow up visits are not recommended after the 19 day period. I understand that I am advised to make a follow up appointment between day 14 and 19 post treatment. If this appointment is not made or kept the effectiveness can not be properly assessed therefore touch ups outside of this time frame may net be permitted and we will wait till your next 3 month visit. The practice will honor the value of their work however, the 2 week follow up must be had in order properly asses and diagnose. The neuromodulation will begin to slowly breakdown over the 3 months so it is common to begin to see movement at about 6 weeks time.

I am aware that the practice of medicine is not an exact science, and I acknowledge that no guarantees have been made to me as to the results of the procedure.

I understand that the success of the procedure is to some extent dependent upon my closely following my practitioner's instructions. I must not perform any vigorous exercise, and I must not massage or manipulate the area (s) of the injections for the 3–4 h post-injection period. I further understand I will not wear a hat or have a facial and or massage for 3 days post injections.

I understand this is an elective procedure and hereby voluntarily consent to treatment with neuromodulation injections for the condition known as: Facial Dynamic Wrinkles. The procedure has been fully explained to me as well as the areas which are "off label."

I have read the above information and understand it. My questions have been answered to my satisfaction.

I accept the risks and complications of the procedure. I understand that no guarantees are implied as to the outcome of the procedure. I certify that I am over the age of 18 and am not under the influence of drugs or alcohol. I have read and understand the pre and post neuromodulation instructions.

I certify if any changes occur in my medical history, I will notify the office. I hereby give my voluntary consent to this procedure, and release Surface Medical Esthetics, PLLC medical staff, and specific technicians from liability associated with the procedure. This consent form is freely and voluntarily executed and shall be binding upon my spouse, relatives, legal representatives, heirs, administrators, successors and assigns.

Patient Name (Print) Patient Signature Date

Provider Name (Print) Provider Signature Date

APPENDIX 5

Neuromodulation pre treatment instructions

It is prudent to follow some simple guidelines before treatment that can make all the **difference between a fair result or great result**, by reducing some possible side effects associated with the injections. We realize this is not always possible; however, **minimizing these risks is always desirable**.

- **Avoid alcoholic beverages** at least **24 h prior** to treatment (alcohol may thin the blood increasing the risk of bruising).
- **Avoid anti-inflammatory/blood thinning medications** ideally, for a period of **two (2) weeks before treatment**. Medications and supplements such as Aspirin, Vitamin E, Gingo Biloba, St. John's Wort, Ibuprofen, Motrin, Advil, Aleve, Vioxx, and other NSAIDS (non-steroidal anti-inflammatory drugs) are all blood thinning and can increase the risk of bruising/swelling after injections.
- **Schedule your neuromodulation appointment at least 2 weeks prior** to a **special event which may be occurring, i.e., wedding, vacation**, etc. It is not desirable to have a very special event occurring and be bruised from an injection.
- You will need to reschedule your appointment if you are on antibiotics for an infection. It is also necessary to reschedule your appointment if you have a fever or had a fever in the last 48 h.

APPENDIX 6

Neuromodulation post treatment instructions

The guidelines to follow post treatment have been followed for years, and are still employed today to prevent the possible side effect of ptosis. These measures should minimize the possibility of ptosis almost 98%.

- No straining, heavy lifting, vigorous exercise for 3–4 h following treatment. It is now known that it takes the toxin approximately 2 h to bind itself to the nerve to start its work, and because we do not want to increase circulation to that area to wash away the neuromodulation from where it was injected. *This waiting period continues to be recommended by most practitioners.*
- Avoid manipulation of area for 3–4 h following treatment. (For the same reasons listed above.) This includes not doing a facial, peel, or micro-dermabrasion after treatment with Neuromodulation for 3 days. A facial, peel, or micro-dermabrasion can be done in same appointment only if they are done before the Neuromodulation.
- Do not lie down or bend over for 3–4 h following treatment. (This instruction has been employed for years by some practitioners.)
- Avoid having a facial or massage for 3 days.
- Do not wear a hat or helmet for 3 days.
- It can take 2–10 days to take effect and 2 weeks for full effect. It is recommended that the patient contact office no later than 3 weeks after treatment if desired effect was not achieved and no sooner than 2 weeks to give toxin time to work.

Makeup may be applied before leaving the office. Some practitioners recommend avoiding Retin-A, Glycolic acid, Vitamin C, and Kinerase *for 24 h* to the treated areas.

I have been counseled in these pre and post-treatment instructions and have been given a written copy of these instructions. All my questions have been answered to my satisfaction.

Patient Name (Printed)	**Patient Signature**	**Date**
Provider Name (Printed)	**Provider Signature**	**Date**

APPENDIX 7

Policy and procedure for neuromodulation administration

Purpose

To ensure safe and effective treatment of patients undergoing Botulinum Toxin A administration at Surface Medical Esthetics, PLLC the following policies and procedures have been developed.

Policy

A medical professional such as a Registered Nurse (RN), Doctor (MD), Advanced Practice Registered Nurse (APRN), Licensed Practical Nurse (LPN) with current state licensure, shall be able to assess, consult and treat patients with Botulinum Toxin A for hyperfunctional lines of the upper face, specifically the forehead, crows feet, and the glabellar region between the eyebrows, following the guidelines set forth.

Setting

The licensed Medical Professional will perform the Administration of Botulinum Toxin A in multi settings and locations, such as but not limited to:

- Surface Medical Esthetics, PLLC at 822 Lafayette Rd suite 4 Hampton NH 03842 authorized by the Medical Director or Advanced Registered Nurse Practitioner as a health care office(s)

All Botulinum Toxin A administration procedures shall be performed in a professional, clean, safe environment, equipped with proper sharps disposal system, and Osha and universal precautions in place.

Supervision

The Medical Professional will function under the general supervision of the Medical Director and/or Advanced Registered Nurse Practitioner who is immediately available for consultation by telecommunication and is physically available as medically necessary.

All adverse reactions such as ptosis, diplopia, lower eyelid retraction, and weakening of the lacrimal pump shall be reported immediately to the Medical Director and/or the Advanced Registered Nurse Practitioner when it occurs. Adverse events will be documented in the patient's chart.

Record keeping

The Medical Professional is responsible for maintaining patient records, including but not limited to patient assessment, signed informed consent of risks, benefits, and potential adverse effects, treatment #, treatment sites, # of injections, solution/concentration used, and the patient response to treatment.

Training/education

The Medical Professional will have specialized training/certification in the Administration of Botulinum Toxin A. A Medical Doctor, Doctor of Osteopath, Advanced Registered Nurse Practitioner experienced in this procedure, or a teaching institution specializing in this procedure, may perform this training/certification. Satisfactory Training will include:

- Mechanism of Action of Botulinum Toxin A
- Basic Theory of Treatment for Cosmetic Purposes
- Facial Anatomy and Muscle Function
- Pharmacokinetics
- Storage, preparation, and dilution of Botulinum Toxin A
- Safety, efficacy, and complication issues
- Contraindications
- Assessment and identification of areas to be treated
- Safe application of injection techniques and return demonstration

Evaluation and competency

Initial evaluation and final determination of satisfactory training and competence shall be by the Medical Director and/or Advanced Registered Nurse Practitioner. On-going evaluation of the competence of the Medical Professional will be done, at a minimum, on a yearly basis at the time of the Medical Professional scheduled company evaluation, and more often, if needed, i.e. indicated by patient dissatisfaction or efficacy issues.

This evaluation will be done by the Medical Director and/or Advanced Registered Nurse Practitioner and shall be documented in the Medical Professionals file.

Authorized personnel documentation

All Medical Professionals who have successfully completed training and posses a certificate of completion and have been determined to be competent to perform the Administration ofBotulinum Toxin A will be documented on the appropriate form and will be kept on file as part of the Standardized Procedure Policy in the Administration office.

Limitations

The Medical Professional will not knowingly treat any patients who are knowingly pregnant, trying to get pregnant, lactating, patients with allergies to human albumin, any patients with significant autoimmune or neurological diseases, i.e., Amyotrophic Lateral Sclerosis (ALS), Lambert-Easton Disorder, Mysthenia Gravis, Multiple Sclerosis, and Parkinson's.

Protocol for administration of Botulinum Toxin A

Medical Professional will:

1. Complete assessment and a medical history questionnaire and take photos at rest and with animation in all positions with all new patients.
2. Patients with a history of allergies to human albumin, patients with significant neurological and autoimmune diseases, pregnant or lactating patients will be denied treatment. *(See Limitations section.)*
3. Upon passing medical screening, patient will be fully informed of risks, benefits, and potential adverse reactions and an informed consent will be signed.
4. Botulinum Toxin A will be stored in freezer (−5°C or lower) until ready for use. Once reconstituted, it must be refrigerated, not refrozen.
5. Botulinum Toxin A will only be reconstituted **just prior to use** and should be used within the first 4 h, but may be refrigerated up to 30 days.
6. **Vacuum will be released**, using a 22 guage needle prior to reconstitution. If no vacuum is present the neuromodulation vial will be sent

back to the manufacturer and a new vial will be used following the same procedure.

7. Botulinum Toxin A (Jeaveau, Botox, Xeomin) will be reconstituted using 2.5–5.0 mL of unpreserved or **preserved saline as a diluent**, resulting in a 4–2 units per 0.1 cc. Dysport will be reconstituted with 1.5–3 mL of unpreserved or preserved saline resulting in 20–10 units per 0.1 mL. A 3–5 cc syringe containing preserved saline is attached to the 18–27 guage needle and **SLOWLY** injected into the vial. Allow the saline to **flow down the sides of the vial**, thus minimizing air bubble formation and not damaging the delicate Botulinum Toxin A (see Appendix 1 for additional acceptable dilution tables for advanced injectors).
8. Botulinum Toxin A is gently drawn up into a 0.3–1 mL syringe using a 18–27 guage needle. The injection is to be administered with a 30 guage-33 gauge—½ in. needle.
9. Patients are injected while in a seated position.
10. Patients are asked to demonstrate dynamically the function of the muscle groups to be injected.
11. Prior to administration the medical professional will map out points of injection according to landmarks and location of muscle belly. The areas of administration will be the corrugator, procerus, frontalis, and orbicularis oculi muscles. (Corrugator and procerus muscles for frown lines, frontalis muscle for horizontal forehead lines, and orbicularis oculi muscle for crow's feet mentalis, masetter, nasalis, pltysma.) Note: Increased toxin dose may be necessary in older and male patients. (See diagrams of each area.)
12. Caution will be taken to administer at least 1 cm above the orbital rim and 1–1.5 cm from the lateral canthus to reduce the chance of Ptosis or other complications. Complications are a rare occurrence. If mild ptosis should occur the nurse will instruct the patient that this will resolve within a few weeks and in the use of OTC, Vasocon to assist in alleviating the ptosis. Ptosis or any other complications will be immediately reported to the Medical Director and/or the Advanced Nurse Practitioner and documented in the patient record along with photos.
13. Syringe is inserted perpendicular to the skin and completed at a depth just beneath the dermis, when injecting the Procerus or Orbicularis Oculi (crow's feet). When injecting the Glabellar or Frontalis muscles, needle will advance to just in front of the periosteum, this is to insure penetration to or into the belly of these muscles. Botulinum Toxin

A 0.025–0.1 mL is injected into all sites according to patient treatment plan.

14. After each injection, a cold compress consisting of a clean 4 × 4 gauze dipped in ice water is applied and gently massaged for a few seconds over the injection site.
15. When procedure is completed the patient will be educated not to rub or manipulate the injection sites, not to lie down for a period of 4 h, no over exertion, and to report any problems or complications to the office immediately. Further the patient should be instructed not to have a massage, facial or wear hats for 3 days.

Development of plan

The Medical Director, Administrator, and the Advanced Practice Registered Nurse have developed this Standardized Procedure and Protocol for the Administration of Botulinum Toxin A by trained Medical Professionals as a comprehensive working model.

This model will be reviewed annually at an annual management meeting and documented in the minutes of the meeting and will be kept in the Administration office.

This Standardized Procedure and Protocol has been approved by:

________________________________ ____________
Nursing Signature Date

________________________________ ____________
Medical Director Date

________________________________ ____________
Administration Signature (THIS IS OPTIONAL) Date

APPENDIX 8

Photo consent form

I consent to and authorize Surface Medical Esthetics, PLLC, its affiliates, agents, representatives and all persons or entities acting with its permission or upon its authority (collectively "Surface Medical Esthetics, corporation") to use my photographs and digital images (the "Material") being taken to evaluate treatment effectiveness, for medical education, training, professional publications or sales purposes. No photographs or digital images revealing my identity will be used without my written consent. If my identity is not revealed, these photographs and digital images may be used, shared, and displayed publicly without my permission. I further grant Surface Medical Esthetics, PLLC the right to incorporate and use the Material in video, print ads, still photographs, catalogs, packaging and package inserts, web site and all other media (the "Advertising"), and to reproduce, exhibit, broadcast, transmit and distribute Advertising containing the Material.

I hereby assign to Surface Medical Esthetics, PLLC all of my right, title and interest in and to the Material, and any rights, including copyrights that may result from the use of any Material in any Advertising.

I further waive any right to inspect or approve the Advertising incorporating the Material and the uses to which it may be applied.

I HAVE READ THE FOREGOING RELEASE AND FULLY UNDERSTAND IT AND AGREE TO THE TERMS OUTLINED.

This release is executed this ________ day of ______________________, 20_____.

____

(Signature)

Print Name: ____

Address: ____

In addition to the above Consent, I further consent to have my identity revealed with full face photos or photos of my body even with identifying markings. I further consent to all the previous language of release of the photos within this consent.

This release is executed this _______ day of _____________________, 20_____.

____

(Signature)

Print Name: ____

APPENDIX 9

Neuromodulation supply list

Supply	Mckesson mms. mckesson.com	Anda Meds andameds.com	Allergan	Galderma	Evolus	Merz
BD 3/10.ML insulin syringe	100 count 37.5 1 = 2.67 Avg use 2–10	100 Count $9.55 1 = $0.96 Avg use 2–10				
Bacteriostatic saline	25 pack = $24.70 30 cc = $0.98 Avg use 1–5 cc	25 Pack = $28.75 30 cc = $1.15 Avg use 1–5 cc				
Botox			$601			
Newtox					$463	
Dysport				540.87		
Xeomin						
18 Guage needles 1.5	100 Pack = $4.95 1 = $0.05	100 Pack = $6.46 1 = $0.07				
3 cc syringes	100 Pack = $6.8 1 = $0.07 Avg use 3–9	100 Pack = $7.87 1 = $0.08 Avg use 3–9				
Gauze (4 × 4)	200 count = $4.72 Avg use 6	200 Count = $2.65 Avg use 6				
Skin marking pen	$1.52	$1.11				
Rubbing alcohol	Bottle $1.70	Bottle $1.70				
Germicidal cleanser	16 FL Ounces $12.07	16 FL Ounces $7.99				
Patient drapes/towels	$0.13	$0.35				
Head band	$1.69 Reusable	$1.69 Reusable				
Gloves	200 Count = $12.69	200 Count = $13.6				
Ice						
Camera						
Mirror						
Necessary forms						
Intake form						
Consent						
Pre and post care						
Treatment record						
Photo release						
Policy and procedure for admin						
SDS						

Glossary of common terms

Blepharoptosis Blepharoptosis is referred to as a droopy lid. Droopy eyelids are common complications if the levator palpebral is inadvertently injected or improper treatment within the infraorbital space. If a patient presents with blepharoptosis, naturally caution should be used when treating the frontalis.

Corrugator crease The corrugator crease is where the skin creases between the medial and lateral corrugator compartments. Perforators from the supratrochlear artery most often clearly outline the crease, a specific landmark for where the supratrochlear artery lies. The supraorbital artery lays within the seat of the lateral corrugator crease.

Corrugator supercilii The corrugator supercilii are two small pyramid-shaped muscles located at the medial portion of each eyebrow. They lay deep to the frontal portion of the frontalis and orbicularis oculi muscles. The length of this muscle spans to and at times across supraorbital creases. This muscle works with the orbicular oculi muscle and the procures in order to pull the eyebrows medially and downward in an attempt to shield the eyes in bright sunlight. When these muscles are contracted, the muscle comes to the surface, and the supratrochlear and supraorbital arteries retract back to the forehead bone.

Depressor anguli oris muscle (DAO) The DAO is a superficial lower facial muscle that depresses or lowers the oral commissures (corners of the mouth). Even though it is noted that the DAO is superficial, it is important to note that the insertion point is deep, and it is imperative to make a note of this anatomical landmark should the injector be treating the platysmas area. It is considered to be a part of the lower lip muscle. Constant and repetitive contraction of these muscles contributes to the cheek lip crease. The DLI is often accidentally hit during this injection. Understand landmarks and safety zones is key to reproducible and esthetically-pleasing outcomes.

Depressor labii inferioris (DLI) The DLI is a deep quadrilateral muscle that originates from the oblique line of the mandible between the menti and the mental foramen. It is continuous with the platysma and is responsible for depressing the lower lip. This muscle is responsible for drawing the lip down and outward laterally a bit with mastication. When active, the muscle is responsible for the emotional show of sorrow, melancholy, and doubt. It is not advised to treat this muscle with neuromodulation as the lip would push up and become rigid. The patient would likely present with oral incompetence.

Depressor septi nasi (DSN) This deep muscle arises from the maxilla's incisive fosa and inserts into the septum and the rear portion of the ala. An active depressor septs muscle can accentuate the nasal tip droop seen in the aging face. It is also responsible for the shortening of the upper lip in animation. The DSN narrows the nostrils and draws the septum down. As we age, the tip of the nose pulls down, and the upper lip area lengthens, and the nares flair. By treating this muscle with neuromodulation, you can raise the tip of the nose, begin to lessen the flair of the nares and shorten the upper lip length. There are two approaches to treatment here with neuromodulators. The injector can perform one deep injection point at the base of the septum or two injection points, one on the right and one on the left of the septum.

Depressor supercilii The depressor supercilli is a depressor muscle that sits laterally to the procerus muscle on both the left and the right. It originates from the orbital rim from the fascia of the nasal bone. Its function is to pull the medial portion of the brows inward and downward. Some literature will describe this muscle as part of the orbicularis oculi muscles instead of being its own structure and function. This particular area is an area that men often do well with treating to hidden the medial brows. However, generally speaking, most women do not want a hidden medial position of the brow.

Dermatochalasis Dermatochalasis is a condition known to medical professionals referring to the laxity or bagging of the upper and lower eyelid region. Neuromodulation can enhance this conditioned aid in the correction. Being mindful of patient selection is essential to this condition.

Diplopia Diplopia is a condition in which the patient develops double vision. Double vision can occur when treating a patient with neuromodulation in the lateral cantonal lines. When the needle is aimed toward the globe, the risk for injection into the extraocular muscles (EOM) increases. Direct or indirect injection into the EOM will cause diplopia.

Dysphagia Dysphagia is the term used for difficulty swallowing. Patients can experience dysphagia after treatment with neuromodulation due to the inadvertent injection or spread into the pharyngeal muscles. Dysphagia is most frequently seen as an adverse event when injection the sternocleidomastoid (SCM) muscles.

Dyspnea Dyspnea is shortness of breath or difficulty breathing. This labored form of breathing can happen as a result of a physical and or emotional trigger. It can be seen in relationship to neuromodulation treatment in regards to the patient's anxiety typically.

Ectropion Ectropion is a condition in which the eyelid turns out. Ectropion is commonly a disease in the elderly population but can be a complication of neuromodulation. Typically it occurs when too much neuromodulation is placed in the infraorbital space. Making sure the patient is a proper candidate for treatment is essential. If the patient presents with scleral show below the iris, they are not candidates for treatment with neuromodulation in the infra-orbital space.

Edema Edema is known as excess fluid trapped in the skin. When a patient has a slower or poorly functioning lymphatic system under the eyes, neuromodulation can cause edema in this area. The orbicularis oculi muscle acts as a mechanical force to squish excess fluid out of the area that the lymphatic system can not handle. Therefore, when muscle activity is reduced due to neuromodulation use, the area can present with edema. Time and lymphatic drainage can help this situation.

Epiphora Epiphora is a medical condition in which excessive tearing or hyperlacrimation. Neuromodulation can help this condition by placing the product into the lacrimal gland. Placing neuromodulation into the gland will prevent the release of acetylcholine, thus decreasing excessive tearing.

Facial nerve The 7th cranial nerve is the facial nerve. This cranial nerve is a motor nerve to the mimetic facial muscles. It is responsible for the movement of the skin and muscles of the face. This nerve's complex course can be predicted based on very specific landmarks seen on each patient or subject.

Festoons Festoons are caused by lax skin and weakening of the orbicularis oculi muscle. Cascading hammock of skin and muscle in the infraorbital rim, which may or may not contain fat. Festoons typically reside from the mid to lateral canthus. Festoons are treated surgically and best left alone regarding dermal filler and neuromodulators in the novice injector.

Frontalis The frontalis muscles are two large fanlike muscles that extend from the eyebrow area to the top of the forehead. The frontalis is where the frontal branch of the facial nerve takes its name. The frontalis muscle is responsible for the elevation of the brows. When elevation of the frontalis muscle occurs, horizontal forehead rhytids appear over time the lines will become etched in static lines at rest. When treating this area, it is smart to take a finger and place it on the forehead where the injection point is planned and then press down toward the brow. The goal is to assess how much skin laxity is present to help the injector determine if this is a good point for an injection of neuromodulation.

Glabellar region The glabellar region is comprised of the procerus and corrugator muscles. This complex region is where the supratrochlear and supraorbital arteries and nerves reside.

Lagopthalmos Lagophthalmos is a condition in which the eyelid can not completely close. Improper treatment with neuromodulation can create lagophthalmos. Use caution when treating patients that have had eyelid surgery and have reduced tissue and scaring in the area as this can lead to improper closing of the eyelid.

Levator labii inferioris (LLI) The LLI is considered to be an accessory muscle to the orbicularis oris muscle. The LLI draws the ala of the nose down and, compresses the nostrils.

Levator labii superioris (LLS) The LLS begins at the maxilla and zygomatic bone above the infraorbital foramen. It is responsible for elevating and everting the upper lip. The LLS can help to raise the angle of the mouth. This muscle also aids in the flair/dilation of the nose.

Levator labii superioris alaeque nasi muscle (LLSAN) The LLSAN is a superficial muscle of the nose and lip, creating facial expression. This muscle stems from the upper part of the maxilla and passes obliquely down and laterally inserting into the alar cartilage and the LLS. There are said to be two slips to the LLSAN; a medial slip and a lateral slip. The lateral slip raises the upper lip and causes it to evert and assists in the dilation of the nostril. The medial portion is responsible for the flair of the alar. This muscle is responsible for a gummy smile in patients. With the proper administration of neuromodulation in the LLSAN it can reduce the gummy smile and slow the deepening of the nasolabial folds development.

Levator palpebrae superioris muscle The levator palpebrae superioris muscle is an eyelid elevator muscle that lies mid pupillary line of the lid and extends up approximately 1 cm above the boney orbital rim. The muscle is often inadvertently injected by new providers when treating the corrugator's lateral portion. When treating the corrugator's lateral portion, one must take great caution to be mindful of depth and position with their injection. If the levator palpebrae muscle becomes paralyzed by neuromodulation injections, the client will not be able to lift their lid.

Malar edema Malar edema is a collection of fluid over the malar eminence below the infraorbital rim. Edema can be temporary or chronic soft tissue swelling and varies in severity.

Malar eminence Malar eminence is the most prominent point on the zygomatic bone. In terms of esthetics, it is a significant land mark to understand as it flattens the landscape of the face changes and ages rapidly. The malar eminence is essential in the beauty industry and should be round and full with natural transition points. The rest of the face is balanced from the malar eminence.

Malar mound Malar mound is soft tissue swelling over the malar eminence resting between the infraorbital rim and mid-cheek. The distinguishing characteristic between malar edema and malar mounds is often not easily discernable. The presence of excess tissue along with bulge is characteristic of a malar mound versus malar edema.

Masseter The masseter is a muscle of mastication and is an extremely powerful muscle. It is responsible for patients who suffer from TMJD. A hyperactive masseter muscle is a broader look to the lower jaw. Think of Angelina Jolie; she has a well-defined jaw with an active masseter, enhancing her unique jaw presentation. When neuromodulation is placed in the masseter muscle, it can create a slimmer jawline and reduce TMJD's pain and discomfort. It is important to note that if the patient has some jowls and or lax skin, treating this area will slim the jawline while enhancing the jowls and exacerbate the lax skin. This muscle is responsible for elevating and protruding the chin.

Mentalis The mentalis (levator Menti) are two deep but small *conical fasciculus* structures that lie on the side of the frenulum of the lower lip. It arises from the incisive fosa of the mandible and then on the descent attaches to the skin on the chin. It is responsible for wrinkling the skin on the chin in an upward fashion. When the mentalis is activated, it raises the lower lip base, thus allowing for protrusion of the lower lip to an everted position to allow for easy drinking. This muscle is also responsible for eliciting emotions of doubt and concern. Over action of this muscle will also create a deep mental crease under the lip.

Nasalis This thin muscle lays across the bridge of the nose and connects with the process. It is at this connection when you can often see a static rhytid across the bridge of the nose. The nasalis muscle draws the tissue up and creates a bunny nose appearance. When active, the nasalis will pull the ala up and, depressors the cartilage of the bridge of the nose. An active nasalis will create horizontal lines on the bridge of the nose and diagonal lines on either side of the nose near the inner canthus.

Orbiculars oculi The orbicularis oculi is a broad, flat, and elliptical muscle that surrounds the orbital rim's circumference and is responsible for the eyelids closure. Much like the orbicularis oris muscle, it is also a sphincter muscle and is ultimately a depressor muscle. The muscle spans several facial regions, which include the eyelid, the cheek, and the forehead. This sphincter muscle's primary function is to act as a depressor muscle. It is partly responsible for the early onset of creasing in the skin at the lateral canthal area and nasalis line.

Orbicularis oris muscle (labi) The orbicularis oris muscle is a superficial yet complex sphincter muscle that encircles the mouth. This muscle is extremely superficial to the upper and lower lips. It is responsible for closing the mouth and protruding the lips. The deep portion of this muscle is responsible for pressing the lip to the teeth. There is an upper and a lower muscle, and then it is further segmented into four quadrants (upper, lower, left, and right).

Photophobia Photophobia is light sensitivity. If injections of neuromodulation position the eyelid to not fully close and the patients lose their ability to shield their eyes from bright light, photophobia will occur.

Platysma The platysma is a superficial muscle of the neck that lay in the subcutaneous tissue and overlaps the sternocleidomastoid. It is considered a broadsheet of muscle with various prominence that covers the majority of the lateral and anterior neck. The platysma fibers cross the clavicle and ascend medially to the side of the neck. The anterior fibers intertwine to the menti while other fibers connect to the mandible and the lower lip. The platysma is responsible for accentuating skin laxity in the aging process. Be careful in the central portion treating bands over the thyroid notch. Just below the thyroid notch sits the laryngeal prominence and below that is the cricothyroid ligament. If a neuromodulator was to affect here, the patient could have significant trouble swallowing. The platysma is responsible for tensing the neck.

Procerus The procerus is a pyramid-shaped muscle located within the glabellar region. It extends from the nasal bone to roughly the middle of the forehead between the eyebrows; this is where it attaches to the frontalis muscle. When injected with neuromodulator, it raises the bridge of the nose and softens the horizontal rhytids on the bridge of the nose. It is responsible for drawing the medial brows down, hence its depressor function.

Ptosis Ptosis is a greek word developed in reference to dropped or fallen. Patients can experience ptosis after neuromodulation if the injections are not precise. Ptosis can occur when treating any muscle that is an elevator. Commonly ptosis is seen in the eyebrow, eyelid, or smile.

Risorius muscle This is a superficial muscle of expression. This muscle pulls the corners of the mouth laterally; it runs through the masseter muscle originating in then fascia. Understanding the anatomical position of this muscle is critical when treating a patient for TMJD. This muscle aids in a number of dynamic expressions such as smiling, grimacing, laughing, and so many others.

Sclera The sclera is the white outermost portion of the part of the eye.

Urticaria Urticaria is also referred to as hives where wheals rise in the skin due to chemical mediators' release. Many factors cause this skin eruption. In terms of urticaria associated with neuromodulation, it would be an IgE mediated drug reaction. Over the counter antihistamine medication is recommended should a patient experience urticaria post-treatment.

Xerophthalmia Xerophthalmia is a condition known as dry eye or aqueous deficiency. Neuromodulation can also be used to treat aqueous deficiency of the eye. If the patient suffers from an aqueous deficiency treating the orbicularis oculi muscle will help reduce contractility and blinking, resulting in a slowed pump motility, improving the eye's dryness.

Zygomatic major This superficial muscle inserts on the lateral aspect of the zygomatic bone posterior to the zygomatic minor muscle. This muscle inserts into the orbicularis oris, the DAO, and the levator angel oris. The zygomatic major muscles draws the angle of the mouth up and out when smiling or laughing. It attaches to the zygomatic bone, and at times it will snuggle under the orbicularis oculi muscle. This muscle is another crucial muscle to avoid when injecting the lateral canthal lines, and the DAO as the smile can be interrupted in an unnatural and displeasing way.

Zygomatic minor The zygomatic minor is a superficial muscle. The ZM elevates the upper lip outward and upward, allowing to show the maxillary teeth. This muscle sits beneath the orbicularis oculi on the zygomatic bone and connects to the subcutaneous tissue of the upper lateral lip muscle, and sits on the lateral portion of the zygomaticus bone. The ZM is essential in the function of smiling. This muscle can be inadvertently injected when the lateral canthal lines are treated in the ill-prepared injector's hands.

Index

Note: Page numbers followed by *f* indicate figures and *t* indicate tables.

D

E

F

MIX
Papier | Fördert gute Waldnutzung
FSC® C083411

Kolibri 360 GmbH
Ferdinand-Jühlke-Straße 7
99095 Erfurt, Deutschland
produktsicherheit@kolibri360.de